VOLUME SIX HUNDRED AND SEVEN

METHODS IN ENZYMOLOGY

Phosphatases

METHODS IN ENZYMOLOGY

Editors-in-Chief

ANNA MARIE PYLE
Departments of Molecular, Cellular and Developmental Biology and Department of Chemistry
Investigator, Howard Hughes Medical Institute
Yale University

DAVID W. CHRISTIANSON
Roy and Diana Vagelos Laboratories
Department of Chemistry
University of Pennsylvania
Philadelphia, PA

Founding Editors

SIDNEY P. COLOWICK and NATHAN O. KAPLAN

VOLUME SIX HUNDRED AND SEVEN

METHODS IN ENZYMOLOGY

Phosphatases

Edited by

KAREN N. ALLEN

Department of Chemistry,
Boston University,
Boston, MA, United States

ELSEVIER

ACADEMIC PRESS

An imprint of Elsevier

Academic Press is an imprint of Elsevier
50 Hampshire Street, 5th Floor, Cambridge, MA 02139, United States
525 B Street, Suite 1650, San Diego, CA 92101, United States
The Boulevard, Langford Lane, Kidlington, Oxford OX5 1GB, United Kingdom
125 London Wall, London, EC2Y 5AS, United Kingdom

First edition 2018

ISBN: 978-0-12-813881-6
ISSN: 0076-6879

For information on all Academic Press publications
visit our website at https://www.elsevier.com/books-and-journals

Publisher: Zoe Kruze
Acquisition Editor: Ashlie Jackman
Editorial Project Manager: Shellie Bryant
Production Project Manager: Divya Krishna Kumar
Cover Designer: Miles Hitchen

Typeset by SPi Global, India

CONTENTS

Section IV
Delineating the Action of Phosphatases in Signaling and Regulation

CONTRIBUTORS

Karen N. Allen
Department of Chemistry, Boston University, Boston, MA, United States

Lesa J. Beamer
Biochemistry Department, University of Missouri, Columbia, MO, United States

Eric J. Berns
Department of Biomedical Engineering, Northwestern University, Evanston, IL, United States

Robert B. Bourret
Department of Microbiology & Immunology, University of North Carolina, Chapel Hill, NC, United States

Nathaniel Tate Burkholder
Department of Molecular Biosciences, The University of Texas at Austin, Austin, TX, United States

Philip A. Cole
Division of Genetics, Brigham and Women's Hospital; Department of Medicine; Department of Biological Chemistry and Molecular Pharmacology, Harvard Medical School, Boston, MA, United States

Qiang Cui
Department of Chemistry and Theoretical Chemistry Institute, University of Wisconsin, Madison, Madison, WI, United States

Darren Demapan
Department of Chemistry and Theoretical Chemistry Institute, University of Wisconsin, Madison, Madison, WI, United States

Daniel R. Dempsey
Division of Genetics, Brigham and Women's Hospital; Department of Medicine; Department of Biological Chemistry and Molecular Pharmacology, Harvard Medical School, Boston, MA, United States

Dong Fang
Department of Chemistry and Theoretical Chemistry Institute, University of Wisconsin, Madison, Madison, WI, United States

Rong Gao
Center for Advanced Biotechnology and Medicine, Rutgers Biomedical and Health Sciences; Department of Biochemistry and Molecular Biology, Rutgers—Robert Wood Johnson Medical School, Piscataway, NJ, United States

Swapnil V. Ghodge
Eshelman School of Pharmacy, University of North Carolina at Chapel Hill, Chapel Hill, NC, United States

Adrian Goldman
Astbury Centre for Structural Molecular Biology, University of Leeds, Leeds, United Kingdom; Faculty of Biological and Environmental Sciences, University of Helsinki, Helsinki, Finland

Mitchell E. Granade
Department of Pharmacology, University of Virginia, Charlottesville, VA, United States

Mark Hannink
University of Missouri, Columbia, MO, United States

Steven P.D. Harborne
Astbury Centre for Structural Molecular Biology, University of Leeds, Leeds, United Kingdom

Sarah A. Harris
Astbury Centre for Structural Molecular Biology, University of Leeds, Leeds, United Kingdom

Thurl E. Harris
Department of Pharmacology, University of Virginia, Charlottesville, VA, United States

Christine M. Harvey
Department of Chemistry, Boston University, Boston, MA, United States

Che-Fan Huang
Department of Chemistry, Northwestern University, Evanston, IL, United States

Seema Irani
Department of Chemical Engineering, The University of Texas at Austin, Austin, TX, United States

Shina Caroline Lynn Kamerlin
Department of Chemistry—BMC, Science for Life Laboratory, Uppsala University, Uppsala, Sweden

Xiya Lu
Department of Chemistry and Theoretical Chemistry Institute, University of Wisconsin, Madison, Madison, WI, United States

Wendy Matthews
Department of Molecular Biosciences, The University of Texas at Austin, Austin, TX, United States

Brenda Medellin
Department of Molecular Biosciences, The University of Texas at Austin, Austin, TX, United States

Milan Mrksich
Department of Chemistry; Department of Biomedical Engineering, Northwestern University, Evanston, IL, United States

Katherine H. O'Toole
Department of Chemistry, Boston University, Boston, MA, United States

Miha Purg
Department of Chemistry—BMC, Science for Life Laboratory, Uppsala University, Uppsala, Sweden

Frank M. Raushel
Department of Chemistry, Texas A & M University, College Station, TX, United States

Daniel Roston
Department of Chemistry and Theoretical Chemistry Institute, University of Wisconsin, Madison, Madison, WI, United States

Scott A. Showalter
Department of Chemistry; Department of Biochemistry and Molecular Biology, Pennsylvania State University, University Park, PA, United States

Ruth E. Silversmith
Department of Microbiology & Immunology, University of North Carolina, Chapel Hill, NC, United States

Kyle M. Stiers
Biochemistry Department, University of Missouri, Columbia, MO, United States

Ann M. Stock
Center for Advanced Biotechnology and Medicine, Rutgers Biomedical and Health Sciences; Department of Biochemistry and Molecular Biology, Rutgers—Robert Wood Johnson Medical School, Piscataway, NJ, United States

Jannik Strauss
Astbury Centre for Structural Molecular Biology, University of Leeds, Leeds, United Kingdom

Tong Su
University of Missouri, Columbia, MO, United States

Lindsey C. Szymczak
Department of Chemistry, Northwestern University, Evanston, IL, United States

Peter Tipton
University of Missouri, Columbia, MO, United States

Roman Tuma
Astbury Centre for Structural Molecular Biology, University of Leeds, Leeds, United Kingdom; Faculty of Science, University of South Bohemia, České Budějovice, Czech Republic

Ainoleena Turku
Drug Research Program, Division of Pharmaceutical Chemistry and Technology, Faculty of Pharmacy, University of Helsinki, Helsinki, Finland

Steven R. Van Doren
Biochemistry Department, University of Missouri, Columbia, MO, United States

Keni Vidilaseris
Faculty of Biological and Environmental Sciences, University of Helsinki, Helsinki, Finland

Matthew A. Watson
Astbury Centre for Structural Molecular Biology, University of Leeds, Leeds, United Kingdom

Craig Wilkinson
Astbury Centre for Structural Biology, University of Leeds, Leeds, United Kingdom

Jia Xu
Biochemistry Department, University of Missouri, Columbia, MO, United States

Yan Jessie Zhang
Department of Molecular Biosciences; Institute for Cellular and Molecular Biology, The University of Texas at Austin, Austin, TX, United States

PREFACE

High-energy phosphate ester bonds are often thought of as the energetic currency of the cell. Phosphoryl transfer is involved in processes ranging from the biosynthesis and catabolism of carbohydrates and polysaccharides, nucleic acids, and cofactors, to the regulation of cellular pathways. In addition, phosphoryl groups on small-molecule metabolites confer important properties—providing excellent "handles" for enzyme binding, trapping metabolic intermediates within the cell, and controlling the reversibility of metabolic pathways. The phosphorylation states of proteins also afford key control points for signal transduction and the regulation of myriad critical physiological processes. For this reason, kinases are extremely important therapeutic targets and there is significant interest in developing diverse approaches to regulate their function through small-molecule inhibitors. We present here the less studied but equally important complement of phosphate ester chemistry, namely the coordinated action of phosphatases. Phosphatases represent control points in pathways and new opportunities for use in chemoenzymatic synthesis. In many cases, such critical players may be functionally unannotated. However, there are many challenges in their study—spurring on the efforts described in this volume. Phosphoryl transfer is, itself, complex chemistry to describe with highly charged phosphate ester substrates and the frequent participation of metal ions in enzyme-catalyzed reactions. Structure/function analysis and specificity are especially thorny problems to study in phosphatases because of the large-scale conformational changes that often accompany binding events. Indeed, these conformational changes can be key in achieving specificity and may also become the rate-limiting step in catalysis. In terms of assaying phosphoryl transfer, the number of direct assays is limited and special considerations must be taken into account to achieve physiologically relevant measurements of activity.

I am excited to present this volume of *Methods in Enzymology* on phosphatases which draws upon approaches from the varied disciplines of computational and organic chemistry, biophysics, biochemistry, bioinformatics, and cell biology to address the structure, mechanism, and regulation of phosphatases. The themes covered in the volume comprise many creative strategies including computational approaches for understanding phosphoryl transfer, means for assessing mechanism, conformational changes and enzyme dynamics, the assignment of enzyme function and specificity, and the regulation of phosphatases. Although each of the methods presented here target-specific enzymes or enzyme families, there is clear application to other phosphatase targets and indeed, in many cases, more generally to other reaction types.

In the first chapter, Purg and Kamerlin tackle the mechanism of diisopropyl fluorophosphatase using the empirical valence bond approach. This computationally efficient method allows them to perform extensive configurational sampling along the entire reaction trajectory of the enzyme. In a complementary discussion, Cui and colleagues apply QM/MM free-energy simulations to elucidate the mechanisms of alkaline phosphatase and myosin. The authors point out that the computational efficiency of this approach is especially helpful in the case of these enzymes, where trends in the transition state or energetics are critical to the interpretation. The two chapters by Goldman and colleagues move the discussion of mechanism into the experimental realm, in an enzyme system in which the hydrolysis of inorganic pyrophosphate is coupled to the pumping of ions across the membrane. The coupled-vectorial processes enacted by these membrane-bound enzymes, previously captured in snapshots by crystallography, are now made dynamic by employing Förster resonance-energy transfer measurements to derive information on distance and the frequency of conformational transitions. The experimental methods presented are enhanced by molecular dynamics and computational modeling, improving both experimental design and data analysis. In the following chapter, bioinformatic approaches on the evolution and differential mechanisms of the pyrophosphatases are enhanced by a hydrolytic activity assay to assess potassium ion dependence. The focus on mechanism is continued by Allen and colleagues, who present methods for obtaining X-ray crystal structures where substrates, products, and intermediate/transition-state analogs are liganded to members of a large superfamily of phosphatases, allowing us to traverse the entire reaction coordinate.

The next four chapters are directed toward the challenges of determining substrate specificity and of defining the enzyme conformational states resulting from ligand binding. The first of these addresses the need for assignment of enzyme function, accelerated by the ever-growing release of new genomic data. In their chapter, Ghodge and Raushel outline a complete template for the assignment of function, using as examples members of the polymerase and histidinol phosphatase family of proteins. The approach leverages the shared structural and mechanistic features of the family to predict substrates and plan targeted mechanistic studies of closely related homologs. Quantitation of the binding affinities of substrates and inhibitors and associated changes in molecular state are tackled in the chapter by Xu and Van Doren. Their described TREND NMR software package fits binding isotherms from titrations detected with NMR spectra. Protein tyrosine phosphatase 1B variants and inhibitor complexes are used to exemplify the application of principal component analysis in order to group and compare NMR-detected enzyme states. Stiers and Beamer define methods to probe the flexibility of the various conformational states associated with phosphorylation of the nucleophile in enzymes in the α-D-phosphohexomutase superfamily and describe means to control and quantify the extent of phosphorylation. The last chapter in this section, by Zhang and colleagues, addresses the question of how to study specificity when the phosphatase is highly selective for *cis*- or *trans*-amide conformations within serine/threonine–proline motifs. Their solution is to use proline-mimicking isosteres that are locked to prevent interconversion.

Phosphatases are integral to cell signaling processes and are thus themselves subject to regulation. The last chapters in this volume describe examples of regulators and regulation and methods to assess them. The contributions from Bourret and Silversmith and from Gao and Stock detail quantitative analyses of bacterial response regulator phosphatases in vitro and in vivo, respectively. The in vivo data combined with mathematical modeling allow for the quantitative assessment of intracellular phosphatase activity of sensor histidine kinases. Complementary in vitro approaches comprise assay methods for chemotaxis two-component signal transduction phosphatases and modifications for those phosphatases not involved in chemotaxis two-component systems, where the timescales differ. The following chapter by Tipton and Hannink delves into the problem of purifying and assaying dimeric and multimeric oligomerization states of a phosphoglycerate mutase family member that regulates mitochondrial dynamics. Purification and physiologically relevant assay methods are also presented in the chapter by Granade and Harris, where the phosphatidic acid phosphatase activity of lipin is measured using small unilamellar vesicles. Mrksich and colleagues describe an approach using mass spectrometry with peptide arrays to study the substrate specificities of two protein tyrosine phosphatases. As the authors highlight, the mass spectrometry method is particularly significant for phosphatases as there are few other existing assays capable of directly measuring dephosphorylation. Finally, Dempsey and Cole showcase the semisynthesis of a C-terminally phosphorylated phosphatase and tensin homolog, PTEN, and develop methods to assess the effect of phosphorylation on protein stability, structure, and enzymatic activity.

Here, the latest methods of interrogating phosphatase chemistry can be found, hopefully benefiting researchers of this multifaceted and important area of enzymology. In conclusion, I wish to thank the authors for their valuable contributions to this volume, detailing approaches spanning from the computer to the cell. I also thank the editorial and production teams at Elsevier for their assistance in bringing this volume to light.

Karen N. Allen
Department of Chemistry,
Boston University,
Boston, MA, United States

SECTION I

Computational Analysis of Enzymatic Phosphoryl Transfer

CHAPTER ONE

Empirical Valence Bond Simulations of Organophosphate Hydrolysis: Theory and Practice

Miha Purg, Shina Caroline Lynn Kamerlin[1]
Department of Chemistry—BMC, Science for Life Laboratory, Uppsala University, Uppsala, Sweden
[1]Corresponding author: e-mail address: lynn.kamerlin@kemi.uu.se

Contents

Abstract

Recent years have seen an explosion of interest in understanding the mechanisms of phosphate ester hydrolysis in biological systems, using a range of computational approaches, each with different advantages and limitations. In this contribution, we present the empirical valence bond (EVB) approach as a powerful tool for modeling biochemical reactivity, using the example of organophosphate hydrolysis by diisopropyl fluorophosphatase as our model reaction. We walk the reader through the protocol for setting up and performing EVB simulations, as well as key technical considerations that need to be taken into account. Finally, we provide examples of the applications of the EVB approach to understanding different experimental observables.

Methods in Enzymology, Volume 607
ISSN 0076-6879
https://doi.org/10.1016/bs.mie.2018.06.007

ABBREVIATIONS

CADEE	computer-aided directed evolution of enzymes
cryoEM	cryoelectron microscopy
DFP	diisopropyl fluorophosphate
DFPase	diisopropyl fluorophosphatase
EVB	empirical valence bond
FEP	free-energy perturbation
LRA	linear response approximation
MC	Monte Carlo
MD	molecular dynamics
NMR	nuclear magnetic resonance
PDB	Protein Data Bank
PON1	serum paraoxonase 1
QM/MM	quantum mechanics/molecular mechanics
SCAAS	surface-constrained all-atom solvent
TS	transition state
TST	transition state theory
US	umbrella sampling
VB	valence bond
vdW	van der Waals

1. INTRODUCTION

Phosphate esters are mechanistically diverse, with even the simplest of phosphate esters (such as phosphate monoester dianions) being potentially hydrolyzed through several different catalytic mechanisms (Kamerlin, Sharma, Prasad, & Warshel, 2013; Lassila, Zalatan, & Herschlag, 2011). While experimental measurements of, for example, linear free-energy relationships (Kirby & Jencks, 1965; Kirby & Varvoglis, 1967), kinetic isotope effects (Hengge, Edens, & Elsing, 1994), or activation entropies (Kirby & Jencks, 1965) (among other markers) can provide some insight into plausible mechanisms, they do not provide atomic resolution on the structures of the transition states (TSs) involved, and thus the mechanistic inferences have to be made based on indirect evidence. While one would assume that this problem can be addressed by high-level calculations to complement the experimental studies, phosphate esters provide particular challenges to theory. These include the high charge of these compounds, the fact that the low-lying d-orbitals of the phosphorus atom can potentially be involved in the reaction, and, in the case of biological phosphoryl transfer, the fact that many of the enzymes that hydrolyze these compounds are metalloenzymes, and correctly describing the properties of metal ions is a major challenge to both quantum and classical mechanical approaches alike (for further discussion, see, e.g., Harvey, 2006; Li & Merz, 2017). As a result of this, unsurprisingly, often contradictory models have been presented in the literature, sometimes even for the hydrolysis of the same compound (see discussion in, e.g., Kamerlin et al., 2013), and therefore even the non-enzymatic hydrolysis of phosphate esters remains mechanistically controversial. Fortunately, however, despite the computational challenges associated with modeling phosphate esters, their biological importance means that there is a wealth of experimental data for benchmarking and validation of computational models (for examples, see Lassila et al., 2011; references cited therein), and, conversely, *provided* that the computational model(s) used can reproduce all relevant experimental data, computation can play, and has played, an important role in the interpretation of existing experimental data (Barrozo, Blaha-Nelson, Williams, & Kamerlin, 2017; Duarte, Åqvist, Williams, & Kamerlin, 2014; Duarte, Barrozo, Åqvist, Williams, & Kamerlin, 2016; Gregersen, Lopez, & York, 2004; López-Canut, Marti, Bertrán, Moliner, & Tuñón, 2009; Plotnikov, Prasad, Chakrabarty, Chu, & Warshel, 2013; Roston & Cui, 2016; Roston, Demapan, & Cui, 2016).

Biological phosphate hydrolysis has been studied using many different computational techniques, including quantum chemical cluster calculations (Chen, Fang, & Himo, 2007; Kim et al., 2008), semiempirical methods (López-Canut, Roca, Bertrán, Moliner, & Tuñón, 2010, 2011), hybrid quantum mechanical/molecular mechanical (QM/MM) approaches (Lu, Ovchinnikov, Demapan, Roston, & Cui, 2017; Roston et al., 2016; Roston & Cui, 2016), as well as various enhanced sampling approaches such as metadynamics and other methods (Rosta, Nowotny, Yang, & Hummer, 2011; Tripathi, Glaves, & Marx, 2017) (these studies are representative but not exclusive examples of the applications of each computational approach to studying phosphate hydrolysis). In our case, our methodology of choice has been the empirical valence bond (EVB) approach of Warshel (1991), due to both its computational speed, and versatility to address a range of different mechanistic questions (Shurki, Derat, Barrozo, & Kamerlin, 2015; Warshel et al., 2006), as discussed in detail in this chapter. While clearly not the only way to address important questions with regard to biological phosphoryl transfer, it has nevertheless been successfully applied by both ourselves (Åqvist & Kamerlin, 2014, 2015; Barrozo, Duarte, Bauer, Carvalho, & Kamerlin, 2015;

Blaha-Nelson, Krüger, Szeler, Ben-David, & Kamerlin, 2017; Purg et al., 2016) and others (Åqvist, Kolmodin, Florian, & Warshel, 1999) to address open questions with regard to questions as diverse as understanding catalytic promiscuity in alkaline phosphatases (Barrozo et al., 2015) to using simulation to understand GTP hydrolysis on the ribosome (Wallin, Kamerlin, & Åqvist, 2013).

In this chapter, we will use as a model system the hydrolysis of diisopropyl fluorophosphate (DFP) by squid diisopropyl fluorophosphatase (DFPase). This enzyme is a six-bladed β-propeller (Fig. 1) that hydrolyzes both organophosphate pesticides such as DFP and nerve agents such as sarin, soman, and tabun (Hartleib & Rüterjans, 2001a). As such, it has been of great interest for engineering for use as a potential biotherapeutic (Melzer et al., 2009) or bioremediation agent (Latifi, Karami, & Khodi, 2015). However, the mechanism of this enzyme and, in particular, the role of a key active-site residue, D229, remained highly controversial, with both general-base catalysis (Xu, Yang, Yu, & Liao, 2016) and nucleophilic substitution mechanisms (Blum, Löhr, Richardt, Rüterjans, & Chen, 2006) having been suggested in the literature (Fig. 1). Interestingly, and despite having almost identical tertiary structure and active-site architecture, no evidence exists for the presence of a covalent intermediate associated with the latter mechanism in the related PON1-catalyzed hydrolysis of paraoxon (Khersonsky & Tawfik, 2005), and it is extremely unlikely that two such similar enzymes hydrolyze their substrates through mechanistically distinct pathways involving essentially the same active-site residues (Purg, Elias, & Kamerlinet al., 2017).

DFPase is an excellent model system for testing the limits of computational studies of enzyme mechanism due to the availability of extensive experimental data on the ion concentration, temperature, and pH dependence of the hydrolysis of DFP (Hartleib & Rüterjans, 2001b), as well as extensive biochemical studies on the effect of mutations on the DFPase activity of this enzyme (Katsemi et al., 2005; Melzer et al., 2009). In addition, DFPase is relatively small and compact (only 314 amino acids) and fairly rigid, which is of benefit to constructing computational models. Our aim in this chapter is *not* to address DFP hydrolysis by DFPase specifically (which we addressed in great detail in a recent study (Purg et al., 2017)), but rather to use this system as a model with which to illustrate both the challenges and the possibilities involved in computational modeling of biological phosphate hydrolysis.

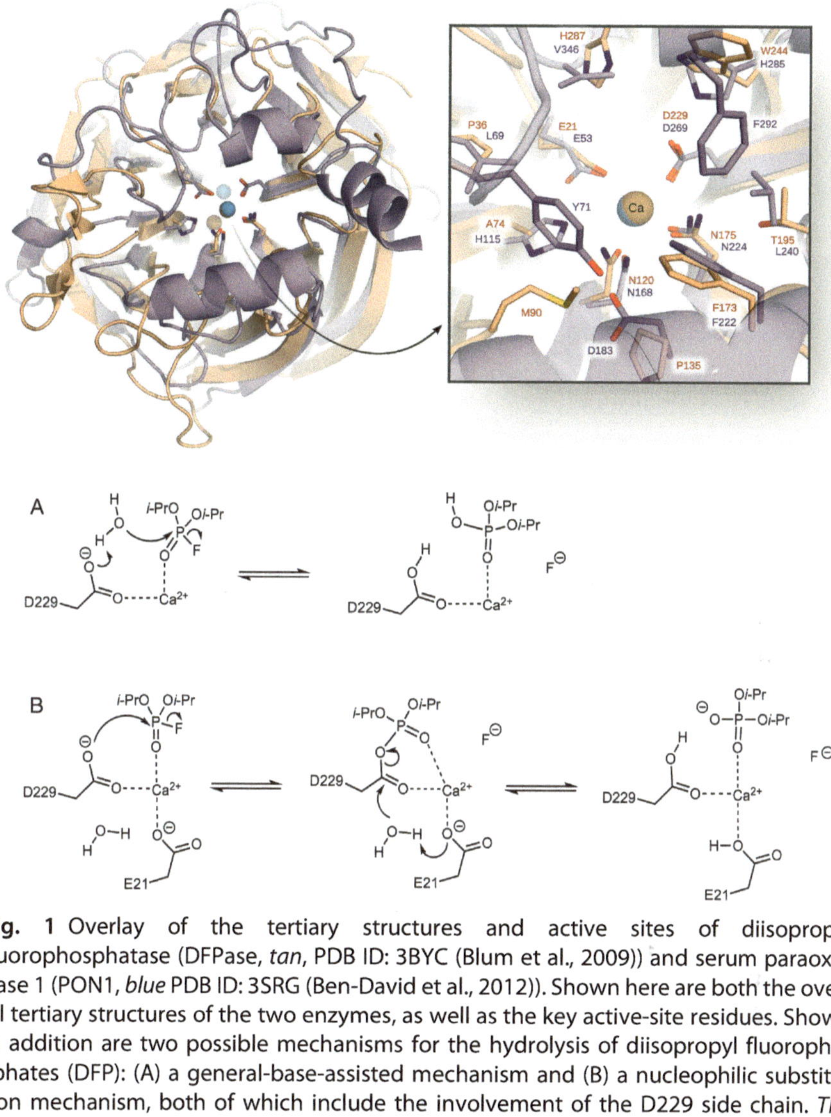

Fig. 1 Overlay of the tertiary structures and active sites of diisopropyl fluorophosphatase (DFPase, *tan*, PDB ID: 3BYC (Blum et al., 2009)) and serum paraoxonase 1 (PON1, *blue* PDB ID: 3SRG (Ben-David et al., 2012)). Shown here are both the overall tertiary structures of the two enzymes, as well as the key active-site residues. Shown in addition are two possible mechanisms for the hydrolysis of diisopropyl fluorophosphates (DFP): (A) a general-base-assisted mechanism and (B) a nucleophilic substitution mechanism, both of which include the involvement of the D229 side chain. *The individual panels for this figure were initially presented in Purg, M., Elias, M., & Kamerlin, S. C. L. (2017). Similar active sites and mechanisms do not lead to cross-promiscuity in organophosphate hydrolysis: Implications for biotherapeutic engineering.* Journal of the American Chemical Society, 139*(48), 17533–17546, reproduced with permission from the American Chemical Society.*

2. THE EMPIRICAL VALENCE BOND APPROACH

Computational enzymology started to take off in the 1970s, based on pioneering work by Warshel, Levitt, and Karplus (Warshel & Karplus, 1972; Warshel & Levitt, 1976) among others (see reviews Friesner & Guallar, 2005; Senn & Thiel, 2009; van der Kamp & Mulholland, 2013). Initial computational studies of biochemical systems were limited by the computational resources available and thus restricted to small or simplified systems. However, as computational power sky-rocketed (and continues to grow), the scope of problems that can be addressed has risen in accordance, and it is now possible not only to address mechanistic questions at atomic detail but also to explore problems as complex as protein synthesis on the ribosome (Xu, Zhang, & Xiang, 2012), to name just one of many examples. However, the challenge of finding the balance between high computational cost and high accuracy in the description of the system remains an ongoing problem, as it is still not feasible to, for example, describe an entire enzyme quantum mechanically, and thus simplifications and approximations need to be used as compromises. It should be noted, however, that it is ill-advised to trust the results of such approximate models, without thorough validation against experiment.

In the context of computational mechanistic enzymology, the system can be simplified by, for example, modeling only the reactive part of the system in a truncated cluster of atoms which are then described fully quantum mechanically (Siegbahn & Himo, 2009), or by including the environment of the reaction (i.e., the rest of the protein and/or solvent) at a lower level of theory via a hybrid QM/MM approach (Senn & Thiel, 2009). Here, the environment can be treated using either explicit all-atom models, coarse-grained models, or a simple dielectric continuum. The resulting model is then applied on a particular set of coordinates of an enzyme–substrate Michaelis complex to, for example, obtain its optimized geometry and energy. Where chemical reactivity is of interest, the typical procedure involves locating also the TS and product or intermediate geometries, and calculating their respective energies. Finally, the activation barrier can, in many cases, be directly related to the experimentally observed kinetic rate constant via transition state theory (TST), although for specific cases nonequilibrium effects and quantum tunneling need to be taken into account and corrected for in the models (Olsson, Mavri, & Warshel, 2006). However, it should be evident that such a static depiction of the system does not adequately portray typical experimental conditions at room temperature. The enormity of the conformational space gives rise to many local energy minima, where choosing a "correct" one is not only impractical but also grossly disregards the dynamical nature of condensed matter. Indeed, the activation enthalpies obtained from static calculations are often highly dependent on the conformation of the system. Although this issue can be somewhat alleviated by considering many conformations (Lonsdale, Harvey, & Mulholland, 2012; Ryde, 2017), for a more complete description, the structural fluctuations of the system should be taken into account, by exploring the thermally accessible microstates of the system along the reaction coordinate. This allows for capturing entropic contributions and, ultimately, for the calculation of reaction free-energy (rather than enthalpy) profiles. Sampling of the configurational space is achieved using either molecular dynamics (MD) or Monte Carlo (MC) simulations and usually is complemented with enhanced sampling techniques such as umbrella sampling (US) (Torrie & Valleau, 1977), replica exchange (Sugita & Okamoto, 1999), or metadynamics (Laio & Parrinello, 2002), in order to sample high-energy regions of a typical reaction landscape.

Here, the EVB approach is unique as a computational approach, in that it uses fully classical force fields in order to provide a QM description of chemical reactivity, within a valence bond (VB) framework (Warshel, 1991; Warshel & Weiss, 1980). This then gives the EVB approach two computational advantages: (1) as it is based on classical force fields, it is very fast, allowing for extensive configurational sampling of the reactive trajectory (currently easily at the cumulatively μs timescale (Purg et al., 2017)), while (2) at the same time, well-parameterized force fields carry a tremendous amount of chemical information, allowing for chemical reactivity to be described in a physically meaningful way.

In the EVB approach, a chemical reaction is described by defining the wave function of the system as a linear combination of VB states (Fig. 2). For the purposes of this chapter, we will describe only a simple two-state system with a single TS; however, in practice, as many states as necessary can be used to describe the chemical reaction, as in standard VB theory, although correctly describing the coupling between the different states becomes increasingly complex as the number of states increases. The ground-state EVB potential function for a two-state system is obtained as the lowest Eigenvalue of the secular equation:

$$E_g = \frac{1}{2}\left(H_{11} + H_{22} - \sqrt{(H_{11} - H_{22})^2 + 4H_{12}^2}\right) \quad (1)$$

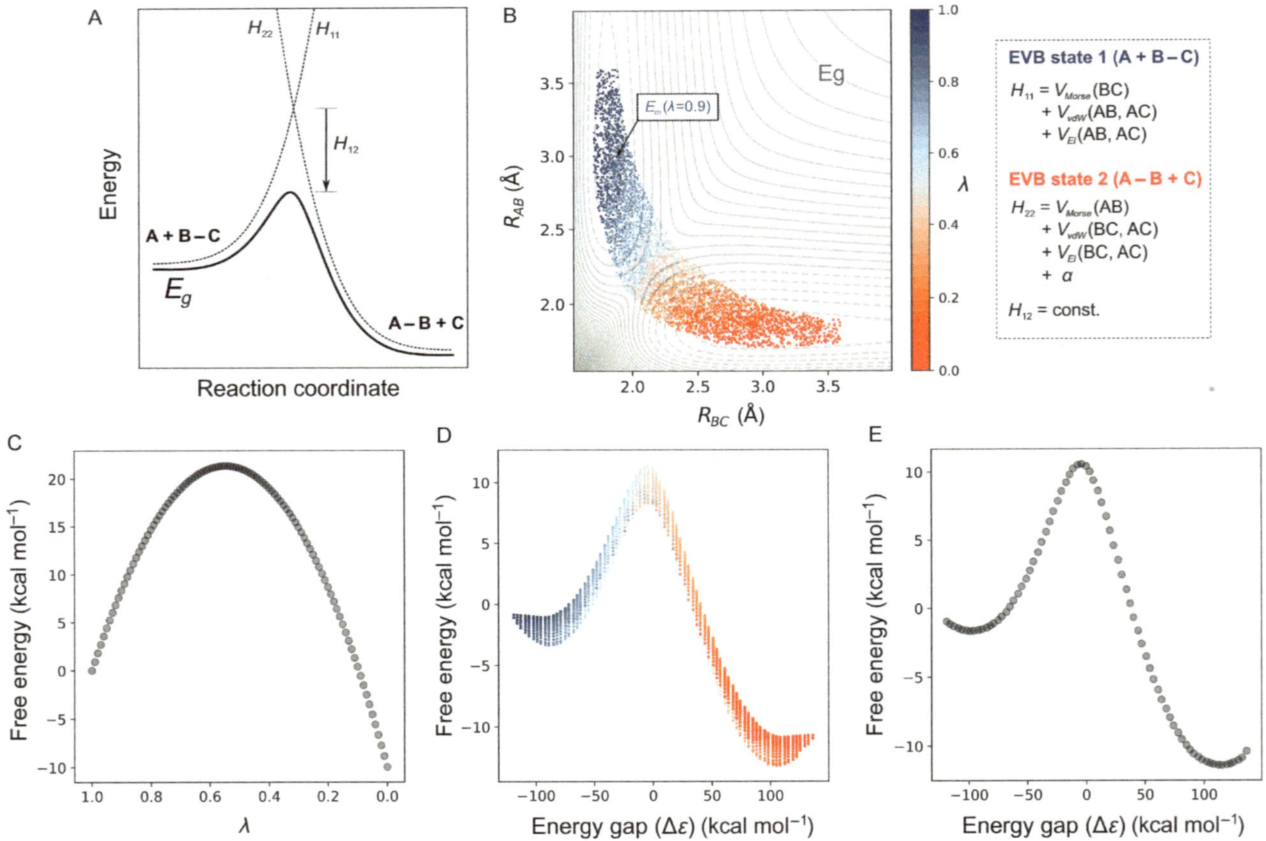

Fig. 2 See legend on opposite page.

The Hamiltonians of the individual VB states H_{11} (reactants) and H_{22} (products) are approximated by analytical potential energy functions, typically using a classical molecular mechanics force field, partitioned into a reactive and solvent/surrounding region (denoted with the superscripts r and s, respectively):

$$H_{ii} = V_i^r + V_i^{rs} + V_i^s \tag{2}$$

A notable exception is the use of Morse potentials in place of Harmonic potentials for atoms involved in bond breaking or forming.

The EVB off-diagonal term, H_{12}, is the phase-independent (Hong, Rosta, & Warshel, 2006) coupling between the two EVB states and essentially describes the energy difference between the intersection of H_{11} and H_{22} and the actual TS on the adiabatic surface, E_g (Fig. 2A). In its simplest form, H_{12} can be described by a geometry-independent constant value, although for greater accuracy, geometry-dependent exponential or Gaussian potentials can be used (see discussion in Chang & Miller, 1990; Schlegel & Sonnenberg, 2006).

Thermal averaging can be addressed via MD simulations directly on E_g, or in conjunction with enhanced sampling techniques to explore high-energy regions (see earlier). More common (and convenient) to the EVB methodology, however, is the so-called mapping approach (Hwang, King, Creighton, & Warshel, 1988; Hwang & Warshel, 1987), often referred to in the literature as FEP/US (free-energy perturbation/umbrella sampling).

Fig. 2 A visualization of the EVB approach as described in Section 2. (A) Schematic representation of E_g (Eq. 1), H_{11} and H_{22} for a concerted substitution reaction (A+B−C→A−B+C) described with two EVB states. (B) Configurations sampled on the mapping potential $E_m(\lambda)$ (color-coded by λ-value) in relation to the actual ground-state potential energy surface E_g (the specific H_{11} and H_{22} are described in the side text box; parameters are based on a realistic system and were simplified for clarity). Note that the points distinctly sample on the E_g minimum energy path (as opposed to many geometry based, e.g., distance, biasing potentials), and that sampling on E_m does not require prior knowledge of either H_{12} or α. (C) Free-energy profile for the alchemical transformation of State 1 ($\lambda = 1$) to State 2 ($\lambda = 0$), obtained with Eq. (7). (D) Free-energy profiles obtained by taking the probability of being at a particular point relative to the ground state ($\Delta G_m(\lambda_k)$), correcting for the free-energy difference of being on the actual ground-state potential E_g, and categorizing the points along the discrete energy-gap reaction coordinate (Eq. 8). Points on the plot are color coded based on λ as in (B). Note that many mapping windows can contribute to a single energy-gap bin. (E) Bin-averaged free-energy profile obtained by weight-averaging the data in (D) (Eq. 9).

Here, the sampling is performed instead on the mapping potential E_m, which is defined as a linear combination of the potentials of the different VB states, i.e.,

$$E_m = \lambda H_{11} + (1 - \lambda) H_{22} \tag{3}$$

The coupling parameter, λ, is used to gradually and discretely drive the system from one VB state to the other, sampling each λ window using classical MD (Fig. 2B).

We now define the reaction coordinate as the "energy gap" between the two VB states: $\xi \equiv H_{11} - H_{22} = \Delta\varepsilon$, and for clarity write out the implicit biasing potential, which is imposed on the EVB potential energy surface (for simplicity, we define H_{12} as constant):

$$E_{bias}(\xi, \lambda) = E_m(\xi, \lambda) - E_g(\xi) = \xi\left(\lambda - \frac{1}{2}\right) + \frac{1}{2}\sqrt{\xi^2 + 4H_{12}^2} \tag{4}$$

The unbiased ground-state probability of being at a particular point on the energy-gap reaction coordinate is obtained from the biased simulation at λ_k via the US expression (Torrie & Valleau, 1977):

$$P_g^k(\xi) = \frac{\left\langle \delta\left(\hat{\xi}(\boldsymbol{r}) - \xi\right) e^{\frac{+E_{bias}^k(\xi)}{RT}} \right\rangle_{m_k}}{\left\langle e^{\frac{+E_{bias}^k(\xi)}{RT}} \right\rangle_{m_k}} \tag{5}$$

where ξ is a given value of the reaction coordinate, $\langle\rangle_{m_k}$ denotes the configurational space sampled on the potential $E_m(\lambda_k)$, and $\delta\left(\hat{\xi}(\boldsymbol{r}) - \xi\right)$ is a Dirac delta function used to discard points where $\hat{\xi}(\boldsymbol{r})$ is *not* equal to ξ. Note that in practice ξ is discretized, and the configurations are binned according to the criteria $|\xi(\boldsymbol{r}) - \xi| \leq (\xi_{max} - \xi_{min})/N_{bin}$.

The PMF (potential of mean force) obtained from configurations sampled on the mapping window λ_k is then evaluated as:

$$\Delta G^k(\xi) = -RT \ln \left\langle \delta\left(\hat{\xi}(\boldsymbol{r}) - \xi\right) e^{\frac{-\left(E_g(\xi) - E_m^k(\xi)\right)}{RT}} \right\rangle_{m_k} - RT \ln \frac{Q_m^k}{Q_g} \tag{6}$$

where the last term is undetermined and represents the free energy associated with introducing the biasing potential (Roux, 1995). As Q_g is constant for all mapping windows and our only interest lies in relative free energies, it can be arbitrarily redefined to a more convenient value, for example, corresponding to the first mapping window—Q_m^0 (Haydock, Sharp, & Prendergast, 1990). The free-energy difference can then be obtained via the FEP formulation (Zwanzig, 1954), as a cumulative sum (Fig. 2C):

$$\Delta G_m(\lambda_k) = \Delta G_m(\lambda_0 \rightarrow \lambda_k) = \sum_{i=0}^{k-1} -RT \ln \left\langle e^{-\frac{E_m(\lambda_{i+1}) - E_m(\lambda_i)}{RT}} \right\rangle_{m_i} \tag{7}$$

This leads to the final FEP/US expression for the free-energy profile (Fig. 2D):

$$\Delta G(\xi, \lambda) = \Delta G_m(\lambda) - RT \ln \left\langle \delta\left(\hat{\xi}(\boldsymbol{r}) - \xi\right) e^{\frac{-\left(E_g(\xi) - E_m(\xi, \lambda)\right)}{RT}} \right\rangle_m \tag{8}$$

Finally, because many mapping windows can contribute to the same reaction coordinate bin, the energies are weight-averaged according to the relative populations (see also the W-PMF technique described by Roux, 1995; Woolf & Roux, 1994):

$$\Delta G(\xi) = \sum_{k=0} \frac{N_k}{N} \Delta G(\xi, \lambda_k) \tag{9}$$

where N_k is the number of points from the λ_k mapping window that contribute to the reaction coordinate at bin ξ, and N is the total number of points that contribute to bin ξ (Fig. 2E).

The energy gap is a particularly powerful choice of reaction coordinate because it projects the full multidimensional conformational space of the system onto a one-dimensional reaction coordinate, allowing the calculations to capture conformational rearrangement along the chemical coordinate, while greatly accelerating the convergence of the free-energy calculations compared to a standard geometric coordinate (Mones, Kulhánek, Simon, Laio, & Fuxreiter, 2009).

A key feature of the EVB approach is that it does not provide absolute free energies, but rather differences in free energy relative to a *reference* state, which can be either the uncatalyzed reaction in aqueous solution (where suitable experimental information or high-level calculations are available) or, for example, the wild-type enzyme against a series of variants (Oanca, Purg, Mavri, Shih, & Stare, 2016). Clearly, parameterization of the EVB force fields to correctly reproduce relevant physicochemical properties is the major and most time-consuming bottleneck in this process. However, once the EVB force fields as well as the associated EVB parameters H_{ij} and the "gas-shift" parameter α (which describes the energies of the VB states at infinite separation) have been appropriately parameterized, it is then possible to use this same parameter set *unchanged* in different environments, which can include different solvents or different enzymes. This is particularly valuable for large-scale computational screening in the advanced stages of computational enzyme design, as we recently illustrated with our computational toolkit CADEE (Amrein et al., 2017), with the limitation that it works only when comparing exactly the same mechanism with the same parameters in different environments, and therefore does not account for potential changes in mechanism upon mutation. In this contribution, we will take the reader through the necessary steps to set up EVB simulations, as well as the potential pitfalls, and show how EVB simulations can be used not only to understand enzymatic reaction mechanisms but also, when coupled to structural bioinformatics approaches, to understand the factors driving the molecular evolution of enzyme functions.

3. EMPIRICAL VALENCE BOND SIMULATIONS IN PRACTICE

In this section, we will describe the general protocol for performing EVB simulations of enzymatic systems using the Q6 simulation package (Bauer et al., 2018), from the initial system setup and EVB parameterization, to mapping of reaction free-energy profiles. We note that while not an a priori requirement for performing EVB simulations in Q6, some steps of the procedure described here will make use of Qtools, an open-source Python toolset and wrapper for Q. As our model system, we will be using the DFPase-catalyzed hydrolysis of DFP via a general-base-assisted mechanism, involving the concerted transfer of a phosphoryl group through a single TS (i.e., a two-state mechanism), with concomitant general-base catalysis by the side chain of D229 to activate the nucleophilic water molecule in the DFPase active site (Fig. 1A) (Purg et al., 2017).

3.1 Obtaining the Relevant Software

In order to perform the simulations described in this work, the reader will need access to the latest version of Q (currently Q6; Bauer et al., 2018) and Qtools (currently version 0.6.0). These are available for free download under GPLv2 and MIT licenses from https://github.com/qusers/Q6 and https://github.com/mpurg/qtools, respectively. Additional files accompanying this chapter have been provided for free download from Github at https://github.com/mpurg/methodsenz.

3.2 System Preparation

3.2.1 Structure Preparation

Reliable starting coordinates of a system are arguably the most important prerequisite for running molecular simulations. A variety of structures obtained by X-ray crystallography, nuclear magnetic resonance (NMR), neutron scattering, or cryoelectron microscopy (cryoEM) studies can be found in the Protein Data Bank (PDB) (Berman et al., 2000). In general, one should approach with care low-resolution (>2.5 Å) X-ray structures, structures with highly mobile or disordered regions (as indicated, for example, by large *B*-factors), large segments of missing residues, or structures of enzymes without bound cofactors and/or other ligands (where necessary to the biological function of the given enzyme).

In particular, while missing segments can be modeled using tools such as Modeller (Webb & Sali, 2014) or Rosetta (Rohl, Strauss, Misura, & Baker, 2004), without some form of validation, the resulting structure will likely be of limited use since the associated errors will be propagated throughout the EVB simulations. Related to this is the fact that amino acids often have alternate locations in the crystal structure due to conformational sampling, and, depending on the computational cost of the simulation approach chosen, not all conformational states will necessarily be sampled during the simulation. Therefore, in cases where more than one amino acid side-chain conformation has been provided in the PDB file, the conformation chosen for the simulations is selected based on occupancy and/or visual inspection of electron-density maps and surrounding interactions.

Following from this, the availability of structures of enzymes in complex with bound ligands in the active site is of tremendous value, due to the insight they can provide into likely positioning of the substrate, and, where both complexes and substrate-free structures are available, of structural changes in the enzyme that accompany the binding of the substrate, substrate or TS analogue, or inhibitor. The same rules as for selecting side-chain rotamers are true for assessing the quality of crystallographic water molecules and bound ligands (Cooper, Porebski, Chruszcz, & Minor, 2011). Alternatively, it is possible, although with more restricted accuracy, to use molecular docking to find suitable chemically reactive binding poses of the substrate from which to initiate the simulations. Finally, a more general assessment of the quality, such as the number of clashes and outliers in Ramachandran plots, can also be performed using the online tool MolProbity (Chen et al., 2010).

In the case of DFPase, several accurate structures are available, including a sub-Ångström resolution X-ray structure (PDB ID: 3O4P (Elias et al., 2013), no ligand), a joint neutron and X-ray structure (PDB ID: 3BYC (Blum et al., 2009), no ligand), as well as an X-ray structure in complex with a substrate analogue (PDB ID: 2GVV (Blum et al., 2006)). As these structures are virtually superimposable, we based our simulations on the neutron-diffraction structure, which was obtained at room temperature, and contains coordinates for the water molecule, which is suggested to act as the nucleophile in the general-base-assisted mechanism (Fig. 1). After obtaining the coordinate file (.pdb) from the PDB, we removed all solvent molecules beyond 20 Å from the catalytic calcium, renamed the remaining *DOD* entries to *HOH*, and exchanged the calcium entries with octahedrally coordinated dummy models (Duarte et al., 2014).

3.2.2 Assigning Correct Protonation States

Although the protonation patterns of amino acid side chains are dynamic, with pK_a reflecting only the ratio between different protonation states, with the exception of specialized methodologies, the protonation states of ionizable residues are fixed throughout classical MD simulations, and thus the protonation patterns of ionizable residues and histidines must be defined in advance of starting the simulation. Correct assignment of the most dominant protonation patterns of ionizable residues is absolutely critical, as incorrect assignment of protonation patterns will lead to both structural instabilities during the simulations as well as producing erroneous results.

There exist several tools for rapid pK_a predictions in proteins and protonation state assignments, including PROPKA (Olsson, Sondergaard, Rostkowski, & Jensen, 2011), the H++ server (Anandakrishnan, Aguilar, & Onufriev, 2012), and MolProbity (Chen et al., 2010). However, these servers mainly provide rough guides, in particular when charged ligands or metal cofactors are present in the active site. They are, nevertheless, very useful for identifying residues that are likely to be significantly shifted from their standard pK_a under physiological conditions. In cases of severe uncertainty, one can make use of more elaborate techniques for predicting pK_a values, such as constant pH simulations, semimacroscopic approaches, FEP, thermodynamic integration, or hybrid QM/MM approaches (for reviews, see, e.g., Kamerlin, Haranczyk, & Warshel, 2008). In addition, when computational resources allow, one can evaluate the effect of a particular protonation state on the reaction free-energy profile directly by simply performing the simulation at both possible protonation states of a given residue. Note that the assignment of neutral histidine residues (i.e., protonation at the δ vs ε nitrogen atoms) could also potentially have a significant impact on the stability of the structure during the simulations and should be visually examined in addition to predictions by any server used.

Finally, as EVB simulations using Q6 are performed in spherical rather than period boundary conditions, all ionizable residues close to the surface of the solvation sphere (i.e., ~15% of the radius, see Section 3.2.5) should be set to their neutral forms during the simulation, to avoid artifacts due to unrealistic dielectric screening (Åqvist, 1996). In the case presented here, the simulation sphere will have a radius of 25 Å centered on the catalytic calcium, which means that we will neutralize all residues beyond 21 Å from the sphere center (for a full list of ionized residues and protonation patterns of histidines, see table S1 in the Supporting Information of Purg et al., 2017). Additionally, we note that, while taking into account the limitations of empirical screening approaches, PROPKA (Olsson et al., 2011) and H++ (Anandakrishnan et al., 2012) analysis of DFPase revealed no likely anomalous protonation states, with all histidines present in their neutral states in the simulation and all other ionizable residues in their standard protonation states at physiological pH. In the Q topologies, ionizable residues have different residue types, depending on whether they are in their ionized (ASP, GLU, ARG, LYS, HIP) or neutral form (ASH, GLH, ARN, LYN, HIE, HID).

3.2.3 Force-Field Parameterization

The Q6 package comes with several force-field topology and parameter files for standard amino acids, including the OPLS-AA (2001 and 2015) (Robertson, Tirado-Rives, & Jorgensen, 2015), AMBER (95 and 14SB) (Maier et al., 2015), CHARMM (v.22) (MacKerell et al., 1998), and GROMOS (87 and 96) (Daura, Mark, & Van Gunsteren, 1998) force fields. We will limit our focus in this chapter to OPLS-AA (2001), as this is the force field we typically use in EVB simulations of phosphate hydrolysis (Blaha-Nelson et al., 2017; Purg et al., 2016). As far as possible, parameters for nonstandard species (such as, in this case, DFP in the different reacting states) should be taken from the force field by considering standard atom types, or from the literature. Failing this, OPLS-AA-compatible parameters for nonstandard species can be obtained using automated assignment tools such as Schrödinger (2017) or LigParGen (Dodda, Cabeza de Vaca, Tirado-Rives, & Jorgensen, 2017). However, as these parameters are often assigned based on functional analogy rather than derived from actual physical properties, the quality of these parameters can be variable and more rigorous parameterization may be necessary.

For our model system, the van der Waals (vdW) and bonding parameters for DFP, and the products diisopropylphosphate (DIP) and fluoride ion (F^-) were assigned via Schrödinger's Macromodel, with partial charges obtained using the RESP protocol (Bayly, Cieplak, Cornell, & Kollman, 1993), following the procedure described later. We conclude that the quality of parameters is satisfactory based on the *high* confidence of the assignments (as reported in the ffld_server output), stable simulations, and energetics in good agreement with known experimental observables.

```
# In directory 0-topol/0-ff/param_dfp

# Obtain parameters for DFP using ffld_server
# (requires schrodinger Maestro)

$SCHRODINGER/utilities/ffld_server -version 14 \
                                   -print_parameters \
                                   -ipdb dfp.pdb > dfp.ffld14

# Convert to Q6 format
q_ffld2q.py dfp.ffld14 dfp.pdb -o DFP

# Calculate RESP charges
# (requires AmberTools and Gaussian)
antechamber -fi pdb -i dfp.pdb -fo gcrt -o dfp_esp.com
g09 dfp_esp.com
antechamber -fi gout -i dfp_esp.log -fo ac -o dfp_resp.ac -nc 0 -c resp
```

The commands above result in a complete set of Q6-compatible topology and parameter files (see the Q6 manual for details). Note that the calculated charges need to be manually transferred from *dfp_resp.ac* to *dfp.lib*. Additionally, since only a single parameter file can be used when generating topologies with Q6, the contents of *dfp.prm* must be transferred into the appropriate sections in the force-field parameter file, in this case *qoplsaa.prm*.

Finally, bonds that are breaking or forming are described in EVB via Morse potentials, for which the parameters, if not available in literature, can often be derived from spectroscopic data (Morse, 1929) or obtained from high-level QM calculations (e.g., Glowacki, Orr-Ewing, & Harvey, 2015). Additionally, due to the steep repulsive nature of the Lennard-Jones potential, nonbonding interactions between reacting atoms are instead modeled using an exponential repulsion potential (see the Q6 manual). These parameters can either be found in the literature (e.g., Warshel, 1991) or fitted to reproduce the shape of the (QM calculated) energy surface, in particular, the region near the TS. Note, however, that their effect essentially cancels out when comparing a reaction in enzyme to a reference reaction in solution (Vagedes, Rabenstein, Åqvist, Marelius, & Knapp, 2000).

3.2.4 Substrate Placement in the Active Site

The starting position of the substrate in the active site during EVB (and other) simulations can have a profound effect on the simulation outcome (see Section 5.1 on simulation convergence and reproducibility), which can be especially problematic if the active site is relatively large, solvent-exposed, and no information about the actual binding pose is known. Computational prediction of the binding pose using molecular docking is an option. If employed, we advise to restrict the search to chemically reactive conformations (e.g., using geometric constraints). Note that the manual placement of the substrate in the active site using chemical intuition, followed by MD equilibration, often produces satisfactory results. In the case of our recent work (Purg et al., 2017), the substrate DFP was placed in the active site (Fig. 3A) in such a way as to optimize (1) the alignment of the nucleophilic water molecule with the oxygen atom of D229 and with the scissile bond (both in terms of the O_{nuc}-P distance and the O_{nuc}-P-F angle), as well as (2) the alignment of the reacting atoms relative to the side chain of the key catalytic residues, and, finally, (3) by comparison to the substrate analogue dicyclopentylphosphoroamidate (DcPPA) found in the DFPase active site (PDB ID: 2GVV (Blum et al., 2006)) (Fig. 3B).

3.2.5 System Solvation and Topology Generation

In the EVB methodology, the system is usually solvated in a spherical droplet of explicit water molecules according to the surface constrained all-atom solvent (SCAAS) model (King & Warshel, 1989) or its variations

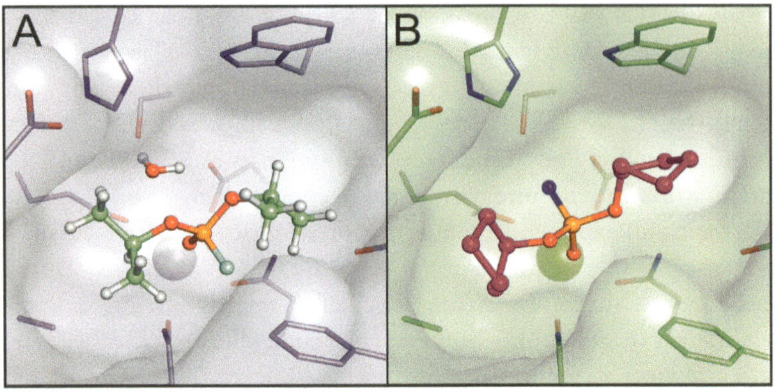

Fig. 3 A structural comparison of (A) diisopropyl fluorophosphate (DFP), and (B) DcPPA in the active site of DFPase (PDB IDs: 2GVV (Blum et al., 2006) and 3BYC (Blum et al., 2009), respectively).

(Marelius, Kolmodin, Feierberg, & Åqvist, 1998). Spherical boundary conditions allow us to limit the size of the system, thus reducing the computational cost, as well as allowing for accurate treatment of long-range electrostatics (Marelius et al., 1998). The choice of the size of the solvation sphere is a trade-off between accuracy and the associated computational cost. All atoms outside the solvation sphere will be restrained to their crystallographic positions as well as excluded from evaluation of nonbonded interactions; thus care must be taken to not exclude any "important" features of the enzyme. In practice, a 20 Å radius is considered to be the lower limit to provide a correct treatment of electrostatics (although spheres of smaller radius have been used successfully in the past when computational resources were more limited Sham & Warshel, 1998). We note also that the appropriateness of the sphere radius can be easily validated by running simulations with different sphere sizes, and increasing the sphere size until the results no longer change (i.e., the simulations converge).

The information stored in Q6 library files (such as atom types, connectivity, or atomic charges) is combined with bonding and nonbonding parameter definitions found in a Q6 parameter file, and atom coordinates stored in a pdb format, into a Q6 topology file using Qprep6 (example input with comments is below). Additionally, the boundary conditions and solvation are defined at this stage, as well as any additional interresidue bonds (e.g., disulfide bridges). Note that as Qprep6 is an interactive command-line utility, the input file must be redirected to standard input:

```
Qprep6 < qprep.inp > qprep.out

Qprep6 input

# read library files
readlib ./0-ff/ohh.lib
readlib ./0-ff/dfp.lib
readlib ./0-ff/ca6.lib
readlib ./0-ff/qoplsaa.lib

# read parameter file (only one)
readprm ./0-ff/qoplsaa_dfp_ca6.prm

# read coordinate file
readpdb dfpase_dfp.pdb

# boundary and solvation
boundary 1 314:C61 25   # sphere, center at 314:C61, radius 25
solvate 314:C61 25 1 HOH  # center at 314:C61, radius 25, grid, tip3p

# write topology and new coordinate file
maketop   pro.top
writetop  pro.top
writepdb  dfpase_dfp_start.pdb y
quit
```

3.2.6 Creating an EVB Parameter File

The EVB force-field parameters for each VB state are defined in the plain text in the EVB parameter file (referred to as an "FEP file" by Q6, as the same file format is used for all free-energy simulations performed by Q6). The relevant VB states used to describe DFPase hydrolysis by DFP are shown in Fig. 4, as well as in the Supporting Information of Purg et al. (2017). As can be seen from the text box below, the FEP file provides the EVB parameters in specific sections, defined by square brackets (*atoms*, *atom_types*, *etc.*). Generating a FEP file requires the comparison of all the parameters that define the reacting atoms in all VB states being considered, and extracting those that change due to the formation and creation of bonds (this can affect all EVB parameter types). To ease the process, scripts for the generation of EVB fep files are provided as part of *Qtools*. Nevertheless, we strongly advise readers to first familiarize themselves with the details of creating the FEP file by hand, before using tools such as *q_makefep.py* (described later). The text box below shows an example of the resulting (truncated and commented) FEP file; for more details, we refer the reader to the Q6 manual.

```
FEP file sample

[FEP]  # define number of states, index offset, ...
states 2

[atoms]  # map topology (.pdb) atom indices to internal indices
1    4841

[atom_types]  # define vdw parameter types (see manual for
   formatting)
dfp_P1  2447.7925  46.7907  43.00  1.5  1730.8507  33.0860  30.974

[change_atoms]  # define changes in vdw parameters (internal
   indices)
1  dfp_P1  dip_P1

[change_charges]  # define changes in charges (internal
   indices)
1  1.2831  1.4156

[soft_pairs]  # use soft-pair potential instead of Lennard-
   Jones
33  27          # (internal indices)

[off_diagonals] # pre-defined off-diagonals
1  2  33  27  0  0

[bond_types]  # bond parameter types (both harmonic and Morse)
1    1312.0    1.250             # (fc, r0)
2     900.0    1.364
5     186.0    1.44    1.574  # (De, b, r0)

[change_bonds]  # bond changes
3454  3455      1    2  # (left: atom pdb indexes)
4841  4849      5    0  # (right: bond_type indexes, 0=no bond)

[angle_types]  # angle parameter types
1     140.000   117.000
2     140.000   108.000

[change_angles]  # angle changes
3451  3454  3455     1  2

# Proper and improper torsions are treated analogously
# (see Q6 manual for details)
```

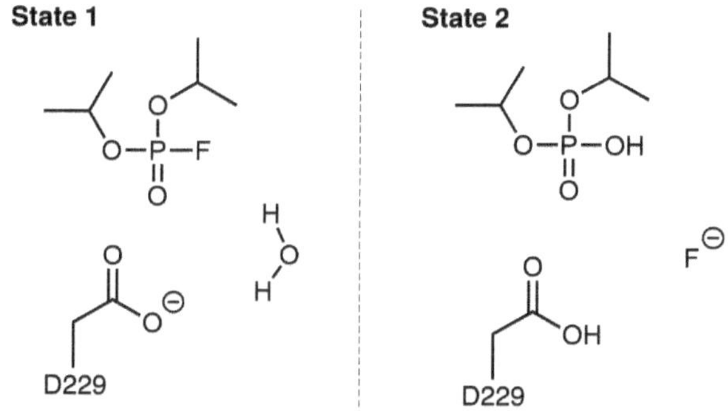

Fig. 4 Valence bond states used to describe the hydrolysis of diisopropyl fluorophosphates (DFP) by DFPase through a general-base mechanism (see Fig. 1A).

The basic requirements to create an FEP file using the *q_makefep.py* utility are:

- Q6 library files (*.lib*) for all species in the system, in all EVB states to be modeled.
- Q6 parameter files (*.prm*) for all species in the system, in all EVB states to be modeled.
- An atomic coordinate file (*.pdb* format) created using *Qprep6*.

Additionally, to identify changes in connectivity during the reaction, *q_makefep.py* requires the user to define a map of each reactive (Q) atom in the atomic coordinate file (defined by the residue index and atom name) to their respective library IDs (residue name and atom name) in each EVB state, in the so-called qmap file. The following command will generate an FEP file for the DFPase model system:

```
# In directory 0-topol/1-fep

# Tip: Use -h flag to display the help message

q_makefep.py -m dfpase_dfp_hoh_asp229.qmap \
             -s dfpase_dfp_start.pdb \
             -f oplsaa \
             -p 0-ff/*prm \
             -l 0-ff/*lib \
             -o dfpase_dfp_hoh_asp229.fep.gen
```

Note, however, that a few parameters such as Morse potentials, soft-pair potentials, and off-diagonal definitions must be added manually by the user. All input files as well as the complete EVB FEP file to model the hydrolysis of DFP by DFPase via a general-base mechanism (*dfpase_dfp_hoh_asp229.fep*) are provided at the following link: https://github.com/mpurg/methodsenz.

3.3 System Equilibration

All MD simulations in Q6 are run using Qdyn6 or its MPI-enabled variant Qdyn6p. A typical Q6 workflow includes some form of equilibration stage prior to actual EVB simulations, comprising several steps, including removing steric clashes, equilibrating the solvent, and finally, equilibrating the solute. An example input file with comments is provided below; for additional information we refer the reader to the official Q6 manual.

Many such input files are typically used during one equilibration protocol, with the restart keyword pointing to the previous simulation's final restart file.

```
Qdyn6 input file

[MD]
steps                    100000      # 100ps
temperature                 300      # 300K
stepsize                      1      # 1fs

[cut-offs]
q_atom                       99      # 99 A

[intervals]
non_bond                     30      # neighbor-list update (30fs)
output                     1000      # output write interval
trajectory                 1000      # trajectory write interval

[files]
topology               N65A.top      # topology
restart            relax_006.re      # input restart file
final              relax_007.re      # output restart file
trajectory        relax_007.dcd      # output trajectory
fep                    N65A.fep      # FEP file

[lambdas]
1.00 0.00                            # lambdas for two states
```

An example of the command-line input for running a simulation with Qdyn6 is:

```
# serial version
Qdyn6 relax.inp > relax.log

# parallel version using 4 MPI threads
mpirun --np 4 Qdyn6p relax.inp > relax.log
```

In the subsequent sections, we describe the equilibration protocol used in our recent study of DFPase (Purg et al., 2017); however, we would like to stress that, while we have tried to develop a fairly universal equilibration protocol for EVB simulations, more elaborate equilibration procedures might be needed for other and more complex systems and therefore this is only a starting point. In addition, the equilibration is not performed at either of the two VB states, but rather with the coupling parameter λ set to 0.5 (denoting an equal mix of the of reactants and products' force fields), to remove the need for positional restraints during the equilibration procedure (see Section 5.2).

3.3.1 Removing Steric Clashes

The starting structure to be used for the simulations (as obtained via X-ray crystallography, neutron scattering, or other techniques) will lie far from the global minima on the potential energy surface that is defined by the chosen force field. That is, the bonding and nonbonding interactions will not be near their equilibrium values as defined by the force field. Additionally, Qprep6 does not position the hydrogen atoms ideally upon their generation. All these factors can lead to the development of large forces during the simulation, resulting in simulation artifacts or "system explosions" (see Section 5.4.1). This can be remedied by running short, low-temperature MD simulations, using an extremely small time step and strong bath coupling, in order to gently ease steric clashes out of the system.

Qdyn6 input files used on the model system discussed in the present contribution can be generated using a helper tool *q_genrelax.py* (in Qtools) and the provided *genrelax.proc*, using the following command (note that the FEP file and topology are required):

```
# In directory 1-min

# Tip: Use -h flag to display the help message

q_genrelax.py genrelax.proc \
              --top   ../0-topol/pro.top \
              --fep   ../0-topol/dfpase_dfp_hoh_asp229.fep \
              --pdb   ../0-topol/dfpase_dfp_start.pdb \
              --rs    run_q.sh
```

The procedure consists of a few short simulations (total 1000 steps with a time step of $\Delta t = 0.1$ fs), followed by a slightly longer simulation (50,000 steps with $\Delta t = 0.5$ fs), all at temperature 0.1 K, using strong bath coupling, restraining the solute atoms to their initial positions with a 20 kcal mol^{-1} harmonic restraint and reseeding the velocities in each step. The whole procedure is then repeated without the positional restraints. The simulation can be run using the provided run_q.sh script.

3.3.2 Thermal Equilibration

The next step in the simulation procedure is to bring the system to the desired simulation temperature, where it is allowed to equilibrate with the heat bath. It is recommended that the solvent be allowed to equilibrate first, in order to prevent unphysical fluctuations of solute atoms, leading the system to stray away from the (minimized) starting structure. This is achieved by applying strong positional restraints on the solute atoms, which are then gradually released in the subsequent steps. In some cases it might also be required to keep small positional (or other geometric) restraints (on the range of 0.5 kcal mol^{-1} Å^{-2}) on the reactive region (Q atoms) to prevent the dissociation of the reacting fragments or the formation of chemically nonreactive (for instance geometrically distorted) starting complexes.

Note: In the beginning of this stage, it is advisable to introduce diversification to the experiment by running many independent simulations with different initial conditions (in practice 30 independent trajectories yield good statistics, with the initial velocities generated via the assignment of different random seeds). This will allow for a good degree of reproducibility of the calculations (see Section 5.1).

Qdyn6 input files for this stage can again be generated using *q_genrelax.py* and the provided input files, once the previous simulation has finished, using the following command:

```
# In directory 2-equil
q_genrelax.py template/genrelax.proc \
              --cont   ../1-min/relax/relax_012.inp \
              --rest   cont_final \
              --pdb    ../0-topol/dfpase_dfp_start.pdb \
              --rs     template/run_q.sh
```

The equilibration procedure begins with the equilibration of the solvent atoms at the desired temperature of 298.15 K, while the solute atoms are restrained to their initial coordinates using 200 kcal mol^{-1} $Å^{-2}$ harmonic restraints on each solute (heavy) atom. This is followed by the gradual release of the positional restraints and, finally, by 10 ns of fully unrestrained equilibration. The simulation can be run using the provided run_q.sh script.

We note that DFPase is a particularly stable system, which is why it was possible to start the equilibration directly at 298.15 K. However, for most systems, an initial heating procedure would be required and we recommend the system to be gradually heated from 0 to 300 K, to avoid instabilities introduced by "system shock" due to sudden heating of the system.

The initial coordinates and velocities are extracted from the final restart file of the previous "minimization" stage; however, to allow for multiple independent trajectories (note that only one is used in this example), the latter are manually reseeded in the first step.

The same coordinates are also used as a reference for positional ("sequence") restraints during the equilibration procedure and are obtained from a copy of the aforementioned restart file (defined by the keyword "restraints" in the "files" section of the input). The alternative would be to omit the "restraints" keyword. In the absence of the "restraints" keyword, the program automatically sets sequence restraints using the coordinates provided in the topology file.

3.4 Simulating Chemical Reactions

A convenient way to simulate chemical reactions in the context of the EVB methodology is the use of the so-called EVB mapping approach (see Section 2). Here, the mapping potential (a linear combination of two or more valence-bond states) is used to sample the phase space along a nongeometric energy-gap reaction coordinate. The imposed bias is removed in postprocessing in the so-called mapping stage, yielding the EVB ground-state reaction free-energy profiles.

A typical system setup would include simulating the *same reaction mechanism* (as defined by the parameters in the FEP file) in both the studied system and a reference system, where the latter is used to calibrate the EVB potential. To obtain a unique fit for the two unknown parameters (H_{12} and α) in a two-state EVB description, we require two known observables—usually these are the activation and reaction free energy of the reference reaction, determined via experimental or high-level QM calculations.

Since the expression for the EVB ground-state potential is introduced only in the postprocessing stage, the calibration process conveniently does not require rerunning the whole reference simulation on the mapping potential each time the recalibration is required, and remapping the energies with the new parameters is sufficient.

3.4.1 Sampling on the Mapping Potential

The coupling parameter, λ, is commonly divided into 51 or 101 equally separated windows, spanning from $\lambda = 1.0$ (reactants) to $\lambda = 0.0$ (products), each window requiring its own Qdyn6 input file and a separate simulation. Note that the above-suggested directionality from reactants to products is not a simulation requirement, and, indeed, all windows can be simulated in parallel and independent of each other. In practice, however, this requires longer equilibration times for each window; thus sequential simulations are preferred to save computational cost. The starting point for a sequential simulation can in principle be at any λ-value, so long as the system has been properly equilibrated at that state. Note that the same procedure should be used for the reference simulation.

Provided the equilibration stage has finished, *q_genfeps.py* can be used to generate Qdyn6 input files for the reaction stage, using the following command:

```
# In directory 3-fep

# Tip: Use -h flag to display the help message

q_genfeps.py template/genfeps.proc \
             ../2-equil/relax/relax_015.inp \
             relax \
             --pdb ../0-topol/dfpase_dfp_start.pdb \
             --repeats 1 \
             --frames 51 \
             --rs run_q.sh
```

The procedure consists of a short equilibration step (100 ps), followed by a total of 51 frames (λ windows) comprised of two sequential trajectories starting from the equilibrated structure at $\lambda = 0.5$, and leading to $\lambda = 1.0$ and $\lambda = 0.0$. Energy values are written out every 10 fs, as is common practice, producing a total of 10,000 energy points per window. The simulation can be run using the provided run_q.sh script.

3.4.2 Obtaining the EVB Free-Energy Profiles

Once the EVB trajectories have been generated, the corresponding EVB free-energy profiles are obtained from the simulations by postprocessing the raw data, via the interactive command-line utility Qfep6, by providing the files containing the energy values during the simulation (energy files, with the .en filename extension) and the calibrated EVB parameters (H_{ij} and α). Additional EVB mapping parameters are required, including k_BT, number of bins, number of points to skip, etc., and these are described in detail in the Q6 manual. Alternatively, to save time, the relevant commands can be written into an input file (example with comments below) and redirected to the standard input:

```
Qfep6 < qfep.inp > qfep.out

Qfep6 input

51                    # number of files/frames
2 0                   # number of states and predefined off-diagonals
0.592       1000      # kT (kcal/mol) and number of points to skip
                        per file
100                   # number of Egap bins
100                   # minimum points per bin
-42.97                # state2 shift (alpha)
1                     # number of off-diagonals
1 2 152.4   0 0 0     # H12: A=const, mu=eta=r0=0
1 -1                  # e1 - e2
fep_050_1.000.en      # list of energy files, sorted by lambda value
fep_049_0.980.en
...
fep_024_0.020.en
fep_025_0.000.en
stop
```

The output from Qfep6 consists of four parts:

- Part 0: Average energies for all EVB states in all files.
- Part 1: FEP summary (Fig. 2C; Eq. 7).
- Part 2: Reaction free-energy summary (Fig. 2D; Eq. 8).
- Part 3: Bin-averaged summary (Fig. 2E; Eq. 9).

We are most interested in part 3, which contains the energy-gap reaction coordinate (the second column of the output) and the normalized bin-averaged free energies (the fourth column of the output).

Qtools includes a command-line tool, *q_mapper.py*, which automatically creates Qfep6 inputs, executes Qfep6, and analyzes the outputs. An example command using the EVB parameters $H_{12} = 152.4$ and $\alpha = -42.97$ is shown below:

```
# Tip: Use -h flag to display the help message
q_mapper.py 152.4 -42.97 --dirs rep_000
```

Additional information about the simulations (LRA contributions, group exclusions, quantum classical path calculations, etc.) can be obtained using the command *q_analysefeps.py*:

```
q_analysefeps.py rep_000
```

This command reports activation, reaction, and FEP free energies as well as exporting most of the raw data to a JSON-formatted file, which can be used with the plotting utility *q_plot.py*:

```
# Run graphical plotting tool (useful for debugging)
q_plot.py qaf.PlotData.json

# Export the data to Grace ascii format
q_plot.py qaf.PlotData.json --export all
```

For further discussion of output using *q_plot.py*, as well as relevant examples, see Section 5.4.3.

3.4.3 Calibrating the EVB Potential

Here, we will consider only the example scenario with two EVB states, where the unknown parameters H_{12} and α can be uniquely determined by fitting the free-energy profile of the reference reaction to known activation ($\Delta G^{\ddagger}$) and reaction (ΔG_0) free energies. In practice, this fitting is an iterative approach of rerunning *Qfep6* with varying values of H_{12} and α until the free energies coincide with reference values.

To save time to not have to find the correct parameters manually, *Qtools* includes the tool *q_automapper.py*, which uses steepest descent to optimize the parameters to reproduce the reference free-energy values with the desired level of precision:

```
# Tip: Use -h flag to display the help message
q_automapper.py 21.2 -2.3 50 0 --dirs rep_000
```

Once the mapping procedure is complete for the reference reaction, the resulting H_{12} and α parameters can be used unchanged to describe the same reaction with the same reacting atoms/parameters in any environment, whether it is in vacuum, other solvents, the active sites of different natural or designed enzyme variants, or even the same reaction occurring in association with the presence of (in)organic catalysts.

4. EMPIRICAL VALENCE BOND SIMULATIONS: SAMPLE RESULTS

In this section, we will provide an overview of the type of results to expect from EVB calculations performed following the protocols described in Sections 2 and 3. For a detailed discussion of the interpretation of these results, see Purg et al. (2017).

As a starting point, a key objective of EVB simulations is to obtain free-energy profiles for different possible reaction mechanisms, in different environments. A comparison of the free-energy profiles for both the nonenzymatic- and DFPase-catalyzed hydrolyses of DFP, via the general-base mechanism shown in Fig. 1, is shown in Fig. 5. These calculations yield average activation free energies of 14.7 and 21.2 kcal mol^{-1} in DFPase and in solution, corresponding to a barrier reduction of 6.5 kcal mol^{-1} upon moving to the DFPase active site, in good agreement with experimental observables. We note that if the system is well parameterized and adequate conformational sampling is performed, the standard deviations between multiple individual trajectories should be small (ideally less than 1 kcal mol^{-1}), and larger standard deviations are indicative of problems with the simulation, whether it is a poor simulation setup, poor parameterization, or insufficient conformational sampling.

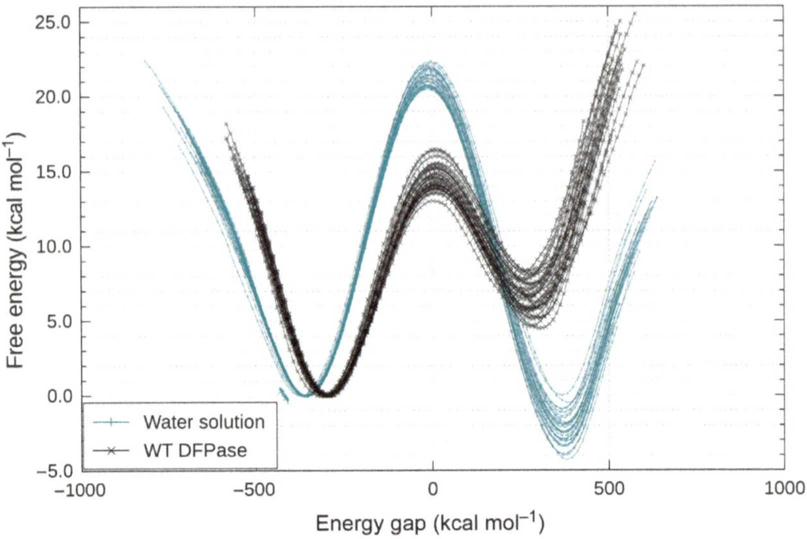

Fig. 5 Actual EVB free-energy profiles for the hydrolysis of diisopropyl fluorophosphate (DFP) in aqueous solution (*blue*) and in the active site of wild-type DFPase (*black*).

One may, of course, also be interested in quantifying the electrostatic or vdW contributions of different amino acids to the calculated activation free energies, for example, to identify hot spots for subsequent mutagenesis in enzyme design studies. There are a number of ways to do this, one of the most straightforward of these being to apply the linear response approximation to the calculated EVB trajectories (for a theoretical background, see, e.g., Lee, Chu, Bolger, & Warshel, 1992; Muegge, Tao, & Warshel, 1997), using *Qcalc6* in combination with *q_calc.py* (*Qtools*) as shown in the example later. The following commands calculate the per-residue contributions of moving from an approximate reactant state ($\lambda = 1.0$) to an approximate TS ($\lambda = 0.5$), for residues indexed 1–312, producing data similar to the plot shown in Fig. 6:

```
q_calc.py gc dfpase_dfp_start.pdb 1 312 --dirs rep_000

# Plot the contributions using q_plot.py
q_plot.py qgc.PlotData.json
# Export the contributions to Grace format
q_plot.py qgc.PlotData.json --export all
```

A more accurate description of the reactant state and TS (as defined by the EVB free-energy profile) can be obtained by locating the mapping

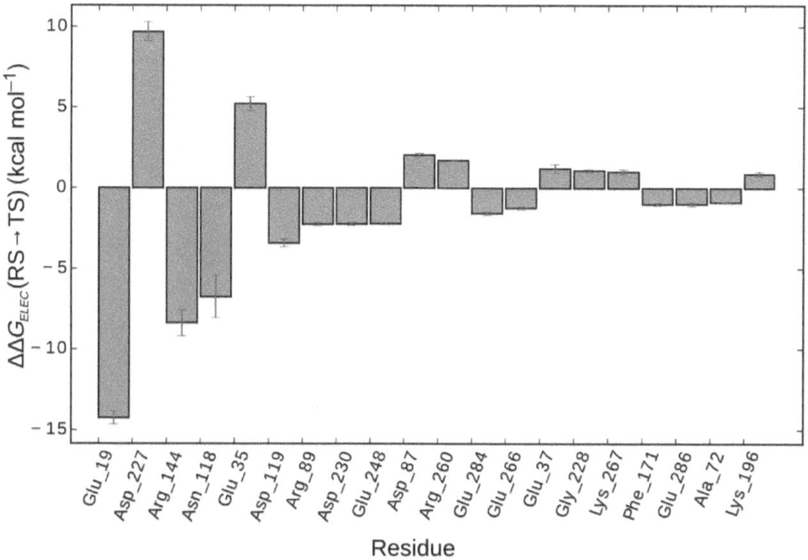

Fig. 6 Electrostatic free-energy contribution of individual residues to the activation free energy for the hydrolysis of diisopropyl fluorophosphate (DFP) by wild-type DFPase, extracted from calculated EVB trajectories using the linear response approximation (Lee et al., 1992). The data shown here is the output from Qtools v.0.6.0.

potential windows with the highest sampling contributions to the appropriate bins (RS, TS) on the energy-gap coordinate (i.e., *Part 2* in the *Qfep6* output). The λ-values associated with these mapping potential windows are then passed to *q_calc.py* via the –lra_l argument (e.g., –lra_l 0.84 0.46).

Clearly, reproducing the catalytic effect of a known enzyme system is a good start; however, it is possible to obtain the right answer for a multitude of wrong reasons, the most obvious being serendipitous cancellation of errors. Therefore, it is critical to challenge the model against all available experimental observables, where simulations are possible. The most obvious of these is to attempt to reproduce known experimental observables for a wide range of enzyme variants, as independent simulations without the need for reparameterization. Fortunately, there exist several studies of DFP hydrolysis by both wild-type DFPase and different DFPase variants (Katsemi et al., 2005; Melzer et al., 2009), providing us with substantial experimental data to test against. For expediency in computational cost, a subset of these were tested in Purg et al. (2017), with the results shown in Fig. 7. As can be seen from this figure, we are able to reproduce the activation free energies of a wide range of enzyme variants,

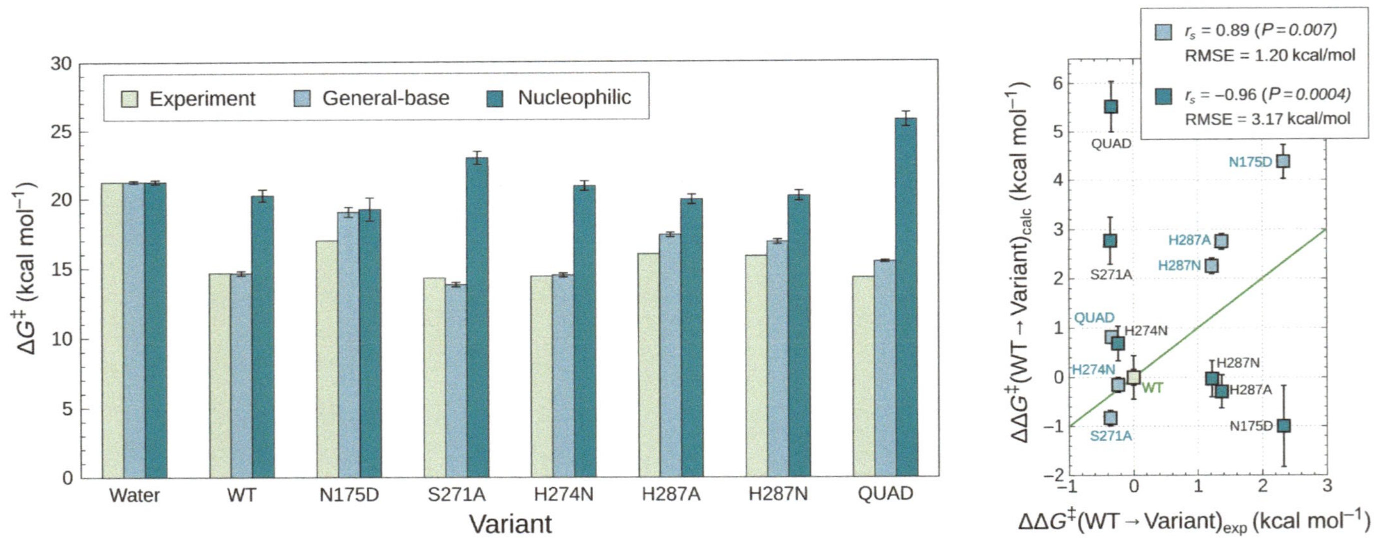

Fig. 7 A comparison of the calculated and experimental free energies for the hydrolysis of diisopropyl fluorophosphate (DFP) by both wild-type DFPase and a range of DFPase variants, via both general-base and nucleophilic substitution mechanisms. "QUAD" denotes an E73D/Y144A/R146A/T195M quadruple mutant form of DFPase. Shown to the *left* are the absolute values, and to the *right* the correlation between the calculated and experimental values for each mechanism. For the original data and more detailed discussion, see Purg et al. (2017). *This figure was originally published in Purg, M., Elias, M., & Kamerlin, S. C. L. (2017). Similar active sites and mechanisms do not lead to cross-promiscuity in organophosphate hydrolysis: Implications for biotherapeutic engineering.* Journal of the American Chemical Society, 139*(48), 17533–17546 and reproduced with permission from the American Chemical Society.*

without the need for subsequent reparameterization, while following experimental trends in activity.

Finally, several studies have shown that computational Arrhenius/Eyring plots are an effective means of distinguishing between different mechanistic possibilities in phosphoryl transfer and other reactions (Åqvist & Kamerlin, 2015; Isaksen, Åqvist, & Brandsdal, 2016; Kazemi & Åqvist, 2015). DFPase is no exception, and we were able to use such plots to discriminate between general-base and nucleophilic substitution mechanisms involving D229. As can be seen from Fig. 8, only the general-base mechanism provides reasonable agreement for experiment, the reasons for which are discussed in Purg et al. (2017). Such plots can be generated effectively by running the EVB simulations at multiple different temperatures. However, a number of factors need to be taken into account: the temperature range and intervals considered the sphere size and associated radius of the "mobile region" in the multilayer solvent model, the number of independent trajectories performed,

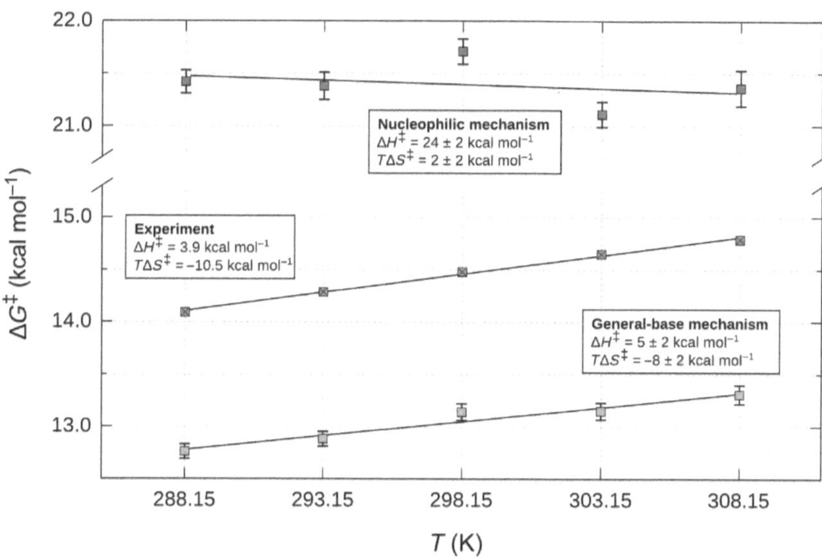

Fig. 8 A comparison of experimental and calculated temperature dependence for the hydrolysis of diisopropyl fluorophosphate (DFP) by wild-type DFPase via both general-base and nucleophilic substitution mechanisms. For the original data and for simulation details, see Purg et al. (2017). *This figure was originally presented in Purg, M., Elias, M., & Kamerlin, S. C. L. (2017). Similar active sites and mechanisms do not lead to cross-promiscuity in organophosphate hydrolysis: Implications for biotherapeutic engineering.* Journal of the American Chemical Society, 139*(48), 17533–17546, and is reproduced with permission from the American Chemical Society.*

and the simulation timescales for the individual trajectories. To obtain meaningful convergent plots of the temperature dependence, typically at least 100–150 simulations per temperature point are necessary.

Clearly, the examples earlier are not the entirety of the type of results that can be obtained by EVB simulations. However, the availability of extensive experimental data including the activities of a wide range of DFPase variants toward DFP, as well as the thermodynamic parameters for the DFPase-catalyzed hydrolysis of DFP, and also the relative stability of the system in simulations make DFP a particularly attractive system for testing methodological protocols for computational enzymology. As can be seen from this data, it is critical not only to rely on a single experimental observable but also to examine all possible available observables when discriminating between different mechanisms. Once the simulations have been performed, the output can be combined with both visual inspection and different types of structural bioinformatics analysis (for example, to obtain insight into structural dynamics, correlated motions, active-site volumes, active-site hydrophobicity, or solvent accessible surface area) to provide further insight into the origins of the observed catalytic effects, and how they are impacted by mutations.

5. TECHNICAL CONSIDERATIONS

5.1 Convergence and Reproducibility

In an ideal scenario, one would perform complete sampling of the configurational space of a biological system in order to obtain fully reproducible and precise (according to the model) observables in a fully convergent fashion. This is not possible, however, due to excessive computational cost and finite computational resources. Therefore, all simulations of biomolecular systems carry approximations and are a compromise between cost and accuracy. Taking into account these limitations, it is important for the user to validate that their simulations are converged and reproducible, for example, by performing simulations of increasingly longer simulation time until the observable of interest (such as an activation free energy) reaches a limiting value. In the case of EVB simulations, this can be done by running simulations with increasingly longer sampling times: 10 ps sampling per window, then 50, 100, or 200 ps, and monitoring how the desired observable changes with increasing simulation time (Fig. 9A). Equilibration time (if inadequate) and sphere size (if too small to correctly screen the charges of the reacting region) can also have a profound effect on simulation outcomes, in particular

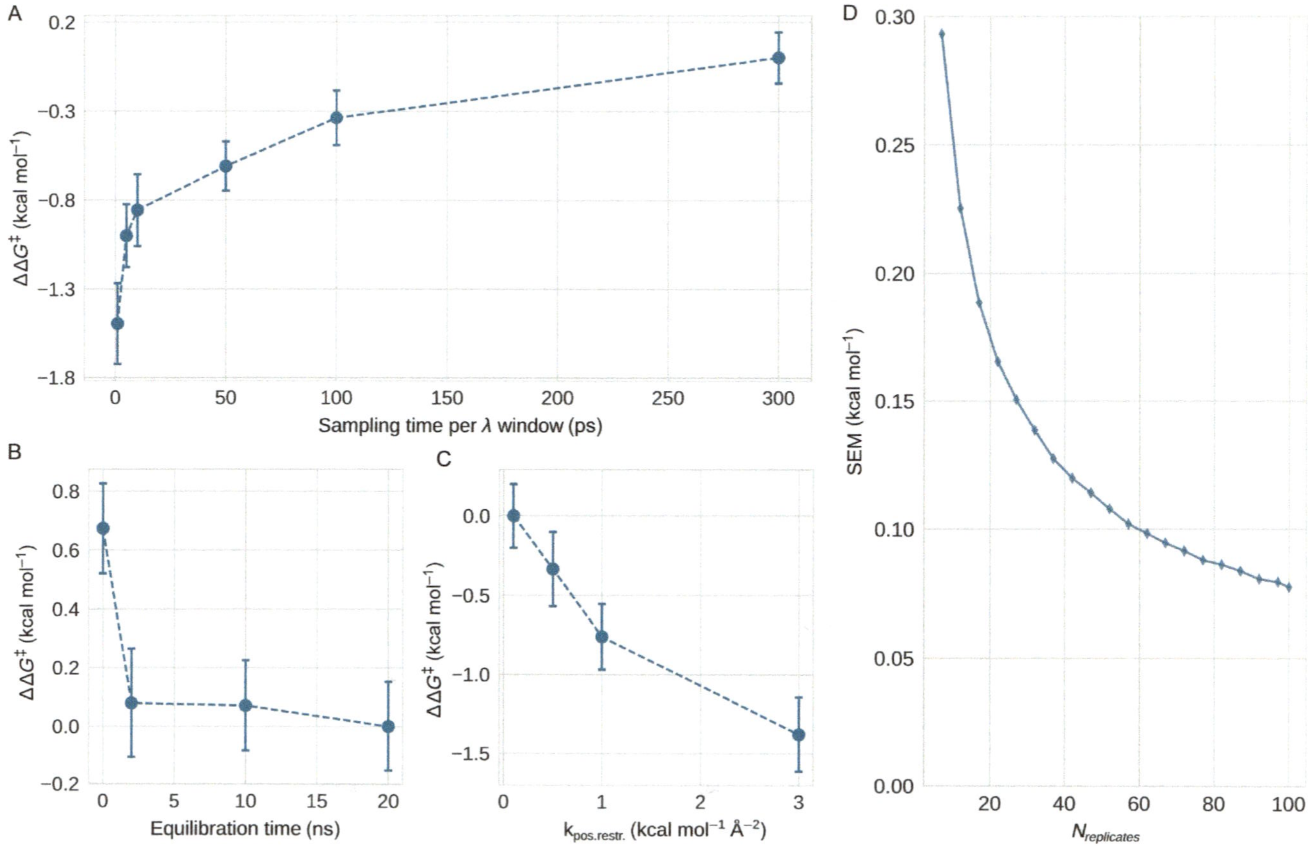

Fig. 9 See legend on next page.

for highly conformationally dynamic systems or reactions with large charge changes along the reaction coordinate. Again, this can be tested by adjusting the equilibration time (Fig. 9B) or sphere size for a test system before starting large-scale simulations.

The lack of fully convergent simulations in conjunction with the fact that classical simulation trajectories are, in general, not reproducible (unless performed on the exact same machine with the same binary, environment, and the same number of MPI threads) demands using a statistical approach to estimate the observables and errors associated with the measurement. In a typical simulation setup, between 10 and 30 independent trajectories will be propagated to obtain better coverage of the conformational space, but for addressing observables where higher precision is required (for example, when studying the temperature dependence of the observed reaction rates), it might be necessary to run 100 or more independent trajectories to obtain statistically meaningful outcomes (Åqvist, Kazemi, Isaksen, & Brandsdal, 2017) (see also Fig. 9D).

5.2 Chemical and Conformational Coordinates

Simulations of chemical reactions should in theory sample space not only along the reaction but also along the entire conformational coordinate which can, depending on the flexibility of the systems involved, drastically increase the complexity of the system and, in many cases, create a need for prohibitively long simulation times to obtain fully convergent results. This

Fig. 9 Various effects on simulation outcome and statistical data as obtained from EVB simulations of the model system described in Section 3. All error-bars shown are standard errors of mean ($N=30$). (A) Dependence of average activation free energy, with respect to sampling time per mapping window (using $t_{equil}=10$ ns, $k_{pos.restr}=0.1$ kcal mol^{-1}), for EVB simulations performed in 51 mapping windows. Shown are values at 1, 5, 10, 50, 100, and 300 ps/frame relative to that obtained at 300 ps/frame simulation time (15.0 kcal mol^{-1}). (B) Dependence of the average activation free energy, with respect to equilibration time (using $t_\lambda=100$ ps, $k_{pos.restr}=0.1$ kcal mol^{-1}). Shown are values at 0, 2, 10, and 20 ns equilibration relative to that obtained at 20 ns equilibration time (14.6 kcal mol^{-1}). (C) Dependence of the average activation free energy, with respect to strength of positional restraints employed on the reactive region during reaction simulations (using $t_{equil}=10$ ns, $t_\lambda=10$ ps). Shown are values obtained using restraints of 0.1, 0.5, 1.0, and 3.0 kcal mol^{-1} Å^{-2}, relative to that obtained using a restraint of 0.1 kcal mol^{-1} Å^{-2} on the reacting atoms (14.2 kcal mol^{-1}). (D) Dependence of the standard error of mean (SEM) on the number of individual replicates. The data was obtained using $t_\lambda=10$ ps, $k_{pos.restr}=0.1$ kcal mol^{-1}, and $t_{equil}=10$ ns (+1 ns per replicate).

can be partially addressed by use of weak restraints on the reactive subsystem, thus restricting the conformational sampling to only a small portion of the conformational phase space and improving the convergence of free-energy profiles along the reaction coordinate for that particular conformation. In practice, this can mean the application of harmonic restraints of up to $0.5\,\text{kcal}\,\text{mol}^{-1}\,\text{Å}^{-2}$ on the reacting atoms in an enzyme active site and up to $1.0\,\text{kcal}\,\text{mol}^{-1}\,\text{Å}^{-2}$ on the reacting atoms for a nonenzymatic reaction. Additional geometric features like donor–acceptor distances or certain angles may also need to be restrained by weak restraints to improve the convergence of the simulations and avoid sampling nonproductive conformational space. Note that the effect of restraints on the underlying potential energy surface has not been quantified in detail, with qualitative analysis revealing possible substantial contributions; thus their use should be as limited as possible and applied with care. Fig. 9C shows the effect of restraints on the activation free energies of the model system described in Section 3. We encourage the reader to review the recently published analysis of sampling-related issues and use of restraints in EVB simulations by Stare (2017).

To properly address the remainder of the somewhat ignored conformational dimension of the reaction's energy landscape, several (restrained) simulations, using different starting conformations, should be performed. These conformations can be obtained via computational methods such as docking or unrestrained MD simulations; however, in practice it is observed that without additional user-imposed geometrical restraints, these methods perform rather poorly due to not being able to discriminate between chemically productive and nonproductive conformations. Manual placement of the substrate in a chemically meaningful reactive conformation, which is driven by chemical intuition, is usually a superior method, albeit introducing a significant level of user bias into the simulation (see Section 5.3 for a possible solution to this problem).

5.3 Over the Mountain Pass or Rappelling Into Valleys?

The use of a biasing potential in EVB simulations involves moving the EVB potential between different reacting states (e.g., from a reactant complex to either the product or different intermediate states), which can be described as "driving over a mountain pass," where the mountain pass represents the reaction TS and the road a conformationally restrained reaction pathway. As explained in Section 5.2, small positional restraints on the reactive region are usually required in order to properly sample the reactive phase space (road) and limit the sampling of the conformational phase space (off-road or alternate road) and thus obtain converged free-energy profiles. These, however, introduce a bias toward a particular starting structure as well as an energy penalty due to pulling the reacting species back to their original position (the reactant's "valley").

A practical remedy to this problem might be to propagate the trajectories from the reaction midpoint (around the TS), which we could instead describe as "rappelling into valleys." The benefit of this approach is that no restraints are needed during the equilibration steps due to the partially bonded character of the TS, thus minimizing the bias from manual (or computational) docking, as well as allowing for unrestricted exploration of the conformational space to ensure diversity between independent replicas. Additionally, the effect of using restraints during the EVB simulations is less pronounced, due to using the midpoint as the reference state against which the restraints are defined.

5.4 Troubleshooting

In this section, we will briefly discuss common problems one might face while performing EVB simulations. Since the solutions to these problems are often not straightforward and comprise many variables, we offer instead general guidelines and strategies as to how to approach these issues, and where to look for possible solutions.

5.4.1 System "Explosions"

There are a number of situations in which the simulation system can "explode" due to a drastic buildup of physically unreasonable large forces and velocities, which lead to a termination of the program. This can happen, for example, when the time step for integration of the equations of motion is too large to adequately sample the fastest motion in the system, which causes the calculated displacements in position to lead to unstable (i.e., high in force) geometries, further increasing the velocities, etc., eventually producing numeric overflows in the program (not-a-number values) or SHAKE failures, as shown in the text box below. We note that, in practice, this problem is most commonly observed when the initial coordinates of the system contain steric clashes and close contacts between atoms, which lead to high forces on the system, and the system has not been properly equilibrated (see Section 3.3). Another source of error could be the use of a physically unreasonable topology file with, for instance, bad connectivity records in the library and/or FEP file, or missing "GAP" or "TER" flags between nonconnected chains in the structure file. The culprit can usually be found by examining the first few "hot atom" warnings in the Qdyn6 output, and by visualizing the associated trajectory, to identify where in the system the problems started. This is best done with a smaller write-out interval, ideally writing output every simulation step until the point the simulation crashes. An example of such a simulation failure is as follows:

```
Qdyn6 output examples

# "Hot atoms"
>>> WARNING: hot atom, i =          4 Temp(i)=1508985.86

# Termination due to detected not-a-number (infinity) value
Total System temperature at step     100 T=       NaN
...
!!!!!!!!!!!!!!!!!!!!!!!!!!!!!!!!!!!!!!!!!!!!!!!!!!!!!!!!!!!!!!!!!!!!!!
ABNORMAL TERMINATION of Q
Terminating due to a detected NaN.
!!!!!!!!!!!!!!!!!!!!!!!!!!!!!!!!!!!!!!!!!!!!!!!!!!!!!!!!!!!!!!!!!!!!!!

# Termination due to detected shake failure
>>> Shake failed, i,j,d,d0 =   1035   1036   0.91627   0.95720
>>> Shake failed, i,j,d,d0 =   1035   1037   0.91627   0.95720
>>> Shake failed, i,j,d,d0 =   1036   1037   2.29098   1.51360
...
!!!!!!!!!!!!!!!!!!!!!!!!!!!!!!!!!!!!!!!!!!!!!!!!!!!!!!!!!!!!!!!!!!!!!!
ABNORMAL TERMINATION of Q
Terminating due to shake failure
!!!!!!!!!!!!!!!!!!!!!!!!!!!!!!!!!!!!!!!!!!!!!!!!!!!!!!!!!!!!!!!!!!!!!!
```

5.4.2 EVB Mapping Failures and Physically Unreasonable Results

The EVB mapping process can fail or produce erroneous results due to several reasons, including trivial ones such as missing energy files (likely a partially failed simulation) or too few data points and too many energy-gap bins, as well as less immediately obvious reasons, such as the use of bad parameters, (even minute) differences between FEP files for different systems, or calibrating the reference reaction to experimental values while simultaneously simulating physically unreasonable reaction pathways (e.g., systems with steric hindrance and/or improper alignment of the reacting species). Visualization of the resulting trajectories using, e.g., VMD (Humphrey,

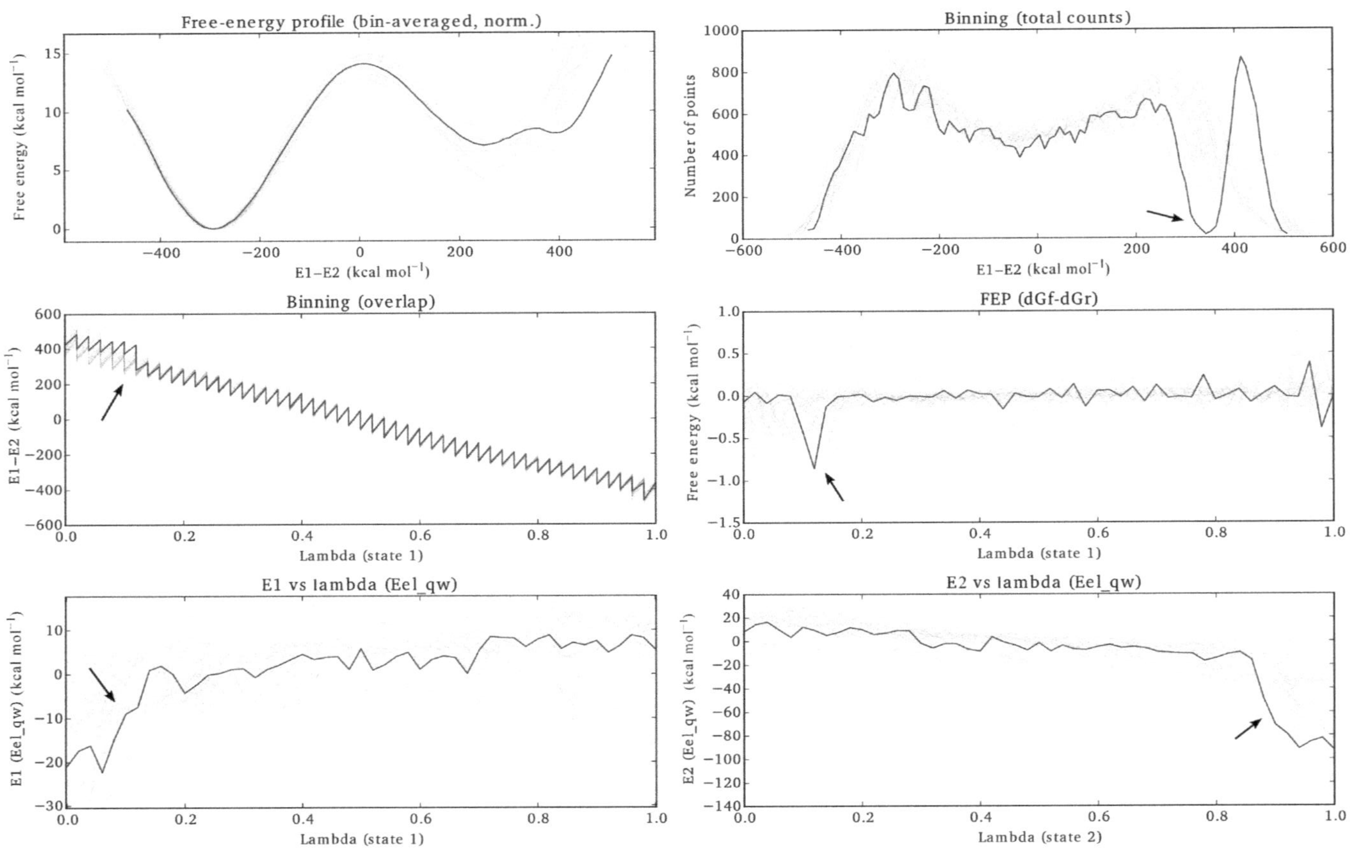

Fig. 10 See legend on opposite page.

Dalke, & Schulten, 1996), PyMOL (DeLano, 2002), or any other popular molecular visualization package is crucial to any simulation troubleshooting, as is measuring all the reactive distances and angles. Potential issues include, as mentioned in Section 5.4.1, poor alignment of the reacting moieties, steric hindrance during the reaction that is unreasonable overcome by the EVB biasing potential, as well as any sudden unexpected geometric changes in the system along the EVB reaction trajectory. We suggest extracting and plotting the energy contributions from the Qfep6 output file using *q_analysefeps.py* and *q_plot.py* (see Section 3.4.2), for both the reference and the nonreference reactions, and comparing them side by side.

5.4.3 "Rough" Free-Energy Profiles

A two-state EVB model is, by design, limited to modeling predefined single-step concerted processes (multistep mechanisms can be modeled in a stepwise fashion). This suggests that a free-energy profile with more than one maximum (TS) and two minima (reactant and product states) is likely a sign of simulation artifacts rather than being physical meaningful. Typical causes are insufficient sampling of rare events (e.g., residue flips, water influx) or sudden unphysical changes (e.g., breaks in distances), resulting in nonconvergent free-energy calculations. An example of such behavior is seen in Fig. 10 where the fluoride-leaving group of DFP (without any imposed distance restraints to the phosphorus atom) samples unwanted regions of phase space (such as completely dissociated conformations), producing a "rough" free-energy profile. Other common scenarios include flipping of a water nucleophile or general-base (an example is shown in Fig. 11; note that the issue here is not due to sampling per se, but rather due to the technical problem of the nucleophile's protons artificially not being chemically equivalent in the EVB framework because of the predefined bonding

Fig. 10 Example analysis of a "rough" free-energy profile (*top left*) obtained from simulations of DFP hydrolysis in DFPase, showing a clear lack of sampling in the product state (*top right*), poor overlap between the mapping potential and the energy-gap reaction coordinate in the product state region (*mid left*) and a spike in the difference between the forward and reverse FEP indicating poor configurational overlap, possibly due to a rare event (*mid right*). Additionally, the *bottom panels* indicate a drastic change in electrostatic interactions between the reactive region (q) and the solvent (w). The effect is significantly larger on the E2 (H_{22}) potential, which suggests that it is the negatively charged leaving group (F^-) that rapidly dissociates into solution, leading to nonconvergent simulations. Analysis of the *P–F* distance during the simulation confirms the nonphysical dissociation event. The plots were obtained with the *Qtools* utilities *q_analysefeps.py* and *q_plot.py* as described in Section 3.4.2.

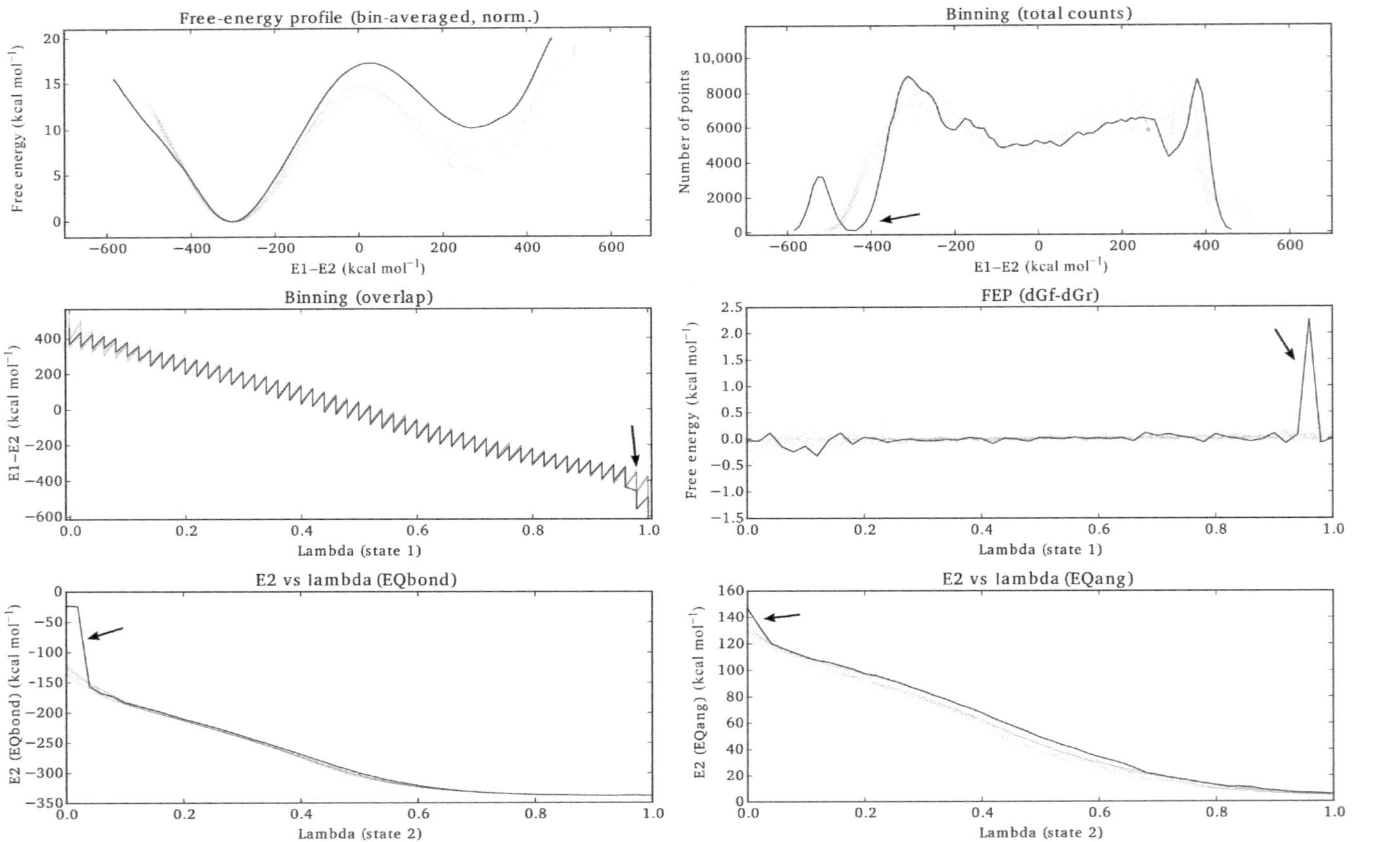

Fig. 11 See legend on opposite page.

patterns), water insertion (solvent molecule between substrate and nucleophile), and dissociation of the reacting species (reactive moieties leave the restrained solvent cage). These challenges are usually tackled by imposing some (justifiable) harmonic geometric restraints to avoid sampling non-reactive regions of phase space and thus adequately describe the desired mechanistic pathway.

5.4.4 Large Variations in the Free-Energy Results

As a final point, we note that large variations in the calculated free energies between individual EVB trajectories can arise due to a lack of complete convergence of the sampling of both the reactive and the conformational (internal and surroundings) coordinates, with large standard deviations often being indicative of insufficient sampling. As a general rule of thumb, one should be careful when interpreting activation free energies with standard deviation greater than 1 kcal mol^{-1}. To assess the reliability of the results, we suggest that one first performs the simulation with several different equilibration and FEP simulation lengths, as described in Section 5.1. The next validation check should be to look at the shape of the distribution of activation free energies to detect outliers as well as possible competing pathways. In such cases, it is often beneficial to compare the individual simulations directly by visualizing the trajectories and examining the energy contributions. Additionally, it might be possible to obtain a correlation with structural features; however, structural features such as residue or loop motions are often coupled to one another, adding significant complexity to the analysis. For an example of such analysis see fig. S3 in the Supplementary Information of Purg et al. (2017).

Fig. 11 Example analysis of an outlying free-energy profile (*top left*) obtained from simulations of DFP hydrolysis by DFPase. The data indicates an issue in the reactant state ($\Delta\varepsilon < -400$, $\lambda_{St.1} > 0.9$, $\lambda_{St.2} < 0.1$) due to the observed break in sampling (*top right*), poor binning overlap (*mid left*), and spikes in ΔG_f–ΔG_r (*mid right*), in comparison to other replicates (shown in *gray*). In contrast to Fig. 10, the electrostatic contributions reveal no anomalies (not shown). However, bonding and angle contributions to the E2 (H_{22}) potential show a rise in energy at $\lambda = 0.98$ not present in E1 (not shown), indicating the issue to be associated with forming bonds and angles (existing only in State 2). Analysis of reactive distances in VMD (Humphrey et al., 1996) reveals the culprit to be a flip of the nucleophile water (swapping hydrogen positions). The reason this produces erroneous results is that the two hydrogen atoms are not considered equivalent within the EVB framework (predefined bond forming/breaking). Nonbonding energies are unaffected due to the symmetric nature of the flip.

6. OVERVIEW AND CONCLUSIONS

The EVB approach has been well documented as a powerful tool for studying enzyme reaction mechanisms, in a broad range of systems (Carvalho et al., 2014; Warshel et al., 2006). In the context of this volume, it has been applied extensively, by both ourselves and others, to study the mechanisms of both enzymatic and nonenzymatic phosphoryl transfer reactions (Carvalho et al., 2014; Kamerlin et al., 2013; Pabis, Duarte, & Kamerlin, 2016). We have presented in this contribution an overview of the EVB approach as applied to the modeling of DFP hydrolysis by DFPase, highlighting key technical considerations and sources of potential errors in the simulations if performed incorrectly. Although several software packages now have the capability to perform EVB simulations, the simulations presented in this chapter are based on the use of our own in-house developed Q simulation package (Marelius et al., 1998), the latest release of which is Q6 (Bauer et al., 2018) under a GPLv2 license. The software is available for free download at https://github.com/qusers/Q6 and provides a broad and comprehensive toolkit for the modeling of phosphoryl transfer and other biochemically important reactions in enzyme active sites, among other features.

REFERENCES

Amrein, B. A., Steffen-Munsberg, F., Szeler, I., Purg, M., Kulkarni, Y., & Kamerlin, S. C. L. (2017). CADEE: Computer-aided directed evolution of enzymes. *IUCrJ*, *4*(1), 50–64.

Anandakrishnan, R., Aguilar, B., & Onufriev, A. V. (2012). H++ 3.0: Automating pK prediction and the preparation of biomolecular structures for atomistic molecular modeling and simulation. *Nucleic Acids Research*, *40*, 537–541.

Åqvist, J. (1996). Calculation of absolute binding free energies for charged ligands and effects of long-range electrostatic interactions. *Journal of Computational Chemistry*, *17*(14), 1587–1597.

Åqvist, J., & Kamerlin, S. C. L. (2014). The conformation of a catalytic loop is central to GTPase activity on the ribosome. *Biochemistry*, *54*(2), 546–556.

Åqvist, J., & Kamerlin, S. C. L. (2015). Exceptionally large entropy contributions enable the high rates of GTP hydrolysis on the ribosome. *Scientific Reports*, *5*, 15817.

Åqvist, J., Kazemi, M., Isaksen, G. V., & Brandsdal, B. O. (2017). Entropy and enzyme catalysis. *Accounts of Chemical Research*, *50*(2), 199–207.

Åqvist, J., Kolmodin, K., Florian, J., & Warshel, A. (1999). Mechanistic alternatives in phosphate monoester hydrolysis: What conclusions can be drawn from available experimental data? *Chemistry & Biology*, *6*(3), R71–R80.

Barrozo, A., Blaha-Nelson, D., Williams, N. H., & Kamerlin, S. C. L. (2017). The effect of magnesium ions on triphosphate hydrolysis. *Pure and Applied Chemistry*, *89*(6), 715–727. https://doi.org/10.1515/pac-2016-1125.

Barrozo, A., Duarte, F., Bauer, P., Carvalho, A. T., & Kamerlin, S. C. L. (2015). Cooperative electrostatic interactions drive functional evolution in the alkaline phosphatase superfamily. *Journal of the American Chemical Society*, *137*(28), 9061–9076.

Bauer, P., Barrozo, A., Purg, M., Amrein, B. A., Esguerra, M., Wilson, P. B., et al. (2018). Q6: A comprehensive toolkit for empirical valence bond and related free energy calculations. *SoftwareX*. https://doi.org/10.1016/j.softx.2017.12.001 (in press).

Bayly, C. I., Cieplak, P., Cornell, W., & Kollman, P. A. (1993). A well-behaved electrostatic potential based method using charge restraints for deriving atomic charges: The RESP model. *The Journal of Physical Chemistry*, *97*(40), 10269–10280.

Ben-David, M., Elias, M., Filippi, J. J., Duñach, E., Silman, I., Sussman, J. L., et al. (2012). Catalytic versatility and backups in enzyme active sites: The case of serum paraoxonase 1. *Journal of Molecular Biology*, *418*, 181–196.

Berman, H. M., Westbrook, J., Feng, Z., Gilliland, G., Bhat, T. N., Weissig, H., et al. (2000). The Protein Data Bank. *Nucleic Acids Research*, *28*, 235–242.

Blaha-Nelson, D., Krüger, D. M., Szeler, K., Ben-David, M., & Kamerlin, S. C. L. (2017). Active site hydrophobicity and the convergent evolution of paraoxonase activity in structurally divergent enzymes: The case of serum paraoxonase 1. *Journal of the American Chemical Society*, *139*, 1155–1167.

Blum, M. M., Löhr, F., Richardt, A., Rüterjans, H., & Chen, J. C. H. (2006). Binding of a designed substrate analogue to diisopropyl fluorophosphatase: Implications for the phosphotriesterase mechanism. *Journal of the American Chemical Society*, *128*(39), 12750–12757.

Blum, M. M., Mustyakimov, M., Rüterjans, H., Kehe, K., Schoenborn, B. P., Langan, P., et al. (2009). Rapid determination of hydrogen positions and protonation states of diisopropyl fluorophosphatase by joint neutron and X-ray diffraction refinement. *Proceedings of the National Academy of Sciences of the United States of America*, *106*, 713–718.

Carvalho, A. T., Barrozo, A., Doron, D., Kilshtain, A. V., Major, D. T., & Kamerlin, S. C. L. (2014). Challenges in computational studies of enzyme structure, function and dynamics. *Journal of Molecular Graphics & Modelling*, *54*, 62–79.

Chang, Y. T., & Miller, W. H. (1990). An empirical valence bond model for constructing global potential energy surfaces for chemical reactions of polyatomic molecular systems. *The Journal of Physical Chemistry*, *94*(15), 5884–5888.

Chen, V. B., Arendall, W. B., Headd, J. J., Keedy, D. A., Immormino, R. M., Kapral, G. J., et al. (2010). MolProbity: All-atom structure validation for macromolecular crystallography. *Acta Crystallographica. Section D, Biological Crystallography*, *66*(1), 12–21.

Chen, S. L., Fang, W. H., & Himo, F. (2007). Theoretical study of the phosphotriesterase reaction mechanism. *The Journal of Physical Chemistry. B*, *111*(6), 1253–1255.

Cooper, D. R., Porebski, P. J., Chruszcz, M., & Minor, W. (2011). X-ray crystallography: Assessment and validation of protein–small molecule complexes for drug discovery. *Expert Opinion on Drug Discovery*, *6*(8), 771–782.

Daura, X., Mark, A. E., & Van Gunsteren, W. F. (1998). Parametrization of aliphatic CH_n united atoms of GROMOS96 force field. *Journal of Computational Chemistry*, *19*(5), 535–547.

DeLano, W. L. (2002). The PyMOL molecular graphics system. http://pymol.org.

Dodda, L. S., Cabeza de Vaca, I., Tirado-Rives, J., & Jorgensen, W. L. (2017). LigParGen web server: An automatic OPLS-AA parameter generator for organic ligands. *Nucleic Acids Research*, *45*(W1), W331–W336.

Duarte, F., Åqvist, J., Williams, N. H., & Kamerlin, S. C. L. (2014). Resolving apparent conflicts between theoretical and experimental models of phosphate monoester hydrolysis. *Journal of the American Chemical Society*, *137*(3), 1081–1093.

Duarte, F., Barrozo, A., Åqvist, J., Williams, N. H., & Kamerlin, S. C. L. (2016). The competing mechanisms of phosphate monoester dianion hydrolysis. *Journal of the American Chemical Society*, *138*(33), 10664–10673.

Elias, M., Liebschner, D., Koepke, J., Lecomte, C., Guillot, B., Jelsch, C., et al. (2013). Hydrogen atoms in protein structures: High-resolution X-ray diffraction structure of the DFPase. *BMC Research Notes*, *6*, 308.

Friesner, R. A., & Guallar, V. (2005). Ab initio quantum chemical and mixed quantum mechanics/molecular mechanics (QM/MM) methods for studying enzymatic catalysis. *Annual Review of Physical Chemistry, 56*, 389–427.

Glowacki, D. R., Orr-Ewing, A. J., & Harvey, J. N. (2015). Non-equilibrium reaction and relaxation dynamics in a strongly interacting explicit solvent: F+ CD3CN treated with a parallel multi-state EVB model. *The Journal of Chemical Physics, 143*(4), 044120.

Gregersen, B. A., Lopez, X., & York, D. M. (2004). Hybrid QM/MM study of thio effects in transphosphorylation reactions: The role of solvation. *Journal of the American Chemical Society, 126*(24), 7504–7513.

Hartleib, J., & Rüterjans, H. (2001a). High-yield expression, purification, and characterization of the recombinant diisopropylfluorophosphatase from *Loligo vulgaris*. *Protein Expression and Purification, 21*(1), 210–219.

Hartlieb, J., & Rüterjans, H. (2001b). Insights into the reaction mechanism of the diisopropyl fluorophosphatase from Loligo vulgaris by means of kinetic studies, chemical modification and site-directed mutagenesis. *Biochimica et Biophysica Acta, 1546*(2), 312–324.

Harvey, J. N. (2006). On the accuracy of density functional theory in transition metal chemistry. *Annual Reports on the Progress of Chemistry, Section C: Physical Chemistry, 102*, 203–226.

Haydock, C., Sharp, J. C., & Prendergast, F. G. (1990). Tryptophan-47 rotational isomerization in variant-3 scorpion neurotoxin. A combination thermodynamic perturbation and umbrella sampling study. *Biophysical Journal, 57*(6), 1269–1279.

Hengge, A. C., Edens, W. A., & Elsing, H. (1994). Transition-state structures for phosphoryl-transfer reactions of p-nitrophenyl phosphate. *Journal of the American Chemical Society, 116*(12), 5045–5049.

Hong, G., Rosta, E., & Warshel, A. (2006). Using the constrained DFT approach in generating diabatic surfaces and off diagonal empirical valence bond terms for modeling reactions in condensed phases. *The Journal of Physical Chemistry. B, 110*(39), 19570–19574.

Humphrey, W., Dalke, A., & Schulten, K. (1996). VMD—Visual molecular dynamics. *Journal of Molecular Graphics, 14*, 33–38.

Hwang, J. K., King, G., Creighton, S., & Warshel, A. (1988). Simulation of free energy relationships and dynamics of SN2 reactions in aqueous solution. *Journal of the American Chemical Society, 110*(16), 5297–5311.

Hwang, J. K., & Warshel, A. (1987). Microscopic examination of free-energy relationships for electron transfer in polar solvents. *Journal of the American Chemical Society, 109*(3), 715–720.

Isaksen, G. V., Åqvist, J., & Brandsdal, B. O. (2016). Enzyme surface rigidity tunes the temperature dependence of catalytic rates. *Proceedings of the National Academy of Sciences of the United States of America, 113*, 7822–7827.

Kamerlin, S. C. L., Haranczyk, M., & Warshel, A. (2008). Progress in ab initio QM/MM free-energy simulations of electrostatic energies in proteins: Accelerated QM/MM studies of pK_a, redox reactions and solvation free energies. *The Journal of Physical Chemistry. B, 113*(5), 1253–1272.

Kamerlin, S. C. L., Sharma, P. K., Prasad, R. B., & Warshel, A. (2013). Why nature really chose phosphate. *Quarterly Reviews of Biophysics, 46*, 1–132.

Katsemi, V., Lücke, C., Koepke, J., Löhr, F., Maurer, S., Fritzsch, G., et al. (2005). Mutational and structural studies of the diisopopylfluorophosphatase from *Loglio vulgaris* shed new light on fthe catalytic mechanism of the enzyme. *Biochemistry, 44*, 9022–9033.

Kazemi, M., & Åqvist, J. (2015). Chemical reaction mechanisms in solution from brute force computational Arrhenius plots. *Nature Communications, 6* (article no. 7293).

Khersonsky, O., & Tawfik, D. S. (2005). Structure-reactivity studies of serum paraoxonase PON1 suggest that its native activity is lactonase. *Biochemistry, 44*, 6371–6382.

Kim, J., Tsai, P. C., Chen, S. L., Himo, F., Almo, S. C., & Raushel, F. M. (2008). Structure of diethyl phosphate bound to the binuclear metal center of phosphotriesterase. *Biochemistry*, *47*(36), 9497–9504.

King, G., & Warshel, A. (1989). A surface constrained all-atom solvent model for effective simulations of polar solutions. *The Journal of Chemical Physics*, *91*, 3647.

Kirby, A. J., & Jencks, W. P. (1965). The reactivity of nucleophilic reagents toward the p-nitrophenyl phosphate dianion. *Journal of the American Chemical Society*, *87*(14), 3209–3216.

Kirby, A. J., & Varvoglis, A. G. (1967). The reactivity of phosphate esters. Monoester hydrolysis. *Journal of the American Chemical Society*, *89*(2), 415–423.

Laio, A., & Parrinello, M. (2002). Escaping free-energy minima. *Proceedings of the National Academy of Sciences of the United States of America*, *99*(20), 12562–12566.

Lassila, J. K., Zalatan, J. G., & Herschlag, D. (2011). Biological phosphoryl-transfer reactions: Understanding mechanism and catalysis. *Annual Review of Biochemistry*, *80*, 669–702.

Latifi, A. M., Karami, A., & Khodi, S. (2015). Efficient surface display of diisopropylfluorophosphatase (DFPase) in *E. coli* for biodegradation of toxic organophosphorus compounds (DFP and Cp). *Applied Biochemistry and Biotechnology*, *177*(3), 624–636.

Lee, F. S., Chu, Z. T., Bolger, M. B., & Warshel, A. (1992). Calculations of antibody-antigen interactions: Microscopic and semi-microscopic evaluation of the free energies of binding of phosphorylcholine analogs to McPC603. *Protein Engineering, Design & Selection*, *5*(3), 215–228.

Li, P., & Merz, K. M., Jr. (2017). Metal ion modeling using classical mechanics. *Chemical Reviews*, *117*(3), 1564–1686.

Lonsdale, R., Harvey, J. N., & Mulholland, A. J. (2012). A practical guide to modelling enzyme-catalysed reactions. *Chemical Society Reviews*, *41*(8), 3025–3038.

López-Canut, V., Marti, S., Bertrán, J., Moliner, V., & Tuñón, I. (2009). Theoretical modeling of the reaction mechanism of phosphate monoester hydrolysis in alkaline phosphatase. *The Journal of Physical Chemistry. B*, *113*(22), 7816–7824.

López-Canut, V., Roca, M., Bertrán, J., Moliner, V., & Tuñón, I. (2010). Theoretical study of phosphodiester hydrolysis in nucleotide pyrophosphatase/phosphodiesterase. Environmental effects on the reaction mechanism. *Journal of the American Chemical Society*, *132*(20), 6955–6963.

López-Canut, V., Roca, M., Bertrán, J., Moliner, V., & Tuñón, I. (2011). Promiscuity in alkaline phosphatase superfamily. Unraveling evolution through molecular simulations. *Journal of the American Chemical Society*, *133*(31), 12050–12062.

Lu, X., Ovchinnikov, V., Demapan, D., Roston, D., & Cui, Q. (2017). Regulation and plasticity of catalysis in enzymes: Insights from analysis of mechanochemical coupling in myosin. *Biochemistry*, *56*, 1482–1497.

MacKerell, A. D., Jr., Bashford, D., Bellott, M. L. D. R., Dunbrack, R. L., Jr., Evanseck, J. D., Field, M. J., et al. (1998). All-atom empirical potential for molecular modeling and dynamics studies of proteins. *The Journal of Physical Chemistry. B*, *102*(18), 3586–3616.

Maier, J. A., Martinez, C., Kasavajhala, K., Wickstrom, L., Hauser, K. E., & Simmerling, C. (2015). ff14SB: Improving the accuracy of protein side chain and backbone parameters from ff99SB. *Journal of Chemical Theory and Computation*, *11*(8), 3696–3713.

Marelius, J., Kolmodin, K., Feierberg, I., & Åqvist, J. (1998). Q: A molecular dynamics program for free energy calculations and empirical valence bond simulations in biomolecular systems. *Journal of Molecular Graphics & Modelling*, *16*(4), 213–225.

Melzer, M., Chen, J. C.-H., Heidenreich, A., Gäb, J., Koller, M., Kehe, K., et al. (2009). Reversed enantioselectivity of diisopropyl fluorophosphatase against organophosphorus

nerve agents by rational design. *Journal of the American Chemical Society*, *131*, 17226–17232.
Mones, L., Kulhánek, P., Simon, I., Laio, A., & Fuxreiter, M. (2009). The energy gap as a universal reaction coordinate for the simulation of chemical reactions. *The Journal of Physical Chemistry. B*, *113*(22), 7867–7873.
Morse, P. M. (1929). Diatomic molecules according to the wave mechanics. II. Vibrational levels. *Physical Review*, *34*(1), 57.
Muegge, I., Tao, H., & Warshel, A. (1997). A fast estimate of electrostatic group contributions to the free energy of protein-inhibitor binding. *Protein Engineering*, *10*, 1363–1372.
Oanca, G., Purg, M., Mavri, J., Shih, J. C., & Stare, J. (2016). Insights into enzyme point mutation effect by molecular simulation: Phenylethylamine oxidation catalyzed by monoamine oxidase A. *Physical Chemistry Chemical Physics*, *18*(19), 13346–13356.
Olsson, M. H., Mavri, J., & Warshel, A. (2006). Transition state theory can be used in studies of enzyme catalysis: Lessons from simulations of tunnelling and dynamical effects in lipoxygenase and other systems. *Philosophical Transactions of the Royal Society B*, *361*(1472), 1417–1432.
Olsson, M. H. M., Sondergaard, C. R., Rostkowski, M., & Jensen, J. H. (2011). PROPKA3: Consistent treatment of internal and surface residues in empirical pKa predictions. *Journal of Chemical Theory and Computation*, 7(2), 525–537.
Pabis, A., Duarte, F., & Kamerlin, S. C. L. (2016). Promiscuity in the enzymatic catalysis of phosphate and sulfate transfer. *Biochemistry*, *55*, 3061–3081.
Plotnikov, N. V., Prasad, B. R., Chakrabarty, S., Chu, Z. T., & Warshel, A. (2013). Quantifying the mechanism of phosphate monoester hydrolysis in aqueous solution by evaluating the relevant ab initio QM/MM free-energy surfaces. *The Journal of Physical Chemistry. B*, *117*(42), 12807–12819.
Purg, M., Elias, M., & Kamerlin, S. C. L. (2017). Similar active sites and mechanisms do not lead to cross-promiscuity in organophosphate hydrolysis: Implications for biotherapeutic engineering. *Journal of the American Chemical Society*, *139*(48), 17533–17546.
Purg, M., Pabis, A., Baier, F., Tokuriki, N., Jackson, C., & Kamerlin, S. C. L. (2016). Probing the mechanisms for the selectivity and promiscuity of methyl parathion hydrolase. *Philosophical Transactions of the Royal Society A*, *374*(2080), 20160150.
Robertson, M. J., Tirado-Rives, J., & Jorgensen, W. L. (2015). Improved peptide and protein torsional energetics with the OPLS-AA force field. *Journal of Chemical Theory and Computation*, *11*(7), 3499–3509.
Rohl, C. A., Strauss, C. E., Misura, K. M., & Baker, D. (2004). Protein structure prediction using Rosetta. *Methods in Enzymology*, *383*, 66–93.
Rosta, E., Nowotny, M., Yang, W., & Hummer, G. (2011). Catalytic mechanism of RNA backbone cleavage by ribonuclease H from quantum mechanics/molecular mechanics simulations. *Journal of the American Chemical Society*, *133*(23), 8934–8941.
Roston, D., & Cui, Q. (2016). Substrate and transition state binding in alkaline phosphatase analyzed by computation of oxygen isotope effects. *Journal of the American Chemical Society*, *138*(36), 11946–11957.
Roston, D., Demapan, D., & Cui, Q. (2016). Leaving group ability observably affects transition state structure in a single enzyme active site. *Journal of the American Chemical Society*, *138*(23), 7386–7394.
Roux, B. (1995). The calculation of the potential of mean force using computer simulations. *Computer Physics Communications*, *91*(1–3), 275–282.
Ryde, U. (2017). How many conformations need to be sampled to obtain converged QM/MM energies? The curse of exponential averaging. *Journal of Chemical Theory and Computation*, *13*(11), 5745–5752.
Schlegel, H. B., & Sonnenberg, J. L. (2006). Empirical valence-bond models for reactive potential energy surfaces using distributed Gaussians. *Journal of Chemical Theory and Computation*, *2*(4), 905–911.

Schrödinger. (2017). *Schrödinger release 2017-1: Macromodel*. New York: Schrödinger, LLC.
Senn, H. M., & Thiel, W. (2009). QM/MM methods for biomolecular systems. *Angewandte Chemie, International Edition*, *48*(7), 1198–1229.
Sham, Y. Y., & Warshel, A. (1998). The surface constraint all atom model provides size independent results in calculations of hydration free energies. *The Journal of Chemical Physics*, *109*(18), 7940–7944.
Shurki, A., Derat, E., Barrozo, A., & Kamerlin, S. C. L. (2015). How valence bond theory can help you understand your (bio)chemical reaction. *Chemical Society Reviews*, *44*(5), 1037–1052.
Siegbahn, P. E., & Himo, F. (2009). Recent developments of the quantum chemical cluster approach for modeling enzyme reactions. *Journal of Biological Inorganic Chemistry*, *14*(5), 643–651.
Stare, J. (2017). Complete sampling of an enzyme reaction pathway: A lesson from gas phase simulations. *RSC Advances*, 7(15), 8740–8754.
Sugita, Y., & Okamoto, Y. (1999). Replica-exchange molecular dynamics method for protein folding. *Chemical Physics Letters*, *314*(1), 141–151.
Torrie, G. M., & Valleau, J. P. (1977). Nonphysical sampling distributions in Monte Carlo free-energy estimation: Umbrella sampling. *Journal of Computational Physics*, *23*(2), 187–199.
Tripathi, R., Glaves, R., & Marx, D. (2017). The GTPase hGBP1 converts GTP to GMP in two steps via proton shuttle mechanisms. *Chemical Science*, *8*(1), 371–380.
Vagedes, P., Rabenstein, B., Åqvist, J., Marelius, J., & Knapp, E. W. (2000). The deacylation step of acetylcholinesterase: Computer simulation studies. *Journal of the American Chemical Society*, *122*(49), 12254–12262.
van der Kamp, M. W., & Mulholland, A. J. (2013). Combined quantum mechanics/molecular mechanics (QM/MM) methods in computational enzymology. *Biochemistry*, *52*(16), 2708–2728.
Wallin, G., Kamerlin, S. C. L., & Åqvist, J. (2013). Energetics of activation of GTP hydrolysis on the ribosome. *Nature Communications*, *4*, 1733.
Warshel, A. (1991). *Computer modeling of chemical reactions in enzymes and solutions*. New York: Wiley.
Warshel, A., & Karplus, M. (1972). Calculation of ground and excited state potential surfaces of conjugated molecules. I. Formulation and parametrization. *Journal of the American Chemical Society*, *94*(16), 5612–5625.
Warshel, A., & Levitt, M. (1976). Theoretical studies of enzymic reactions: Dielectric, electrostatic and steric stabilization of the carbonium ion in the reaction of lysozyme. *Journal of Molecular Biology*, *103*(2), 227–249.
Warshel, A., Sharma, P. K., Kato, M., Xiang, Y., Liu, H., & Olsson, M. H. (2006). Electrostatic basis for enzyme catalysis. *Chemical Reviews*, *106*(8), 3210–3235.
Warshel, A., & Weiss, R. M. (1980). An empirical valence bond approach for comparing reactions in solutions and in enzymes. *Journal of the American Chemical Society*, *102*, 6218–6226.
Webb, B., & Sali, A. (2014). Protein structure modeling with *MODELLER*. In *Protein structure prediction* (pp. 1–15). Springer.
Woolf, T. B., & Roux, B. (1994). Conformational flexibility of o-phosphorylcholine and o-phosphorylethanolamine: A molecular dynamics study of solvation effects. *Journal of the American Chemical Society*, *116*(13), 5916–5926.
Xu, C., Yang, L., Yu, J.-G., & Liao, R.-Z. (2016). What roles do the residue Asp229 and the coordination variation of calcium play of the reaction mechanism of the diisopropylfluorophosphatase? A DFT investigation. *Theoretical Chemistry Accounts*, *135*, 138.
Xu, J., Zhang, J. Z., & Xiang, Y. (2012). Ab initio QM/MM free energy simulations of peptide bond formation in the ribosome support an eight-membered ring reaction mechanism. *Journal of the American Chemical Society*, *134*(39), 16424–16429.
Zwanzig, R. W. (1954). High-temperature equation of state by a perturbation method. I. nonpolar gases. *The Journal of Chemical Physics*, *22*(8), 1420–1426.

CHAPTER TWO

Analysis of Phosphoryl-Transfer Enzymes with QM/MM Free Energy Simulations

Daniel Roston, Xiya Lu, Dong Fang, Darren Demapan, Qiang Cui[1,2]
Department of Chemistry and Theoretical Chemistry Institute, University of Wisconsin, Madison, Madison, WI, United States
[1]Corresponding author: e-mail address: qiangcui@bu.edu

Contents

Abstract

We discuss the application of quantum mechanics/molecular mechanics (QM/MM) free energy simulations to the analysis of phosphoryl transfers catalyzed by two enzymes: alkaline phosphatase and myosin. We focus on the nature of the transition state and the issue of mechanochemical coupling, respectively, in the two enzymes. The results illustrate unique insights that emerged from the QM/MM simulations, especially concerning the interpretation of experimental data regarding the nature of enzymatic transition states and coupling between global structural transition and catalysis in the active site. We also highlight a number of technical issues worthy of attention when applying QM/MM free energy simulations, and comment on a number of technical and mechanistic issues that require further studies.

[2] Current Address: Department of Chemistry, Boston University, Boston, MA, United States.

Methods in Enzymology, Volume 607
ISSN 0076-6879
https://doi.org/10.1016/bs.mie.2018.05.005

1. INTRODUCTION

Phosphoryl transfer represents arguably one of the most important classes of biological transformations (Alberts et al., 1994; Cleland & Hengge, 2006; Kamerlin, Sharma, Prasad, & Warshel, 2013; Knowles, 1980; Lassila, Zalatan, & Herschlag, 2011; Westheimer, 1987). It is involved in many key biological processes such as bioenergy transduction (e.g., ATP hydrolysis in biomolecular motors), signal transduction (e.g., GTP hydrolysis in G-proteins and phosphorylation/dephosphorylation by kinases/phosphatases), and genome processing (e.g., DNA/RNA synthesis in DNA/RNA polymerases); there are ~2000 protein kinases and ~1000 phosphatases in the human genome. Therefore, disruption or impairment of phosphoryl-transfer reaction may significantly perturb the function of the proteins involved and lead to serious diseases such as cancer (Campisi & di Fagagna, 2007; Kiaris & Spandidos, 1995; Lange, Takata, & Wood, 2011; Roberts & Der, 2007; Shaw & Cantley, 2006). Indeed, protein kinases and phosphatases are among the most important drug targets (Cohen, 2002; Collins & Workman, 2006; Davies, Reddy, Caivano, & Cohen, 2000; Garber, 2001; Robertson, 2007; Schwartz & Murray, 2009; Zhang, 2002).

Motivated by such considerations, extensive efforts have been paid to studying the mechanism of phosphoryl-transfer reactions in both small molecules and proteins in the past few decades. These investigations have targeted both the underlying chemical processes (e.g., whether the reaction involves any intermediate and the nature of the transition state) and strategies that proteins employ to regulate the rate of phosphoryl transfers. In an excellent review article, rather up to date findings from extensive experimental studies, especially concerning the chemical mechanism of biological phosphoryl transfers and nature of transition state, were summarized (Lassila et al., 2011). In recent years, as computational hardware and methodologies continue to improve, computational studies have become increasingly effective at analyzing the mechanism of reaction mechanisms in biomolecules. Specifically for phosphoryl-transfer reactions, hybrid quantum mechanics/molecular mechanics (QM/MM) type of computations (Brunk & Rothlisberger, 2015; Friesner & Guallar, 2005; Garcia-Viloca, Gao, Karplus, & Truhlar, 2004; Hu & Yang, 2008; Kamerlin, Haranczyk, & Warshel, 2009; Monard & Merz, 1999; Riccardi et al., 2006; Senn & Thiel, 2009) has been used by several research groups (Åqvist & Kamerlin, 2016; Carvalho, Szeler, Vavitsas, Åqvist, & Kamerlin, 2015; Duarte, Amrein, &

Kamerlin, 2013; Genna, Vidossich, Ippoliti, Carloni, & De Vivo, 2016; Grigorenko et al., 2007; Hayashi et al., 2012; Hou & Cui, 2012, 2013; Kamerlin et al., 2013; Kiani & Fischer, 2014, 2016; McCullagh, Saunders, & Voth, 2014; McGrath, Kuo, Hayashi, & Takada, 2013; Mlynsky et al., 2014; Pabis, Duarte, & Kamerlin, 2016; Rosta, Kamerlin, & Warshel, 2008; Roston & Cui, 2016a, 2016b; Roston, Demapan, & Cui, 2016) to provide mechanistic insights into a broad set of enzymes that catalyze different phosphoryl transfers. These studies underlined subtleties in the interpretation of experimental data, which include both "direct" observations such as crystal structures and "indirect" observables such as kinetic isotope effects (KIEs), activation entropy, and (linear) free energy relationships.

In this contribution, we discuss our recent studies of phosphoryl transfers catalyzed by several enzymes using QM/MM methodologies developed in our group. In addition to highlighting the unique contribution of these QM/MM studies to the understanding of detailed catalytic mechanism, another aim is to summarize our QM/MM methodologies regarding both strength and remaining limitations. We also comment on future development and application of computational studies targeting biological phosphoryl transfers.

2. BACKGROUND ON COMPUTATIONAL METHODS

Enzymes feature motions that span a broad range of temporal and spatial scales. Identifying which motions make an essential contribution to the chemical reaction that an enzyme catalyzes remains a major challenge in the field of enzymology (Nashine, Hammes-Schiffer, & Benkovic, 2010). In principle, all motions that occur at a time scale faster than the chemical reaction, which for natural enzymes is commonly in the range of milliseconds, should be properly averaged to define the thermodynamic and kinetic properties of the catalyzed reaction (Cui & Karplus, 2003; Zhou, 2010). In practice, it remains a major challenge to conduct an adequate canonical average over the configurations implicated in these motions with either direct or enhanced sampling methods. Meeting this challenge is particularly important to the study of phosphoryl transfers as many relevant enzymes are known to be particularly flexible and therefore rich in μs–ms motions. Our general aim, therefore, is to develop QM/MM methods that are able to reach the μs time scale in the near future using modest computational resources, while maintaining a level of accuracy adequate for addressing the relevant mechanistic questions.

2.1 QM/MM Potential Function

Two levels of QM/MM potential functions are used in our studies. The semiempirical type (Christensen, Kubar, Cui, & Elstner, 2016; Thiel, 1996) of QM/MM methods is computationally efficient and thus can be used in direct molecular dynamics and free energy simulations (Gao et al., 2006; Pu, Gao, & Truhlar, 2006; Riccardi et al., 2006). QM/MM methods based on ab initio or density functional theory (DFT) (Brunk & Rothlisberger, 2015; Hu & Yang, 2008) are much more expensive (about a factor of 10^3 or more) and therefore are used in either minimization or free energy path type of calculations (see Section 2.2). Because the coupling between QM and MM atoms and subtleties associated with the QM/MM partitioning have been well described in previous work (Gao, Amara, Alhambra, & Field, 1998; König, Hoffmann, Frauenheim, & Cui, 2005; Reuter, Dejaegere, Maigret, & Karplus, 2000; Senn & Thiel, 2009), we focus here on the discussion of the QM potential (Cui, 2016). The proper size of the QM region in QM/MM simulations of enzymes has been an interesting subject of discussion and debate in the recent literature (Flaig, Beer, & Ochsenfeld, 2012; Hu, Soderhjelm, & Ryde, 2011; Jindal & Warshel, 2016; Kulik, Zhang, Klinman, & Martinez, 2016); in our own experience, provided that the QM/MM boundary is carefully treated (König et al., 2005; Reuter et al., 2000; Roston & Cui, 2016a), a QM region involving 100–150 atoms is often adequate for many mechanistic analyses, especially when sampling is properly done to allow adequate response of the MM region to charge redistribution of the QM region during the chemical reaction (Riccardi et al., 2006).

2.1.1 Semiempirical QM/MM Methods

There are two popular types of semiempirical QM models: the more traditional neglect of diatomic differential overlap (NDDO) methods (Thiel, 1996) such as AM1, PM3, and the more recent OM2/3 models (Dral, Wu, Sporkel, Koslowski, & Thiel, 2016), and the density functional tight binding model (DFTB) (Elstner et al., 1998; Gaus, Cui, & Elstner, 2014). With the use of a minimal basis and approximations of integrals (or Hamiltonian matrix elements), these methods are computationally efficient and are 2–3 orders of magnitude faster than ab initio or DFT calculations. Therefore, these methods can be routinely used in direct MD-based simulations at the 10–100 ns scale, with a QM region of about <200 atoms.

For larger QM regions, linear-scaling methods have been developed (Liu et al., 2001; Nishizawa, Nishimura, Kobayashi, Irle, & Nakai, 2016) although the application to realistic enzyme systems has been limited, due in part to the limited accuracy of semiempirical methods for noncovalent interactions in biomolecules, as compared to well-parameterized classical force fields; as discussed in Christensen et al. (2016), systematic improvement of semiempirical methods for biomolecular structure and interaction is an important area of research.

Specifically in the context of phosphoryl-transfer reactions, both NDDO type of method (AM1-d) (Nam, Cui, Gao, & York, 2007) and the third-order variant of DFTB (DFTB3 (Gaus, Cui, & Elstner, 2011)) have been specifically adjusted for describing phosphoryl-transfer reactions (Gaus, Lu, Elstner, & Cui, 2014). The fact that specific parameterizations are required suggests that approximations in both methods make them less robust for describing the complexity of the underlying potential energy surfaces for phosphoryl transfers (Florian & Warshel, 1998). This is not due to the strong participation of *d* orbitals, as both AM1-d and DFTB3 explicitly include *d* orbitals in the basis set and a natural bonding orbital (NBO (Reed, Curtiss, & Weinhold, 1988; Weinhold & Landis, 2005)) analysis of relevant structures reveals that the participation of the *d* orbitals is very limited (Table 1) at both DFTB3 and DFT levels. Identifying the origins for the limited transferability of current semiempirical methods for phosphoryl-transfer reactions is of major interest (Mlynsky et al., 2014). For specific applications, it is always important to test the semiempirical method using active-site model systems against high-level QM methods, as we show in Section 2.1.2 using alkaline phosphatase (AP) as an example.

Table 1 Natural Bonding Orbitals for P–O Bonds in a Protonated Dimethyl Phosphate From DFTB3/3OB and B3LYP/aug-cc-pVTZ Calculations

	DFTB3		B3LYP	
Bond	**P**	**O**	**P**	**O**
P–O_1	0.53 $sp^{3.41}d^{0.02}$	0.85 $sp^{3.22}$	0.47 $sp^{3.34}d^{0.10}$	0.88 $sp^{2.47}$
P–O_2	0.60 $sp^{2.14}d^{0.00}$	0.80 $sp^{2.94}$	0.51 $sp^{2.07}d^{0.04}$	0.86 $sp^{2.18}$
P–O_3	0.54 $sp^{3.28}d^{0.02}$	0.84 $sp^{2.81}$	0.47 $sp^{3.29}d^{0.10}$	0.89 $sp^{2.60}$
P–O_4	0.54 $sp^{3.47}d^{0.02}$	0.84 $sp^{3.17}$	0.46 $sp^{3.31}d^{0.11}$	0.89 $sp^{2.79}$

Another major issue for applying semiempirical methods to phosphoryl-transfer systems is the treatment of metal ions, which are often featured in the active site of the relevant enzymes. Currently available NDDO methods are usually not sufficiently reliable for metal ions, especially transition-metal ions. DFTB, which is based on DFT rather than Hartree–Fock, appears more promising for the description of metal ions; specifically for DFTB3, parameterization has been done for several main group ions such as Na^{+}, Mg^{2+}, and Ca^{2+} (Kubillus, Kubař, Gaus, Řezáč, & Elstner, 2015; Lu, Gaus, Elstner, & Cui, 2015), as well as for simpler transition-metal ions such as Zn^{2+} (Lu et al., 2015) and $Cu^{+/2+}$ (Gaus et al., 2015). The performance of DFTB3 is most impressive for structural properties of metal compounds, while the energetics are less accurate, especially for highly charged ligands, due presumably to the limited description of polarization and charge-transfer effects. Even with the current set of parameters, nevertheless, DFTB models have been shown to be highly effective for the description of ligand binding to zinc-enzymes (Pecina et al., 2017); systematic improvement of DFTB models for more complex transition-metal ions remains a topic of major interest.

2.1.2 Ab Initio and DFT QM/MM Methods

Ab initio and DFT methods are generally more robust than semiempirical methods. Due to computational cost, DFT methods are often chosen in QM/MM calculations for enzymes, although energies can be further corrected based on higher-level methods, including QM/QM′ embedding methods in which the central region is described with a highly correlated QM method such as CCSD(T) (Claeyssens et al., 2006), while the rest of the QM atoms are described with DFT (Bennie et al., 2016; Manby, Stella, Goodpaster, & Miller, 2012). When transition-metal ions are involved, doing such a local correction is worthwhile as it is not straightforward to identify the most reliable density functional for the metal center of interest (Cramer & Truhlar, 2009; Jiang, Laury, Powell, & Wilson, 2012).

Another issue of concern is the contribution from long-range dispersion interactions, which are missing in most popular GGA-based DFT methods (Grimme, Hansen, Brandenburg, & Bannwarth, 2016); for large QM regions, long-range dispersion can be important and neglect of the contribution can affect both the structure (a comparison is made in Table 2) and the reaction energetics (see Table 3 as an example). Fortunately, empirical dispersion models have been developed for both DFT (Grimme et al., 2016) and semiempirical models (Brandenburg, Hochheim, Bredow, & Grimme,

Table 2 Key Structural Parameters (in Å) in an Active-Site Model of AP in the Gas Phase (see Fig. 1) From DFTB3 and B3LYP Calculations

	B3LYP/6-31+G(d,p)			DFTB3/3OB		
Substrate	$O^{nuc}\cdots P$	$P\cdots O^{lg}$	δ	$O^{nuc}\cdots P$	$P\cdots O^{lg}$	δ
$MpNPP^-$	2.22	1.99	+0.23	2.21	1.91	+0.30
$MmNPP^-$	2.08	1.96	+0.12	2.14	1.86	+0.28
MPP^-	1.99	2.06	−0.07	1.86	2.14	−0.28

Diester phosphate substrates are studied. MpNPP, methyl *p*-nitrophenyl phosphate; MmNPP, methyl *m*-nitrophenyl phosphate; MPP, methyl phenyl phosphate. "δ" is the difference between the breaking ($P\cdots O^{lg}$) and forming ($O^{nu}\cdots P$) bonds.

Table 3 Calculated Activation Barrier (in kcal/mol) for Diester Hydrolysis in the Active-Site Model of AP in the Gas Phase (see Fig. 1)

		Single Points at DFTB3 Geometries			Single Points at B3LYP Geometries		
Substrate	**DFTB3**	**B3LYP**	**M06**	**MP2**	**B3LYP**	**M06**	**MP2**
$MpNPP^-$	8.8	9.2 (4.2[a])	5.7 (3.1)	7.6	12.0 (6.7)	6.8 (6.1)	8.8
$MmNPP^-$	9.7	9.2 (9.4)	6.8 (7.5)	13.6	15.3 (12.0)	13.2 (12.8)	13.2
MPP^-	20.0	16.2 (9.1)	11.6 (7.1)	9.6	20.1 (11.6)	13.0 (10.8)	10.4

[a]The number with parentheses is obtained with corresponding DFT functional plus D3 dispersion. The geometries are optimized either at DFTB3/3OB or B3LYP/6-31+G(d,p) level. Single point energy calculations are carried out at B3LYP, M06, and MP2 method at two levels of geometries using a larger basis set of 6-311++G(d,p).

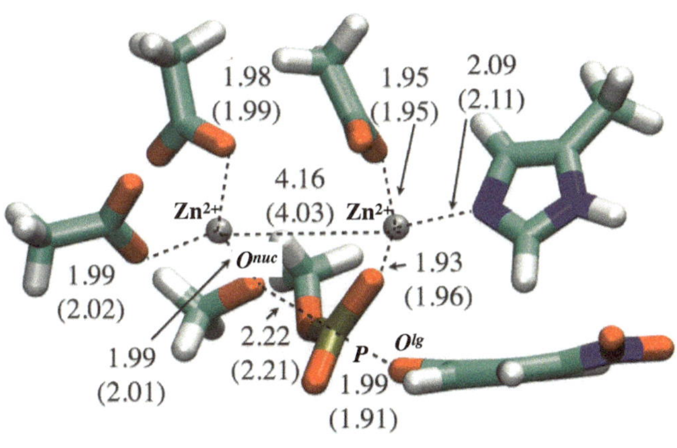

Fig. 1 An active-site model for alkaline phosphatase (AP), with methyl *p*-nitrophenyl phosphate (MpNPP) as a substrate. Key geometrical parameters (in Å) are shown to compare DFTB3/3OB (with parentheses) and B3LYP/6-31+G(d,p); for additional comparisons, see Tables 2 and 3.

2014; Elstner, Hobza, Frauenheim, Suhai, & Kaxiras, 2001; Tuttle & Thiel, 2008) and they appear to be effective in many practical applications.

2.2 Free Energy Computations With QM/MM Potentials

The topic of free energy simulations has been covered in many recent review articles (Barducci, Bonomi, & Parrinello, 2011; Pohorille, Jarzynski, & Chipot, 2010), including that focus on QM/MM potential functions (Brunk & Rothlisberger, 2015; Kamerlin et al., 2009; Lu et al., 2016). Below, we touch on a few points particularly relevant to the study of biological phosphoryl-transfer reactions.

2.2.1 Direct Sampling: Metadynamics and String Methods

With semiempirical QM/MM potentials, it is feasible to conduct direct sampling of the free energy surface using techniques such as umbrella sampling (Torrie & Valleau, 1977), metadynamics (Barducci et al., 2011), and string methods (Weinan & Vanden-Eijnden, 2010). One specific feature associated with phosphoryl transfers is that multiple reaction pathways are likely available and relatively similar in free energy profiles (Florian & Warshel, 1998; Yang & Cui, 2009). For example, the competition between dissociative and associative mechanisms, which differ in terms of whether a pentavalent intermediate exists or not, remains debated in several key enzymes (Lassila et al., 2011); moreover, as discussed in Section 3.2 using myosin as an example, multiple water molecules and/or protein sidechains might be involved in mediating proton transfers during the phosphoryl-transfer reaction. Therefore, care has to be exercised to select reaction coordinates (or referred to as collective variables) that drive the free energy simulations; inappropriate choice of reaction coordinates projects configurations of distinct chemical structures into the same region on the reduced conformational space, thus leading to a less physically meaningful free energy surface.

The string suite of methods has become popular in recent years and has been applied to enzymatic reactions (Rosta, Nowotny, Yang, & Hummer, 2011); these methods represent the reaction pathway with a string in a reduced space spanned also by a set of collective variables. Because the string is a one-dimensional quantity, the number of collective variables used in such calculations can be substantially larger than in metadynamics simulations, which aim to construct a multidimensional free energy *surface*. Nevertheless, it is important to emphasize that there are multiple variants of the string method, which treat thermal fluctuations in different ways with different underlying approximations (Maragliano, Fischer, Vanden-Eijnden, & Ciccotti, 2006). For example, in the variant of the string method applied to the myosin example (Lu, Ovchinnikov, Roston, Demapan, & Cui, 2017) in Section 3.2, the entropic contribution associated with thermal fluctuations of the collective variables is treated approximately compared to the finite temperature string approach (Maragliano et al., 2006; Ovchinnikov, Karplus, & Vanden-Eijnden, 2011); therefore, the number of collective variables used to parameterize the string should be carefully selected. Moreover, in both metadynamics and string calculations, there are a number of key parameters that influence the convergence and accuracy of these calculations; an example in the string calculations is the force constant used to control the degree of sampling orthogonal to the path, which determines the accuracy of the mean force (Lu et al., 2016); another issue is the overlap of the configurations sampled along the string, and conducting Hamiltonian replica exchange is useful in this regard (Rosta et al., 2011). Therefore, it is important to recognize that these methods are not entirely "black boxes" and the key parameters need to be chosen carefully based on test calculations.

2.2.2 Minimum Free Energy Path Approach for Ab Initio QM/MM

With ab initio/DFT-based QM/MM potentials, it is usually too expensive to conduct sufficient sampling for biological applications. The most practical approach is the minimum free energy path (MFEP) method pioneered by Hu and Yang (2008), in which the fluctuations of the QM and MM atoms are decoupled; the MFEP is then defined as the steepest descent path on the potential of mean force for the QM atoms that connects the reactant and product basins. Similar to the variant of the string approach applied below to myosin, the thermal fluctuations of the QM atoms orthogonal to the MFEP are treated approximately and estimated using a harmonic model. Moreover, to compute the mean force on the QM atoms in a cost-effective manner, the MM degrees of freedom are sampled using MD simulations in which the QM atoms are replaced with a set of classical atoms whose charge distribution is determined using a mean-field model. Using a small phosphate molecule in solution, we found (Lu et al., 2016) that the mean-field approximation leads to an adequate estimate of the mean force with errors on the order of 0.5–1 kcal/mol Å^{-1}; a reweighting scheme has been proposed in the literature (Hayashi et al., 2017; Kosugi & Hayashi, 2012) and might reduce the magnitude of the error. For biomolecular applications, one potential concern is that the QM degrees of freedom are treated with minimization in the MFEP search, thus large amplitude motions of the QM atoms (e.g., rearrangement of water molecules, see discussion of myosin, or change of hydration (Liang, Swanson, Peng, Wikström, & Voth, 2016; Lu et al., 2016)) during the reaction might be missed. Therefore, for complex chemical transformations, it is beneficial to compare results from the MFEP approach and direct free energy simulations using an inexpensive QM/MM potential; once subtleties associated with sampling for the specific system in hand are better understood, the energetics can be further improved by either MFEP calculations using a high-level QM/MM potential, or other multilevel free energy approaches discussed in the next section.

2.2.3 Multilevel QM/MM Free Energy Calculations

The importance of integrating different levels of QM/MM simulations can be highlighted using the example of proton transfer in human carbonic anhydrase (CAII), a well-studied system both experimentally (Silverman & McKenna, 2007) and computationally (Maupin, McKenna, Silverman, & Voth, 2009; Riccardi, Yang, & Cui, 2010). With extensive DFTB/MM free energy simulations using a collective variable that describes the translocation of the excess proton, we found that the reaction proceeds through a set of stepwise proton transfers (Riccardi, Koenig, Guo, & Cui, 2008); this aspect of the reaction was also observed in MS-EVB-based free energy simulations (Maupin et al., 2009). Using a B3LYP/MM potential and MFEP, however, the transition state was observed to be highly concerted in nature (see Fig. 2). Evidently, it is worth further exploring strategies that integrate extensive sampling and higher accuracy, which are offered by semiempirical and ab initio/DFT QM/MM methods, respectively.

One strategy in this context is multilevel QM/MM free energy simulations, in which the reaction pathway is explored at the low-level (e.g., semiempirical QM/MM) with umbrella sampling or the string method; then the relative free energies of key stationary points (e.g., transition state and reactant/product) are improved by conducting high-level QM/MM calculations (Marti, Moliner, & Tuñón, 2005; Polyak, Benighaus, Boulanger, & Thiel, 2013). The simplest "correction" is to compute the potential energy difference between low- and high-level QM/MM potentials along the minimum energy path, although it is difficult to evaluate whether such a correction remains transferable as thermal fluctuations of

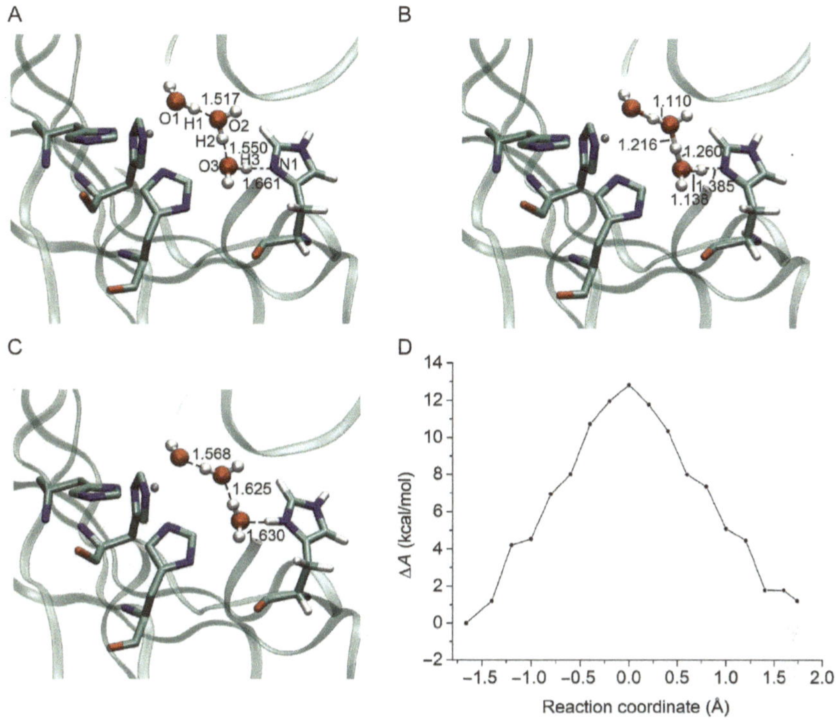

Fig. 2 Optimized structures for the reactant (A), transition state (B), and product (C) for the proton transfer in human carbonic anhydrase using the B3LYP/6-31+G(d,p)// CHARMM MFEP approach; the computed free energy profile is shown in (D). The QM region includes the zinc ions, its ligands, His 64 (proton acceptor), and the two bridging water molecules.

the system are taken into consideration. To evaluate the free energy difference between different levels of QM/MM potentials, it is necessary to conduct MD sampling for the relevant stationary points (e.g., by constraining the value of the reaction coordinate to specific values). The degree of sampling required for such a free energy perturbation to reach reasonable convergence depends on the distribution overlap between the two levels of QM/MM potentials (Boresch & Woodcock, 2017; Lu et al., 2016). In practice, it is likely that some degree of sampling at the high QM/MM level is required (König, Hudson, Boresch, & Woodcock, 2014; Lu et al., 2016; Plotnikov, Kamerlin, & Warshel, 2011), although innovative strategies [e.g., machine learning (Shen, Wu, & Yang, 2016)] may be required to minimize the computational cost; this remains an active area of research.

2.3 Other Experimental Observables

In addition to the activation free energy barrier, other experimental observables should be computed to allow a systematic comparison to experiments; considering the various approximations in computational methods, agreement with experimental activation free energy alone does not provide a compelling support for the computational model. These include, for example, various spectroscopic data such as infrared, Raman and nuclear magnetic resonance chemical shifts, free energy relationships, and KIEs. For the computation of these experimental observables, QM/MM models are uniquely useful as compared to reactive force field models, such as MS-EVB; while it is possible to parameterize VB type of models for energetics, it is generally difficult to parameterize such models for other properties such as variation of vibrational frequency and charge along the reaction pathway. As computational hardware and methodologies continue to improve, QM/MM calculations for these experimental observables can be conducted for increasingly realistic enzyme models with less approximation. For example, although KIEs were largely computed using models based on (generalized) normal mode analysis (Gao & Truhlar, 2002), it has become feasible to conduct path integral simulations that treat nuclear quantum effects beyond the harmonic model (Major & Gao, 2007; Wang, Ceriotti, & Markland, 2014). As discussed in Section 3.1 using AP as an example, QM/MM calculations can help highlight subtleties associated with the interpretation of various experimental observables.

3. CASE STUDIES

In this section, using two enzymes that we have studied in recent work, we illustrate the application of various QM/MM approaches to the analysis of phosphoryl-transfer mechanisms.

3.1 Alkaline Phosphatase: Phosphoryl-Transfer Transition State

AP has attracted much attention in recent years in the area of enzymology because it is one of the remarkable examples of catalytic promiscuity (Jonas & Hollfelder, 2009; O'Brien & Herschlag, 1999a; Pabis et al., 2016): it catalyzes the hydrolysis of phosphate monoesters with an impressive proficiency of 10^{27}, but it is also able to catalyze the hydrolysis of phosphate diesters, triesters, and sulfate monoesters with respectable efficiency as well.

Such a level of promiscuity makes it possible to study the catalysis with a broad range of substrates, providing fairly unique data, such as free energy relationships (Nikolic-Hughes, Rees, & Herschlag, 2004; Zalatan & Herschlag, 2006), that shed light onto the nature of the phosphoryl-transfer transition state and factors that dictate the catalytic efficiency and selectivity. The large body of diverse experimental data (Andrews, Deng, & Herschlag, 2011; Andrews, Fenn, & Herschlag, 2013; Bobyr et al., 2011, 2012; Catrina et al., 2007; Lassila & Herschlag, 2008; Nikolic-Hughes, O'Brien, & Herschlag, 2005; Nikolic-Hughes et al., 2004; O'Brien & Herschlag, 1998, 1999b, 2001, 2002; O'Brien, Lassila, Fenn, Zalatan, & Herschlag, 2008; Peck, Sunden, Andrews, Pande, & Herschlag, 2016; Sunden, Peck, Salzman, Ressl, & Herschlag, 2015; Zalatan et al., 2007; Zalatan, Fenn, Brunger, & Herschlag, 2006; Zalatan, Fenn, & Herschlag, 2008; Zalatan & Herschlag, 2006) also provides an opportunity to test computational methods as well as popular models used for interpreting experimental data.

3.1.1 Nature of Transition State and Free Energy Relationships

Free energy relationships are widely used in the physical organic literature (Greig, 2010; Williams, 2003) (also see Florian, Aqvist, & Warshel, 1998; Rosta et al., 2008). For the phosphoryl-transfer reaction in AP, the nucleophile is a protein residue, Ser102 (see Fig. 3). The free energy relationships were thus evaluated with different substrate leaving groups for a series of phosphate monoesters and diesters (Zalatan & Herschlag, 2006); for the cognate substrate phosphate monoesters, the measurements were done with the R166S variant to ensure that the chemical step is rate limiting. The Brönsted slope β_{lg} was measured to be -0.85 ± 0.13 and -0.95 ± 0.08 for the monoesters (O'Brien & Herschlag, 2002) and diesters (Zalatan & Herschlag, 2006), respectively; these are similar in magnitude (following the consideration of binding contribution to the observed β_{lg} (O'Brien & Herschlag, 2002)) to the values measured in solution, -1.23 and -0.94 ± 0.05, for aryl monoesters and diesters, respectively (Lassila et al., 2011; Zalatan & Herschlag, 2006). Therefore, the interpretation (Zalatan & Herschlag, 2006) was that the phosphoryl-transfer transition state in AP resembles that in solution, which has been established to be loose and synchronous in nature for monoesters and diesters, respectively (Lassila et al., 2011). In other words, in contrast to the traditional picture of enzyme catalysis, the AP active site appears to be able to stabilize different types of transition states, despite the substantially different charge distributions expected for these transition states.

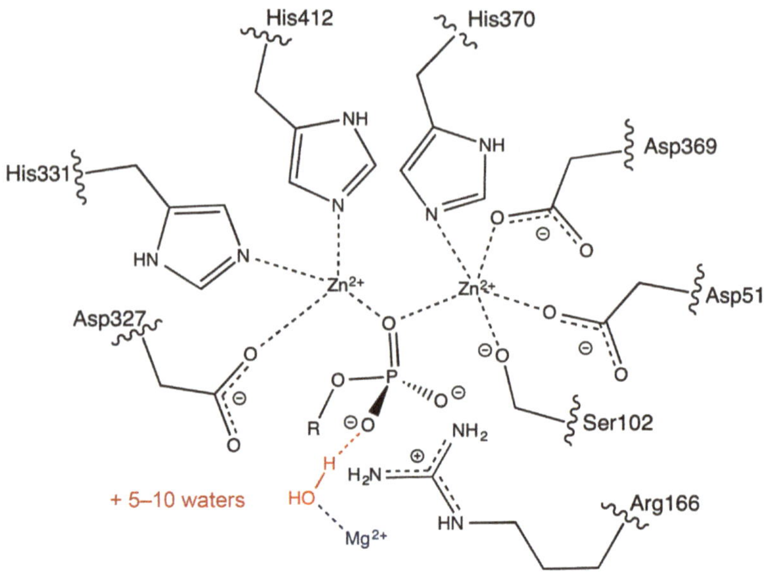

Fig. 3 A schematic illustration for the active site of alkaline phosphatase; this also illustrates the minimal QM region used in QM/MM simulations (Roston & Cui, 2016b; Roston et al., 2016).

In a loose transition state, there is substantial bond cleavage of the P–O^{lg} bond and little bond formation with the nucleophile; as a result, the formal charge associated with the PO_3 moiety is expected to be lower in the transition state as compared to the reactant state (Fig. 4A). In the canonical model for the phosphoryl-transfer transition state in AP (Nikolic-Hughes et al., 2005, 2004; Zalatan et al., 2008), the PO_3 moiety interacts with several cationic species in the active site: two zinc ions and the highly conserved Arg166. Thus, the reduction of the formal charge of PO_3 would imply that Arg166 is, in fact, "anticatalytic" in the chemical step. This rather unexpected inference suggests that the nature of the transition state in AP deserves further scrutiny; either the electronic structure or the binding mode of the reactive moieties, or both, deviate from the canonical model that emerged from experimental studies.

Motivated by these considerations, we have conducted QM/MM free energy simulations (Roston et al., 2016) for a series of monoesters and diesters in wild-type AP using the latest formulation of the DFTB model (DFTB3 (Gaus et al., 2011)/3OB (Gaus, Goez, & Elstner, 2012; Gaus, Lu, et al., 2014)); the computational efficiency of the DFTB3 model was essential to the extensive sampling required for the AP system,

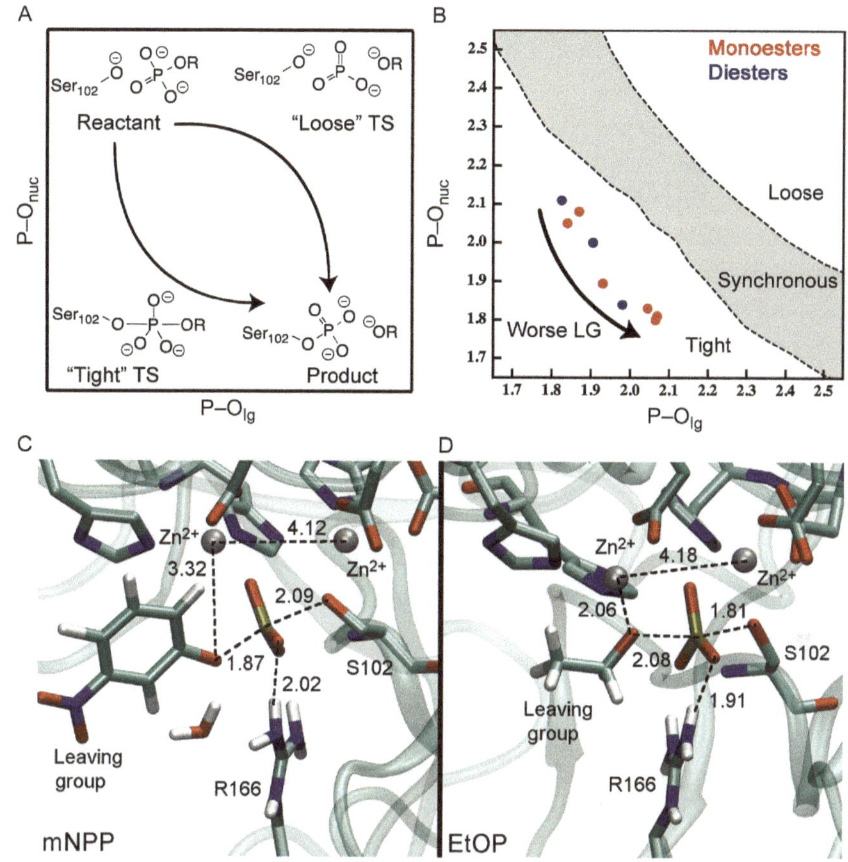

Fig. 4 Results from DFTB3/MM simulations for AP with multiple substrates. (A) Formal charges associated with the reaction pathways that involve a loose and tight transition state, respectively. (B) Values for the P–O^{nuc} and P–O^{lg} distances in the transition state computed for a series of mono- and diester phosphate substrates; the boundary between "tight," "synchronous," and "loose" transition states is defined approximately based on bond-order considerations (Roston et al., 2016); (C) and (D) Snapshots for the transition state from simulations for phosphate monoesters with a good (m-NP) and a poor (EtO) leaving group, respectively.

where catalytic promiscuity stems in part from the open nature of the active site, which is highly accessible to solvent molecules (Hou & Cui, 2013). For each substrate, adaptive umbrella sampling simulations were conducted to construct the two-dimensional free energy surface for the phosphoryl-transfer reaction; the natural reaction coordinates are the breaking and forming P–O bonds associated with the leaving group and nucleophile (Ser102). The QM region contains approximately

125 atoms (Fig. 3) and the free energy simulations include more than 10 ns of sampling for each substrate.

Several features from the QM/MM free energy simulations are worth noting. First, for the series of monoesters studied, the nature of the transition state is tight as reflected by a significant degree of bond formation with the nucleophile (Fig. 4B). As a result, partial charges on the PO_3 moiety (Mulliken charge and NBO charge analyses led to similar trends) are more negative in the transition state than in the reactant state (Roston et al., 2016); this trend supports the catalytic role of the conserved Arg166. Second, the degree of bond cleavage in the transition state depends on the nature of the leaving group (Roston et al., 2016); the Wiberg bond order changes from ~0.7 for a decent leaving group (*meta*-nitrophenyl, mNP) to ~0.4 for a poor leaving group (ethyl oxide). In other words, there is a significant shift in the nature of the phosphoryl-transfer transition state as a function of the leaving group, which suggests that the free energy relationships for the leaving group should contain a significant degree of curvature; a close examination of the experimental data reported in previous literature (Zalatan & Herschlag, 2006) indeed indicates notable deviation from linearity for both phosphate monoesters and diesters. Interpretation of the free energy relationships data without considering the curvature would conclude that the transition state features significant bond cleavage, which is supported by our calculations only for the poor leaving groups. Third, the binding mode of the reactive moieties in the transition state also varies with the leaving group (Fig. 4C and D). For a decent leaving group such as mNP, as there is little charge development on the leaving group in the transition state, the mNP-O fragment does not interact strongly with the active-site zinc ions and is stabilized largely by solvent molecules; the PO_3 also interacts only with one zinc ion. For a poor leaving group such as ethyl oxide, which features significant bond cleavage in the transition state, the EtO^- fragment is bound to the zinc ion and the PO_3 is coordinated with both zinc ions, similar to the canonical model adopted in the experimental literature (Nikolic-Hughes et al., 2005, 2004; Zalatan et al., 2008).

Therefore, the computational studies point to a model in which the bonding nature and binding mode of the transition state vary significantly depending on the leaving group; for additional discussion of tests and predictions of the model, see the original publication (Roston et al., 2016). It should be pointed out that the possibility of transition-state variation with respect to perturbation has been recognized in the literature (Greig, 2010; Williams, 2003); e.g., it was stated that "in principle," all free energy relationships should be curved. In practice, however, it is usually assumed that deviation from linearity is insignificant (Jencks, 1985) and thus curvature is taken as evidence for a change in the rate limiting step or a change in mechanism. Our analysis for AP suggests otherwise and highlights that experiments that rely on systematic perturbations, either through mutation or change of substrate, should be interpreted with care, as perturbations may lead to nontrivial changes in the properties of interest (e.g., nature of the transition state in the WT enzyme).

Another implication of the computational results, as emphasized in a recent study (Roston & Cui, 2016a), is that crystal structures with transition-state analogs may not accurately reflect the structure and binding mode of the transition state. For AP, for example, crystal structures with vanadate (Bobyr et al., 2011) and tungstate (Peck et al., 2016) suggested a binding mode that appears to be a better representation of the transition state for a substrate with a poor leaving group, in which the PO_3 is coordinately simultaneously with the two zinc ions and Arg166.

3.1.2 Nature of Transition State and Kinetic Isotope Effects

KIE measurement is a powerful tool for interrogating the properties of the transition state (Cleland & Hengge, 2006). Over the years, a set of empirical rules have been established for the interpretation of KIEs. For phosphoryl-transfer transitions states (Hengge, 2002), for example, a loose transition state is usually associated with a large and normal ^{18}O KIE for the bridging oxygen and an inverse ^{18}O KIE for the nonbridging oxygens. Although tremendously valuable, these empirical rules were established by considering the contributions from a very limited number of vibrational modes that involve the substituted isotopes. In the enzyme active site, the reactive fragment is bound with multiple interactions, thus many "vibrational modes" might be perturbed due to an isotope substitution and therefore complicate the interpretation of KIEs. Another important issue is that the measurements are on V/K, which includes the binding step.

For AP, KIE measurements for two phosphate monoesters led to results that were not simple to interpret (Zalatan et al., 2007). In the WT enzyme, an aryl leaving group (*para*-nitrophenyl, pNP) leads to a normal but very small ^{18}O KIE (1.0003) for the bridging oxygen and an inverse nonbridging KIE; in the R166S variant, where the chemical step is fully rate limiting, the bridging oxygen KIE is slightly larger (1.0091). With an alkyl leaving group (*m*-nitrobenzyl, mNB), the bridging oxygen KIE is substantially larger in both the WT (1.0072) and R166S variant (1.0199). Those results were interpreted (Zalatan et al., 2007) to suggest that the transition state is loose for both substrates, although the magnitude of bridging oxygen KIE is reduced due to strong interactions with the active site; the larger bridging oxygen KIEs for the alkyl leaving group were thought to arise from the intrinsically larger KIEs for alkyl phosphates than aryl phosphates, even with the same degree of P–O^{lg} cleavage.

The computational efficiency of the DFTB3/MM approach allowed us to compute the ^{18}O KIEs (Roston & Cui, 2016b) with both the bridging and nonbridging oxygen substitutions using a path-integral-based free energy perturbation approach (Major & Gao, 2007). The calculation assumes that the location of the transition state does not vary due to isotope substitution, while the vibrational contribution to the free energy is evaluated with anharmonicity included. The results led to several interesting observations. First, the QM/MM KIE results generally captured the trends observed in the experimental measurements; i.e., a normal bridging oxygen KIE and an inverse nonbridging oxygen KIE, and that alkyl leaving groups have larger bridging oxygen KIEs than aryl leaving groups. Therefore, the KIE results generally support the qualitative trends in the computed transition-state structures for the set of substrates studied. Second, comparison of computed equilibrium and KIEs indicated that the origin of the difference between aryl and alkyl leaving groups was not the intrinsic difference between the corresponding substrates, but rather due to the different degrees of bond cleavage in the transition state, further supporting our model that the nature of the transition state varies with different leaving groups. Third, with an active-site model, DFTB3 and B3LYP calculations give consistent results but they do not capture the experimental trends. This observation highlights the importance of explicitly including the enzyme contribution and that interactions with the enzyme environment may make substantial contributions to KIEs. In fact, path integral calculations found rather significant binding isotope effects for the bridging oxygen in the ground state, suggesting that the substrate structure is deformed toward the transition state upon binding to the enzyme active site. For additional discussion, see Roston and Cui (2016b).

3.1.3 Remaining Questions and Methodological Challenges

Taken together, the DFTB3/MM free energy and KIE calculations led to a different model for the phosphoryl-transfer reaction in AP as compared to previous experimental studies. Although the model appears consistent with experimental data published to date, and it led to a number of testable predictions explicitly summarized in Roston et al. (2016), we note that several features of AP catalysis remain to be better understood.

First of all, while the DFTB3/MM simulations thoroughly explored the free energy surface near the transition-state region, the calculations were not able to reliably predict the activation free energy. As the substrate moves further away from the nucleophile, solvent molecules penetrate into the active site; as a result, the free energy surface for the reactant region was found to be too broad to allow a clear definition of the Michaelis complex, making it difficult to reliably evaluate the activation free energy and therefore an explicit evaluation of the free energy relationships. Although the substrate is expected to bind weakly to the active site due to its solvent accessible nature, the difficulty in defining the Michaelis complex in the simulations highlights the importance of balancing QM-QM and QM/MM interactions at the enzyme/solvent interface (Roston & Cui, 2016a). Further improvement of DFTB3 and DFTB3/MM for noncovalent interactions is clearly warranted (Goyal et al., 2014).

Second, recent experimental studies in the Herschlag lab identified nonadditive contributions from residues in the active site (Sunden et al., 2015). It was argued that understanding such cooperative interactions is important for defining features that dictate the catalytic proficiency and selectivity of enzymes. While several cooperative interactions are straightforward to rationalize due to the spatial proximity of the relevant residues, the underlying mechanism for other nonadditive effects requires further explanation. Along this line, QM/MM simulations are expected to make useful contributions, although the issue with defining the Michaelis complex needs to be resolved first.

Finally, although trends in the DFTB3/MM free energy and KIE calculations are robust for both mono- and diester substrates, it is worth evaluating the transition-state properties using ab initio/DFT-based QM/MM simulations. As indicated by the active-site model comparison mentioned in Section 2.1.2, for example, while DFTB3 captures the key trends in DFT calculations (also see Roston & Cui, 2016b), it appears to overestimate the dependence of transition-state geometries on the nature of the leaving group (Table 2); moreover, because the QM region is fairly large (125–250 atoms), dispersion interactions likely make a notable contribution, as hinted by the difference between DFT and MP2 results for the active-site models (Table 3).

3.2 Myosin: Regulation of the Timing of ATP Hydrolysis

ATP and GTP hydrolysis reactions power many biomolecular machines in cells. Although it has become increasingly clear that the remarkable structural transitions in those systems are usually driven by nucleotide binding or dissociation rather than the chemical (hydrolysis) step per se (Gao & Karplus, 2004; Vale & Milligan, 2000; Yu, Ma, Yang, & Cui, 2007), the question of what controls the timing of the hydrolysis step remains relevant; if hydrolysis could occur independent of large structural transitions, the efficiency of energy transduction would be compromised. In this context, there are two possible models: in one model, hydrolysis of nucleotides is strictly controlled by the nearby residues, and conformational transition of the active site is strongly coupled with global structural transitions; i.e., global structural changes are the *consequence* of active-site events. In the other model, global structural changes explicitly impact the efficiency of ATP hydrolysis, presumably by influencing the structural and dynamical properties of the active site; i.e., global structural changes are the *cause* of active-site events. The goals of our work are to establish which model is applicable to realistic biomolecular motors and to reveal the underlying molecular mechanism. We choose to focus on the molecular motor myosin due to the rich experimental background in terms of both structural studies and biochemical/kinetic analysis (Geeves & Holmes, 2005; Sweeney & Houdusse, 2010).

3.2.1 Multiple Reaction Pathways

To probe how conformational properties impact the hydrolysis activity, we first need to establish the hydrolysis mechanism. We do so by performing QM/MM calculations for ATP hydrolysis in the crystal structure for the prepowerstroke state; this structure (Fig. 5A) was obtained with ADP·vanadate in the active site (Smith & Rayment, 1996), thus regarded as the state in which ATP is efficiently hydrolyzed.

The mechanism of ATP hydrolysis in myosin has been analyzed by numerous experimental (Málnási-Csizmadia et al., 2001; Malnasi-Csizmadia et al., 2007; Onishi, Mochizuki, & Morales, 2004; Onishi, Ohki, Mochizuki, & Morales, 2002) and computational (Grigorenko et al., 2007; Kiani & Fischer, 2014; Li & Cui, 2004; Lu et al., 2017; Schwarzl, Smith, & Fischer, 2006; Yang & Cui, 2009; Yang, Yu, & Cui, 2008) investigations. For the purpose of this chapter, it is worth noting that debates remain regarding the identity of a catalytic base: one mechanism involves the γ phosphate of ATP as the ultimate base, although the proton transfer from the lytic water

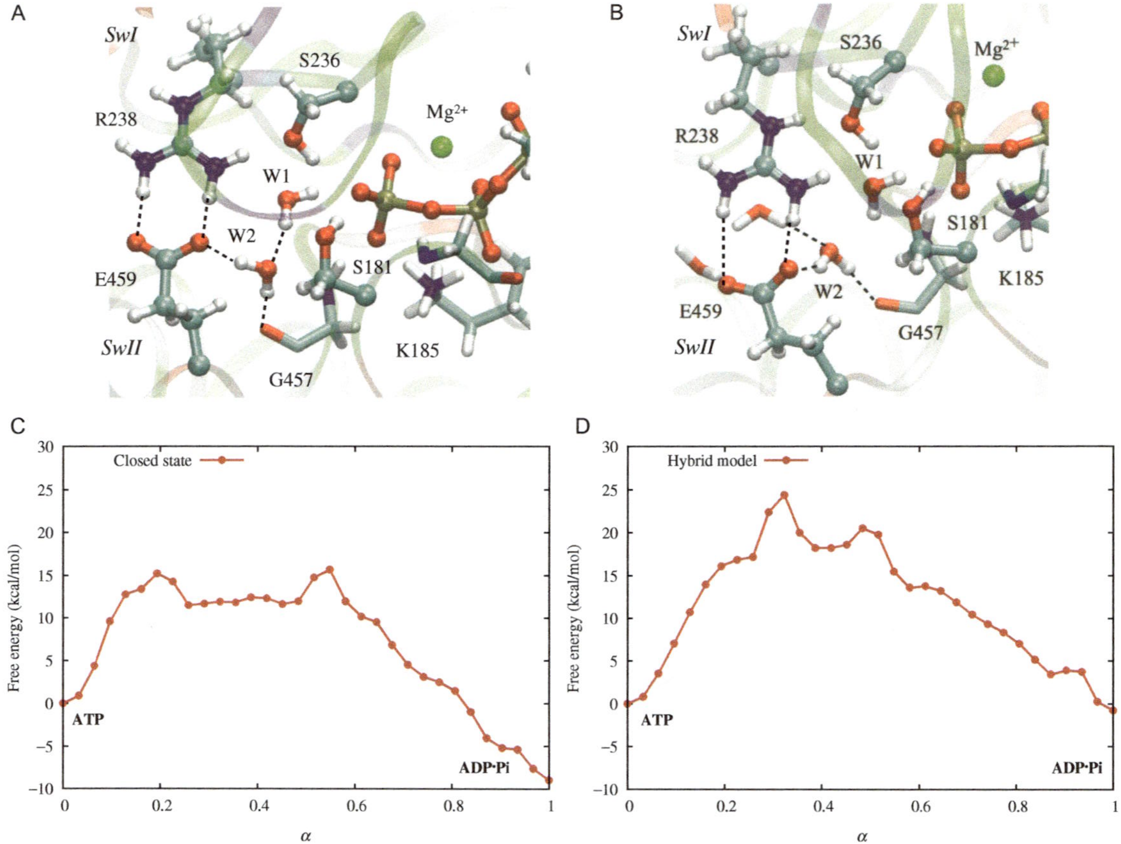

Fig. 5 Active-site structure of myosin and computed free energy profile for ATP hydrolysis. (A) The prepowerstroke state based on the crystal structure of the myosin motor domain complexed with ADP·vanadate (Smith & Rayment, 1996); (B) A computational "hybrid" model, in which the active-site protein residues are restrained to adopt the structure of the prepowerstroke state, while the rest adopts the postrigor crystal structure (Bauer et al., 2000). (C) and (D) Free energy profiles computed for the two structural models based on DFTB3/MM string calculations (Lu et al., 2017). Note that although the active-site structures are very similar in the two models, the hydrolysis free energy profiles are very different.

is likely mediated by active-site water molecules and/or polar side chains, while another mechanism involves a highly conserved glutamate (Glu459 in *Dictyostelium discoideum* myosin). While glutamate serving as the base for ATP/GTP hydrolysis has been supported in several other molecular motors such as F_1-ATPase (Hayashi et al., 2012) and kinesin (McGrath et al., 2013), this role in myosin is not expected as Glu459 is engaged in a signature salt-bridge with Arg238; this interaction has been deemed essential to the closure of the active site and also expected to lower the pK_a of Glu459, making it less likely to act as a base. Different computations, however, led to different conclusions (Grigorenko et al., 2007; Kiani & Fischer, 2014; Lu et al., 2017), and mutational studies involving this salt-bridge (Onishi et al., 2002) also led to results that were not straightforward to interpret (see below).

From a technical perspective, the mechanism of ATP hydrolysis is not straightforward to analyze using standard umbrella sampling methods. While the nucleophilic attack is well described by the antisymmetric stretch coordinate involving the lytic water oxygen, γ phosphorus (P^{γ}) and the bridging oxygen between the β and γ phosphates, proton transfer from the lytic water may occur through multiple routes and implicate different water molecules and amino acids in the active site. Coordinates that explicitly specify the identity of the "proton relay groups" are too restrictive and not appropriate for free energy simulations; coordinates that are collective in nature, such as those based on the center of excess charge (Koenig et al., 2006), are suitable for describing long-range proton transfers in which pathways involving rapidly exchanging water molecules are averaged over. However, coordinates that are highly collective in nature may also map very different structures onto the same region of the reaction coordinate space and thus compromises the physical meaningfulness of the free energy surface (see Fig. 6 for an example). Therefore, the strategy found effective in Lu et al. (2017) was to use metadynamics simulations to explore the reaction mechanism with delocalized collective variables; once a complete reaction pathway is sampled (it does not need to feature a converged free energy profile), the structures are fed into string calculations, which are able to use many more structural variables to uniquely define the pathway for meaningful free energy calculations.

Using this strategy, five distinct reaction pathways were identified for ATP hydrolysis in the prepowerstroke state in essentially an automated fashion. The free energy profiles in Fig. 7 indicate different barriers but very consistent exothermicity, supporting the adequate convergence of the simulations; in terms of the amount of required sampling, the convergence

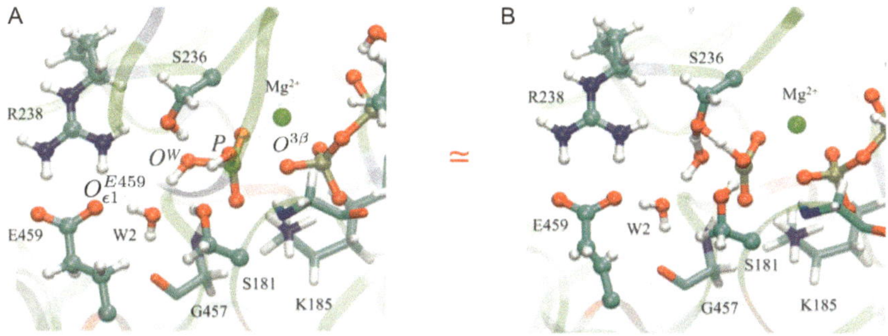

Fig. 6 Two active-site structures of the myosin motor domain have rather different bonding structures but very similar collective variables defined in metadynamics simulations (Lu et al., 2017) (CV1 is the antisymmetric stretch involving the breaking (P–$O^{3\beta}$) and forming (O^{W}-P) P–O bonds; CV2 is the antisymmetric stretch involving the location of the excess charge relative to P^{γ} and carboxylate oxygen of Glu459, $O_{\epsilon 1}^{E459}$). As a result, they are mapped into the same region in the collective variable space, compromising the physical meaningfulness of the computed free energy surface.

behavior shown for one pathway indicates that on the order of 1 ns per image is required (Fig. 8), which is only feasible at the moment with semiempirical-based QM/MM simulations.

Although the calculations do not have the required accuracy (see discussion in Section 3.1.3) to reliably rank the three low-barrier pathways, analysis of energetics along the multiple pathways indicates that the barrier is low when the hydronium is directly stabilized by Glu459. Indeed, among the low-free energy pathways, Glu459 either explicitly acts as the catalytic base (in the E459-protonation path) or participates indirectly by electrostatically stabilizing the hydronium during proton transfer (e.g., in the W2 only pathway). These observations are consistent with the significant reduction of ATPase activity in variants in which Glu459 is replaced by an alanine (Onishi et al., 2002). The result that the E459-protonation pathway features a low-free energy barrier is not contradictory to the expectation that the salt-bridge interaction with Arg238 lowers the pK_a of Glu459; proton transfer to Glu459 was indeed found to be an uphill process, leading to a transient intermediate that is ~10 kcal/mol above the reactant state (Fig. 7A). Nevertheless, once the hydroxide is generated, the subsequent nucleophilic attack has only a modest barrier. Moreover, as illustrated in Fig. 9, active-site water molecules reorient to accommodate changes in protonation pattern throughout the reaction; these nonmonotonic changes in orientation are not straightforward to sample in minimization type of studies, which might explain the observation that the Glu459 pathway was found high in barrier

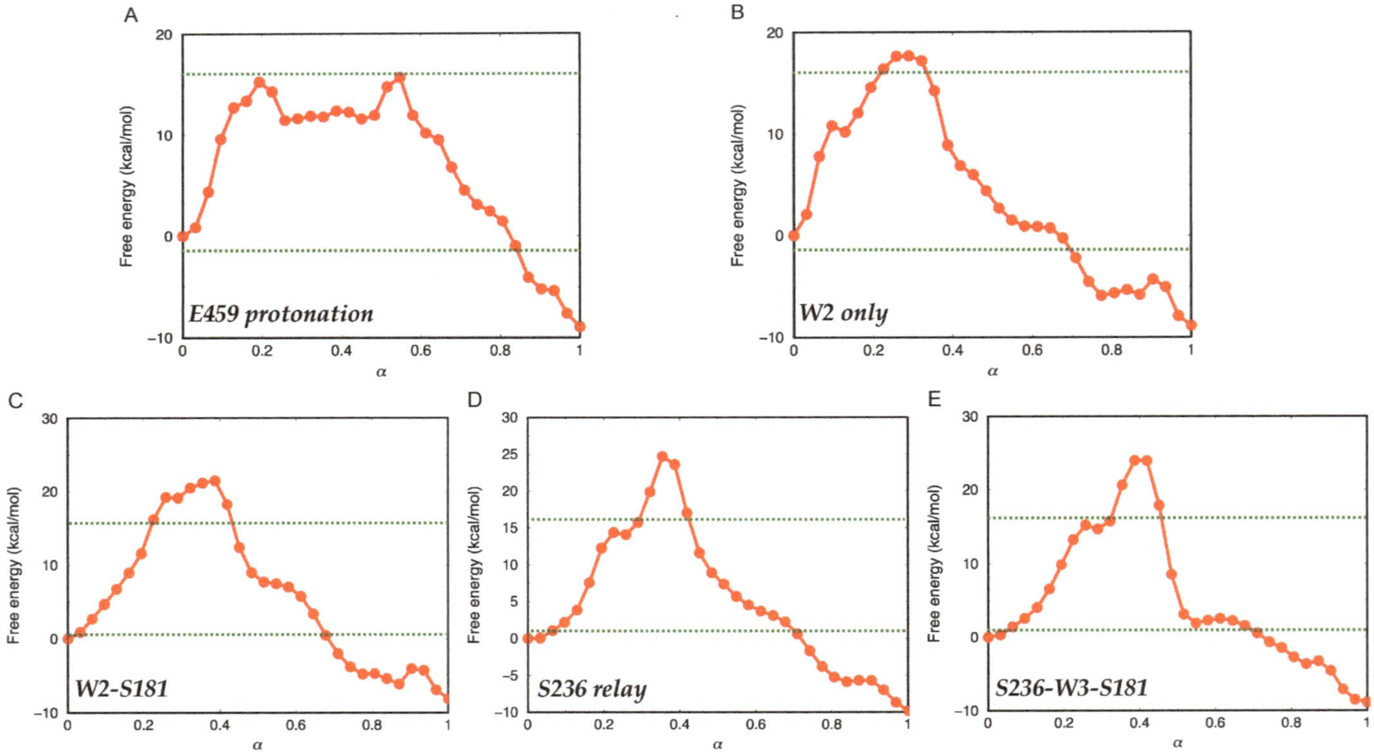

Fig. 7 Free energy profiles computed for distinct reaction mechanisms of ATP hydrolysis in myosin using DFTB3/MM string simulations. Note that independent simulations lead to consistent exothermicities, supporting the convergence of the free energy simulations. As discussed in the text and in Lu et al. (2017), the exothermicity is overestimated with the DFTB3/MM approach. Experimental activation barrier and exothermicity (Geeves & Holmes, 1999, 2005) are indicated with *dotted lines*.

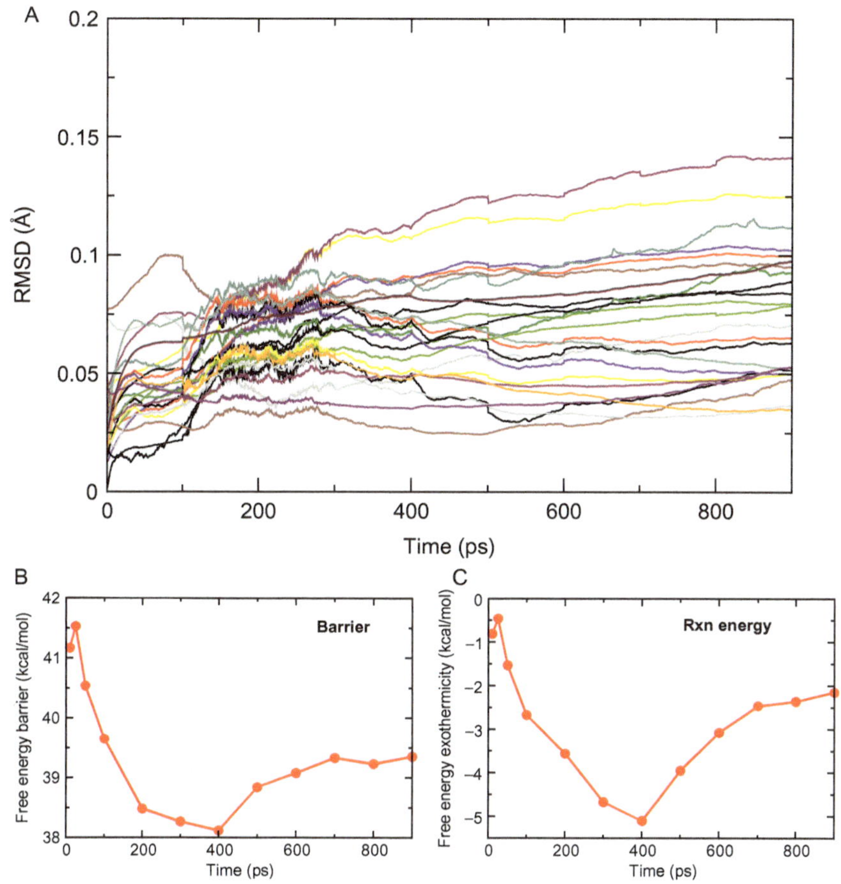

Fig. 8 Example of convergence of DFTB3/MM string simulations for ATP hydrolysis in myosin. The results shown are for the hydrolysis in the postrigor state, which features a very high barrier and low exothermicity as expected. In (A), the root-mean-square-differences (RMSD) in the collective variables relative to the initial guess are shown as functions of the sampling time per image in the string simulations. In (B,C), the convergence behaviors of the free energy barrier and reaction exothermicity are shown with respect to the amount of sampling time per image.

in the minimum energy path-based calculations (Kiani & Fischer, 2014). Such a comparison highlights the importance of including proper thermal sampling in the study of complex chemical transformations in enzymes.

3.2.2 Impact of Global Structural Transitions

To explore the roles of global structural transitions on the ATPase activity, we studied the hydrolysis reaction in a hybrid computational model, in which the active-site conformation in the postrigor crystal structure (Bauer et al., 2000) (which was obtained with ATP bound and clearly

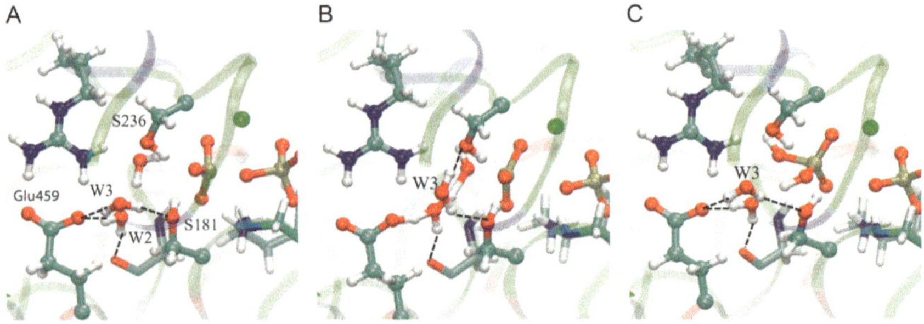

Fig. 9 Water molecules in the active site rearrange during the hydrolysis reaction along the E459-protonation pathway (Lu et al., 2017).

reflected a hydrolysis-incompetent state) was changed to that in the prepowerstroke structure through biased molecular dynamics simulations. As shown in Fig. 5B, the active site in the resulting hybrid structure closely resembles that in the prepowerstroke crystal structure (Fig. 5A), in terms of all amino acid side chains that are in direct contact with γ phosphate and the key salt-bridge between the *SwI* and *SwII* motifs. If ATP hydrolysis were strictly controlled by the "first coordination shell", free energy profiles of hydrolysis computed with these two models are expected to be very similar. Rather surprisingly, the computed free energy profiles differ significantly (compare Figs. 5C and D); in the hybrid model, the hydrolysis is less exothermic by almost 10 kcal/mol, and the free energy barrier is also higher by $\sim$ 9 kcal/mol. Evidently, hydrolysis is very much influenced by residues beyond the first and second coordination shell.

A close inspection of the active site in the two models indicates that water molecules therein behave differently. In the prepowerstroke conformation, the active site is well shielded from the bulk solvent and water molecules near the γ phosphate are engaged in a set of stable hydrogen-bonding interactions. In the hybrid "closed postrigor" model, although *SwII* is displaced to close the active site and establish more extensive interactions with ATP (e.g., through the main chain NH of Gly457), the other structural elements (e.g., the N-terminal loop of the relay helix) remain in the postrigor conformation (Yang et al., 2008). As a result, the active site remains, in fact, accessible to water; i.e., water molecules in the active site are in contact with the interfacial solvent molecules and more disordered compared to those in the prepowerstroke state (compare Figs. 5A and B). Another string calculation was conducted to estimate the free energy cost associated with reorganizing the active-site water molecules into configurations consistent with those in the prepowerstroke state; the penalty was found to be

~3 kcal/mol, which is significant but not the dominant contribution to the difference (~9 kcal/mol) between the free energy barriers in the two models. While further analysis is clearly warranted, the results support a model in which global structural transitions are important to catalysis; without these changes, the active site is not well set up for efficient chemistry even if key protein residues appear to be properly positioned; the effects may involve perturbation of active-site water orientations and reorganization energy, which contribute to the free energy of activation (Warshel et al., 2006). These considerations are reminiscent of the recurring observations from studies of natural enzymes and artificially evolved enzymes that remote mutations (Goodey & Benkovic, 2008; Schulenburg, Stark, Kunzle, & Hilvert, 2015; Tracewell & Arnold, 2009; Wang, Goodey, Benkovic, & Kohen, 2006), collectively, may lead to a significant impact on the chemical step in the active site.

3.2.3 Remaining Questions and Methodological Challenges

Numerous issues remain to be clarified. From a technical point of view, the accuracy of DFTB3 for multistep phosphoryl-transfer reactions needs to be improved. As shown in Fig. 7, the hydrolysis reaction in the prepowerstroke state is exothermic by almost 10 kcal/mol; the experimental data suggested that the hydrolysis is nearly thermoneutral (Geeves & Holmes, 1999). Studies of simple model reactions indicated that the error in exothermicity can be qualitatively understood (Lu et al., 2017), yet the issue of more concern is whether the DFTB3/MM approach captures the proper sequence of events between proton transfer and nucleophilic attack (Mlynsky et al., 2014); this affects the nature of the rate limiting transition state and therefore any factors that stabilize that transition state. Although gas phase models support the use of DFTB3 (Lu et al., 2017), more systematic calibration using solution phase model reactions will be valuable.

In terms of mechanistic details, the role of Glu459 needs to be further understood. In particular, mutation experiments involving the Arg238-Glu459 salt-bridge revealed several unexplained observations (Onishi et al., 2004, 2002). First, although this salt-bridge has been regarded as a key signature for the recovery stroke in myosin, mutation of either Arg238 or Glu459 did not abolish ATP binding or the recovery stroke; determining the structure and solvation of the active site in those variants will be informative. Second, while mutation of Glu459 into an alanine was, as expected, found to significantly reduce the ATP activity, a double mutant in which Arg238 and Glu459 were exchanged also had very low

ATPase activity; whether this was due to significantly perturbed active-site water properties or altered electrostatics remains to be clarified. Considering the similarity in active-site structure among many biomolecular motors and GTPases, insights into these questions are likely transferable to other enzymes that employ nucleotide hydrolysis to drive other processes.

Concerning the role of global structural transitions, it is of interest to better understand the properties of active sites that are, on the one hand, essential to catalysis, and on the other hand, tightly coupled to global structural changes. In this regard, in addition to the average position of key residues (e.g., the population of configurations that are most conducive to catalysis), it is likely that their fluctuations are also important due to contribution to the reorganization energy (Warshel et al., 2006). A quantitative understanding beyond the continuum electrostatic analysis performed in Lu et al. (2017) (also see Swiderek, Tuñón, Moliner, & Bertran, 2015) requires QM/MM methodologies that are able to compute reorganization energies with well-defined decomposition of the QM region into chemically meaningful diabatic states (Mo & Gao, 2000). Moreover, structural transitions distant from the active site may also influence the activation free energy through entropic factors, as discussed in the context of temperature adaptation of enzymes (Aqvist, Kazemi, Isaksen, & Brandsdal, 2017; Isaksen, Aqvist, & Brandsdal, 2016). A reliable determination of entropic contributions, however, requires extensive sampling, for which further simplification of the QM model (e.g., with an empirical valence bond model) is likely productive, as demonstrated by Åqvist and coworkers in the analysis of several enzymes (Carvalho et al., 2015; Kazemi, Himo, & Aqvist, 2016).

4. CONCLUDING REMARKS

In this contribution, we discussed the application of QM/MM free energy simulations to two phosphoryl-transfer systems. The main goal is to illustrate the applicability of QM/MM methods, including some of the technical issues worthy of attention when applied to realistic enzymes (as opposed to active-site models) and the remaining limitations of the DFTB3-based methodologies. Although the DFTB3/MM method should be regarded as semiquantitative in nature, its computational efficiency compared to ab initio/DFT-based QM/MM calculations makes it particularly attractive in problems where trends are of most interest; in the case of AP, it is the trend in the nature of the transition state as the leaving group varies, and in the case of myosin, it is the trend in energetics with either different routes of proton transfers or with different structural models of the active site. Whenever possible, it is always valuable to supplement DFTB3/MM simulations with ab initio/DFT-based calculations, either with active-site models or QM/MM calculations of the realistic enzyme model through the MFEP approach. Ultimately, models that emerge from the computational analysis need to be tested with experiments, by either comparing to published data in the literature or explicitly documenting predictions (Roston et al., 2016) to stimulate new experimental tests. Along this line, the analysis of AP highlights that the interpretation of experimental data is not always straightforward, and calibrated computational models can be informative in identifying subtleties in the analysis.

In addition to the "Remaining Questions and Methodological Challenges" already summarized earlier, it is important to analyze processes other than the chemical step in phosphoryl-transfer enzymes because very often it is not the chemical step that controls the rate or functional specificity. For example, while the mechanism of large-scale conformational transitions has started to be explored at an atomic level for several kinases (Meng, Shukla, Pande, & Roux, 2016; Meng, Pond, & Roux, 2017; Shukla, Meng, Roux, & Pande, 2014; Zheng & Cui, 2018), it remains challenging to identify factors that dictate the rate of such transitions. Further progress along these lines is essential to obtaining a global understanding of phosphoryl-transfer enzymes, especially regarding their evolution and new strategies that regulate their activity for biomedical and biotechnological applications.

ACKNOWLEDGMENTS

D.R. was partially supported by an NIH NRSA Post-doctoral Fellowship (1F32GM112371-01A1), and D.D. was supported by an NIH Pre-doctoral Fellowshop (T32 GM008293). This research was also supported by an NIH grant (R01GM106443) and an XSEDE allocation (TG-MCB110014) to Q.C. Computational resources from the Extreme Science and Engineering Discovery Environment (XSEDE), which is supported by NSF grant number OCI-1053575, are greatly appreciated; computations are also partly supported by the National Science Foundation through a major instrument grant (CHE-0840494).

REFERENCES

Alberts, B., Bray, D., Lewis, J., Raff, M., Roberts, K., & Watson, J. D. (1994). Molecular biology of the cell. Garland Publishing, Inc.

Andrews, L. D., Deng, H., & Herschlag, D. (2011). Isotope-edited FTIR of alkaline phosphatase resolves paradoxical ligand binding properties and suggests a role for ground-state destabilization. *Journal of the American Chemical Society*, *133*, 11621–11631.

Andrews, L. D., Fenn, T. D., & Herschlag, D. (2013). Ground state destabilization by anionic nucleophiles contributes to the activity of phosphoryl transfer enzymes. *PLOS Biology*, *11*, e1001599.

Åqvist, J., & Kamerlin, S. C. L. (2016). Conserved motifs in different classes of GTPases dictate their specific modes of catalysis. *ACS Catalysis*, *6*, 1737–1743.

Aqvist, J., Kazemi, M., Isaksen, G. V., & Brandsdal, B. O. (2017). Entropy and enzyme catalysis. *Accounts of Chemical Research*, *50*, 199–207.

Barducci, A., Bonomi, M., & Parrinello, M. (2011). Metadynamics. *WIREs Computational Molecular Science*, *1*, 826–843.

Bauer, C. B., Holden, H. M., Thoden, J. B., Smith, R., & Rayment, I. (2000). X-ray structures of the Apo and MgATP-bound states of *Dictyostelium discoideum* myosin motor domain. *Journal of Biological Chemistry*, *275*, 38494–38499.

Bennie, S. J., van der Kamp, M. W., Pennifold, R. C. R., Stella, M., Manby, F. R., & Mulholland, A. J. (2016). A projector-embedding approach for multiscale coupled-cluster calculations applied to citrate synthase. *Journal of Chemical Theory and Computation*, *12*, 2689–2697.

Bobyr, E., Lassila, J. K., Wiersma-Koch, H. I., Fenn, T. D., Lee, J. J., Nikolic-Hughes, I., … Herschlag, D. (2011). High-resolution analysis of Zn2+ coordination in the alkaline phosphatase superfamily by EXAFS and x-ray crystallography. *Journal of Molecular Biology*, *415*, 102–117.

Bobyr, E., Lassila, J. K., Wiersma-Koch, H. I., Fenn, T. D., Lee, J. J., Nikolic-Hughes, I., … Herschlag, D. (2012). High-resolution analysis of Zn2+ coordination in the alkaline phosphatase superfamily by EXAFS and x-ray crystallography. *Journal of Molecular Biology*, *415*, 102–117.

Boresch, S., & Woodcock, H. L. (2017). Convergence of single-step free energy perturbation. *Molecular Physics*, *115*, 1200–1213.

Brandenburg, J. G., Hochheim, M., Bredow, T., & Grimme, S. (2014). Low-cost quantum chemical methods for noncovalent interactions. *The Journal of Physical Chemistry Letters*, *5*, 4275–4284.

Brunk, E., & Rothlisberger, U. (2015). Mixed quantum mechanical/molecular mechanical molecular dynamics simulations of biological systems in ground and electronically excited states. *Chemical Reviews*, *115*, 6217–6263.

Campisi, J., & di Fagagna, F. (2007). Cellular senescence: When bad things happen to good cells. *Nature Reviews Molecular Cell Biology*, *8*, 729–740.

Carvalho, A. T. P., Szeler, K., Vavitsas, K., Åqvist, J., & Kamerlin, S. C. L. (2015). Modeling the mechanisms of biological GTP hydrolysis. *Archives of Biochemistry and Biophysics*, *582*, 80–90.

Catrina, I., O'Brien, P. J., Purcell, J., Nikolic-Hughes, I., Zalatan, J. G., Hengge, A. C., & Herschlag, D. (2007). Probing the origin of the compromised catalysis of E-coli alkaline phosphatase in its promiscuous sulfatase reaction. *Journal of the American Chemical Society*, *129*, 5760–5765.

Christensen, A. S., Kubar, T., Cui, Q., & Elstner, M. (2016). Semi-empirical quantum mechanical methods for non-covalent interactions for chemical and biochemical applications. *Chemical Reviews*, *116*, 5301–5337.

Claeyssens, F., Harvey, J. N., Manby, F. R., Mata, R. A., Mulholland, A. J., Ranaghan, K. E., … Werner, H. J. (2006). High-accuracy computation of reaction barriers in enzymes. *Angewandte Chemie International Edition*, *45*, 6856–6859.

Cleland, W. W., & Hengge, A. C. (2006). Enzymatic mechanisms of phosphate and sulfate transfer. *Chemical Reviews*, *106*, 3252–3278.

Cohen, P. (2002). Protein kinases - The major drug targets of the twenty-first century? *Nature Reviews Drug Discovery*, *1*, 309–315.

Collins, I., & Workman, P. (2006). Design and development of signal transduction inhibitors for cancer treatment: Experience and challenges with kinase targets. *Current Signal Transduction Therapy*, *1*, 13–23.

Cramer, C. J., & Truhlar, D. G. (2009). Density functional theory for transition metals and transition metal chemistry. *Physical Chemistry Chemical Physics*, *11*, 10757–10816.

Cui, Q. (2016). Quantum mechanical methods in biochemistry and biophysics. *The Journal of Chemical Physics*, *145*, 140901.

Cui, Q., & Karplus, M. (2003). Catalysis and specificity in enzymes: A study of triosephosphate isomerase (*TIM*) and comparison with methylglyoxal synthase (*MGS*). *Advances in Protein Chemistry*, *66*, 315–372.

Davies, S. P., Reddy, H., Caivano, M., & Cohen, P. (2000). Specificity and mechanism of action of some commonly used protein kinase inhibitors. *Biochemical Journal*, *351*, 95–105.

Dral, P. O., Wu, X., Sporkel, L., Koslowski, A., & Thiel, W. (2016). Semiempirical quantum-chemical orthogonalization-corrected methods: Benchmarks for ground-state properties. *Journal of Chemical Theory and Computation*, *12*, 1097–1120.

Duarte, F., Amrein, B. A., & Kamerlin, S. C. L. (2013). Modeling catalytic promiscuity in the alkaline phosphatase superfamily. *Physical Chemistry Chemical Physics*, *15*, 11160–11177.

Weinan, E., & Vanden-Eijnden, E. (2010). Transition-path theory and path-finding algorithms for the study of rare events. *Annual Review of Physical Chemistry*, *61*, 391–420.

Elstner, M., Hobza, P., Frauenheim, T., Suhai, S., & Kaxiras, E. (2001). Hydrogen bonding and stacking interactions of nucleic acid base pairs: A density-functional-theory based treatment. *The Journal of Chemical Physics*, *114*, 5149–5155.

Elstner, M., Porezag, D., Jungnickel, G., Elsner, J., Haugk, M., Frauenheim, T., … Seifert, G. (1998). Self-consistent-charge density-functional tight-binding method for simulations of complex materials properties. *Physical Review B*, *58*(11), 7260–7268.

Flaig, D., Beer, M., & Ochsenfeld, C. (2012). Convergence of electronic structure with the size of the QM region: Example of QM/MM NMR shieldings. *Journal of Chemical Theory and Computation*, *8*, 2260–2271.

Florian, J., Aqvist, J., & Warshel, A. (1998). On the reactivity of phosphate monoester dianions in aqueous solution: Bronsted linear free-energy relationships do not have an unique mechanistic interpretation. *Journal of the American Chemical Society*, *120*, 11524–11525.

Florian, J., & Warshel, A. (1998). Phosphate ester hydrolysis in aqueous solution: Associative versus dissociative mechanisms. *The Journal of Physical Chemistry B*, *102*, 719–734.

Friesner, R. A., & Guallar, V. (2005). Ab intio QM and QM/MM methods for studying enzyme catalysis. *Annual Review of Physical Chemistry*, *56*, 389–427.

Gao, J. L., Amara, P., Alhambra, C., & Field, M. J. (1998). A generalized hybrid orbital (GHO) method for the treatment of boundary atoms in combined QM/MM calculations. *The Journal of Physical Chemistry A*, *102*, 4714–4721.

Gao, J. L., Ma, S. H., Major, D. T., Nam, K., Pu, J. Z., & Truhlar, D. G. (2006). Mechanisms and free energies of enzymatic reactions. *Chemical Reviews*, *106*, 3188–3209.

Gao, J. L., & Truhlar, D. G. (2002). Quantum mechanical methods for enzyme kinetics. *Annual Review of Physical Chemistry*, *53*, 467–505.

Gao, Y., & Karplus, M. (2004). Biomolecular motors: The F_1-ATPase paradigm. *Current Opinion in Structural Biology*, *14*, 250–259.

Garber, K. (2001). Rapamycin's resurrection: A new way to target the cancer cell cycle. *Journal of the National Cancer Institute*, *93*, 1517–1519.

Garcia-Viloca, M., Gao, J., Karplus, M., & Truhlar, D. G. (2004). How enzymes work: Analysis by modern rate theory and computer simulations. *Science*, *303*, 186–195.

Gaus, M., Cui, Q., & Elstner, M. (2011). DFTB-3rd: Extension of the self-consistent-charge density-functional tight-binding method SCC-DFTB. *Journal of Chemical Theory and Computation*, *7*, 931–948.

Gaus, M., Cui, Q., & Elstner, M. (2014). Density functional tight binding (DFTB): Application to organic and biological molecules. *WIREs Computational Molecular Science*, *4*, 49–61.

Gaus, M., Goez, A., & Elstner, M. (2012). Parametrization and benchmark of DFTB3 for organic molecules. *Journal of Chemical Theory and Computation*, *9*, 338–354.

Gaus, M., Jin, H., Demapan, D., Christensen, A. S., Goyal, P., Elstner, M., & Cui, Q. (2015). DFTB3 parametrization for copper: The importance of orbital angular momentum dependence of Hubbard parameters. *Journal of Chemical Theory and Computation*, *11*, 4205–4219.

Gaus, M., Lu, X., Elstner, M., & Cui, Q. (2014). Parameterization of DFTB3/3OB for sulfur and phosphorus for chemical and biological applications. *Journal of Chemical Theory and Computation*, *10*, 1518–1537.

Geeves, M. A., & Holmes, K. C. (1999). Structural mechanism of muscle contraction. *Annual Review of Biochemistry*, *68*, 687–728.

Geeves, M. A., & Holmes, K. C. (2005). The molecular mechanism of muscle contraction. *Advances in Protein Chemistry*, *71*, 161–193.

Genna, V., Vidossich, P., Ippoliti, E., Carloni, P., & De Vivo, M. (2016). A self-activated mechanism for nucleic acid polymerization catalyzed by DNA/RNA polymerases. *Journal of the American Chemical Society*, *138*, 14592–14598.

Goodey, N. M., & Benkovic, S. J. (2008). Allosteric regulation and catalysis emerge via a common route. *Nature Chemical Biology*, *4*, 474–482.

Goyal, P., Qian, H. J., Irle, S., Lu, X., Roston, D., Mori, T., … Cui, Q. (2014). Feature article: Molecular simulation of water and hydration effects in different environments: Challenges and developments for DFTB based models. *The Journal of Physical Chemistry B*, *118*, 11007–11027.

Greig, I. R. (2010). The analysis of enzymic free energy relationships using kinetic and computational models. *Chemical Society Reviews*, *39*, 2272–2301.

Grigorenko, B. L., Rogov, A. V., Topol, I. A., Burt, S. K., Martinez, H. M., & Nemukhin, A. V. (2007). Mechanism of the myosin catalyzed hydrolysis of ATP as rationalized by molecular modeling. *Proceedings of the National Academy of Sciences of the United States of America*, *104*, 7057–7061.

Grimme, S., Hansen, A., Brandenburg, J. G., & Bannwarth, C. (2016). Dispersion-corrected mean-field electronic structure methods. *Chemical Reviews*, *116*, 5105–5154.

Hayashi, S., Uchida, Y., Hasegawa, T., Higashi, M., Kosugi, T., & Kamiya, M. (2017). QM/MM geometry optimization on extensive free-energy surfaces for examination of enzymatic reactions and design of novel functional properties of proteins. *Annual Review of Physical Chemistry*, *68*, 135–154.

Hayashi, S., Ueno, H., Shaikh, A. R., Umemura, M., Kamiya, M., Ito, Y., … Noji, H. (2012). Molecular mechanism of ATP hydrolysis in F-1-ATPase revealed by molecular simulations and single-molecule observations. *Journal of the American Chemical Society*, *134*(20), 8447–8454. https://doi.org/10.1021/ja211027m.

Hengge, A. C. (2002). Isotope effects in the study of phosphoryl and sulfuryl transfer reactions. *Accounts of Chemical Research*, *35*, 105–112.

Hou, G. H., & Cui, Q. (2012). QM/MM analysis suggests that alkaline phosphatase (AP) and nucleotide pyrophosphatase/phosphodiesterase (NPP) slightly tighten the transition state for phosphate diester hydrolysis relative to solution: Implication for catalytic promiscuity in the AP superfamily. *Journal of the American Chemical Society*, *134*, 229–246.

Hou, G. H., & Cui, Q. (2013). Stabilization of different types of transition states in a single enzyme active site: QM/MM analysis of enzymes in the alkaline phosphatase superfamily. *Journal of the American Chemical Society*, *135*, 10457–10469.
Hu, H., & Yang, W. T. (2008). Free energies of chemical reactions in solution and in enzymes with ab initio quantum mechanics/molecular mechanics methods. *Annual Review of Physical Chemistry*, *59*, 573–601.
Hu, L. H., Soderhjelm, P., & Ryde, U. (2011). On the convergence of QM/MM energies. *Journal of Chemical Theory and Computation*, *7*, 761–777.
Isaksen, G. V., Aqvist, J., & Brandsdal, B. O. (2016). Enzyme surface rigidity tunes the temperature dependence of catalytic rates. *Proceedings of the National Academy of Sciences of the United States of America*, *113*, 7822–7827.
Jencks, W. P. (1985). A primer for the BEMA HAPOTHLE—An empirical approach to the characterization of changing transition-state structure. *Chemical Reviews*, *85*, 511–527.
Jiang, W., Laury, M. L., Powell, M., & Wilson, A. K. (2012). Comparative study of single and double hybrid density functionals for the prediction of 3d transition metal thermochemistry. *Journal of Chemical Theory and Computation*, *8*, 4102–4111.
Jindal, G., & Warshel, A. (2016). Exploring the dependence of QM/MM calculations of enzyme catalysis on the size of the QM region. *Journal of Physical Chemistry B*, *120*, 9913–9921.
Jonas, S., & Hollfelder, F. (2009). Mapping catalytic promiscuity in the alkaline phosphatase superfamily. *Pure and Applied Chemistry*, *81*, 731–742.
Kamerlin, S. C. L., Haranczyk, M., & Warshel, A. (2009). Progress in ab initio QM/MM free-energy simulations of electrostatic energies in proteins: Accelerated QM/MM studies of pK(a), redox reactions and solvation free energies. *The Journal of Physical Chemistry B*, *113*, 1253–1272.
Kamerlin, S. C. L., Sharma, P. K., Prasad, R. B., & Warshel, A. (2013). Why nature really chose phosphate. *Quarterly Reviews of Biophysics*, *46*, 1–132.
Kazemi, M., Himo, F., & Aqvist, J. (2016). Enzyme catalysis by entropy without circe effect. *Proceedings of the National Academy of Sciences of the United States of America*, *113*, 2406–2411.
Kiani, F. A., & Fischer, S. (2014). Catalytic strategy used by the myosin motor to hydrolyze ATP. *Proceedings of the National Academy of Sciences of the United States of America*, *111*(29), E2947–E2956. https://doi.org/10.1073/pnas.1401862111.
Kiani, F. A., & Fischer, S. (2016). Comparing the catalytic strategy of ATP hydrolysis in biomolecular motors. *Physical Chemistry Chemical Physics*, *18*, 20219–20233.
Kiaris, H., & Spandidos, D. A. (1995). Mutations of Ras genes in human tumors. *International Journal of Oncology*, *7*, 413–421.
Knowles, J. R. (1980). Enzyme catalyzed phosphoryl transfer reactions. *Annual Review of Biochemistry*, *49*, 877–919.
Koenig, P., Ghosh, N., Hoffman, M., Elstner, M., Tajkhorshid, E., Frauenheim, T., & Cui, Q. (2006). Towards theoretical analysis of long-range proton transfer kinetics in biomolecular pumps. *The Journal of Physical Chemistry A*, *110*, 548–563.
König, G., Hudson, P. S., Boresch, S., & Woodcock, H. L. (2014). Multiscale free energy simulations: An efficient method for connecting classical MD simulations to QM or QM/MM free energies using Non-Boltzmann Bennett reweighting schemes. *Journal of Chemical Theory and Computation*, *10*, 1406–1419.
König, P. H., Hoffmann, M., Frauenheim, T., & Cui, Q. (2005). A critical evaluation of different QM/MM frontier treatments with SCC-DFTB as the QM method. *The Journal of Physical Chemistry B*, *109*, 9082–9095.
Kosugi, T., & Hayashi, S. (2012). QM/MM reweighting free energy SCF for geometry optimization on extensive free energ surface of enzymatic reaction. *Journal of Chemical Theory and Computation*, *8*, 322–334.

Kubillus, M., Kubař, T., Gaus, M., Řezáč, J., & Elstner, M. (2015). Parameterization of the DFTB3 method for Br, Ca, Cl, F, I, K, and Na in organic and biological systems. *Journal of Chemical Theory and Computation*, *11*(1), 332–342.

Kulik, H. J., Zhang, J. Y., Klinman, J. P., & Martinez, T. J. (2016). How large should the QM region be in QM/MM calculations? The case of catechol O-methyltransferase. *Journal of Physical Chemistry B*, *120*, 11381–11394.

Lange, S. S., Takata, K., & Wood, R. D. (2011). DNA polymerases and cancer. *Nature Reviews Cancer*, *11*, 96–110.

Lassila, J. K., & Herschlag, D. (2008). Promiscuous sulfatase activity and thio-effects in a phosphodiesterase of the alkaline phosphatase superfamily. *Biochemistry*, *48*, 12853–12859.

Lassila, J. K., Zalatan, J. G., & Herschlag, D. (2011). Biological phosphoryl transfer reactions: Understanding mechanism and catalysis. *Annual Review of Biochemistry*, *80*, 669–702.

Li, G. H., & Cui, Q. (2004). Mechanochemical coupling in myosin: A theoretical analysis with molecular dynamics and combined QM/MM reaction path calculations. *The Journal of Physical Chemistry B*, *108*, 3342–3357.

Liang, R. B., Swanson, J. M. J., Peng, Y. X., Wikström, M., & Voth, G. A. (2016). Multiscale simulations reveal key features of the proton-pumping mechanism in cytochrome c oxidase. *Proceedings of the National Academy of Sciences of the United States of America*, *113*, 7420–7425.

Liu, H. Y., Elstner, M., Kaxiras, E., Frauenheim, T., Hermans, J., & Yang, W. T. (2001). Quantum mechanics simulation of protein dynamics on long timescale. *Protein I: Structure, Function, and Genetics*, *44*, 484–489.

Lu, X., Fang, D., Ito, S., Okamoto, Y., Ovchinnikov, V., & Cui, Q. (2016). QM/MM free energy simulations: Recent progress and challenges. *Molecular Simulation*, *42*, 1056–1078.

Lu, X., Gaus, M., Elstner, M., & Cui, Q. (2015). Parameterization of DFTB3/3OB for magnesium and zinc for chemical and biological applications. *The Journal of Physical Chemistry B (Jorgensen Festschrift)*, *119*, 1062–1082.

Lu, X., Ovchinnikov, V., Roston, D., Demapan, D., & Cui, Q. (2017). Regulation and plasticity of catalysis in enzymes: Insights from analysis of mechanochemical coupling in myosin. *Biochemistry*, *56*, 1482–1497.

Major, D. T., & Gao, J. L. (2007). An integrated path integral and free-energy perturbation-umbrella sampling method for computing kinetic isotope effects of chemical reactions in solution and in enzymes. *Journal of Chemical Theory and Computation*, *3*, 949–960.

Málnási-Csizmadia, A., Pearson, D. S., Kovács, M., Woolley, R. J., Geeves, M. A., & Bagshaw, C. R. (2001). Kinetic resolution of a conformational transition and the ATP hydrolysis step using relaxation methods with a *Dictyostelium* myosin II mutant containing a single tryptophan residue. *Biochemistry*, *40*, 12727–12737.

Malnasi-Csizmadia, A., Toth, J., Pearson, D. S., Hetenyi, C., Nyitray, L., Geeves, M. A., … Kovacs, M. (2007). Selective perturbation of the myosin recovery stroke by point mutations at the base of the lever arm affects ATP hydrolysis and phosphate release. *Journal of Biological Chemistry*, *282*, 17658–17664.

Manby, F. R., Stella, M., Goodpaster, J. D., & Miller, T. F., III. (2012). A simple, exact density-functional-theory embedding scheme. *Journal of Chemical Theory and Computation*, *8*, 2564–2568.

Maragliano, L., Fischer, A., Vanden-Eijnden, E., & Ciccotti, G. (2006). String method in collective variables: Minimum free energy paths and isocommittor surfaces. *The Journal of Chemical Physics*, *125*, 024106.

Marti, S., Moliner, V., & Tuñón, I. (2005). Improving the QM/MM description of chemical processes: A dual level strategy to explore the potential energy surface in very large systems. *Journal of Chemical Theory and Computation*, *1*, 1008–1016.

Maupin, C. M., McKenna, R., Silverman, D. N., & Voth, G. A. (2009). Elucidation of the proton transport mechanism in human carbonic anhydrase II. *Journal of the American Chemical Society*, *131*, 7598–7608.

McCullagh, M., Saunders, M. G., & Voth, G. A. (2014). Unraveling the mystery of ATP hydrolysis in actin filaments. *Journal of the American Chemical Society*, *136*, 13053–13058.

McGrath, M. J., Kuo, I. F. W., Hayashi, S., & Takada, S. (2013). Adenosine triphosphate hydrolysis mechanism in kinesin studied by combined quantum-mechanical/molecular-mechanical metadynamics simulations. *Journal of the American Chemical Society*, *135*(24), 8908–8919. https://doi.org/10.1021/ja401540g.

Meng, Y., Shukla, D., Pande, V. S., & Roux, B. (2016). Transition path theory analysis of c-Src kinase activation. *Proceedings of the National Academy of Sciences of the United States of America*, *113*, 9193–9198.

Meng, Y. L., Pond, M. P., & Roux, B. (2017). Tyrosine kinase activation and conformational flexibility: Lessons from Src-family tyrosine kinases. *Accounts of Chemical Research*, *50*, 1193–1201.

Mlynsky, V., Banas, P., Sponer, J., van der Kamp, M. W., Mulholland, A. J., & Otyepka, M. (2014). Comparison of ab initio, DFT, and semiempirical QM/MM approaches for description of catalytic mechanism of hairpin ribozyme. *Journal of Chemical Theory and Computation*, *10*, 1608–1622.

Mo, Y. R., & Gao, J. L. (2000). Ab initio QM/MM simulations with a molecular orbital-valence bond (MOVB) method: Application to an S(N)2 reaction in water. *Journal of Computational Chemistry*, *21*, 1458–1469.

Monard, G., & Merz, K. M. (1999). Combined quantum mechanical/molecular mechanical methodologies applied to biomolecular systems. *Accounts of Chemical Research*, *32*, 904–911.

Nam, K., Cui, Q., Gao, J., & York, D. M. (2007). A specific reaction parameterization for the AM1/d hamiltonian for transphosphorylation reactions. *Journal of Chemical Theory and Computation*, *3*, 486–504.

Nashine, V. C., Hammes-Schiffer, S., & Benkovic, S. J. (2010). Coupled motions in enzyme catalysis. *Current Opinion in Chemical Biology*, *14*, 644–651.

Nikolic-Hughes, I., O'Brien, P. J., & Herschlag, D. (2005). Alkaline phosphatase catalysis is ultrasensitive to charge sequestered between the active site zinc ions. *Journal of the American Chemical Society*, *127*, 9314–9315.

Nikolic-Hughes, I., Rees, D. C., & Herschlag, D. (2004). Do electrostatic interactions with positively charged active site groups tighten the transition state for enzymatic phosphoryl transfer? *Journal of the American Chemical Society*, *126*, 11814–11819.

Nishizawa, H., Nishimura, Y., Kobayashi, M., Irle, S., & Nakai, H. (2016). Three pillars for achieving quantum mechanical molecular dynamics simulations of huge systems: Divide-and-conquer, density-functional tight-binding, and massively parallel computation. *Journal of Computational Chemistry*, *37*, 1983–1992.

O'Brien, P. J., & Herschlag, D. (1998). Sulfatase activity of E-coli alkaline phosphatase demonstrates a functional link to arylsulfatases, an evolutionarily related enzyme family. *Journal of the American Chemical Society*, *120*, 12369–12370.

O'Brien, P. J., & Herschlag, D. (1999b). Does the active site arginine change the nature of the transition state for alkaline phosphatase-catalyzed phosphoryl transfer? *Journal of the American Chemical Society*, *121*, 11022–11023.

O'Brien, P. J., & Herschlag, D. (1999a). Catalytic promiscuity and the evolution of new enzymatic activities. *Chemistry & Biology*, *4*, R91–R105.

O'Brien, P. J., & Herschlag, D. (2001). Functional interrelationships in the alkaline phosphatase superfamily: Phosphodiesterase activity of Escherichia coli alkaline phosphatase. *Biochemistry*, *40*, 5691–5699.

O'Brien, P. J., & Herschlag, D. (2002). Alkaline phosphatase revisited: Hydrolysis of alkyl phosphates. *Biochemistry*, *41*, 3207–3225.

O'Brien, P. J., Lassila, J. K., Fenn, T. D., Zalatan, J. G., & Herschlag, D. (2008). Arginine coordination in enzymatic phosphoryl transfer: Evaluation of the effect of Arg166 mutations in Escherichia coli alkaline phosphatase. *Biochemistry*, *47*, 7663–7672.

Onishi, H., Mochizuki, N., & Morales, M. F. (2004). On the myosin catalysis of ATP hydrolysis. *Biochemistry*, *43*, 3757–3763.

Onishi, H., Ohki, T., Mochizuki, N., & Morales, M. F. (2002). Early stages of energy transduction by myosin: Roles of Arg in switch I, of Glu in switch II, and of the salt-bridge between them. *Proceedings of the National Academy of Sciences of the United States of America*, *99*, 15339–15344.

Ovchinnikov, V., Karplus, M., & Vanden-Eijnden, E. (2011). Free energy of conformational transition paths in biomolecules: The string method and its application to myosin VI. *The Journal of Chemical Physics*, *134*, 085103.

Pabis, A., Duarte, F., & Kamerlin, S. C. L. (2016). Promiscuity in the enzymatic catalysis of phosphate and sulfate transfer. *Biochemistry*, *55*, 3061–3081.

Pecina, A., Haldar, S., Fanfrlik, J., Meier, R., Rezac, J., Lepsik, M., & Hobza, P. (2017). SQM/COSMO scoring function at the DFTB3-D3H4 level: Unique identification of native protein-ligand poses. *Journal of Chemical Information and Modeling*, *57*, 127–132.

Peck, A., Sunden, F., Andrews, L. D., Pande, V. S., & Herschlag, D. (2016). Tungstate as a transition state analog for catalysis by alkaline phosphatase. *Journal of Molecular Biology*, *428*, 2758–2768.

Plotnikov, N. V., Kamerlin, S. C. L., & Warshel, A. (2011). Paradynamics: An effective and reliable model for ab Initio QM/MM free-energy calculations and related tasks. *The Journal of Physical Chemistry B*, *115*, 7950–7962.

Pohorille, A., Jarzynski, C., & Chipot, C. (2010). Good practices in free-energy calculations. *The Journal of Physical Chemistry B*, *114*, 10235–10253.

Polyak, I., Benighaus, T., Boulanger, E., & Thiel, W. (2013). Quantum mechanics/molecular mechanics dual Hamiltonian free energy perturbation. *The Journal of Chemical Physics*, *139*, 064105.

Pu, J. Z., Gao, J. L., & Truhlar, D. G. (2006). Multidimensional tunneling, recrossing, and the transmission coefficient for enzymatic reactions. *Chemical Reviews*, *106*, 3140–3169.

Reed, A. E., Curtiss, L. A., & Weinhold, F. (1988). Intermolecular interactions from a natural bond orbital, donor-acceptor viewpoint. *Chemical Reviews*, *88*, 899–926.

Reuter, N., Dejaegere, A., Maigret, B., & Karplus, M. (2000). Frontier bonds in QM/MM methods: A comparison of different approaches. *The Journal of Physical Chemistry A*, *104*(8), 1720–1735.

Riccardi, D., Koenig, P., Guo, H., & Cui, Q. (2008). Proton transfer in carbonic anhdyrase is controlled by electrostatics rather than the orientation of the acceptor. *Biochemistry*, *47*, 2369–2378.

Riccardi, D., Schaefer, P., Yang, Y., Yu, H., Ghosh, N., Prat-Resina, X., … Cui, Q. (2006). Feature Article: Development of effective quantum mechanical/molecular mechanical (QM/MM) methods for complex biological processes. *The Journal of Physical Chemistry B*, *110*, 6458–6469.

Riccardi, D., Yang, S., & Cui, Q. (2010). Proton transfer function of carbonic anhydrase: Insights from QM/MM simulations. *Biochimica et Biophysica Acta*, *1804*, 342–351.

Roberts, P. J., & Der, C. J. (2007). Targeting the Raf-MEK-ERK mitogen-activated protein kinase cascade for the treatment of cancer. *Oncogene*, *26*, 3291–3310.

Robertson, J. G. (2007). Enzymes as a special class of therapeutic target: Clinical drugs and modes of action. *Current Opinion in Structural Biology*, *17*, 674–679.

Rosta, E., Kamerlin, S. C. L., & Warshel, A. (2008). On the interpretation of the observed linear free energy relationship in phosphate hydrolysis: A thorough computational study of phosphate diester hydrolysis in solution. *Biochemistry*, *47*, 3725-2735.

Rosta, E., Nowotny, M., Yang, W., & Hummer, G. (2011). Catalytic mechanism of RNA backbone cleavage by ribonuclease H from quantum mechanics/molecular mechanics simulations. *Journal of the American Chemical Society*, *133*, 8934–8941.

Roston, D., & Cui, Q. (2016a). QM/MM analysis of transition states and transition state analogues in metalloenzymes. *Methods in Enzymology*, *557*, 213–250.

Roston, D., & Cui, Q. (2016b). Substrate and transition state binding in alkaline phosphatase exhibited by computational analysis of isotope effects. *Journal of the American Chemical Society*, *138*, 11946–11957.

Roston, D., Demapan, D., & Cui, Q. (2016). Leaving group ability affects transition state structure for phosphoryl transfer in a single enzyme active site. *Journal of the American Chemical Society*, *138*, 7386–7394.

Schulenburg, C., Stark, Y., Kunzle, M., & Hilvert, D. (2015). Comparative laboratory evolution of ordered and disordered enzymes. *Journal of Biological Chemistry*, *290*, 9310–9320.

Schwartz, P. A., & Murray, B. W. (2009). Protein kinase biochemistry and drug discovery. *Bioorganic Chemistry*, *39*, 192–210.

Schwarzl, S. M., Smith, J. C., & Fischer, S. (2006). Insights into the chemomechanical coupling of the myosin motor from simulation of its ATP hydrolysis mechanism. *Biochemistry*, *45*, 5830–5847.

Senn, H. M., & Thiel, W. (2009). QM/MM methods for biomolecular systems. *Angewandte Chemie International Edition*, *48*, 1198–1229.

Shaw, R. J., & Cantley, L. C. (2006). Ras, PI(3)K and mTOR signalling controls tumour cell growth. *Nature*, *441*, 424–430.

Shen, L., Wu, J. H., & Yang, W. T. (2016). Multiscale quantum mechanics/molecular mechanics simulations with neural networks. *Journal of Chemical Theory and Computation*, *12*, 4934–4946.

Shukla, D., Meng, Y. L., Roux, B., & Pande, V. S. (2014). Activation pathway of Src kinase reveals intermediate states as targets for drug design. *Nature Communications*, *5*, 3397.

Silverman, D. N., & McKenna, R. (2007). Solvent-mediated proton transfer in catalysis by carbonic anhydrase. *Accounts of Chemical Research*, *40*, 669–675.

Smith, C. A., & Rayment, I. (1996). X-ray structure of the Magnesium(II) · ADP · Vanadate complex of the *Dictyostelium discoideum* myosin motor domain to 1.9 Å resolution. *Biochemistry*, *35*, 5404–5417.

Sunden, F., Peck, A., Salzman, J., Ressl, S., & Herschlag, D. (2015). Extensive site-directed mutagenesis reveals interconnected functional units in the alkaline phosphatase active site. *eLife*, *4*, e06181.

Sweeney, H. L., & Houdusse, A. (2010). Structural and functional insights into the myosin motor mechanism. *Annual Review of Biophysics*, *39*, 539–557.

Swiderek, K., Tuñón, I., Moliner, V., & Bertran, J. (2015). Protein flexibility and preorganization in the design of enzymes. the kemp elimination catalyzed by HG3.17. *ACS Catalysis*, *5*, 2587–2595.

Thiel, W. (1996). Perspectives on semiempirical molecular orbital theory. *Advances in Chemical Physics*, *93*, 703–757.

Torrie, G. M., & Valleau, J. P. (1977). Non-physical sampling distributions in Monte-Carlo free-energy estimation: Umbrella sampling. *Journal of Computational Physics*, *23*, 187–199.

Tracewell, C. A., & Arnold, F. H. (2009). Directed enzyme evolution: Climbing fitness peaks one amino acid at a time. *Current Opinion in Chemical Biology*, *13*, 3–9.

Tuttle, T., & Thiel, W. (2008). OMx-D: Semiempirical methods with orthogonalization and dispersion corrections. Implementation and biochemical application. *Physical Chemistry Chemical Physics*, *10*, 2159–2166.
Vale, R. D., & Milligan, R. A. (2000). The way things move: Looking under the hood of molecular motor proteins. *Science*, *288*, 88–95.
Wang, L., Ceriotti, M., & Markland, T. E. (2014). Quantum fluctuations and isotope effects in ab initio descriptions of water. *The Journal of Chemical Physics*, *141*, 104502.
Wang, L., Goodey, N. M., Benkovic, S. J., & Kohen, A. (2006). Coordinated effects of distal mutations on environmentally coupled tunneling in dihydrofolate reductase. *Proceedings of the National Academy of Sciences of the United States of America*, *103*, 15753–15758.
Warshel, A., Sharma, P. K., Kato, M., Xiang, Y., Liu, H. B., & Olsson, M. H. M. (2006). Electrostatic basis for enzyme catalysis. *Chemical Reviews*, *106*, 3210–3235.
Weinhold, F., & Landis, C. R. (2005). Valency and bonding. Cambridge University Press.
Westheimer, F. H. (1987). Why nature chose phosphates. *Science*, *235*, 1173–1178.
Williams, A. (2003). Free energy relationships in organic and bio-organic chemistry. Cambridge, UK: Royal Society of Chemistry.
Yang, Y., & Cui, Q. (2009). The hydrolysis activity of adenosine triphosphate in myosin: A theoretical analysis of anomeric effects and the nature of transition state. *The Journal of Physical Chemistry A*, *113*, 12439–12446. PMCID: PMC2783400.
Yang, Y., Yu, H., & Cui, Q. (2008). Extensive conformational changes are required to turn on ATP hydrolysis in myosin. *Journal of Molecular Biology*, *381*, 1407–1420.
Yu, H., Ma, L., Yang, Y., & Cui, Q. (2007). Mechanochemical coupling in myosin motor domain, I. Equilibrium active site simulations. *PLOS Computational Biology*, *3*, 0199. PMC1796662.
Zalatan, J. G., Catrina, I., Mitchell, R., Grzyska, P. K., O'Brien, P. J., Herschlag, D., & Hengge, A. C. (2007). Kinetic isotope effects for alkaline phosphatase reactions: Implications for the role of active-site metal ions in catalysis. *Journal of the American Chemical Society*, *129*, 9789–9798.
Zalatan, J. G., Fenn, T. D., Brunger, A. T., & Herschlag, D. (2006). Structural and functional comparisons of nucleotide pyrophosphatase/phosphodiesterase and alkaline phosphatase: Implications for mechanism and evolution. *Biochemistry*, *45*, 9788–9803.
Zalatan, J. G., Fenn, T. D., & Herschlag, D. (2008). Comparative enzymology in the alkaline phosphatase superfamily to determine the catalytic role of an active-site metal ion. *Journal of Molecular Biology*, *384*, 1174–1189.
Zalatan, J. G., & Herschlag, D. (2006). Alkaline phosphatase mono- and diesterase reactions: Comparative transition state analysis. *Journal of the American Chemical Society*, *128*, 1293–1303.
Zhang, Z. Y. (2002). Protein tyrosine phosphatases: Structure and function, substrate specificity, and inhibitor development. *Annual Review of Pharmacology and Toxicology*, *42*, 209–234.
Zheng, Y. Q., & Cui, Q. (2018). Multiple pathways and timescales for conformational transitions in apo-adenylate kinase. *Journal of Chemical Theory and Computation*, *14*, 1716–1726.
Zhou, H. X. (2010). Rate theories for biologists. *Quarterly Reviews of Biophysics*, *43*, 219–293.

SECTION II

Structure, Mechanism and Evolution of Phosphatases

CHAPTER THREE

Defining Dynamics of Membrane-Bound Pyrophosphatases by Experimental and Computational Single-Molecule FRET

Steven P.D. Harborne*, Jannik Strauss*, Ainoleena Turku†, Matthew A. Watson*, Roman Tuma*,‡, Sarah A. Harris*, Adrian Goldman*,§,1

*Astbury Centre for Structural Molecular Biology, University of Leeds, Leeds, United Kingdom
†Drug Research Program, Division of Pharmaceutical Chemistry and Technology, Faculty of Pharmacy, University of Helsinki, Helsinki, Finland
‡Faculty of Science, University of South Bohemia, České Budějovice, Czech Republic
§Faculty of Biological and Environmental Sciences, University of Helsinki, Helsinki, Finland
[1]Corresponding author: e-mail address: a.goldman@leeds.ac.uk

Contents

Methods in Enzymology, Volume 607
ISSN 0076-6879
https://doi.org/10.1016/bs.mie.2018.04.017

Abstract

Membrane-bound pyrophosphatases couple the hydrolysis of inorganic pyrophosphate to the pumping of ions (sodium or protons) across a membrane in order to generate an electrochemical gradient. This class of membrane protein is widely conserved across plants, fungi, archaea, and bacteria, but absent in multicellular animals, making them a viable target for drug design against protozoan parasites such as *Plasmodium falciparum*. An excellent understanding of many of the catalytic states throughout the enzymatic cycle has already been afforded by crystallography. However, the dynamics and kinetics of the catalytic cycle between these static snapshots remain to be elucidated.

Here, we employ single-molecule Förster resonance energy transfer (FRET) measurements to determine the dynamic range and frequency of conformations available to the enzyme in a lipid bilayer during the catalytic cycle. First, we explore issues related to the introduction of fluorescent dyes by cysteine mutagenesis; we discuss the importance of residue selection for dye attachment, and the balance between mutating areas of the protein that will provide useful dynamics while not altering highly conserved residues that could disrupt protein function. To complement and guide the experiments, we used all-atom molecular dynamics simulations and computational methods to estimate FRET efficiency distributions for dye pairs at different sites in different protein conformational states. We present preliminary single-molecule FRET data that points to insights about the binding modes of different membrane-bound pyrophosphatase substrates and inhibitors.

ABBREVIATIONS

ALEX	alternating laser excitation
BCA	bicinchoninic acid assay
$C_{12}E_8$	octaethylene glycol monododecyl ether
DDM	dodecyl-β-maltopyranoside
DTT	dithiothreitol
FRET	Förster resonant energy transfer
IDP	imidodiphosphate
MD	molecular dynamics
mPPase	membrane-bound pyrophosphatase
MWCO	molecular weight cutoff
PDB	protein data bank
P_i	inorganic phosphate
PMSF	phenylmethanesulfonyl fluoride
POPC	phosphatidylcholine
POPE	phosphatidylethanolamine
PP_i	inorganic pyrophosphate
RASP	recurrence analysis of single particles
SC-Leu	synthetic-complete leucine-dropout media
smFRET	single-molecule FRET

TCEP	Tris(2-carboxyethyl)phosphine
TmPPase	mPPase from *Thermotoga maritima*
VrPPase	mPPase from *Vigna radiate*
YPD	yeast extract, peptone, and dextrose media

1. INTRODUCTION

Membrane-bound pyrophosphatases (mPPases) are found in plants, fungi, archaea, and bacteria and couple the hydrolysis of inorganic pyrophosphate (PP_i) to the pumping of sodium ions or protons, generating an electrochemical potential across a membrane (Fig. 1A). Evolutionarily distinct from soluble pyrophosphatases, mPPases are often found in specialized organelles such as vacuoles in plants or acidocalcisomes in *protozoan* parasites, where they play a role in the acidification of these compartments. The absence of mPPases from higher eukaryotes makes them an attractive drug target against certain bacterial infections and against *protozoan* parasites such as *Plasmodium falciparum* (Shah, Vidilaseris, Xhaard, & Goldman, 2016). The mPPases of different species differ with respect to the ion they pump, either sodium, protons, or both; they display varying dependence on potassium ions. Although the molecular details of these mechanisms also remain somewhat unresolved, they have been discussed and reviewed at length elsewhere (Baykov, Malinen, Luoto, & Lahti, 2013; Luoto, Baykov, Lahti, & Malinen, 2013; Luoto, Nordbo, Malinen, Baykov, & Lahti, 2015).

1.1 mPPase Structure and Mechanism

During one catalytic turnover, an mPPase must bind its substrate (PP_i), pump an ion(s) across the membrane, hydrolyze PP_i, and release $2P_i$ as product. We already have a good understanding of the relationship between the three-dimensional protein structure of mPPases and enzyme function due to numerous crystal structure snapshots of the protein throughout the catalytic cycle (Kellosalo et al., 2012; Li et al., 2016; Lin et al., 2012). There has been extensive structural work on two mPPases: a potassium dependent, sodium pumping mPPase from *Thermotoga maritima* (TmPPase) and a potassium dependent, proton pumping mPPase from *Vigna radiata* (VrPPase).

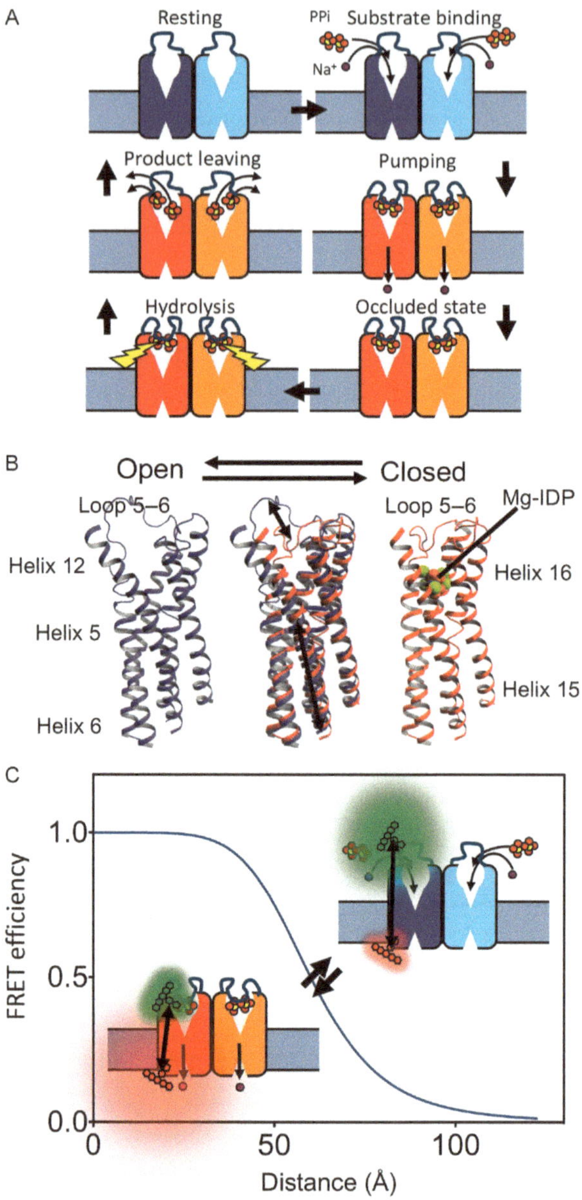

Fig. 1 Using FRET to measure distance fluctuations during the catalytic cycle. (A) Scheme for the proposed enzymatic cycle of membrane-bound PPases. (B) Structural changes in TmPPase between resting state (*blue*; PDB: 4AV3) (Kellosalo, Kajander, Kogan, Pokharel, & Goldman, 2012) and substrate analogue imidodiphosphate (IDP)-bound state (*red*; PDB: 5LZQ) (Li et al., 2016). For clarity, only the helices lining the central pore (helices 5, 6, 12, 15, and 16) are shown. (C) The relationship between distance and FRET efficiency between Alexa Fluor 488 and 594 in relation to movement of the 5–6 loop of mPPase.

There are four distinct TmPPase structures and two distinct VrPPase structures: TmPPase has been determined in a resting state (TmPPase:Ca:Mg; PDB: 4AV3) (Kellosalo et al., 2012), a substrate analogue imidodiphosphate (IDP)-bound state (TmPPase:IDP:Na; PDB: 5LZQ) (Li et al., 2016), a phosphate analogue tungstate (WO_4)-bound state (TmPPase:WO_4; PDB: 5LZR) (Li et al., 2016), and a product-bound state (TmPPase:2P_i; PDB: 4AV6) (Kellosalo et al., 2012). VrPPase has been determined in a substrate analogue IDP-bound state (VrPPase:IDP; PDB: 4A01) and a single phosphate-bound state (VrPPase:P_i; PDB: 5GPJ) (Lin et al., 2012).

Coupled with evidence from various biophysical assays, the crystal structures suggest that binding of substrate to the active site of the mPPase provides energy to drive a conformational change resulting in the pumping of an ion (Li et al., 2016; Shah et al., 2017). In order for the enzyme to return to the resting state it must then hydrolyze PP_i before releasing phosphate. In other words, substrate binding drives ion pumping and is then followed by PP_i hydrolysis, rather than substrate hydrolysis providing the energy for pumping (Fig. 1A). It is unclear at what point the ion associates with the enzyme in the cycle, presumably preceding or concomitantly with the binding of PP_i. However, once the ion has entered the system we have hypothesized that proton pumping occurs by a Grotthuss chain mechanism (Li et al., 2016; Tsai, Kellosalo, Sun, & Goldman, 2014), whereas sodium pumping must occur by a slightly different mechanism. Indeed, the TmPPase:IDP:Na crystal structure contains a bound sodium ion (Li et al., 2016). Regardless of which ion is pumped, the process is mediated by the position of key residues that undergo movements during the cycle due to the twisting, bending, and repositioning of transmembrane helices 5, 6, 12, and 16 (Fig. 1B) (Li et al., 2016). Furthermore, both the crystal structures and some preliminary molecular dynamics (MD) simulations we have performed (Shah et al., 2017) highlight the importance of the loop between the fifth and sixth transmembrane helices (loop 5–6) to the catalytic mechanism. Loop 5–6 is closed over the enzyme active site in the substrate analogue-bound states. However, in the absence of substrate there is a lack of electron density in this region of the crystal structure, suggesting that the loop is disordered. All-atom MD simulations (Shah et al., 2017) confirmed that loop 5–6 is highly dynamic in a substrate-free resting-state model but constrained over the active site in the presence of the substrate (or substrate analogue). If the substrate is removed from the active site at the start of the simulations, the loop immediately becomes more dynamic and is no longer constrained over the active site (Shah et al., 2017). We hypothesize that the closing of loop 5–6 over the enzyme active site is directly coupled to ion pumping as part of the binding change mechanism.

1.2 Defining Dynamic Motions in mPPases by FRET

Quantitative information about the rapid and dynamic changes associated with the enzymatic process is lacking. Therefore, several unresolved issues persist relating to the catalytic cycle. For example, it is unclear whether the rate-limiting step(s) in the enzymatic cycle is the binding of substrate, the pumping of ions, the hydrolysis of PP_i, or the release of product. Thus, by monitoring the dynamic motions in mPPase, we will be able to explore rates associated with each step and test the proposed sequence of events.

One powerful technique for measuring distance changes between parts of a protein with Ångstrom precision is single-molecule Förster resonance energy transfer (smFRET) (Fig. 1C). Pairs of fluorescent dye molecules chosen to have an overlap between their emission and absorption spectra can exchange energy in an efficient and highly distance-dependent nonradiative process. Upon direct photoexcitation of the donor, excitation is either emitted or transferred to the second dye, the acceptor. The acceptor then returns to its ground state by emission of a lower energy (red shifted) photon. The donor and acceptor dyes emit at different wavelengths (colors), and the relative intensities of emission at the two colors are primarily sensitive to the distance between the dyes, although relative orientation and local environment also play a role. However, the latter two factors are minimized by attaching dyes to the protein via a flexible linker that allows rapid rotational averaging. At single-molecule level, additional information can be gained about the relative populations of different conformations of the protein by plotting the distribution of FRET efficiencies. Furthermore, kinetic and dynamic parameters can be inferred from time-resolved analysis of the data.

The attachment of the fluorescent dyes to the protein can be accomplished by a reactive maleimide group on the dye that specifically reacts with sulfhydryl groups allowing us to selectively label a part of the protein by site-directed mutagenesis of wild-type residues to cysteine. We describe a simulation-guided residue selection process we have used in order to maximize chances of observing the dynamic changes in the protein we wish to see, and we highlight some potential pitfalls we have encountered.

Finally, we present preliminary experimental smFRET data that afford new insights into the binding modes of different mPPase substrates and inhibitors.

2. COMBINING EXPERIMENTAL AND COMPUTATIONAL METHODS FOR OPTIMIZING THE ATTACHMENT OF FLUORESCENT DYES FOR SINGLE-MOLECULE FRET

To observe movements of helix-12, loops 5–6 and 13–14 during the catalytic cycle (Fig. 1B), cysteine residues were introduced for the attachment of FRET labels. For successful FRET experiments, positioning of the fluorescent dyes is extremely important, and a number of factors were considered when placing cysteine residues. As the distance dependency of the efficiency of FRET is nonlinear (Fig. 1C), the FRET pair ought to be separated by between 40 and 60 Å in order to ensure good but highly variable FRET efficiency across the range of the protein conformations seen in the available high-resolution structures (Fig. 1B and C).

We sought to place one label in a highly dynamic region of the protein, and a second at a relatively static residue. Furthermore, we desired a design that afforded protein that remained active once cysteine mutations had been introduced and labeled with the dyes. Usually these mutations are introduced into a cysteine-free variant, but this is not always possible. The introduced cysteine residues needed to be solvent accessible to ensure good labeling efficiency. Finally, to detect signal changes relating to the movement of dye pairs toward or away from each other, the magnitude of the movement needed to be large enough to produce change in FRET efficiency above error in measurement (due primarily to shot noise, which occurs in photon counting optical devices due to the discrete nature of photons), but also background caused by thermal fluctuations in the protein conformation. Therefore, the dyes needed to be introduced at positions that undergo the largest expected conformational changes during the catalytic cycle.

It should also be noted that for the FRET experiments we used the same thermophilic orthologue of mPPase used for crystallographic study (TmPPase). The reason for this is twofold: first, there are well-established protocols for expression and purification, and TmPPase is easy to handle in the lab due to high protein stability (Kellosalo, Kajander, Palmgren, Lopéz-Marqués, & Goldman, 2011). Second, the crystal structures provide a meaningful starting point for MD simulations.

2.1 Choosing Positions for Mutagenesis From Sequence and Structure

The strategy we implement in choosing positions for labeling avoids highly conserved residues that, when mutated or labeled, are likely to disrupt protein function, folding, or stability. We did this by constructing a comprehensive multisequence alignment to include orthologues from highly diverse branches of evolution. The need for the labeled residues to be solvent accessible, positioned at appropriate distances from each other, and result in the largest distance changes between positions in different states was addressed by visual inspection of the selected sites relative to their position in three-dimensional models of the protein. In this case, there are numerous crystal structures of TmPPase in several different conformational states. From previous work (Li et al., 2016; Shah et al., 2017), it was known that helix-12, loops 5–6 and 13–14 are the most dynamic elements and therefore should encompass desirable sites for fluorescent label attachment.

2.1.1 Equipment and Software

- Software for multiple alignments, e.g., Jalview (Waterhouse, Procter, Martin, Clamp, & Barton, 2009).
- Software for protein structure viewing and manipulation, e.g., PyMol (Schrödinger, LLC, 2015).

2.1.2 Procedure

1. Construct a multiple sequence alignment of orthologues of your protein of interest (see Notes).
2. Choose potential residues for dye labeling by selecting residues with low conservation (and therefore most likely to tolerate mutagenesis without affecting protein function and folding). It may be possible to select potential labeling sites by looking for positions where cysteine residues are frequently found in orthologues. Furthermore, residues with similar properties could also be targeted, for example, serine or threonine residues (see Notes).
3. For membrane proteins, select a number of residues on each side of the membrane.
4. Visualize structural models of your protein of interest using PyMol and locate potential sites selected in the previous step (see Notes). Exclude residues that are buried in secondary structure, and in the case of membrane proteins, likely to be covered by membrane or detergent micelle.

5. Using the PyMol distance wizard or by custom script, construct a distance matrix between all remaining potential sites (see Fig. 2A–C for TmPPase example). Furthermore, if structures are available in several states, a matrix of the differences in distances between states should be constructed. This will help select residues that change in distance the most between states (see Notes).
6. The final selection of residues should include 5–10 sites on either side of the membrane that are positioned on average 40–60 Å from opposing residues (see Notes).

2.1.3 Notes

- It is important to create an alignment with a broad representation of different orthologues of your protein of choice in order to understand conservation across distinct evolutionary branches.
- Due to the limited regions of TmPPase considered for labeling, we did not find any positions frequently substituted to cysteine, threonine, or serine in other mPPase orthologues.
- For the crystal structure of the TmPPase in the substrate analogue-bound (closed) form, all putative sites selected for labeling were present in the model, so a distance matrix for this state could be constructed. However, residues of loops 5–6 and 13–14 were missing in the crystal structure of the apo (open) form, so it was not possible to construct a matrix of differences in distance between states. Instead, we used a loop building and MD approach to estimate the size of changes in distance between states for this enzyme (see Sections 2.7 and 3.2).
- In TmPPase, accessible native cysteine residues needed to be removed to avoid off-target labeling of the protein by the dyes. Preliminary mutagenesis and labeling of TmPPase indicated that three of the four native cysteine residues (C300, C474, and C559) could be mutated to serine without detriment to enzymatic function (J. Kellosalo, unpublished), with the remaining cysteine (C183) buried in the center of the membrane-spanning region, and therefore unlikely to be labeled. A similar strategy must be adopted for any protein to be subjected to FRET analysis, and the number of native cysteine residues may also be used to guide choice of orthologue.
- Two further mutations were already present in the TmPPase sequence used for the structural work (V352L and S395G), which increase the stability of the protein during purification and do not have a detrimental effect on enzyme activity. These residues were excluded from consideration for mutations to cysteine.

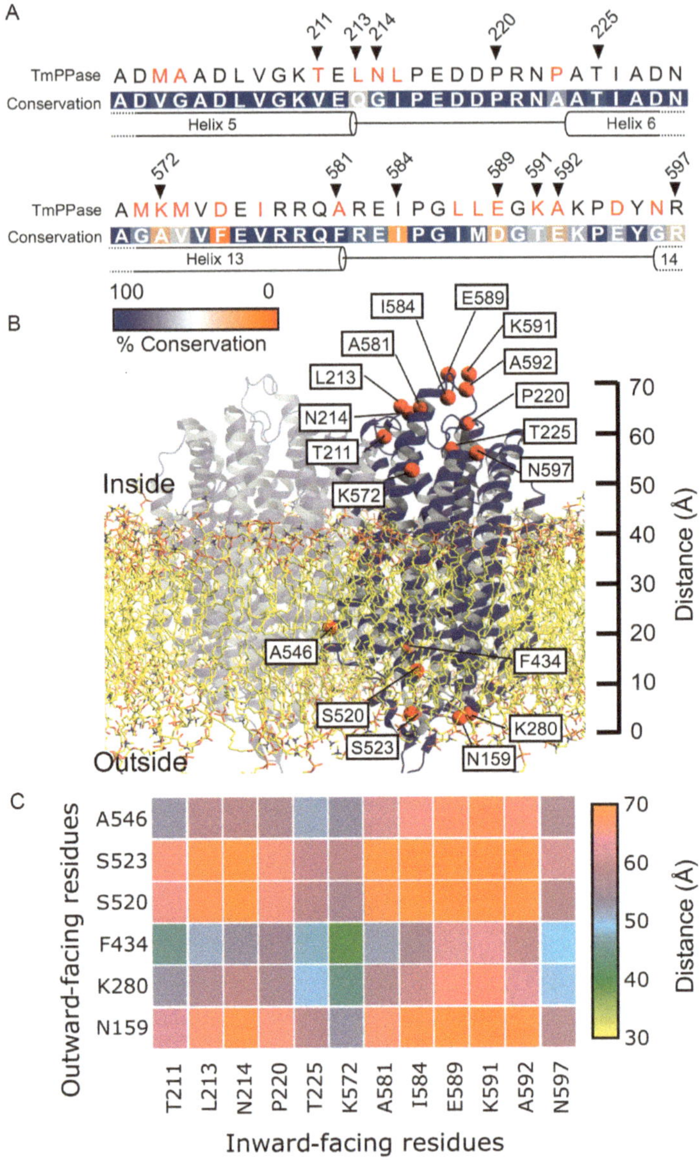

Fig. 2 Positions chosen for cysteine mutagenesis in TmPPase. (A) TmPPase sequence for loops 5–6 and 13–14 is displayed in comparison to the consensus and conservation derived from an alignment of 177 distantly related sequences. Residues are colored *red* in the TmPPase sequence if they differ from the overall consensus. Coloring of the consensus sequence indicates the conservation of the residue at that position. *Arrows* indicate the positions that were chosen for mutation. (B) Positions chosen for mutations (*red spheres*) mapped onto the models for IDP-bound state of TmPPase. (C) Heatmap of the distance matrix between Cα atoms of residues on the inside face, and those on the outside face in TmPPase resting-state structure.

- Even avoiding highly conserved residues for mutations, we expected a proportion of the mutations to be detrimental to protein function. Therefore, we advise that a large number of potential sites are tested initially.
- Our preliminary selection of 18 residues for cysteine mutagenesis comprised the inward-facing residues (dynamic residues): T211, L213, N214, P220, T225, K572, A581, I584, E589, K591, A592, N597, and the outward-facing residues (static residues): N159, K280, F434, S520, S523, A546 (Fig. 2A–C).

2.2 Expression, Harvest, and *S. cerevisiae* Membrane Preparation

The procedure outlined in Sections 2.2.1–2.2.4 is specific for the expression and purification of our target protein TmPPase from *S. cerevisiae* as previously reported (Kellosalo et al., 2011). However, this procedure can be substituted with a suitably optimized one for the expression and purification of other proteins of interest.

2.2.1 Equipment and Consumables

- 2.5-L baffled flasks, e.g., TunAir shake flasks (Sigma-Aldrich, catalog number: Z710822).
- Shaking incubator, e.g., INNOVA44 (New Brunswick, catalog number: M1282-0000).
- Equipment for yeast cell lysis, e.g., beadbeater (BioSpec, catalog number: 1107900).
- 0.5–0.75 mm diameter glass beads (Thistle Scientific, catalog number: 11079105).
- High-speed centrifuge, e.g., Sorvall RC6.
- Ultracentrifuge and rotor, e.g., Beckman coulter Ti 45.
- 40-mL Glass dounce homogenizer (Wheaton, catalog number: 357546).

2.2.2 Buffers and Reagents

- Yeast extract, peptone, dextrose (YPD) medium (1% (w/v) yeast extract, 2% (w/v) tryptone, 2% (v/v) D-glucose).
- Synthetic-complete leucine-dropout (SC-Leu) medium: 2% (v/v) D-glucose, 1.35 g/L yeast nitrogen base (without amino acids), 0.5 m*M* tryptophan, 0.4 m*M* histidine, 0.2 m*M* adenine, 0.2 m*M* uracil.

- Phenylmethanesulfonyl fluoride (PMSF) (Sigma-Aldrich, catalog number: P7626-1G).
- Pepstatin A (Sigma-Aldrich, catalog number: P5318-25MG).
- 1 *M* MES–NaOH pH 6.5.
- 1 *M* Tris–HCl pH 7.5.
- 1 *M* dithiothreitol (DTT).
- 1 *M* Tris(2-carboxyethyl)phosphine (TCEP).
- 25% (w/v) Dodecyl-β-maltopyranoside (DDM) (Anatrace, catalog number: D310-25-GM).
- 0.5 *M* EDTA pH 7.5.
- 1 *M* $MgCl_2$.
- 1 *M* NaCl.
- 3.5 *M* KCl.
- 1 *M* Na_2HPO_4.
- 100 m*M* sodium pyrophosphate pH 8.
- 100% glycerol.
- Breaking buffer: 200 m*M* Tris–HCl pH 7.5, 40% (w/v) glycerol, 10 m*M* EDTA pH 7.5, 2 m*M* DTT, 0.2 m*M* PMSF, 0.2 m*M* Pepstatin A.
- Tris-buffered glycerol (TBG): 10 m*M* Tris–HCl pH 7.5, 10% (w/v) glycerol, 5 m*M* EDTA.
- Resuspension buffer: 50 m*M* MES pH 6.5, 20% (w/v) glycerol, 50 m*M* KCl, 5 m*M* $MgCl_2$, 1.3 m*M* DTT, 0.34 m*M* PMSF, 0.2 m*M* Pepstatin A.

2.2.3 Procedure

1. For each variant, use a single colony of *S. cerevisiae* (see Notes) bearing the correct plasmid (see Notes) to inoculate 5–10 mL SC-Leu medium primary precultures. Incubate at 30°C overnight (>18 h) with shaking at 200 rpm.
2. Use primary precultures to inoculate 1 L of SC-Leu glucose medium secondary precultures at a starting $OD_{600\ nm}$ of 0.05. Incubate cultures at 30°C overnight (>18 h) with shaking at 200 rpm.
3. Use the 1 L secondary precultures to inoculate 10 L of YPD medium. Use 1 L of total culture in each 2.5-L full-baffle shake flask. Incubate at 30°C for 8–24 h (see Notes) with shaking at 225 rpm.
4. Harvest the cells by centrifugation at 3000 × *g* for 20 min.
5. Water wash the cells twice in one-tenth the total culture volume of H_2O with centrifugation at 3000 × *g* for 20 min between each wash.
6. Pelleted cells can be stored at −80°C until further use (see Notes).
7. Resuspend the final cell pellet in 50 mL breaking buffer.

8. Assemble the beadbeater (large chamber ~300 mL) (see Notes) filled with approximately 170 g of 0.5 mm diameter glass beads and 150 mL of resuspended cells. Top the chamber up with breaking buffer until 90%–95% full. Fill the outside reservoir with ice-saturated water.
9. Activate the beadbeater 25 times for 30 s each time. Allow 1-min cool down between each activation.
10. Recover the cell lysate from beads in 50-mL falcon tubes. Wash the glass beads two to three times with approximately 50 mL of TBG and combine with the cell lysate.
11. Add 5 mL of 3.5 *M* KCl per 40 mL of cell lysate (see Notes).
12. Centrifuge the lysate to remove particulate material in a benchtop centrifuge (3000 × *g*, 5 min, 4°C).
13. Recover the supernatant and centrifuge at 235,000 × *g* for 1 h at 4°C to recover membranes.
14. Discard the supernatant, recover, and resuspend the membrane pellet in 15 mL resuspension buffer. Use a glass homogenizer to assist resuspension.
15. Carry out a bicinchoninic acid assay (BCA) assay to determine total membrane protein concentration (see Notes).
16. Snap-freeze the membranes suspension in liquid nitrogen and store at −80°C until use for purification.

2.2.4 Notes

- It is essential to use a freshly transformed single colony picked from a plate, as the plasmid is not stable if the colony is kept at −80°C in glycerol stocks that are warmed up for expression.
- We use the *S. cerevisiae* strain BJ1991 (*MATα, pep4-3, prbl-1122, ura3-52, leu2, trpl*) (Ernst & Chan, 1985).
- We transformed the plasmid pMP-TmPPase CSSS with the desired site-directed point mutations into *S. cerevisiae* using the lithium acetate/PEG/DNA method (Gietz & Schiestl, 2007).
- The pMP-TmPPase plasmid contains a high copy number pUC-derived origin of replication and an ampicillin selectable marker for cloning in *E. coli* (Kellosalo et al., 2011; Palmgren & Christensen, 1993; Villalba, Palmgren, Berberián, Ferguson, & Serrano, 1992). Furthermore, the pMP plasmid has a 2-μ origin of replication and *LEU2* selectable marker for yeast, with the TmPPase gene under the constitutive *PMAI* gene promoter for high levels of protein expression (Kellosalo et al., 2011; Palmgren & Christensen, 1993; Villalba et al., 1992).

- It is important to inoculate rich media with a high density of cells from the selective minimal media. This ensures that the yeast cells undergo fewer doubling events in the rich media. Fewer doubling events in rich media means there is less chance that the yeast cells will eliminate the plasmid or the gene, as they are under no selective pressure to keep it in rich media, and our data indicate that expression of the gene is deleterious for the organism. Incubation time in rich media is protein dependent and should be experimentally determined.
- In our experience, sonication and cell disruptor devices yield significantly lower amounts of active protein.
- We typically obtain 100–200 g of cells from 10 L of rich yeast culture.
- Potassium is added to the samples in excess as potassium is a key cofactor. Without potassium, the protein is less active and less stable (López-Marqués et al., 2005).
- From 10 L of culture, we typically recover 200–300 mg of total membrane protein as determined by BCA assay.

2.3 Solubilization and Purification of TmPPase Cysteine Variants

2.3.1 Equipment and Consumables

- Water bath.
- 50K MWCO concentrator, e.g., VivaSpin 20, 6, or 2 (GE Healthcare, catalog numbers: 28932362, 28932318, 28932257).

2.3.2 Buffers and Reagents

- 1 *M* MES–NaOH pH 6.5.
- 1 *M* Tris–HCl pH 7.5.
- 1 *M* DTT.
- 1 *M* TCEP.
- 25% (w/v) DDM (Anatrace, catalog number: D310 25-GM).
- 0.5 *M* EDTA pH 7.5.
- 1 *M* $MgCl_2$.
- 1 *M* NaCl.
- 3.5 *M* KCl.
- 1 *M* Na_2HPO_4.
- 100 m*M* sodium pyrophosphate pH 8.
- 100% Glycerol.
- 1 *M* imidazole pH 7.5.
- Nickel affinity resin (Ni Sepharose® 6FF, GE Healthcare, catalog number: GE17-5318-02).

- cOmplete EDTA-free protease inhibitor tablet (Roche, catalog number: 5056489001).
- Empty 20-mL bed volume plastic column (e.g., Econo-Pac® Chromatography Columns, Bio-Rad, catalog number: 7321010)
- Resuspension buffer: 50 m*M* MES pH 6.5, 20% (w/v) glycerol, 50 m*M* KCl, 5 m*M* $MgCl_2$, 1.3 m*M* DTT, 0.34 m*M* PMSF, 0.2 m*M* Pepstatin A.
- Solubilization buffer: 50 m*M* MES pH 6.5, 20% (w/v) glycerol, 5% (w/v) DDM, protease inhibitor tablet.
- Wash buffer: 50 m*M* MES pH 6.5, 20% (w/v) glycerol, 50 m*M* KCl, 5 m*M* $MgCl_2$, 20 m*M* imidazole, 1 m*M* Na_2HPO_4, 0.2% DDM, 1 m*M* TCEP.
- Elution buffer: 50 m*M* MES pH 6.5, 20% (w/v) glycerol, 50 m*M* KCl, 5 m*M* $MgCl_2$, 400 m*M* imidazole, 1 m*M* Na_2HPO_4, 0.2% DDM, 1 m*M* TCEP.
- Storage buffer: 20% (v/v) glycerol, 50 m*M* HEPES pH 7.4, 50 m*M* KCl, 5 m*M* $MgCl_2$, 0.02% DDM, 1 m*M* TCEP.

2.3.3 Procedure

1. Thaw yeast membranes and turn on the water bath (for preheating to 75°C).
2. Mix solubilization buffer with yeast membranes to a final total membrane protein concentration between 5 and 10 mg/mL.
3. Place the solubilization suspension in the 75°C water bath for 1.0–1.5 h. Mix by inversion every 10 min to ensure the samples remain in suspension (see Notes).
4. Separate insoluble from soluble fraction by centrifugation at 4000 × *g* for 5 min at 20°C (room temperature).
5. Prepare and preequilibrate 4 mL of nickel resin slurry by washing first in 40 mL H_2O, then in 40 mL resuspension buffer with 0.2% (w/v) DDM. This step can be achieved either using an empty column or by centrifugation to collect nickel resin at 3000 × *g* for 5 min at 4°C. Finally, resuspend the nickel resin in 32 mL of resuspension buffer and 6 mL of 3.5 *M* KCl.
6. Mix the solubilized membrane sample with the preequilibrated nickel resin to bind in batch at 37°C for 2 h with mixing.
7. Apply the resin slurry to an empty 20-mL plastic column and collect a sample of the flow-through for SDS-PAGE analysis.
8. Wash the column with 20 mL of wash buffer and collect a sample of the flow-through for SDS-PAGE analysis.

9. Wash the column with another 20 mL of wash buffer and collect a sample of the flow-through for SDS-PAGE analysis.
10. Elute the protein with 20 mL of elution buffer, collect the flow-through.
11. Concentrate the eluted protein in a 50-K MWCO VivaSpin 20 (approximately 2 h at 4°C, 3000 × *g*) to approximately 2.5 mL. This concentrates the protein by approximately 10-fold.
12. Dilute the sample to a total volume of 20 mL in storage buffer. This will dilute the imidazole concentration to approximately 40 m*M*.
13. Concentrate the sample once again in a 50-K MWCO VivaSpin 20 (approximately 2 h at 4°C, 3000 × *g*) to 2–10 mL (aim for 0.5–5 mg/mL).
14. Snap-freeze aliquots in liquid nitrogen.
15. Analyze the collected samples of the purification by SDS-PAGE.

2.3.4 Notes

- TmPPase is solubilized at high temperatures in an adaptation of the "Hot-Solve" method (López-Marqués et al., 2005). The majority of protein, apart from thermostable TmPPase, will denature and can easily be removed by centrifugation. For new proteins, the temperature of solubilization will need to be determined experimentally. The method is not suitable for mesophilic proteins.

2.4 Activity of Cysteine Variants

The activity of the mPPases can be measured using a colorimetric assay to detect P_i produced from the hydrolysis of pyrophosphate (Baginski, Foà, & Zak, 1967; Kellosalo et al., 2011). This assay would also be suitable for other enzymatic reactions that produce P_i, for example, those catalyzed by ATP-hydrolyzing enzymes. If the protein of interest cannot be assayed by accumulation of P_i, this activity assay should be replaced with a suitably robust assay to indicate the effect of cysteine mutagenesis on protein activity.

2.4.1 Equipment

- Water bath.
- Multichannel pipettes, e.g., Pipetman 8-channel 2–20 μL (Gilson, catalog number: F14401) and Pipetman 8-channel 20–200 μL (Gilson, catalog number: F14403).
- Suitable 96-well plate reader that can read absorption at $A_{860\ \mathrm{nm}}$, e.g., Tecan Spark 20M plate reader (Tecan) (see Notes).

2.4.2 Buffers and Reagents

- 0.5 *M* HCl.
- 1 *M* Tris–HCl pH 8.
- 1 *M* $MgCl_2$.
- 1 *M* NaCl.
- 1 *M* DTT.
- 1 *M* KCl.
- 1 *M* Na_2HPO_4.
- 100 m*M* sodium pyrophosphate pH 8.
- 25% (w/v) DDM (Anatrace, catalog number: D310-25-GM).
- L-α-Phosphatidyl choline from soybean lecithin (Sigma-Aldrich, catalog number: 429415-100GM).
- L-Ascorbic acid (Sigma-Aldrich, catalog number: A5960).
- Ammonium molybdate tetrahydrate (Sigma-Aldrich, catalog number: A7302).
- Sodium (meta)arsenite (Sigma-Aldrich, catalog number: S7400).
- Sodium citrate tribasic (Sigma-Aldrich, catalog number: S4641).
- Glacial acetic acid.
- Reaction buffer: 60 m*M* Tris–HCl pH 8, 2.4 m*M* $MgCl_2$, 100 m*M* KCl, 20 m*M* NaCl.

2.4.3 Procedure

1. Prepare lipids by sonicating 30 mg/mL soybean lecithin in 20 m*M* Tris–HCl pH 8, 1 m*M* DTT until the solution turns clear. This stock can be aliquoted and frozen at −20°C for future use.
2. Solubilize lipid by the addition of 54 μL of 25% DDM to 120 μL of lipid stock at 55°C for 15 min.
3. Prepare the sodium arsenite solution by dissolving 5 g of sodium arsenite and 5 g of sodium citrate tribasic in 100 mL H_2O, add 5 mL of glacial acetic acid, mix, and adjust to 250 mL with H_2O.
4. Prepare solution A by mixing 0.3 g of ascorbic acid in 10 mL of ice-cold 0.5 *M* HCl.
5. Prepare solution B by mixing 70 mg of ammonium heptomolybdate in 1 mL of ice-cold H_2O.
6. Prepare phosphate standards: from the 1 *M* Na_2HPO_4 stock, make a 5-m*M* working stock. Make 0, 2.5, 5, 10, 15 nmol P_i standards by adding 0, 2.5, 5, 10, 15 nmol Na_2HPO_4 to a total of 100 μL of H_2O, respectively.
7. Prepare protein samples for the assay by mixing 4.2 μL of the sample with 5.8 μL of lipid/detergent mix for relipidation.

8. Prepare the assay plate: add 10 μL of each P_i standard in triplicate. Add 15 μL of 4.5 × reaction buffer and 20 μL of diluted sample in triplicate. We suggest several different dilutions of the sample, e.g., 10-fold, fivefold, and twofold dilution in H_2O.
9. Mix solutions A and B on ice. The resulting solution should be yellow and transparent. If a color change has occurred, the solution must be remade.
10. Preheat the samples for 5 min (put the plate in the water bath preheated to 70°C).
11. Add 10 μL of 2 m*M* sodium pyrophosphate to each well using multichannel pipette return to water bath for 5 min.
12. Place the plate on ice for 10 min.
13. Add 60 μL of solution A/B using a multichannel pipette.
14. Leave the plate on ice for a further 10 min.
15. Add 90 μL of sodium arsenite solution using multichannel pipette.
16. Incubate at room temperature for 1 h.
17. Measure the absorbance at $A_{860\ \mathrm{nm}}$ using a suitable plate reader.

2.4.4 Notes

- Any suitable plate reader can be used, but must be able to read absorption at $A_{860\ \mathrm{nm}}$.
- We have found that TmPPase must be relipidated following purification for maximal enzyme activity (Kellosalo et al., 2011).

2.5 Labeling of TmPPase Cysteine Variants

2.5.1 Equipment

- Desiccator, e.g., Vacufuge plus (Eppendorf, catalog number: 5305000509).
- UV–vis spectrophotometer, e.g., DS-11 + (DeNovix, catalog number: DS-11 +).

2.5.2 Buffers and Reagents

- 1 *M* HEPES pH 7.4.
- Alexa Fluor 488 maleimide (Invitrogen, catalog number: A10254).
- Alexa Fluor 594 maleimide (Invitrogen, catalog number: A10256).
- Anhydrous DMSO (Invitrogen, catalog number: D12345).
- Nickel affinity resin (Nickel Sepharose® 6FF, GE Healthcare, catalog number: GE17-5318-02).

- Empty micro-biospin chromatography columns (Bio-Rad, catalog number: 732-6204).
- Labeling buffer A: 20% (v/v) glycerol, 50 m*M* HEPES pH 7.4, 50 m*M* KCl, 5 m*M* $MgCl_2$, 0.02% DDM.
- Labeling buffer B: 20% (v/v) glycerol, 50 m*M* HEPES pH 7.4, 50 m*M* KCl, 5 m*M* $MgCl_2$, 0.02% DDM, 400 m*M* imidazole.

2.5.3 Procedure

1. Prepare stocks of fluorescent dyes by resuspension to 10 m*M* in dry DMSO (155 or 122 μL for Alexa Fluor 488 and Alexa Fluor 594, respectively) and aliquot into 10 μL portions in 0.3-mL PCR tubes. These aliquots can be stored in a dry environment below −70°C (ideally liquid nitrogen).
2. Thaw one aliquot of each dye, mix in a 1:1 ratio of Alexa Fluor 488 to Alexa Fluor 594, and desiccate in a 1.5-mL Eppendorf tube (2 h under vacuum at 40°C).
3. Add concentrated protein stock (between 10 and 30 μL of 2.5–10 mg/mL protein)
4. Incubate for 18 h with shaking at 37°C in the dark (wrap tubes in tin foil to block out the light).
5. Dilute the sample to 500 μL with labeling buffer A.
6. Batch bind to 100 μL of preequilibrated nickel resin for 30 min at 50°C with mixing at 1000 rpm.
7. Collect the resin and discard the supernatant by centrifugation through an empty micro-biospin column.
8. Wash the resin with a total of 4 mL labeling buffer A (in 0.5 mL aliquots) to remove the majority of the unattached label.
9. Elute protein using 0.2 mL labeling buffer B.
10. Determine the protein concentration and labeling efficiency by measuring absorbance between 200 and 800 nm (see Fig. 3; Table 1, for example). The contribution of dyes to absorbance at $A_{280\ \mathrm{nm}}$ needs to be accounted for using Eq. (1). Weightings of 0.11 and 0.56 are specific to the absorbance of Alexa Fluor 488 and 594 at $A_{280\ \mathrm{nm}}$, respectively.

$$\text{Protein concentration} = \frac{[A_{280} - (A_{494} \times 0.11) - (A_{590} \times 0.56)] \times \text{dilution factor}}{203{,}000} \qquad (1)$$

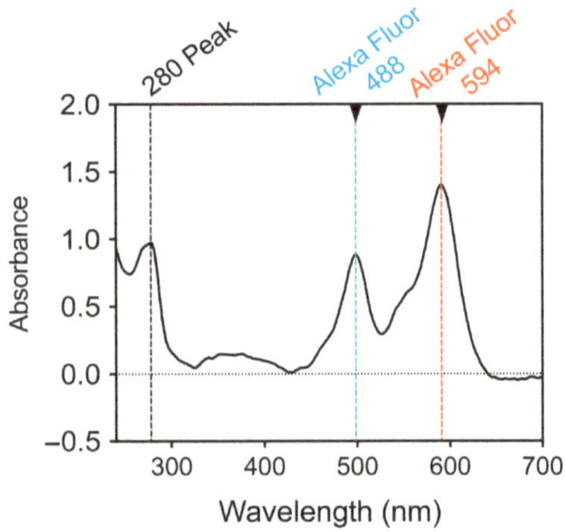

Fig. 3 Example absorption scan of TmPPase labeled with Alexa Fluor 488 and 594.

Table 1 Example Dye Labeling Efficiency for TmPPase

Protein	Peak at 280 nm (AU)	0.961
	Adjusted 280 nm peak (protein only) (AU)	0.079
	Estimated protein concentration (μ*M*)	109
Alexa Fluor 488	Peak at 494 nm (AU)	0.661
	Concentration (μ*M*)	91
	Contribution to $A_{280\ \mathrm{nm}}$ (AU)	0.07
	Estimated labeling efficiency (% total cysteines)	41.6
Alexa Fluor 594	Peak at 590 nm (AU)	1.69
	Concentration (μ*M*)	149
	Contribution to $A_{280\ \mathrm{nm}}$ (AU)	0.88
	Estimated labeling efficiency (% total cysteines)	68.5

AU, absorbance units.

2.6 Reconstitution of Labeled TmPPase Into Liposomes

For the smFRET-ALEX experiments, we reconstitute the protein into liposomes. Liposome reconstitution provides a more native-like environment for the protein than detergent micelles. Furthermore, by greatly increasing the size of the diffusing species, liposome reconstitution reduces the rates of rotational and translational diffusion of the emitting species in solution.

Therefore, the mean incident time of an emitter in the confocal volume of the experiment for observation of a smFRET signal is increased leading to a greater signal-to-noise ratio and the possibility to extract kinetics on the millisecond time scale, e.g., by recurrence analysis of single particles (RASP) (Hoffmann et al., 2011). The procedure in Sections 2.6.1–2.6.3 outlines the steps required for the successful reconstitution of TmPPase. However, the lipid, detergent, BioBead additions, and buffer conditions are protein specific and may need extensive optimization for new proteins.

2.6.1 Equipment and Consumables

- Carousel mixer, e.g., SB3 adjustable speed rotator (Stuart, catalog number: 1143996)
- SM2 BioBeads (Bio-Rad, catalog number: 1523920).
- Desalting column, e.g., Centripure P25 desalting column (Generon, catalog number: CP-0108-Z025.0-001).
- Empty micro-biospin chromatography columns (Bio-Rad, catalog number: 732-6204).
- 500- to 1000-μL syringes, e.g., gastight 1700 syringes (Hamilton, catalog number: 203320).
- Extruder and syringe support (Avanti, catalog number: 610023).
- 0.4 and 0.1 μm nuclepore track-etched membrane filters (Whatman, catalog numbers: WHA800309 and WHA800282, respectively).
- Teflon membrane support disks (Avanti, catalog number: 610025).
- Simple fluorimeter, e.g., QFX (DeNovix, catalog number: QFX).

2.6.2 Buffers and Reagents

- 1 *M* $MgCl_2$.
- 1 *M* KCl.
- 1 *M* Tris–HCl pH 8.
- Octaethylene glycol monododecyl ether ($C_{12}E_8$; Anatrace, catalog number: O3304ML).
- L-α-Phosphatidyl choline from soybean lecithin (Sigma-Aldrich, catalog number: 429415-100GM).
- Assay buffer: 20 m*M* Tris–HCl pH 8, 50 m*M* KCl, 2 m*M* $MgCl_2$.

2.6.3 Procedure

1. Prepare a starting solution such that, accounting for the volume of protein added in step 2, concentrations of reagents will equal 4 mg/mL L-α-phosphatidyl choline (from soybean lecith), 20 m*M* Tris–HCl

Table 2 Example Volumes Used for Labeled TmPPase Liposome Reconstitution

Component	Stock Conc.	Final Conc.	Volume to Add (μL)
L-α-Phosphatidyl choline	30 mg/mL	4 mg/mL	66.67
Tris–HCl pH 8	1 *M*	20 m*M*	10
KCl	3.5 *M*	50 m*M*	7.5
MgCl	1 *M*	2 m*M*	1
$C_{12}E_8$	25% (v/v)	1% (v/v)	20
Alexa Fluor 488 on dual labeled TmPPase	117 μ*M*	5 μ*M*	21.36
H_2O			373.4

pH 8, 50 m*M* KCl, 2 m*M* $MgCl_2$, and 1% (v/v) $C_{12}E_8$ in a total volume of 600 μL (see Table 2, for example). Vortex to mix until the turbid lipid is completely solubilized (indicated by the solution becoming completely transparent).

2. Dilute labeled protein in this solution to a final Alexa Fluor 488 concentration of 5 μ*M*.
3. Mix with inversion for 1 h at 4°C.
4. For detergent removal, weigh four portions of BioBeads at 30 mg (wet weight) and four portions of BioBeads at 60 mg.
5. Add one 30 mg portion of BioBeads to the lipid/detergent/protein mixture and mix with inversion for 20 min at 4°C.
6. Repeat step 5 with the remaining 30 mg portions of BioBeads, followed by the 60 mg portions of BioBeads. Mix between additions by inversion for 20 min at 4°C.
7. Following the final BioBead addition, mix at 4°C overnight.
8. Remove BioBeads by centrifugation through an empty micro-biospin column.
9. Desalt the proteoliposomes into assay buffer using a preequilibrated desalting column. The sample will be diluted to approximately 1 mL. Liposome elution from the desalting column can be monitored by measuring fluorescence from the Alexa Fluor dyes in the fractions.
10. Assemble the extrusion apparatus and pass the desalted liposomes 11 times through a 0.4-μm pore filter, then 11 times through a 0.1-μm pore filter, to ensure proteoliposomes have a uniform size distribution.

2.7 All-Atom Molecular Dynamics Simulations

Proteins in solution are subject to thermal noise, which can have dramatic effects on protein shape and conformation. We require knowledge of the nature and magnitude of these conformational changes as they need to be accounted for when analyzing FRET data. We used all-atom MD simulations to determine the size and amplitude of these thermal fluctuations, allowing us to better refine our model to account for motions we observe experimentally. Furthermore, due to the numerous crystal structures of TmPPase available in different conformational states, coupled with MD simulations, it is possible to estimate the size of conformational changes (i.e., distance between a pair of residues), and whether these will be observable over the thermal fluctuations in the protein. The MD simulations are computationally expensive; therefore, they were only used when we had available a selection of double-cysteine variants that were active, could be labeled with dyes, and that remained active once labeled.

2.7.1 Equipment and Software

- Access to parallel computing (simulations were run on 64 processors of the Leeds ARC2 supercomputer at the University of Leeds, which contains 380 CPUs comprising 3040 cores of standard compute, furnished with InfiniBand fast interconnects for efficient parallel processing), or GPUs, e.g., 2–4 NVIDIA GTX 1080T graphics cards.
- Software for homology modeling and loop building, e.g., Rosetta (Mandell, Coutsias, & Kortemme, 2009), Modeller (Webb & Sali, 2014), or iTasser (Yang et al., 2015).
- Protein structure viewing software, e.g., VMD (Humphrey, Dalke, & Schulten, 1996), Chimera (Pettersen et al., 2004), or PyMol (Schrödinger, LLC, 2015).
- We use a PMEMD and CUDA version of the AMBER molecular dynamics software (Le Grand, Götz, & Walker, 2013); however, NAMD and GROMACS can also be used (Pronk et al., 2013).
- CHARMM Molecular Modelling Builder (Brooks et al., 2009).

2.7.2 Procedure

1. Generate input coordinates for simulations by modeling missing loops and residues (for TmPPase in the resting state, loops 5–6 and 13–14 were built into the structure above the active site in an orientation parallel to the protomer channel).

2. Remove solvent molecules and any bound inhibitor not required in the simulations (for TmPPase, resting state, inhibitor-bound state, and the inhibitor-bound state after removal of the ligands (inhibitor, three Mg^{2+}, and Na^{+}) were used).
3. Embed the models in a membrane consisting of a 50% mixture of phosphatidylethanolamine (POPE) and phosphatidylcholine (POPC) lipids. Insure that the box is sufficiently large to ensure at least 8 Å between the protein and the edge of the simulation box (in CHARMM Molecular Modelling Builder).
4. Solvate the protein and membrane with TIP3P water molecules and 0.1 *M* KCl using the CHARMM Molecular Modelling Builder, and output a PDB file, following the online AMBER tutorial (available at http://ambermd.org/tutorials/advanced/tutorial16/).
5. Use the XLEaP module of AMBER (Le Grand et al., 2013) to construct the molecular topology and coordinate files for MD simulation. Use the AMBER lipid14 force field (Dickson et al., 2014) for the membrane, and the ff14SB force field (Maier et al., 2015) for the protein.
6. Perform a standard equilibration according to the AMBER tutorial (available at http://ambermd.org/tutorials/advanced/tutorial16/) that follows the steps of minimization, first heating, fixing the lipids, second heating, fixing the lipids, holding to equilibrate periodic box dimensions, and finally adding a constant pressure.
7. Subject models to >100 ns of MD simulation at constant temperature and pressure using the AMBER suite of programs (see Notes).
8. For data analysis, use the PTRAJ module of AMBER.
9. For trajectory visualization use VMD.
10. Trajectories can be analyzed to plot changes in distance between Cα atoms of pairs of residues selected for mutagenesis in order to understand the magnitude of fluctuations in distance due to thermal noise (see Fig. 4A and B for TmPPase example).

2.7.3 Notes

- MD simulations of TmPPase were run for >250 ns, with approximately 1000 frames per nanosecond (Fig. 4B). Fluctuations between Cα positions of cysteine pairs provide insight into normal thermal fluctuations and observable conformational changes in the enzyme between the open and closed state.

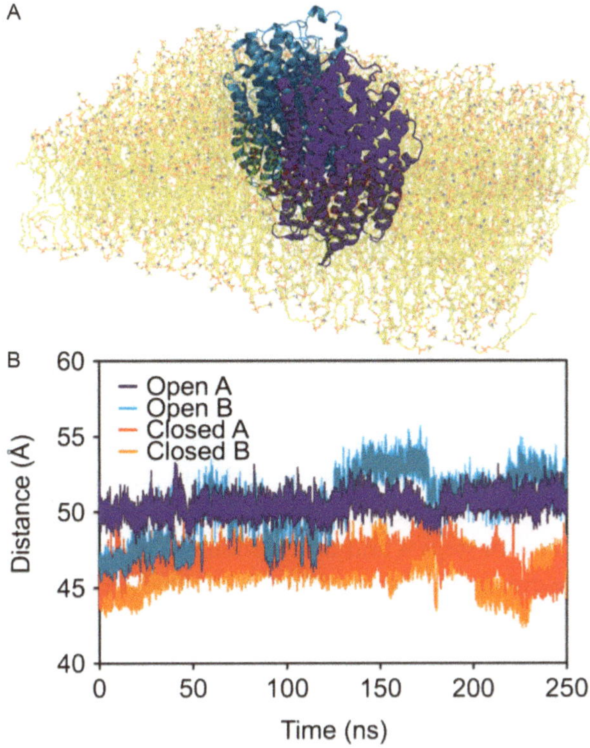

Fig. 4 All-atom molecular dynamics simulation of TmPPase in a lipid membrane. (A) A representative frame from all-atom MD simulations of a TmPPase dimer in the resting (open) state. (B) Plot of the distance fluctuations between Cα atoms of residues 211 and 546 in both open (*navy/marine*) and closed (*red/orange*) states in both chain A (*navy/red*) and B (*marine/orange*) over 250 ns of simulation. Initial starting models for all-atom MD simulations were based on resting-state (TmPPase:Ca:Mg; PDB: 4AV3) and closed-state (TmPPase:IDP:Na; PDB: 5LZQ) structures equilibrated in a POPC:POPE bilayer.

2.8 Computational Methods to Account for Dye Flexibility

The dyes include a flexible linker, and therefore will have access to a positional range in each protein conformation. We need to discern between noise introduced by flexible dyes and genuine structural fluctuations associated with conformational changes in observations of FRET signal. We use computational methods to account for the linker flexibility and its effect on dye position. As output models from MD simulations were used we were able to systematically explore all of the possible rotamer states and dye positions relative to the protein and the membrane.

2.8.1 Equipment

- PyMol (Schrödinger, LLC, 2015) (Available at: https://pymol.org/).
- MtsslWizard (PyMol Plugin) (Hagelueken, Ward, Naismith, & Schiemann, 2012).
- Python dependencies for MtsslWizard (NumPy (van der Walt, Colbert, & Varoquaux, 2011), WX, Matplotlib (Hunter, 2007), PIL)

2.8.2 Procedure

1. Install MtsslSuite plugin for PyMol containing the MtsslWizard (instructions available at http://www-hagelueken.thch.uni-bonn.de/) (Hagelueken et al., 2012).
2. Open protein model in PyMol.
3. Activate MtsslWizard through the menu bar.
4. Select residue to be labeled and select dye to use as label.
5. Select rotamer search option (see Notes) and perform the calculation.
6. Repeat process in steps 4 and 5 for second dye position.
7. Measure the distance distribution between the dyes and output the result for analysis (e.g., Fig. 5A–D).

2.8.3 Notes

- It is recommended that a "painstaking" search (100,000 tries) is conducted with "loose" constraints in order to find enough rotamers for a realistic representation of the distribution of distances available to the dyes.
- For TmPPase, we carried out the procedure to label both the inhibitor-bound and resting state structures that had been output from MD simulations with modeled lipid bilayer, to indicate the potential difference in distributions we might observe between states.
- The panels in Fig. 5 show an example of a double variant that was discounted as a viable option for smFRET. The accessible volume of the dyes was taken into account, and it was found that the two dyes were too far (producing low FRET), and their distance distributions were not distinct enough between the two states, meaning that it is unlikely that informative data could be extracted from a dye pair attached to these positions.

2.9 Single-Molecule FRET Using Alternating Laser Excitation

The labeled proteins reconstituted into liposomes exhibit clear FRET signals. The low proteoliposome concentration (approximately 70 pmol)

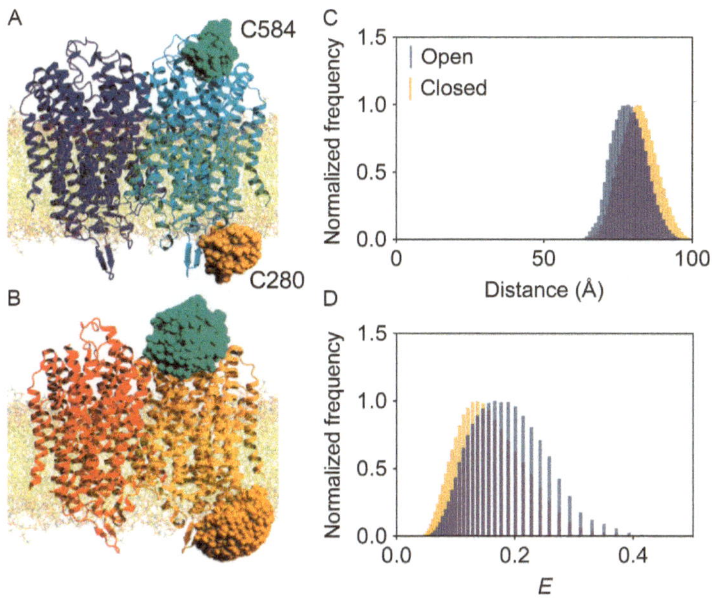

Fig. 5 Accounting for accessible volume of the dyes (space fill) by computation. Models of TmPPase with FRET dyes added to residues K280C and I584C in both (A) open and (B) closed states. Histograms of (C) calculated distances and (D) calculated FRET efficiencies (*E*).

together with the low confocal volume minimizes the chance that observed signals correspond to more than one vesicle. The fluorescence is collected in separate channels for the donor and acceptor in a form of short bursts (millisecond duration) resulting from passage of individual vesicles containing labeled protein. Anticorrelated changes to donor and acceptor signals indicate the occurrence of FRET, and the ratio of intensities averaged over a burst allow calculation of the FRET efficiency. It is important to note that protein samples labeled at two cysteines with a mixture of two dyes will also exhibit substantial amount of species with one type of dye attached. This is effectively dealt with by alternating laser excitation (ALEX) (Kapanidis et al., 2005) on a microsecond time scale (i.e., shorter than diffusion time through the confocal volume ~1 ms) which allows the experimenter to estimate dye stoichiometry and filter out donor-only and acceptor-only species. The observations from many bursts, i.e., many molecules, are combined into histograms, and the peaks can be fitted to normal distributions, if necessary. Conditions can be varied and their influence upon the relative positions and populations reveals the conformational behavior of the protein.

2.9.1 Equipment

- Python version 2.7 or above (Python Software Foundation. Python Language Reference. Available at http://www.python.org).
- FRETBursts open-source python package (Ingargiola, Lerner, Chung, Weiss, & Michalet, 2016).
- Dependencies for FRETBursts (including SciPy (van der Walt et al., 2011), NumPy (van der Walt et al., 2011), Pandas (McKinney, 2010), Matplotlib (Hunter, 2007)). FRETBursts and dependencies can easily be installed and managed with Anaconda graphical user interface (Available at http://www.anaconda.com).
- ALEX apparatus, originally invented by Kapanidis et al. (2005) and implemented with modifications as described in Sharma et al. (2014).
- Imaging chambers (Lab-Tek II Chambered coverglass, Nunc, catalog number: 155360)

2.9.2 Procedure

- Proteoliposome samples for smFRET-ALEX must be made fresh on the day of use (see Notes).
- 100 μL of sample (diluted to 50–100 p*M* of the Alexa Fluor 488 dye) in the presence or absence of substrate or inhibitor are added to an imaging chamber on an oil immersion objective and focused approximately 30 μm above the plane of the coverslip. For short and/or preliminary data collection a drop of 50–100 μL on a clean 0.17-mm cover slip may suffice instead of the rather expensive chamber.
- smFRET bursts are captured in batches of 5 min files, with fresh sample added after every third collection to counteract the issue of liposomes fusing and sedimenting, and insure loss of enzymatic activity was minimized. Typically, 1–2 h of data were required to capture the required number of bursts to perform robust statistical analyses. As more bursts are needed for kinetic analysis by, e.g., RASP, longer data collection is likely.
- Data collected as batches of single-spot μs-ALEX measurement files (SM file format) should be merged together and converted to photon-hdf5 format using FRETBursts python package.
- For each dataset, calculate background, perform corrections, and perform burst search and burst selection using FRETBursts python package.

Dependent on the labeling efficiency, bursts should be filtered to exclude contributions from singly labeled proteins (see Notes).

- Extract histograms from filtered observations and fit multiple Gaussian distributions.

2.9.3 Notes

- Proteoliposomes should be prepared and used fresh to avoid liposome fusing caused by extended incubation on ice or freeze-thawing.
- Due to the "shotgun" labeling procedure used here, perfect 1:1 labeling of sites is unachievable as the chemistry for attachment of both dyes is the same. Therefore, even for well-labeled sample, as little as 10% of the sites may possess the correct, alternately labeled pattern that provides meaningful measurements. Burst contributions from singly labeled species and free dyes were filtered out using the FRETBursts python package which requires data collected on ALEX setup.

3. SUMMARY AND CONCLUSIONS

To understand the motions of helix-12 and loops 5–6 and 13–14 in TmPPase during catalysis, cysteine residues were added to these elements to allow the attachment of a FRET label. Furthermore, a relatively static residue 40–60 Å distant from these elements was replaced with cysteine so that a second label could be added. This allowed us to monitor changes in distance between these two sites.

3.1 Nonfunctional Cysteine Variants of Loop 5–6

Prior to implementing our rational approach for selecting cysteine mutations, some positions of TmPPase loop 5–6 were targeted for mutation based on satisfying distance constraints alone. Residues of loop 5–6 were targeted for mutation and labeling at residues L215, R221, and N222 as these residues were at an ideal distance (40–60 Å) from F434. These double variants (L215C:F434C, R221C:F434C, and N222:F434C) were expressed, purified, and pyrophosphatase activity tested (Fig. 6A and B). Mutation of any one of these three residues (L215, R221, and N222) resulted in a completely inactive enzyme, whereas the single variant F434C remained fully active. When this result is rationalized using structural information (Fig. 6B), it is clear that these residues of loop 5–6 play a central role in

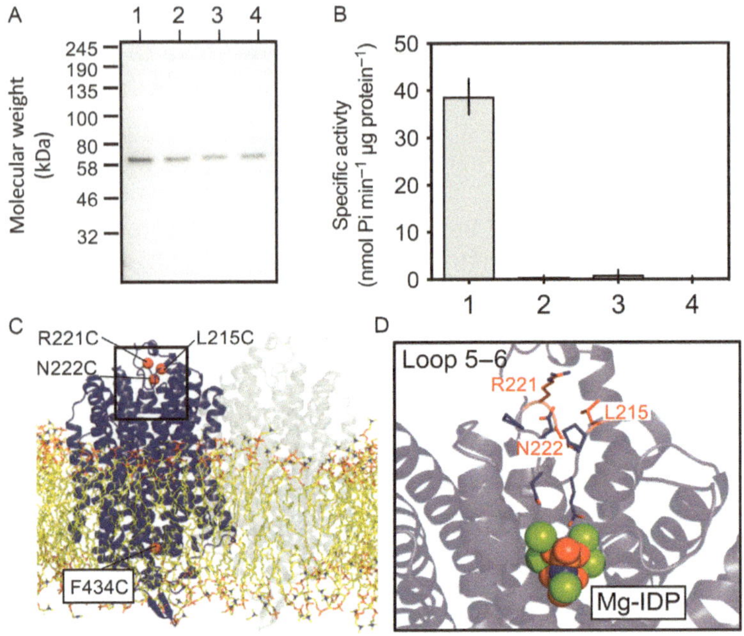

Fig. 6 Nonfunctional double-cysteine variants. (A) SDS-PAGE of purified TmPPase single-cysteine variant 1: F434C, and double-cysteine variant 2: F434C:L215C; 3: F434C:R221C; 4: F434C:N222C. (B) Pyrophosphatase activity of TmPPase single- and double-cysteine variants (as in (A)). (C) Positions (*red spheres*) of cysteine mutations in the IDP-bound TmPPase structure. (D) Close-up view of mutated residues in loop 5–6 that gave rise to inactive enzyme (*red*).

the binding and catalysis of PP_i hydrolysis. Although not all of these residues are in contact with the substrate (Fig. 6C and D), they are still highly conserved across mPPase orthologues (Fig. 2), and so they probably play an important structural or mechanistic role associated with substrate binding or the conformational change that leads to ion pumping. Thus, mutating them, unsurprisingly, leads to an inactive enzyme. This finding required us to perform a more thorough investigation of potential sites for attaching dyes. Although certain positions appeared to be ideal from a distance and positional point of view, labeling these positions would create meaningless observations as the enzyme is not active.

3.2 Improved Strategy for Selecting Sites for Cysteine Mutagenesis

We adapted our strategy as outlined in the procedure above. Importantly, we aimed to avoid selection of well-conserved residues, which we identified

Table 3 Effect of Point Mutations and Labeling on TmPPase Expression and Activity

Mutation	Position	B&W Numbering	Outcome
T211C	Loop 5–6	5.70	Half PP_i hydrolysis activity
L213C	Loop 5–6	5.73	Did not express
N214C	Loop 5–6	5.74	Was not cloned
P220C	Loop 5–6	6.27	Inactive for PP_i hydrolysis
T225C	Loop 5–6	6.33	Was not cloned
K572C	Loop 13–14	13.56	Active with and without label
A581C	Loop 13–14	13.65	Did not express
I584C	Loop 13–14	13.68	Half PP_i hydrolysis activity
E589C	Loop 13–14	14.28	Did not express
K591C	Loop 13–14	14.30	Was not cloned
A592C	Loop 13–14	14.31	Did not express
N597C	Loop 13–14	14.36	Was not cloned
N159C	Loop 4–5	4.79	Did not express
K280C	TMH7	7.33	Active with and without label
F434C	TMH10	10.68	Active with and without label
S520C	TMH12	12.71	Active with and without label
S523C	Loop 12–13	12.73	Did not express
A546C	TMH13	13.30	Active with and without label

B&W, Ballesteros and Weinstein numbering (Ballesteros & Weinstein, 1995; Tsai et al., 2014); *TMH*, transmembrane helix.

by a multisequence alignment. Furthermore, we selected a larger number of initial sites for mutations to increase the odds of finding an active enzyme variant.

Three of the total 18 mutations chosen as potential labeling sites failed at the cloning stage (Table 3). Of the remaining 15 single-cysteine TmPPase variants, 9 were successfully expressed and purified (Fig. 7A; Table 3), of which 7 (3 inward facing and 4 outward facing) were still active for PP_i hydrolysis (Fig. 7B; Table 3). Eight different combinations of TmPPase double-cysteine mutations were made, five of which retained activity when labeled with dyes (Fig. 7C).

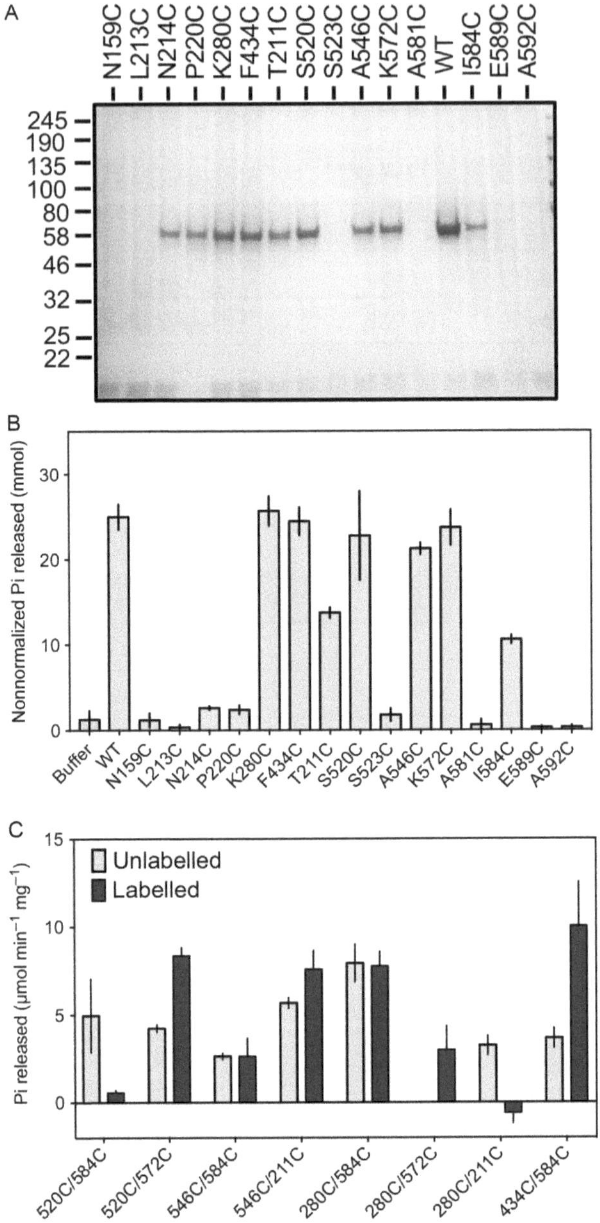

Fig. 7 Expression and activity of labeled and unlabeled cysteine variants. (A) SDS-PAGE of purified TmPPase single-cysteine variants. (B) Pyrophosphatase activity of TmPPase single-cysteine variants. (C) Pyrophosphatase activity of labeled (*black*) and unlabeled (*gray*) TmPPase double-cysteine variants.

3.3 Using MD Simulations and Computational Modeling to Assist Selection

With eight potential labeling pairs, we prioritized experimental testing of these constructs to those that would produce observable differences in dye position between different conformational states of the protein. Furthermore, we considered the size of "molecular breathing" in the protein due to thermal fluctuation in comparison to the size of the conformational change on substrate binding.

We used all-atom MD simulations of the enzyme in both a resting (open) and an inhibitor-bound (closed) conformation. This allowed an observation of the distribution of distances between Cα atoms of residue pairs in both simulated states, and for each protomer (Fig. 8A–H). By this analysis, the mutated pairs A546C:T211C (Fig. 8D) and K280C:T211C (Fig. 8G) appear to provide the largest change in the position of the maxima in the inter-Cα distance distribution between the two states. However, it should be noted that the K280C:T211C construct was ruled out due to the fact that it was no longer enzymatically active once labeled (Fig. 7C).

It was also possible to account for the freedom and flexibility of the dye molecules tethered at each of the specific sites on TmPPase in different states; FRET dyes were added to the models in silico and possible rotamers of the dyes were sampled (Fig. 9A–C). This method provided normal distributions of distances between two dyes, taking into account the freedom and flexibility of the dyes, and also specific factors such as steric hindrance created by

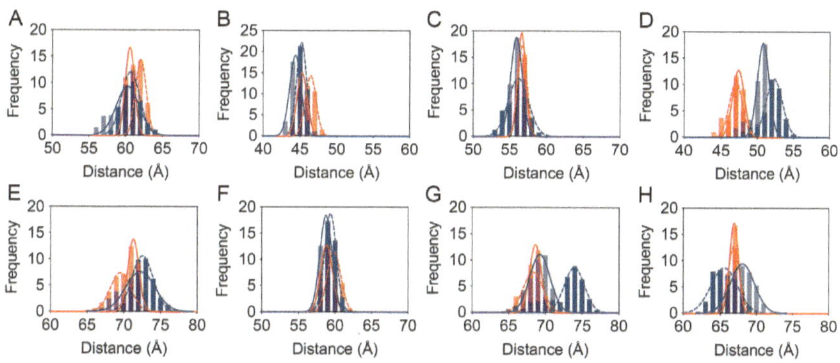

Fig. 8 Distances between pairs of mutation sites for both open (*blue*) and closed (*red*) states in both chain A (*solid line*) and B (*dashed line*). Histograms of distance fluctuations between Cα atoms of TmPPase residue pairs (A) S520C:I584C, (B) S520C:K572C, (C) A546C:I584C, (D) A546C:T211C, (E) K280C:I584C, (F) K280C:K572C, (G) K280C:T211C, and (H) F434C:I584C.

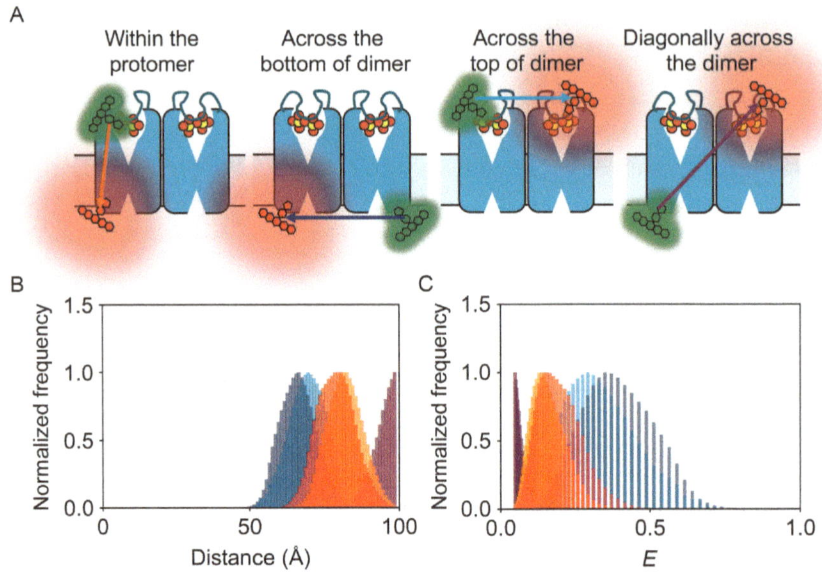

Fig. 9 All theoretical distance distributions of a double-labeled TmPPase dimer (K280C, I584C). (A) Scheme of possible FRET distances that could be measured in TmPPase dimers. Histograms of (B) distances and (C) calculated FRET efficiencies (*E*) corresponding to situations in (A), as calculated using MtsslWizard to understand freedom of movement of FRET dyes through rotation. Colors of the histograms represent distances between FRET pairs within a promoter (*orange*), across the dimer horizontally (*blue*), or across the dimer diagonally (*purple*).

surrounding environment (e.g., protein or lipid bilayer). Furthermore, using this analysis we were able to calculate distance distributions for dye pairs that cross the dimer interface (Fig. 9A–C). Some double variants were ruled out as good candidates for smFRET, as cross talk over the dimer interface would have theoretical FRET efficiencies that overlap severely with the signal we wish to observe (e.g., the TmPPase double-cysteine variant K280C: I584C—Fig. 9A–C). This analysis indicates that, for some potential FRET pairs, a heterodimer of labeled:unlabeled TmPPase may be required; otherwise histograms will be too complex to deconvolute meaningful observations.

3.4 Labeling TmPPase Double Variant and Observing smFRET in the Presence of Substrate or Inhibitors

To observe conformational changes, the TmPPase double variant A546C: I584C was dual labeled with Alexa Fluor 488 and 594 with a labeling efficiency of ~50%. We confirmed the protein retained 89 (±18)% activity after labeling. The labeled protein was reconstituted into liposomes, and we assume a 50:50, inward:outward-facing mixture of enzyme orientations in the proteoliposomes. Future efforts to improve the quality of the data could include the use of lipid-bilayer nanodiscs (Bayburt & Sligar, 2009) instead of proteoliposomes as this would render all active sites accessible to freely diffusing substrate or inhibitors. Single-molecule FRET was observed using ALEX.

FRET efficiency increases when two dyes are close to one another and decreases when two dyes move apart (Fig. 1B). Therefore, the FRET efficiency inversely correlates to the distance between the two dyes. In the above example, on the addition of substrate at a concentration equivalent to the K_m (10 μ*M* PP_i), there is a clear change in the shape of the histogram, and a decrease in mean FRET efficiency from 0.7 to 0.6, indicating that the dyes move apart (Fig. 10). In contrast, on the addition of a known inhibitor IDP, a PP_i substrate analogue, a second distribution at an approximate mean FRET efficiency of 0.76 appears. This indicates that the dyes move toward each other. Another inhibitor, etidronic acid (Eti) was also added, which resulted in a broad distribution around a mean FRET efficiency of 0.63.

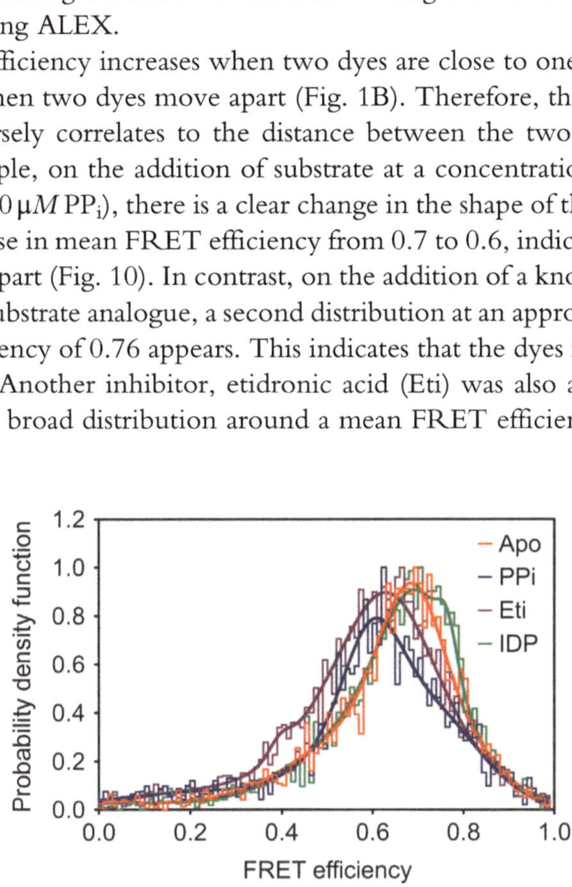

Fig. 10 The smFRET-ALEX analysis of double-labeled TmPPase (A546C:I584C) in the absence of substrate (unliganded; *red*), in the presence of 10 μ*M* PP_i (*blue*), 2.5 m*M* etidronic acid (eti; *purple*), or 2.5 m*M* imidodiphosphate (IDP; *green*). Histograms were produced by binning the FRET observations after processing the raw data using FRETBursts open-source python package. The sum of three Gaussian distributions was used to fit the shape of the histograms in each case. Due to the fact that substrates and inhibitors were only added to the outside of the liposomes, it is possible that not all enzyme active sites were accessible. Therefore, the enzyme could not be locked into a single state by saturating with substrate or inhibitor, and observed signals may have arisen from a mixture of several states.

We have shown that it is possible to measure changes in the conformation between different states of TmPPase in liposomes using smFRET. The size of these observed changes, although small, are in good agreement with our MD modeling of the loop movement. Intriguingly, we do not observe the same behavior in loop movement for the PP_i as for the IDP-bound case. We hypothesize that we are observing states during PP_i hydrolysis that have not been previously observed in crystal structures, and that our model of the enzyme mechanism should now be improved to include these previously undefined conformations. Indeed, the acquisition of dynamic information during catalytic turnover on a single-molecule basis is inaccessible by crystallography. Thus, the FRET approach we describe here is complementary to our previous work.

ACKNOWLEDGMENT

This work was supported by a Grant from BBSRC (BB/M021610/1). A.T. was supported by a grant from the Jane and Aatos Erkko foundation and Integrative Life Science Doctoral Program. J.S. was supported by a Grant from the European Union through a Marie-Skłodowsk-Curie Action Innovative Training Network (Grant agreement No. 722687).

REFERENCES

Baginski, E. S., Foà, P. P., & Zak, B. (1967). Microdetermination of inorganic phosphate, phospholipids, and total phosphate in biologic materials. *Clinical Chemistry*, *13*(4), 326–332.

Ballesteros, J. A., & Weinstein, H. (1995). Integrated methods for the construction of three-dimensional models and computational probing of structure-function relations in G protein-coupled receptors. *Methods in Neurosciences*, *25*, 366–428. Retrieved from, https://www.sciencedirect.com/science/article/pii/S1043947105800497.

Bayburt, T. H., & Sligar, S. G. (2009). Membrane protein assembly into Nanodiscs. *FEBS Letters*, *584*(9), 1721–1727. https://doi.org/10.1016/j.febslet.2009.10.024.

Baykov, A. A., Malinen, A. M., Luoto, H. H., & Lahti, R. (2013). Pyrophosphate-fueled Na^+ and H^+ transport in prokaryotes. *Microbiology and Molecular Biology Reviews: MMBR*, 77(2), 267–276. https://doi.org/10.1128/MMBR.00003-13.

Brooks, B. R., Brooks, C. L., Mackerell, A. D., Nilsson, L., Petrella, R. J., Roux, B., et al. (2009). CHARMM: The biomolecular simulation program. *Journal of Computational Chemistry*, *30*(10), 1545–1614. https://doi.org/10.1002/jcc.21287.

Dickson, C. J., Madej, B. D., Skjevik, Å. A., Betz, R. M., Teigen, K., Gould, I. R., et al. (2014). Lipid14: The Amber lipid force field. *Journal of Chemical Theory and Computation*, *10*(2), 865–879. https://doi.org/10.1021/ct4010307.

Ernst, J. F., & Chan, R. K. (1985). Characterization of *Saccharomyces cerevisiae* mutants supersensitive to aminoglycoside antibiotics. *Journal of Bacteriology*, *163*(1), 8–14.

Gietz, R. D., & Schiestl, R. H. (2007). High-efficiency yeast transformation using the LiAc/SS carrier DNA/PEG method. *Nature Protocols*, *2*(1), 31–34. https://doi.org/10.1038/nprot.2007.13.

Hagelueken, G., Ward, R., Naismith, J. H., & Schiemann, O. (2012). MtsslWizard: In silico spin-labeling and generation of distance distributions in PyMOL. *Applied Magnetic Resonance*, *42*(3), 377–391. https://doi.org/10.1007/s00723-012-0314-0.

Hoffmann, A., Nettels, D., Clark, J., Borgia, A., Radford, S. E., Clarke, J., et al. (2011). Quantifying heterogeneity and conformational dynamics from single molecule FRET of diffusing molecules: Recurrence analysis of single particles (RASP). *Physical Chemistry Chemical Physics*, *13*(5), 1857. https://doi.org/10.1039/c0cp01911a.

Humphrey, W., Dalke, A., & Schulten, K. (1996). VMD: Visual molecular dynamics. *Journal of Molecular Graphics*, *14*(1), 33–38. https://doi.org/10.1016/0263-7855(96)00018-5.

Hunter, J. D. (2007). Matplotlib: A 2D graphics environment. *Computing in Science & Engineering*, *9*(3), 90–95. https://doi.org/10.1109/MCSE.2007.55.

Ingargiola, A., Lerner, E., Chung, S. Y., Weiss, S., & Michalet, X. (2016). FRETBursts: An open source toolkit for analysis of freely-diffusing single-molecule FRET. *PLoS One*, *11*(8), 1–27. https://doi.org/10.1371/journal.pone.0160716.

Kapanidis, A. N., Laurence, T. A., Lee, N. K., Margeat, E., Kong, X., & Weiss, S. (2005). Alternating-laser excitation of single molecules. *Accounts of Chemical Research*, *38*(7), 523–533. https://doi.org/10.1021/ar0401348.

Kellosalo, J., Kajander, T., Kogan, K., Pokharel, K., & Goldman, A. (2012). The structure and catalytic cycle of a sodium-pumping pyrophosphatase. *Science*, *337*(6093), 473–476. https://doi.org/10.1126/science.1222505.

Kellosalo, J., Kajander, T., Palmgren, M. G., Lopéz-Marqués, R. L., & Goldman, a. (2011). Heterologous expression and purification of membrane-bound pyrophosphatases. *Protein Expression and Purification*, *79*(1), 25–34. https://doi.org/10.1016/j.pep.2011.05.020.

Le Grand, S., Götz, A. W., & Walker, R. C. (2013). SPFP: Speed without compromise—A mixed precision model for GPU accelerated molecular dynamics simulations. *Computer Physics Communications*, *184*(2), 374–380. https://doi.org/10.1016/j.cpc.2012.09.022.

Li, K.-M., Wilkinson, C., Kellosalo, J., Tsai, J.-Y., Kajander, T., Jeuken, L. J. C., et al. (2016). Membrane pyrophosphatases from *Thermotoga maritima* and *Vigna radiata* suggest a conserved coupling mechanism. *Nature Communications*, 7, 13596. https://doi.org/10.1038/ncomms13596.

Lin, S.-M., Tsai, J.-Y., Hsiao, C.-D., Huang, Y.-T., Chiu, C.-L., Liu, M.-H., et al. (2012). Crystal structure of a membrane-embedded H^+-translocating pyrophosphatase. *Nature*, *484*(7394), 399–403. https://doi.org/10.1038/nature10963.

López-Marqués, R. L., Pérez-Castiñeira, J. R., Buch-Pedersen, M. J., Marco, S., Rigaud, J.-L., Palmgren, M. G., et al. (2005). Large-scale purification of the proton pumping pyrophosphatase from *Thermotoga maritima*: A "Hot-Solve" method for isolation of recombinant thermophilic membrane proteins. *Biochimica et Biophysica Acta (BBA)—Biomembranes*, *1716*(1), 69–76. https://doi.org/10.1016/j.bbamem.2005.08.004.

Luoto, H. H., Baykov, A. A., Lahti, R., & Malinen, A. M. (2013). Membrane-integral pyrophosphatase subfamily capable of translocating both Na^+ and H^+. *Proceedings of the National Academy of Sciences of the United States of America*, *110*(4), 1255–1260. https://doi.org/10.1073/pnas.1217816110.

Luoto, H. H., Nordbo, E., Malinen, A. M., Baykov, A. A., & Lahti, R. (2015). Evolutionarily divergent, Na^+-regulated H^+-transporting membrane-bound pyrophosphatases. *Biochemical Journal*, *467*(2), 281–291. https://doi.org/10.1042/BJ20141434.

Maier, J. A., Martinez, C., Kasavajhala, K., Wickstrom, L., Hauser, K. E., & Simmerling, C. (2015). ff14SB: Improving the accuracy of protein side chain and backbone parameters from ff99SB. *Journal of Chemical Theory and Computation*, *11*(8), 3696–3713. https://doi.org/10.1021/acs.jctc.5b00255.

Mandell, D. J., Coutsias, E. A., & Kortemme, T. (2009). Sub-angstrom accuracy in protein loop reconstruction by robotics-inspired conformational sampling. *Nature Methods*, *6*(8), 551–552. https://doi.org/10.1038/nmeth0809-551.

McKinney, W. (2010). In *Data structures for statistical computing in Python. Proceedings of the 9th Python in Science Conference, 1697900(Scipy)* (pp. 51–56). Retrieved from, http://conference.scipy.org/proceedings/scipy2010/mckinney.html.

Palmgren, M. G., & Christensen, G. (1993). Complementation in situ of the yeast plasma membrane H^+-ATPase gene pma1 by an H^+-ATPase gene from a heterologous species. *FEBS Letters*, *317*(3), 216–222. https://doi.org/10.1016/0014-5793(93)81279-9.

Pettersen, E. F., Goddard, T. D., Huang, C. C., Couch, G. S., Greenblatt, D. M., Meng, E. C., et al. (2004). UCSF Chimera—A visualization system for exploratory research and analysis. *Journal of Computational Chemistry*, *25*(13), 1605–1612. https://doi.org/10.1002/jcc.20084.

Pronk, S., Páll, S., Schulz, R., Larsson, P., Bjelkmar, P., Apostolov, R., et al. (2013). GROMACS 4.5: A high-throughput and highly parallel open source molecular simulation toolkit. *Bioinformatics*, *29*(7), 845–854. https://doi.org/10.1093/bioinformatics/btt055.

Schrödinger, LLC. (2015). *The {PyMOL} Molecular Graphics System, Version~1.8*.

Shah, N. R., Vidilaseris, K., Xhaard, H., & Goldman, A. (2016). Integral membrane pyrophosphatases: A novel drug target for human pathogens? *AIMS Biophysics*, *3*(1), 171–194. https://doi.org/10.3934/biophy.2016.1.171.

Shah, N. R., Wilkinson, C., Harborne, S. P. D., Turku, A., Li, K.-M., Sun, Y.-J., et al. (2017). Insights into the mechanism of membrane pyrophosphatases by combining experiment and computer simulation. *Structural Dynamics*, *4*(3), 32105. https://doi.org/10.1063/1.4978038.

Sharma, A., Leach, R. N., Gell, C., Zhang, N., Burrows, P. C., Shepherd, D. A., et al. (2014). Domain movements of the enhancer-dependent sigma factor drive DNA delivery into the RNA polymerase active site: Insights from single molecule studies. *Nucleic Acids Research*, *42*(8), 5177–5190. https://doi.org/10.1093/nar/gku146.

Tsai, J.-Y., Kellosalo, J., Sun, Y.-J., & Goldman, A. (2014). Proton/sodium pumping pyrophosphatases: The last of the primary ion pumps. *Current Opinion in Structural Biology*, *27*, 38–47. https://doi.org/10.1016/j.sbi.2014.03.007.

van der Walt, S., Colbert, S. C., & Varoquaux, G. (2011). The NumPy array: A structure for efficient numerical computation. *Computing in Science & Engineering*, *13*(2), 22–30. https://doi.org/10.1109/MCSE.2011.37.

Villalba, J. M., Palmgren, M. G., Berberián, G. E., Ferguson, C., & Serrano, R. (1992). Functional expression of plant plasma membrane H^+-ATPase in yeast endoplasmic reticulum. *The Journal of Biological Chemistry*, *267*(17), 12341–12349. Retrieved from http://www.ncbi.nlm.nih.gov/pubmed/1534807.

Waterhouse, A. M., Procter, J. B., Martin, D. M. A., Clamp, M., & Barton, G. J. (2009). Jalview version 2—A multiple sequence alignment editor and analysis workbench. *Bioinformatics*, *25*(9), 1189–1191. https://doi.org/10.1093/bioinformatics/btp033.

Webb, B., & Sali, A. (2014). Comparative protein structure modeling using modeller. In *Current protocols in bioinformatics: Vol. 47* (pp. 5.6.1–5.6.32). Hoboken, NJ: John Wiley & Sons, Inc. https://doi.org/10.1002/0471250953.bi0506s47.

Yang, J., Yan, R., Roy, A., Xu, D., Poisson, J., & Zhang, Y. (2015). The I-TASSER suite: Protein structure and function prediction. *Nature Methods*, *12*(1), 7–8. https://doi.org/10.1038/nmeth.3213.

CHAPTER FOUR

A Simple Strategy to Determine the Dependence of Membrane-Bound Pyrophosphatases on K^+ as a Cofactor

Jannik Strauss*, Craig Wilkinson*, Keni Vidilaseris†, Steven P.D. Harborne*[,1], Adrian Goldman*[,†,1]

*Astbury Centre for Structural Biology, University of Leeds, Leeds, United Kingdom
†Faculty of Biological and Environmental Sciences, University of Helsinki, Helsinki, Finland
[1]Corresponding authors: e-mail address: s.harborne@leeds.ac.uk; a.goldman@leeds.ac.uk

Contents

Abstract

Membrane-bound pyrophosphatases (mPPases) couple pyrophosphate hydrolysis to H^+ and/or Na^+ pumping across membranes and are found in all domains of life except for multicellular animals including humans. They are important for development and stress resistance in plants. Furthermore, mPPases play a role in virulence of human pathogens that cause severe diseases such as malaria and African sleeping sickness.

Sequence analysis, functional studies, and recently solved crystal structures have contributed to the understanding of the mPPase catalytic cycle. However, several key mechanistic features remain unknown. During evolution, several subgroups of mPPases

Methods in Enzymology, Volume 607
ISSN 0076-6879
https://doi.org/10.1016/bs.mie.2018.04.018

differing in their pumping specificity and cofactor dependency arose. mPPases are classified into one of five subgroups, usually by sequence analysis. However, classification based solely on sequence has been inaccurate in several instances due to our limited understanding of the molecular mechanism of mPPases. Thus, pumping specificity and cofactor dependency of mPPases require experimental confirmation. Here, we describe a simple method for the determination of K^+ dependency in mPPases using a hydrolytic activity assay. By coupling these dependency studies with site-directed mutagenesis, we have begun to build a better understanding of the molecular mechanisms of mPPases. We optimized the assay for thermostable mPPases that are commonly used as model systems in our lab, but the method is equally applicable to mesophilic mPPases with minor modifications.

ABBREVIATIONS

B&W Ballesteros and Weinstein
ChPPase mPPase from *Carboxydothermus hydrogenoformans*
DDM dodecyl-β-maltopyranoside
DTT dithiothreitol
EDTA ethylenediaminetetraacetic acid
IDP imidodiphosphate
mPPase membrane-bound pyrophosphatase
PaPPase mPPase from *Pyrobaculum aerophilum*
PDB protein data bank
P_i inorganic phosphate
PMSF phenylmethylsulfonyl fluoride
PP_i inorganic pyrophosphate
SCD-Leu synthetic complete leucine-dropout media
sPPase soluble pyrophosphatase
TBG TRIS-buffered glycerol
TmPPase mPPase from *Thermotoga maritima*
TRIS tris(hydroxymethyl)aminomethane
VrPPase mPPase from *Vigna radiata* (mung bean)
YPD yeast extract, peptone, and glucose media

1. INTRODUCTION

Membrane-bound pyrophosphatases (mPPases) couple the hydrolysis of inorganic pyrophosphate (PP_i) to the generation of an electrochemical gradient by the translocation of Na^+ and/or H^+ across a membrane (Baltscheffsky, Von Stedingk, Heldt, & Klingenberg, 1966; Luoto, Baykov, Lahti, & Malinen, 2013; Malinen, Belogurov, Baykov, & Lahti, 2007; Moyle, Mitchell, & Mitchell, 1972). mPPases are found in a variety of organisms including archaea, bacteria, plants, and protozoan parasites—but not in multicellular animals (Baltscheffsky et al., 1966; Bucke, 1970; Drozdowicz et al., 1999; Karlsson, 1975; Scott et al., 1998). These pumps are complex in terms of ion specificity and dependence on cofactors for maximal enzyme activity leading to their classification into five main subgroups (Table 1).

Classification has been primarily based on functional characterization of the pumped ion and dependence on cationic cofactors such as Mg^{2+}, Na^+, or K^+ for ion transport or hydrolytic activity (Baltscheffsky et al., 1966; Luoto, Nordbo, Malinen, Baykov, & Lahti, 2015; Malinen et al., 2007; Moyle et al., 1972; Nordbo, Luoto, Baykov, Lahti, & Malinen, 2016). Sequence alignments and phylogenetic studies further elucidated the evolutionary origin of mPPases and revealed sequence determinants linked to classification parameters such as K^+ dependence or ion transport specificity (Belogurov & Lahti, 2002; Luoto, Belogurov, Baykov, Lahti, & Malinen, 2011; Luoto et al., 2015; Nordbo et al., 2016). Using site-directed mutagenesis, putative key residues were evaluated and the molecular mechanism of these enzymes was explored (Belogurov & Lahti, 2002; Luoto et al., 2011). Even after half a century, the classification of mPPases continues to be dynamic; recently, it was discovered that certain dual-pumping mPPases lose H^+-translocation ability at Na^+ concentrations above physiological levels (Nordbo et al., 2016).

1.1 Distribution, Biological Role, and Significance of mPPases

Na^+-pumping and dual-pumping mPPases are exclusive to prokaryotes, whereas H^+-pumping mPPases are found in all domains of life (Luoto et al., 2013). Not only the distribution but also the cellular localization of mPPases varies as the eukaryotic enzyme is predominately localized in organelles such as the vacuole in plants or the acidocalcisome in parasites rather than in the cell membrane (Rea et al., 1992; Scott et al., 1998). However, the biological role of mPPases in resistance to abiotic stress conditions (e.g., cold, drought, and hypoxia in plants) is preserved across all subfamilies and in all types of organisms (García-Contreras, Celis, & Romero, 2004; Lander, Cordeiro, Huang, & Docampo, 2016; Li et al., 2005; Liu et al., 2014; Pasapula et al., 2011; Yoon, Kim, & Kim, 2013). mPPases are less important for the regulation of intracellular P_i level than soluble PPases (sPPases), and their biological function is broader (Kajander, Kellosalo, & Goldman, 2013). Overexpression and knockout experiments in several species demonstrated their role in plant development and parasite osmo- and pH-regulation in addition to stress resistance (Lander et al., 2016; Li et al., 2005; Liu et al., 2014).

Table 1 mPPase Classification

Family	Pumping Specificity	$K^{12.46}$	K^+ Dep.	Occurrence	Key Glutamate $E^{6.53}$	Key Glutamate $E^{6.57}$	Key Glutamate $E^{5.43}$	Other	Example
I	Na^+	✗	✓	(Archae)bacteria	✓			Evolutionary ancestor	*T. maritima* (TmPPase)
IIa	H^+	✗	✓	Plants, parasites		✓			*V. radiata* (VrPPase)
IIb		✗	✓	*F. johnsoniae*			✓	$E^{5.43}$ occupies position similar to $E^{6.57}$ in IIb	*F. johnsoniae* (FjPPase)
III		✓	✗	Archae, bacteria, plants	✓				*P. aerophilum* (PaPPase)
IVa	Na^+/H^+	✗	✓	Bacteria esp. *Bacteroides*	✓			Na^+ pumping retained above 10 m*M* NaCl	*B. vulgatus* (BvPPase)
IVb		✗	✓	Bacteria esp. *Clostridioides*	✓			Na^+ pumping lost above 10 m*M* NaCl	*C. leptum* (CpPPase)
V	H +	✓	✓	Bacteria	✓			Na^+ pumping lost above 0.1 m*M* NaCl	*C. limicola* (ClPPase)

Due to the presence of mPPases in several pathological microorganisms such as *Plasmodium falciparum* and *Trypanosoma brucei*, and their crucial role in host biology, these pumps are an attractive target for drug design. Using X-ray crystallography, the structures of two mPPases from different subgroups have been solved in different catalytic stages providing the basis for structure-guided drug design (Kellosalo, Kajander, Kogan, Pokharel, & Goldman, 2012; Li et al., 2016; Lin et al., 2012). By combining structural data, molecular dynamic simulations, and electrometric studies, the molecular mechanism of the catalytic cycle and ion selectivity was explored and has been reviewed in detail (Li et al., 2016; Shah et al., 2017).

1.2 The Evolution and Structure of mPPases

Here, we will briefly introduce the molecular mechanism and evolution of K^+ dependency and ion selectivity. More detailed information about the classification and identification of key residues in mPPases can be found in the respective original literature (Asaoka, Segami, & Maeshima, 2014; Belogurov & Lahti, 2002; Hirono & Maeshima, 2009; Hirono, Nakanishi, & Maeshima, 2007; Kellosalo et al., 2012; Lee et al., 2011; Li et al., 2016; Lin et al., 2012; Luoto et al., 2011, 2015; Mimura, Nakanishi, Hirono, & Maeshima, 2004; Nakanishi, Saijo, Wada, & Maeshima, 2001; Nordbo et al., 2016; Pan et al., 2011; Schultz & Baltscheffsky, 2003; Zhen, Kim, & Rea, 1997). To allow for a simple comparison of residues between species, we label according to the Ballesteros and Weinstein (B&W) nomenclature (Ballesteros & Weinstein, 1995) in the format $R^{N1.N2}$ (R is the amino acid, N1 is the transmembrane helix, and N2 specifies the offset from a well-conserved amino acid in the middle of the respective helix—Table 2).

mPPases consist of 16 transmembrane helices (TMH), which form an inner and an outer ring. At the center of the inner ring, there are four main functional regions: the hydrolytic center, the coupling funnel, the ion gate, and the exit channel (Fig. 1) (Kellosalo et al., 2012; Lin et al., 2012). Our understanding of the catalytic cycle of mPPases places ion pumping prior to PP_i hydrolysis in the sequence of events (binding change mechanism) as indicated by electrometric studies using the SURFE^{2}R technique (Li et al., 2016).

During one catalytic turnover, substrate in form of Mg_2PP_i binds on the cytoplasmic side in the hydrolytic center leading to a conformational change that drives ion pumping and release of Na^+ and/or H^+ from the exit channel on the luminal/periplasmic side. PP_i is then hydrolyzed to return the enzyme to the resting state with phosphates being released stepwise (Fig. 1). Protons

Table 2 Residue Numbering in mPPases

B&W	TmPPase[a]	VrPPase[b]	PaPPase[c]
1.50	F17	F25	Y20
2.50	K55	K94	R58
3.50	S87	S153	S96
4.50	G130	G194	G138
5.50	R191	R242	R183
6.50	D243	D294	D235
7.50	G294	G335	A271
8.50	L321	L361	I294
9.50	G369	G411	G333
10.50	S416	S458	S380
11.50	D458	D500	D439
12.50	K499	K541	K480
13.50	V566	V597	V554
14.50	M611	M642	F599
15.50	A649	A680	A637
16.50	K707	K742	K691

[a]mPPase from *Thermotoga maritima*.
[b]mPPase from *Vigna radiata*.
[c]mPPase from *Pyrobaculum aerophilum*.

are likely to be translocated in a Grotthuss-chain mechanism, whereas Na^+ pumping occurs via a defined ion-binding site, as seen in the IDP-bound structure of TmPPase (PDB: 5LZQ). However, the molecular mechanism defining ion selectivity is still largely unknown.

mPPases are evolutionarily ancient, with Na^+-pumping mPPases thought to have evolved first, followed by divergence of H^+- and dual-pumping mPPases along individual evolutionary lines from this ancestor (Nordbo et al., 2016). As the structural similarity (C_α root-mean-square deviation = 0.862 Å of IDP-bound VrPPase (PDB: 4A01) to TmPPase (PDB:5LZQ)) and sequence identity (>50% identity) between mPPases are high, small changes in protein sequence may account for differences found between subgroups (Baykov, Malinen, Luoto, & Lahti, 2013; Luoto et al., 2013, 2015; Shah et al., 2017).

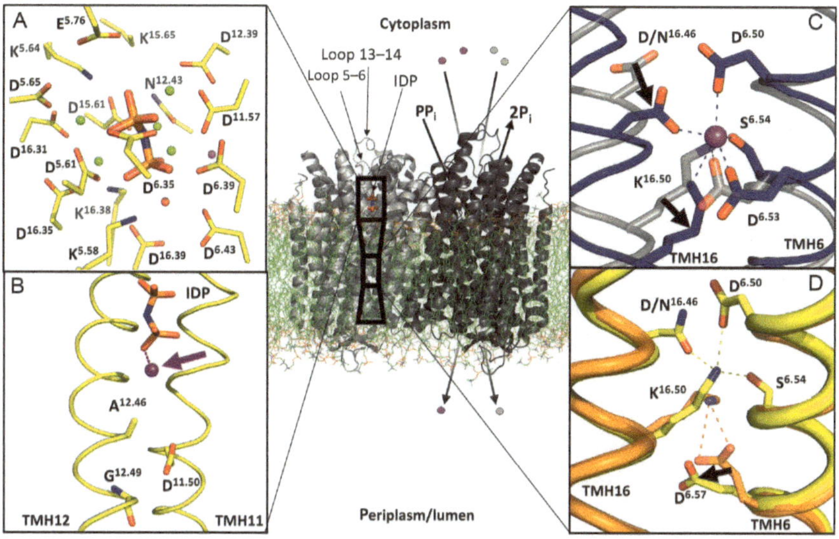

Fig. 1 Structural overview of IDP-bound mPPase dimer embedded in lipid bilayer with its reaction mechanism schematically shown. *Black boxes* indicate location of hydrolytic center, coupling funnel, ion gate, and exit channel (from cytoplasmic to periplasmic/luminal side). Residue labeling according to the B&W nomenclature. Ions are represented by *spheres*: Na^+ (*purple*) and H^+ (*gray*). (A) Detailed view of the VrPPase (*orange*) hydrolytic center with nucleophilic water (*red*) and IDP bound in a conserved environment of residues and metal ions (Mg^{2+}, *green* and K^+, *purple*). (B) Hydrolytic center of K^+-dependent VrPPase (*yellow*) in the IDP-bound state with focus on residues $A^{12.46}$ and $G^{12.49}$ defining the K^+ requirement. K^+ shown as *purple sphere*. *Arrow* indicates where ε-NH_3 group of $K^{12.46}$ is thought to be located in K^+-independent mPPases. (C) Ion gate of Na^+-pumping TmPPase in product-bound (*gray*) and IDP-bound (*blue*) states with major conformational changes indicated by *arrows*. (D) Ion gate of H^+-pumping VrPPase in product-bound (*orange*) and IDP-bound (*yellow*) states with major conformational changes indicated by *arrows*.

For example, the transition from Na^+-pumping to K^+-dependent H^+-pumping mPPases may only require repositioning of a single semiconserved residue. K^+-dependent H^+-pumping mPPases have a Glu at position 6.57, rather than position 6.53, moving the carboxylate group one turn down in TMH6, allowing it to act as a H^+ acceptor and abolishing the Na^+-binding site (Fig. 1C and D). Similarly, the evolution of K^+ dependence may have only required the replacement of key residues at positions 12.46 and 12.49 as described later.

Multiple sequence alignments of K^+-dependent and K^+-independent H^+-pumping mPPases indicated that positions 12.46 and 12.49 were key to defining a requirement for K^+ for maximal activity. In K^+-dependent enzymes, those positions contained Ala and Ala/Gly, respectively, whereas in the K^+-independent ones they were replaced by Lys or Thr, respectively (Fig. 2) (Belogurov & Lahti, 2002). Studies further demonstrated that the $A^{12.46}K$ mutation by itself was sufficient to abolish the requirement for K^+ in the K^+-dependent H^+-pumping mPPase of *Carboxydothermus hydrogenoformans* (ChPPase). In contrast, $A^{12.49}T$ single variants only showed a threefold reduced affinity for K^+ but still required it for maximal enzymatic activity (Belogurov & Lahti, 2002). In K^+-dependent mPPases, the conserved PP_i-binding site consists of 3 Lys, 10 Asp, and 1 Asn (Fig. 1A). Together with three additional Mg^{2+} ions, K^+ coordinates the substrate Mg_2PP_i in a metal cage, affecting the enzymatic activity of mPPases (Kellosalo et al., 2012; Lin et al., 2012). In K^+-independent enzymes, it has been hypothesized that K^+ is substituted by the positively charged amine group of $K^{12.46}$ (Fig. 1B). Indeed, our 3.8 Å structure of the K^+-independent PaPPase indicates occupation of the K^+-binding site by $K^{12.46}$ supporting this idea (Wilkinson, C., unpublished). However, phylogenetic studies have revealed an evolutionary distant subfamily of mPPases requiring low K^+ concentrations for maximal enzymatic activity despite the presence of $K^{12.46}$, which suggests it may not be the sole determinant of K^+ independency (Luoto et al., 2015).

Although categorizing mPPases into subgroups based on the presence or absence of certain key residues as described earlier has largely been successful, there have been several exceptions. This indicates that our models are not yet robust. In order to improve our ability to reliably categorize mPPases based on sequence alone, we still must determine the mechanistic properties of different mPPases experimentally.

1.3 Determination of K^+ Dependence in mPPases

The enzymatic activity of mPPases can be assessed in several ways. Either the hydrolytic activity or ion pumping can be measured. Both types of assay have been used in mutational studies previously and provide complementary information, as hydrolytic activity and ion pumping can be uncoupled by mutagenesis (Asaoka et al., 2014; Hirono et al., 2007; Pan et al., 2011). Assays of hydrolytic activity use the specific quantification of inorganic phosphate (P_i) by a colorimetric assay as a measure of the catalysis of PP_i. Assays of ion pumping use fluorescent probes (e.g., acridine orange) (Scott et al., 1998), quantification of accumulated Na^+ isotopes (Na^{22}) (Malinen et al., 2007), or current measurements on a membrane-coated gold chip

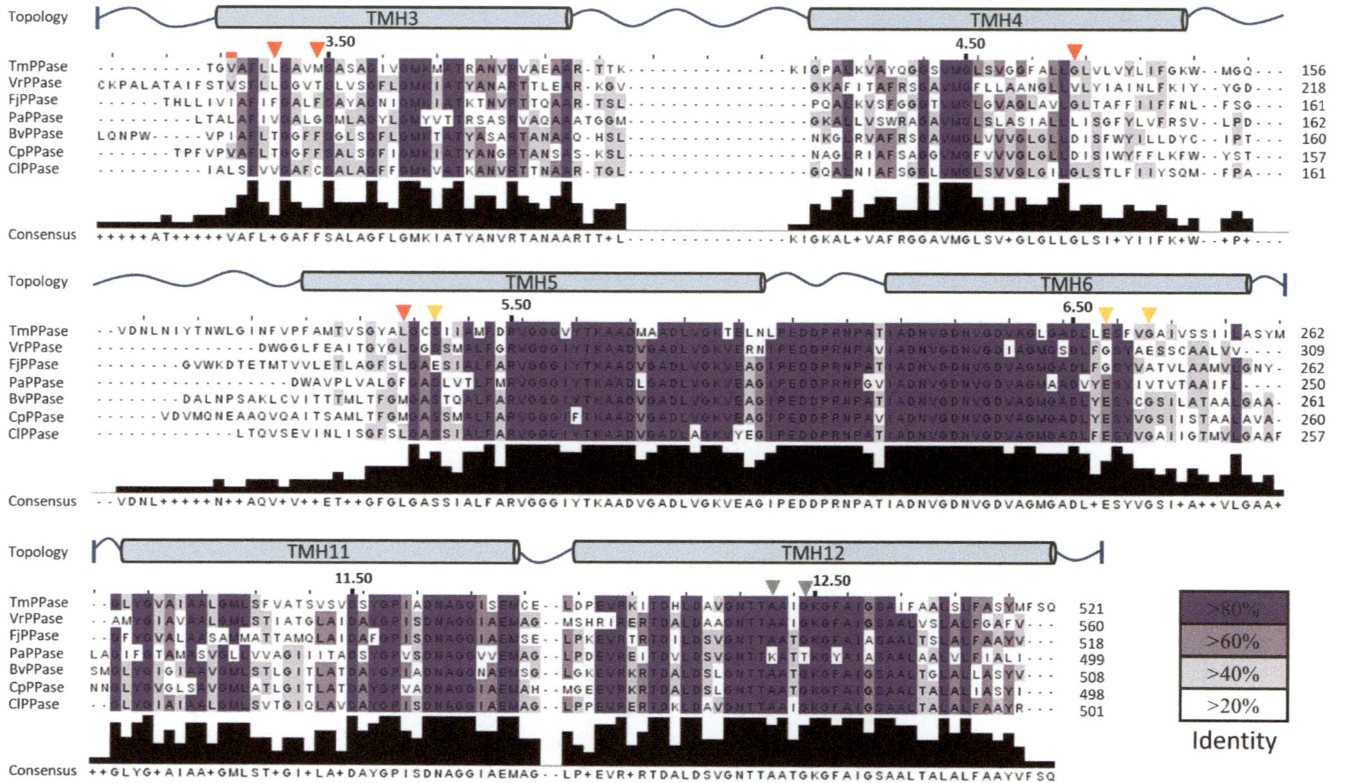

Fig. 2 Sequence alignment segments (TMH3-6 and TMH11-12) of a representative subset of mPPases. Topology is shown based on TmPPase structure. B&W labeling is included for conserved residue 50 of each helix. Position of key residues is indicated by *colored triangles*: *red*, dual-pumping signature; *orange*, semiconserved Glu; *gray*, K^+ dependence.

with immobilized and mPPase-containing liposomes (SURFE^{2}R) (Li et al., 2016) to determine the ion translocation. However, for a simple and fast determination of K^+ dependency, measurement of ion pumping is inferior to measurement of hydrolytic activity, due to the need for different techniques (dependent on which ion is pumped) or the requirement of special equipment.

Here, we describe a rapid quantitative P_i assay that allows us to test the requirement of K^+ for maximal protein activity with mPPase orthologs. It can be used to compare wild-type and site-directed variant enzyme in the presence and absence of K^+. This provides a way to quickly evaluate sequence-based predictions of K^+ dependence providing us with a greater understanding of the underlying molecular mechanism involved. Additionally, a variation of this protocol can be used for inhibitor testing on mPPases in a cheap, simple, and robust manner as described elsewhere (Vidilaseris, Kellosalo, & Goldman, 2018). The K^+-independent H^+-pumping PaPPase is used as model system as its expression and purification protocols are well established and the enzyme is stable at high temperature. This thermostability allows its solubilization at high temperatures, which has the added bonus of denaturing most contaminant protein, preventing their interference in the assay. With slight adjustment, the assay is also amenable to mesophilic mPPases: protein is solubilized at low temperature instead of high temperature, and contaminating sPPases are inhibited with fluoride.

In the assay, we use the optimal reaction conditions for the mPPase in question (temperature, cofactors, and lipid). Either purified protein, which has to be relipidated, or crude membrane extract containing the expressed mPPase is used. PP_i hydrolysis is then allowed to proceed for a defined time at native-like reaction temperatures before a temperature drop halts the enzymatic reaction. P_i quantification follows a protocol from Baginski and Zak and utilizes the colorimetric reaction of P_i with molybdate (Fig. 3) (Baginski, Foà, & Zak, 1967).

Upon substrate hydrolysis, P_i instantaneously reacts with molybdate to form an activated complex. In the presence of ascorbic acid, the complex is reduced resulting in a color change (yellow to blue), which is used for quantification in combination with a set of P_i standards at known concentrations (Fig. 3). Excess molybdate is complexed by addition of a citrate–arsenite solution. The color change is linear only when arsenite and citrate are present, as these chemicals prevent unfavorable side reactions (Baginski et al., 1967). The total P_i amount released during the activity assay is

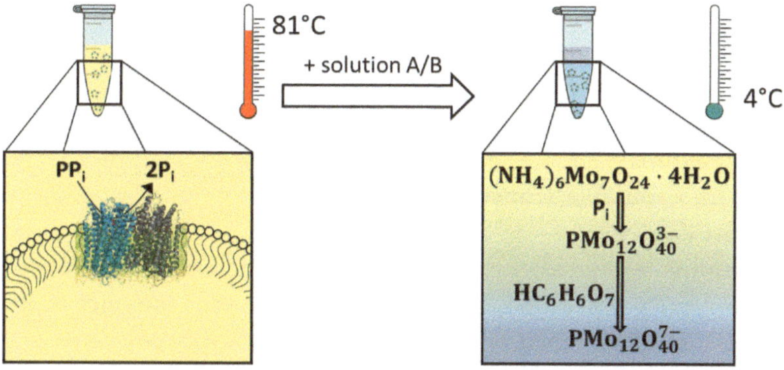

Fig. 3 Schematic drawing of the colorimetric P_i-release assay upon substrate addition to relipidated sample. Ions are represented as *spheres* with Na^+ in *yellow* and H^+ in *purple*.

calculated based on absorbance measurement at 860 nm and comparison to a P_i standard curve. To allow for an accurate comparison between the presence and absence of K^+, the specific activity must be calculated (Eq. 1).

$$\text{Activity}_{\text{spec.}} = \frac{n_{P_i}}{t_{\text{reaction}} \cdot m_{\text{mPPase}}} \tag{1}$$

n_{P_i} : averaged phosphate nanomoles released during hydrolytic assay (nmol),
t_{reaction} : reaction time (min), and
m_{mPPase} : amount of mPPase in activity assay (mg)

2. PRELIMINARY SEQUENCE-BASED DETERMINATION OF K^+ DEPENDENCY

Preliminary determination of the K^+ dependency in mPPases, based on the $A/K^{12.46}$ and $G/A/T^{12.49}$ dichotomy (occurrence of $G^{12.49}$ vastly outnumbers $A^{12.49}$ in K^+-dependent mPPases), is useful to get a first indication of how K^+ might affect the maximal enzymatic activity. This information is important for protein solubilization and purification, where the presence of cofactors during these processes can significantly improve the quality and yield of protein obtained. Sequence analysis of our target protein PaPPase revealed the presence of $K^{12.46}$ and $T^{12.49}$, indicating that it does not require K^+ (Fig. 2). However, the discovery of evolutionarily distant K^+-dependent H^+-PPases that contain $K^{12.46}$ highlights limitations in predicting ion selectivity based on sequence analysis only (Luoto et al., 2015).

2.1.1 Equipment and Software

- Software for multiple alignments, e.g., Jalview (Waterhouse, Procter, Martin, Clamp, & Barton, 2009)

2.1.2 Procedure

1. Perform a multiple sequence alignment of orthologs of your protein of interest (see *Notes*).
2. Determine residue identity at positions 12.46 and 12.49 (B&W nomenclature) and predict the K^+ dependence based on the presence or absence of Lys and Thr at each position, respectively (see *Notes*).

2.1.3 Notes

- It is important to create an alignment covering a sufficient range of orthologs of your protein of interest to allow for an accurate comparison of residue or motif conservation. The alignment of 42 varied sequences will include proteins with 68%–98% sequence identity.
- An alignment of 42 mPPase sequences from all mPPase subgroups shows the presence of $K^{12.46}$ and $T^{12.49}$ in PaPPase, indicating that it does not require K^+ for maximal activity (Fig. 2; seven sequences displayed).

2.2 PaPPase Expression in *Saccharomyces cerevisiae* and Membrane Extraction

Later, we describe the expression and membrane preparation of the thermostable PaPPase in yeast. As standard, we express the protein in yeast in batches of 10 L of media, typically yielding protein in excess of that needed for testing protein activity by hydrolytic assay. The expression scale can be adjusted for different mPPases if required.

2.2.1 Equipment and Consumables

- L-baffled flasks, e.g., TunAir shake flasks (Sigma-Aldrich, catalogue number: Z710822)
- Shaking incubator, e.g., INNOVA44 (New Brunswick, catalogue number: M1282-0000)
- Equipment for yeast cell lysis, e.g., BeadBeater (BioSpec, catalogue number: 1107900)
- 0.5 mm Diameter glass beads (ThistleScientific, catalogue number: 11079105)
- High-speed centrifuge, e.g., Avant J-26 XP and rotor, e.g., JLA 8.1000
- Ultracentrifuge and rotor, e.g., Ti 45, Beckman coulter
- 40-mL glass dounce homogenizer (Wheaton, catalogue number: 357546)

2.2.2 Buffers and Reagents

- YPD medium (1% (w/v) yeast extract, 2% (w/v) tryptone, 2% (v/v) D-glucose)
- 1.5 × SCD-Leu: 3% (v/v) D-glucose, 2 g/L yeast nitrogen base (without amino acids), 0.75 m*M* tryptophan, 0.6 m*M* histidine, 0.3 m*M* adenine, 0.3 m*M* uracil
- Phenylmethanesulfonyl fluoride (PMSF) (Sigma-Aldrich, catalogue number: P7626-1G).
- Pepstatin A (Sigma-Aldrich, catalogue number: P5318-25MG)
- 1 *M* MES–NaOH pH 6.5
- 1 *M* Tris–HCl pH 7.5
- 1 *M* dithiothreitol (DTT)
- 25% (w/v) dodecyl-β-maltopyranoside (DDM) (Anatrace, catalogue number: D310-25-GM)
- 0.5 *M* EDTA pH 7.5
- 1 *M* $MgCl_2$
- 1 *M* NaCl
- 1 *M* KCl
- 1 *M* Na_2HPO_4
- 100 m*M* sodium pyrophosphate pH 8
- 100% glycerol
- Lysis buffer: 200 m*M* Tris–HCl pH 7.5, 40% (w/v) glycerol, 10 m*M* EDTA pH 7.5, 2 m*M* DTT, 0.2 m*M* PMSF
- TBG: 200 m*M* Tris–HCl pH 7.5, 10% (w/v) glycerol, 5 m*M* EDTA
- Resuspension buffer: 50 m*M* MES pH 6.5, 20% (w/v) glycerol, 50 m*M* KCl, 5 m*M* $MgCl_2$, 1.3 m*M* DTT, 0.34 m*M* PMSF, 0.2 m*M* Pepstatin A

2.2.3 Procedure

1. Use individual colonies of *S. cerevisiae* containing the mPPase expression plasmid (see *Notes*) to inoculate three 7.5 mL 1 × SCD-Leu medium primary precultures. Incubate for 24 h at 30°C with shaking at 200 rpm.
2. Inoculate the three 850 mL 1 × SCD-Leu secondary precultures using the primary precultures grown overnight. Incubate secondary precultures for 24 h at 30°C with shaking at 200 rpm.
3. Using the secondary precultures, inoculate 10 × 750 mL of 1.5 × SCD-Leu medium each in a 2.5-L full-baffle shake flask. Incubate at 30°C for 10 h with shaking at 200 rpm (see *Notes*).
4. Harvest the cells by centrifugation at 3000 × *g* for 20 min.
5. Resuspend cell pellet in H_2O to 10% of the original culture volume and spin down cells at 3000 × *g* for 20 min. Repeat washing step twice.

6. Remove supernatant and weigh the obtained cell mass. Pellets can be stored at −80°C until further use (see *Notes*).
7. Resuspend the final cell pellet in 1 mL lysis buffer per 2 g of dry cell weight.
8. Assemble the BeadBeater (smallest chamber fitting the whole cell suspension volume) filled with approximately 1.5 g ice-cold 0.5 mm diameter glass beads per 1 mL of cell suspension. If the BeadBeater chamber is not filled completely, top up with TBG until full. Fill the outside reservoir with ice water.
9. Lyse the cells by 12 × 60 s activations with 60 s cooldown periods between each activation. Replace some water in the outside reservoir with fresh ice after six cycles (see *Notes*).
10. Recover the cell lysate from beads in 40-mL aliquots transferred into 50-mL falcon tubes. Wash the glass beads two to three times with approximately 50 mL of TBG and combine with the cell lysate.
11. Add 5 mL of 3.5 *M* KCl per 40 mL of cell lysate.
12. Centrifuge the lysate to remove cell debris in a benchtop centrifuge (4000 × *g*, 10 min, 4°C).
13. Recover the membrane-containing supernatant and collect membranes at 235,000 × *g* for 1 h at 4°C.
14. Discard the supernatant, resuspend the membrane pellets in a total of 4 mL resuspension buffer, and transfer membrane suspension into 15-mL Falcon tubes. Wash ultracentrifugation tubes with another 4 mL of resuspension buffer and combine with membrane suspension. Use a dounce glass homogenizer to assist resuspension (see *Notes*).
15. Carry out a Bradford assay to determine total membrane protein concentration (see *Notes*).
16. Snap freeze the membranes in liquid nitrogen and store at −80°C until further use.

2.2.4 Notes

- We use the BJ1991 strain of *S. cerevisiae* (*MATα, pep4-3, prbl-1122, ura3-52, leu2, trpl*) (Ernst & Chan, 1985) for heterologous expression of PaPPase. The plasmid pMP649-PaPPase with N-terminally 6xHis-tagged PaPPase and a GG-linker separating the tag and the gene was used for transformation into *S. cerevisiae* by the lithium acetate/PEG/DNA method (Gietz & Schiest, 2007). pMP649 plasmid contains a high copy number pUC derived origin of replication and an ampicillin-resistance selection marker for cloning in *E. coli* as well as a *LEU2* selectable marker for yeast (Kellosalo, Kajander, Palmgren, Lopéz-Marqués, & Goldman, 2011; Palmgren & Christensen, 1993; Villalba, Palmgren, Berberián, Ferguson, & Serrano, 1992). The pMP649 plasmid contains a *2μ* origin of replication. The PaPPase gene is under the control of a constitutive *PMA1* gene promoter for high protein expression levels in our pMP649-PaPPase plasmid (Kellosalo et al., 2011; Palmgren & Christensen, 1993; Villalba et al., 1992).

- The incubation time and media type for expression are protein dependent and should be experimentally determined.
- We typically obtain 50–100 g of cells from 10 L of SCD-Leu yeast culture.
- Ensure the samples are kept cold throughout the lysis process in order to minimize protease activity.
- If extracted membrane samples are to be used directly for activity assay measurements, ensure that the resuspension buffer is free from K^+. K^+-free buffers do not usually require additional treatments, such as chelation, in order to remove contaminating K^+ as only millimolar concentrations affect the mPPase activity (Belogurov & Lahti, 2002).
- Typically, 1–3 g of total membrane protein is recovered, as determined by the Bradford assay.

2.3 Solubilization and Purification of PaPPase Using the "Hot-Solve" Method

2.3.1 Equipment and Consumables

- Water bath
- Spectrophotometer, e.g., DS-11 (DeNovix)
- 50K MWCO concentrator, e.g., VivaSpin 20 (GE Healthcare, catalogue number: 28932362)
- Buffer exchange columns, e.g., Micro Bio-Spin® Columns (Bio-Rad, catalogue number: 7326221)

2.3.2 Buffers and Reagents

- 1 *M* MES–NaOH pH 6.5
- 1 *M* Tris–HCl pH 7.5
- 1 *M* DTT
- 25% (w/v) DDM (Anatrace, catalogue number: D310-25-GM)
- 0.5 *M* EDTA pH 7.5

- 1 *M* $MgCl_2$
- 1 *M* NaCl
- 3.5 *M* KCl
- 100% Glycerol
- 1 *M* Imidazole pH 7.5
- Nickel affinity resin (Ni Sepharose® 6FF, GE Healthcare, catalogue number: GE17-5318-02).
- Complete EDTA-free protease inhibitor tablet (Roche, catalogue number: 5056489001).
- Resuspension buffer: 50 m*M* MES pH 6.5, 20% w/v glycerol, 50 m*M* KCl, 5 m*M* $MgCl_2$, 1.3 m*M* DTT, 0.34 m*M* PMSF, 0.2 m*M* Pepstatin A.
- Solubilization buffer: 50 m*M* MES pH 6.5, 20% w/v glycerol, 5.34% w/v DDM, 1 m*M* Na_2PP_i, protease inhibitor tablet.
- Wash buffer: 50 m*M* MES pH 6.5, 20% w/v glycerol, 50 m*M* KCl, 5 m*M* $MgCl_2$, 20 m*M* imidazole, 0.02% DDM, 1 m*M* DTT, 0.2 m*M* PMSF, 0.2 m*M* Pepstatin A.
- Elution buffer: 50 m*M* MES pH 6.5, 3.5% w/v glycerol, 50 m*M* KCl, 5 m*M* $MgCl_2$, 400 m*M* imidazole pH 6.5, 0.02% DDM, 1 m*M* DTT, 0.2 m*M* PMSF, 0.2 m*M* Pepstatin A.
- Assay buffer: 50 m*M* MES pH 6.5, 3.5% w/v glycerol, 5 m*M* $MgCl_2$, 0.02% DDM, 1 m*M* DTT, 0.2 m*M* PMSF, 0.2 m*M* Pepstatin A.

2.3.3 Procedure

1. Preheat the water bath to 75°C and thaw yeast membranes on ice.
2. Dilute yeast membranes to a final concentration of 7.2 mg/mL using resuspension buffer (see *Notes*).
3. Preheat solubilization buffer and diluted membranes for 15 min in the 75°C water bath.
4. Mix 10 mL of solubilization buffer with 30 mL of diluted membranes in 50-mL Falcon tubes.
5. Place the combined sample in the 75°C water bath for 1.5 h. Mix by inversion at 5 min intervals to ensure the samples remain in suspension (see *Notes*).
6. Separate the insoluble material by centrifugation at 4000 × *g* for 5 min at room temperature.
7. Transfer supernatant to new 50-mL Falcon tubes and repeat centrifugation at 4000 × *g* for 5 min prior to incubation with nickel affinity beads to remove any residual insoluble material.

8. Prepare 0.5 mL nickel affinity resin slurry per 40 mL of solubilized membrane sample. Remove excess ethanol, wash nickel affinity resin slurry in 40 mL water, and equilibrate in 10 × column volume (CV) resuspension buffer with 0.02% (w/v) DDM. Finally, resuspend the nickel affinity resin in 32 mL of resuspension buffer and 6 mL of 3.5 *M* KCl for every 40 mL of solubilized membrane sample.
9. Mix the solubilized membrane sample with the preequilibrated nickel affinity beads and batch bind at 37°C for 1.5–2 h, shaking at 200 rpm to maintain the beads in solution.
10. Apply the sample to a 20 mL purification column and collect a small aliquot of the flow-through (FT) for SDS-PAGE analysis (see *Notes*).
11. Wash the column with 2 × CV of Wash buffer and collect a small aliquot of the flow-through (W1) for SDS-PAGE analysis.
12. Wash the column with another 2 × CV of Wash buffer and collect a small aliquot of the flow-through (W2) for SDS-PAGE analysis.
13. Elute the protein with 2 × CV ofElution buffer, collect the flow-through (E1), and store on ice.
14. Concentrate the eluted protein in a 50 K MWCO VivaSpin 20 (4°C, 3000 × *g*) to approximately 10 mg/mL (analyzing samples via A_{280} readings during the concentrating steps—see *Notes*).
15. Once concentrated, a buffer exchange is carried out using Micro Bio-Spin® Columns to remove imidazole and K^+. Follow the manufacturer's instructions and elute protein in assay buffer (see *Notes*).
16. Analyze the collected purification samples by SDS-PAGE.
17. Snap freeze aliquots in liquid nitrogen and store protein at −80°C until further use.

2.3.4 Notes

- Total membrane protein concentration is based on results obtained via the Bradford assay. Typically, mPPase solubilization and purification from 0.5 to 1 mL thawed membrane sample yields amounts of protein suitable for testing the K^+ dependency.
- PaPPase is solubilized at high temperatures according to the "Hot-Solve" method described previously (López-Marqués et al., 2005). Using high temperatures for solubilization leads to denaturation of most contaminant proteins, typically resulting in a pure sample after affinity chromatography. The optimal solubilization temperature for the target protein must be empirically determined. Mesophilic proteins cannot be solubilized

using the "Hot-Solve" method, but alternative methods have been used to purify the H^+-pumping mPPase from the *Vigna radiata* (Li et al., 2016; Vidilaseris, K. unpublished).

- Detergent can be exchanged during the purification by substituting DDM in both the Wash and the Elution buffer.
- Buffer exchange can also be performed by repetitive concentration and subsequent dilution of sample in imidazole-free Elution buffer or dialysis. For new mPPases, it is recommended to prepare a sample in K^+-containing (50 m*M* KCl) assay buffer.
- Lower protein concentrations can be used as well but may require adjustment of the sample volume added to the reaction mixture (described in Section 2.4.3.)

2.4 Quantitative Phosphate Assay

The phosphate assay can be used to determine the K^+ dependency of mPPases from all subgroups in a medium-throughput manner by comparing the hydrolytic activity in the presence or absence of K^+.

2.4.1 Equipment

- PCR machine capable of accepting 8-tube PCR strips, e.g., MasterCycler (Eppendorf, catalogue number: 6321000019)
- Heat block, e.g., ThermoMixer C (Eppendorf, catalogue number: 5382000015)
- Multichannel pipettes operating between 10 and 200 μL, e.g., Pipetman 8-channel 2–20 μL (Gilson, catalogue number: F14401) and Pipetman 8-channel 20–200 μL (Gilson, catalogue number: F14403)
- 96-Well plate reader with absorbance measurement at 860 nm, e.g., Tecan Spark 20M plate reader (Tecan) (see *Notes*)

2.4.2 Buffers and Reagents

- 0.5 *M* HCl
- 1 *M* Tris–HCl pH 8
- 1 *M* $MgCl_2$
- 1 *M* NaCl
- 1 *M* DTT
- 1 *M* KCl
- 5 m*M* Na_2HPO_4
- 100 m*M* Sodium pyrophosphate pH 8
- 25% (w/v) DDM (Anatrace, catalogue number: D310-25-GM)

- L-α-Phosphatidyl choline from soybean lecithin (Sigma-Aldrich, catalogue number: 429415-100GM)
- Ascorbic acid
- Ammonium heptamolybdate
- Sodium arsenite
- Sodium citrate tribase
- Glacial acetic acid
- 4 × Reaction buffer: 80 m*M* Tris–HCl pH 8, 5 m*M* $MgCl_2$, 20 m*M* NaCl, 1 m*M* NaF, ±120 m*M* KCl, ±protein sample

2.4.3 Procedure

1. Prepare lipids for protein relipidation by sonicating 30 mg/mL soybean lecithin in 20 m*M* Tris–HCl pH 8 and 1 m*M* DTT until the solution turns clear. A stock of prepared lipids can be aliquoted and frozen at −80°C for future use.
2. For protein relipidation, heat 40 μL of sonicated lipids at 55°C for 15 min and allow to cool to room temperature. Mix 2 μL of concentrated protein with 40 μL of prepared lipids and top up to 100 μL using assay buffer (see *Notes*).
3. Dissolve 5 g of sodium arsenite and 5 g of sodium citrate tribasic in 100 mL H_2O, add 5 mL of glacial acetic acid, and top-off to 250 mL with H_2O after mixing.
4. Prepare solution A by dissolving 0.3 g of ascorbic acid in 10 mL of ice-cold 0.5 *M* HCl (see *Notes*).
5. Prepare solution B by dissolving 70 mg of ammonium heptamolybdate in 1 mL of ice-cold H_2O.
6. Prepare samples by mixing 1 μL of relipidated protein sample with 60 μL of 4 × reaction buffer with and without K^+ and adjusting to a total volume of 240 μL using water. Also include a negative control-containing assay buffer instead of protein sample (see *Notes*).
7. Prepare 0, 12.5, 25, 50, and 75 nmol P_i standards by replacing the protein sample with 0, 2.5, 5, 10, and 15 μL of a 5 m*M* Na_2HPO_4 stock, respectively.
8. Transfer 40 μL of reaction mixture containing relipidated protein to 8-well PCR strips in triplicate. Additionally, add 40 μL of the negative control and P_i standards in triplicate to determine the background P_i levels and standard curve, respectively.
9. Prewarm the reaction mixtures to 81°C for 5 min, adding one strip to the PCR machine every 20 s.

10. Combine solutions A and B on ice, which should form a yellow and transparent mixture. If there is a color change prior to conducting the assay, the solutions cannot be used and must be remade.
11. After 5 min of preincubation, add 10 μL of 10 m*M* sodium pyrophosphate to each PCR strip using a multichannel pipette, one strip every 20 s, and incubate for 5 min at 81°C.
12. Place each PCR strip on ice, one strip every 20 s, for 10 min to stop the reaction.
13. Add 60 μL of solution A/B mixture to each strip, one strip every 20 s, using a multichannel pipette and leave on ice for an additional 10 min.
14. Add 90 μL of sodium arsenite solution to each strip, one strip every 20 s, using a multichannel pipette to stop the color change (see *Notes*).
15. Transfer contents of each PCR strip to a 96-well plate and incubate the plate for 1 h at room temperature
16. Measure the absorbance at A_{860nm} using a suitable plate reader.

2.4.4 Notes

- Any plate reader capable of absorption measurement at 860 nm can be used.
- mPPase relipidation following protein purification is required for maximal enzyme activity (Kellosalo et al., 2011). Lipid solubilization can be supported by substitution of 18 μL assay buffer with 25% DDM. However, detergent addition may reduce the maximal enzymatic activity, as seen for PaPPase, in which case it should not be added. Instead of purified protein, extracted membrane samples can be used. Prepare a serial dilution of extracted membrane samples. Ensure that no K^+ has been added to buffers used to generate the extracted membrane samples. Specific activity results cannot be obtained due to the absence of accurate protein concentration data, nevertheless K^+ dependency can still be determined.
- Ascorbic acid should be dissolved thoroughly and kept on ice throughout the procedure to prevent premature color change.
- Depending on the protein concentration obtained in purification, the relipidated protein sample volume added to the reaction buffer may need to be adjusted so that the measurements fall within the linear range of the standard curve.
- The presence of the sPPase inhibitor NaF in the reaction buffer is only required if working with mesophilic proteins or membrane extractions below denaturing temperatures.

- The optimal reaction temperature is protein dependent and must be adjusted for different mPPases.
- Sodium arsenite is toxic and should be handled with extra care at all times. Do not allow contact with skin and use fume hood when working with powder.

2.5 Data Analysis

To allow accurate evaluation of the effect of K^+, the raw data need to be processed and normalized to the exact reaction time and enzyme amount used in the P_i-release assay. The specific activities of samples with or without K^+ are compared to determine the K^+ dependency.

2.5.1 Procedure

- Average the triplicates of each sample and P_i standard and subtract the negative control to eliminate background signal.
- Make a calibration curve by plotting the normalized A_{860} values against the correlated P_i amount (nmol) of the standards. Use linear regression to obtain a function describing the trendline ($f(x) = mx + b$) (see *Notes*).
- Determine phosphate amount (nmol) of the sample by substituting $f(x)$ with the normalized A_{860} absorbance value and solving the equation for x.
- Calculate the specific activity (Eq. 1) and determine K^+ dependence by comparing specific activities in K^+-containing and -lacking reaction mixtures (see *Notes*).

2.5.2 Notes

- The coefficient of determination (R^2) should be above 0.95 to allow accurate calculation of the phosphate concentration in the sample.
- The specific activity of our target protein PaPPase (wild type) was not affected by the presence or absence of K^+ in the reaction mixture. Therefore, we were able to confirm our sequence-based prediction that PaPPase is K^+ independent. We also found that single-site variants $K^{12.46}A$ and $T^{12.49}A$ had comparable hydrolytic activities to wild-type enzyme in the presence or absence of K^+. Only the double site-directed variant enzyme showed a significant decrease in the hydrolytic activity in the absence of K^+ (Fig. 4).

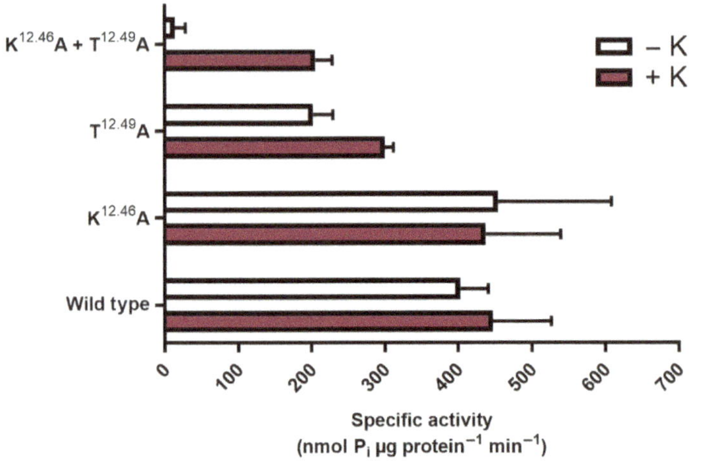

Fig. 4 Hydrolytic activity assay of PaPPase wild-type and site-directed variant enzyme in the presence and absence of potassium. Error bars denote standard error of the mean.

3. SUMMARY AND CONCLUSION

Although sequence analysis greatly supports classification of mPPases, recurring identification of enzymes that do not fit the current classification scheme highlights the need for experimental evaluation. Additionally, the molecular model explaining K^+ dependence is based on mutational studies of a single protein (ChPPase). In this protein, the key change $A^{12.49}T$ is relatively rare, as Gly usually occupies this position in K^+-dependent mPPases (Belogurov & Lahti, 2002). To allow for definitive conclusions, mutational studies should be repeated with a representative subset of mPPases. Both the evaluation of sequence-based classification and analysis of mPPase variants studying K^+ dependency can be addressed using our P_i-quantification assay. We present a simple, low-cost, and rapid method to determine mPPase dependence on K^+ without the need for specialized equipment or the use of radioactive substances. We show PaPPase wild-type and site-directed variant protein as a case study. The assay was optimized for the use with thermostable enzymes at high temperatures, which are particularly useful as model system to study mPPases. Comments have been added where changes to the protocol are necessary if working with nonthermostable or nonpurified protein.

The mutation of key residues involved in K^+ dependency (12.46 and 12.49) aimed to reverse-engineer evolutionary steps and reintroduce K^+ dependency in PaPPase, thereby supporting conclusions of previous mutational studies on ChPPase. This would increase confidence in the general applicability of our current mechanistic model of K^+ dependency. Wild-type PaPPase showed hydrolytic activity without K^+ supplementation at levels similar to those obtained in the presence of K^+; hence, its sequence-based classification as K^+-independent enzyme due to the presence of $K^{12.46}$ could be confirmed. According to the mechanistic model of K^+ dependency, replacement of Lys with Ala by itself ($K^{12.46}A$) should be sufficient to introduce a requirement for K^+ as the amine group substituting in substrate coordination is removed. However, PaPPase single-site variants ($K^{12.46}A$) failed to introduce K^+ dependence. Only double site-directed variants ($K^{12.46}A$ and $T^{12.49}A$) showed K^+ dependence, indicating a more important role of residue $T^{12.49}$, e.g., via stabilizing interactions that may maintain the integrity of the K^+-binding site even in the absence of its ligand. To confirm this hypothesis, the P_i-quantification assay must be repeated in the absence of any monovalent cation that might be able to occupy the K^+-binding site. However, if there is no evidence of other cations rescuing hydrolytic activity in the absence of K^+, a more complex model of K^+ dependency including both key residues (12.46 and 12.49) may be required in order to be applicable to mPPases of all subgroups.

Here, we were able to reverse the steps evolution has made toward K^+ independency by mutating two key residues that were previously identified in large mPPase sequence alignments and evaluated the mutations using a rapid P_i-quantification assay. This supports the hypothesis that alteration to a few key residues in mPPases during evolution was sufficient to radically change enzyme characteristics such as cofactor dependency.

ACKNOWLEDGMENTS

This project has received funding from the BBSRC (BB/M021610/1) and the Leeds 110 Anniversary Scholarship, the Academy of Finland (1286429 and 308105) and the Erkko Foundation.

REFERENCES

Asaoka, M., Segami, S., & Maeshima, M. (2014). Identification of the critical residues for the function of vacuolar H^+-pyrophosphatase by mutational analysis based on the 3D structure. *Journal of Biochemistry*, *156*(6), 333–344. https://doi.org/10.1093/jb/mvu046.

Baginski, E. S., Foà, P. P., & Zak, B. (1967). Microdetermination of inorganic phosphate, phospholipids, and total phosphate in biologic materials. *Clinical Chemistry*, *13*(4), 326–332.

Ballesteros, J. A., & Weinstein, H. (1995). Integrated methods for the construction of three-dimensional models and computational probing of structure-function relations in G protein-coupled receptors. *Methods in Neurosciences*, *25*, 366–428. https://doi.org/10.1016/S1043-9471(05)80049-7.

Baltscheffsky, H., Von Stedingk, L. V., Heldt, H. W., & Klingenberg, M. (1966). Inorganic pyrophosphate: Formation in bacterial photophosphorylation. *Science*, *153*(3740), 1120–1122. https://doi.org/10.1126/science.153.3740.1120.

Baykov, A. A., Malinen, A. M., Luoto, H. H., & Lahti, R. (2013). Pyrophosphate-fueled Na^+ and H^+ transport in prokaryotes. *Microbiology and Molecular Biology Reviews*, 77(2), 267–276. https://doi.org/10.1128/MMBR.00003-13.

Belogurov, G. A., & Lahti, R. (2002). A lysine substitute for K^+. *The Journal of Biological Chemistry*, *277*(51), 49651–49654. https://doi.org/10.1074/jbc.M210341200.

Bucke, C. (1970). The distribution and properties of alkaline inorganic pyrophosphatase from higher plants. *Phytochemistry*, *9*(6), 1303–1309. https://doi.org/10.1016/S0031-9422(00)85323-6.

Drozdowicz, Y. M., Lu, Y.-P., Patel, V., Fitz-Gibbon, S., Miller, J. H., & Rea, P. A. (1999). A thermostable vacuolar-type membrane pyrophosphatase from the archaeon *Pyrobaculum aerophilum*: Implications for the origins of pyrophosphate-energized pumps. *FEBS Letters*, *460*(3), 505–512. https://doi.org/10.1016/S0014-5793(99)01404-0.

Ernst, J. F., & Chan, R. K. (1985). Characterization of *Saccharomyces cerevisiae* mutants supersensitive to aminoglycoside antibiotics. *Journal of Bacteriology*, *163*(1), 8–14. Retrieved from https://www.ncbi.nlm.nih.gov/pmc/articles/PMC219073/.

García-Contreras, R., Celis, H., & Romero, I. (2004). Importance of *Rhodospirillum rubrum* H^+-pyrophosphatase under low-energy conditions. *Journal of Bacteriology*, *186*(19), 6651–6655. https://doi.org/10.1128/JB.186.19.6651.

Gietz, R. D., & Schiest, R. H. (2007). High-efficiency yeast transformation using the LiAc/SS carrier DNA/PEG method. *Nature Protocols*, *2*(1), 31–34. https://doi.org/10.1007/978-1-4939-1363-3_1.

Hirono, M., & Maeshima, M. (2009). Functional enhancement by single-residue substitution of *Streptomyces coelicolor* A3(2) H^+-translocating pyrophosphatase. *The Journal of Biochemistry*, *146*(5), 617–621. https://doi.org/10.1093/jb/mvp114.

Hirono, M., Nakanishi, Y., & Maeshima, M. (2007). Essential amino acid residues in the central transmembrane domains and loops for energy coupling of *Streptomyces coelicolor* A3(2) H^+-pyrophosphatase. *Biochimica et Biophysica Acta—Bioenergetics*, *1767*(7), 930–939. https://doi.org/10.1016/j.bbabio.2007.03.014.

Kajander, T., Kellosalo, J., & Goldman, A. (2013). Inorganic pyrophosphatases: One substrate, three mechanisms. *FEBS Letters*, *587*(13), 1863–1869. https://doi.org/10.1016/j.febslet.2013.05.003.

Karlsson, J. (1975). Membrane-bound potassium and magnesium ion-stimulated inorganic pyrophosphatase from roots and cotyledons of sugar beet (*Beta Vulgaris* L.). *Biochimica et Biophysica Acta—General Subjects*, *399*(2), 356–363. https://doi.org/10.1016/0304-4165(75)90264-0.

Kellosalo, J., Kajander, T., Kogan, K., Pokharel, K., & Goldman, A. (2012). The structure and catalytic cycle of a sodium-pumping pyrophosphatase. *Science*, *337*(6093), 473–476. https://doi.org/10.1126/science.1222505.

Kellosalo, J., Kajander, T., Palmgren, M. G., Lopéz-Marqués, R. L., & Goldman, A. (2011). Heterologous expression and purification of membrane-bound pyrophosphatases. *Protein Expression and Purification*, *79*(1), 25–34. https://doi.org/10.1016/j.pep.2011.05.020.

Lander, N., Cordeiro, C., Huang, G., & Docampo, R. (2016). Polyphosphate and acidocalcisomes. *Biochemical Society Transactions*, *44*(1), 1–6. https://doi.org/10.1042/BST20150193.

Lee, C.-H., Pan, Y.-J., Huang, Y.-T., Liu, T.-H., Hsu, S.-H., Lee, C.-H., et al. (2011). Identification of essential lysines involved in substrate binding of vacuolar H+-pyrophosphatase. *The Journal of Biological Chemistry*, *286*(14), 11970–11976. https://doi.org/10.1074/jbc.M110.190215.

Li, K.-M., Wilkinson, C., Kellosalo, J., Tsai, J.-Y., Kajander, T., Jeuken, L. J. C., et al. (2016). Membrane pyrophosphatases from *Thermotoga maritima* and *Vigna radiata* suggest a conserved coupling mechanism. *Nature Communications*, 7(13596), 1–11. https://doi.org/10.1038/ncomms13596.

Li, J., Yang, H., Peer, W. A., Richter, G., Blakeslee, J., Bandyopadhyay, A., et al. (2005). Arabidopsis H^+-PPase AVP1 regulates auxin-mediated organ development. *Science*, *310*(5754), 121–125. https://doi.org/10.1126/science.1115711.

Lin, S.-M., Tsai, J.-Y., Hsiao, C.-D., Huang, Y.-T., Chiu, C.-L., Liu, M.-H., et al. (2012). Crystal structure of a membrane-embedded H^+-translocating pyrophosphatase. *Nature*, *484*, 399–404. https://doi.org/10.1038/nature10963.

Liu, J., Pace, D., Dou, Z., King, T. P., Guidot, D., Li, Z.-H., et al. (2014). A vacuolar-H^+-pyrophosphatase (TgVP1) is required for microneme secretion, host cell invasion, and extracellular survival of *Toxoplasma gondii*. *Molecular Microbiology*, *93*(4), 698–712. https://doi.org/10.1111/mmi.12685.A.

López-Marqués, R. L., Pérez-Castiñeira, J. R., Buch-Pedersen, M. J., Marco, S., Rigaud, J.-L., Palmgren, M. G., et al. (2005). Large-scale purification of the proton pumping pyrophosphatase from *Thermotoga maritima*: A "Hot-Solve" method for isolation of recombinant thermophilic membrane proteins. *Biochimica et Biophysica Acta—Biomembranes*, *1716*(1), 69–76. https://doi.org/10.1016/j.bbamem.2005.08.004.

Luoto, H. H., Baykov, A. A., Lahti, R., & Malinen, A. M. (2013). Membrane-integral pyrophosphatase subfamily capable of translocating both Na^+ and H^+. *Proceedings of the National Academy of Sciences of the United States of America*, *110*(4), 1255–1260. https://doi.org/10.1073/pnas.1217816110.

Luoto, H. H., Belogurov, G. A., Baykov, A. A., Lahti, R., & Malinen, A. M. (2011). Na^+-translocating membrane pyrophosphatases are widespread in the microbial world and evolutionarily precede H^+-translocating pyrophosphatases. *The Journal of Biological Chemistry*, *286*(24), 21633–21642. https://doi.org/10.1074/jbc.M111.244483.

Luoto, H. H., Nordbo, E., Malinen, A. M., Baykov, A. A., & Lahti, R. (2015). Evolutionarily divergent, Na^+-regulated H^+-transporting membrane-bound pyrophosphatases. *Biochemical Journal*, *467*(2), 281–291. https://doi.org/10.1042/BJ20141434.

Malinen, A. M., Belogurov, G. A., Baykov, A. A., & Lahti, R. (2007). Na^+-pyrophosphatase: A novel primary sodium pump. *Biochemistry*, *46*(30), 8872–8878. https://doi.org/10.1021/bi700564b.

Mimura, H., Nakanishi, Y., Hirono, M., & Maeshima, M. (2004). Membrane topology of the H^+-pyrophosphatase of *Streptomyces coelicolor* determined by cysteine-scanning mutagenesis. *The Journal of Biological Chemistry*, *279*(33), 35106–35112. https://doi.org/10.1074/jbc.M406264200.

Moyle, J., Mitchell, R., & Mitchell, P. (1972). Proton-translocating pyrophosphatase of *Rhodospirillum rubrum*. *FEBS Letters*, *23*(2), 233–236. https://doi.org/10.1016/0014-5793(72)80349-1.

Nakanishi, Y., Saijo, T., Wada, Y., & Maeshima, M. (2001). Mutagenic analysis of functional residues in putative substrate-binding site and acidic domains of vacuolar H^+-pyrophosphatase. *The Journal of Biological Chemistry*, *276*(10), 7654–7660. https://doi.org/10.1074/jbc.M009743200.

Nordbo, E., Luoto, H. H., Baykov, A. A., Lahti, R., & Malinen, A. M. (2016). Two independent evolutionary routes to Na^+/H^+ cotransport function in membrane pyrophosphatases. *Biochemical Journal*, *473*(19), 3099–3111. https://doi.org/10.1042/BCJ20160529.

Palmgren, M. G., & Christensen, G. (1993). Complementation in situ of the yeast plasma membrane H^+-ATPase gene *pma1* by an H^+-ATPase gene from a heterologous species. *FEBS Letters*, *317*(3), 216–222. https://doi.org/10.1016/0014-5793(93)81279-9.

Pan, Y. J., Lee, C. H., Hsu, S. H., Huang, Y. T., Lee, C. H., Liu, T. H., et al. (2011). The transmembrane domain 6 of vacuolar H^+-pyrophosphatase mediates protein targeting and proton transport. *Biochimica et Biophysica Acta—Bioenergetics*, *1807*(1), 59–67. https://doi.org/10.1016/J.BBABIO.2010.10.006.

Pasapula, V., Shen, G., Kuppu, S., Paez-Valencia, J., Mendoza, M., Hou, P., et al. (2011). Expression of an Arabidopsis vacuolar H^+-pyrophosphatase gene (*AVP1*) in cotton improves drought- and salt tolerance and increases fibre yield in the field conditions. *Plant Biotechnology Journal*, *9*(1), 88–99. https://doi.org/10.1111/j.1467-7652.2010.00535.x.

Rea, P. A., Kim, Y., Sarafian, V., Poole, R. J., Davies, J. M., & Sanders, D. (1992). Vacuolar H^+-translocating pyrophosphatases: A new category of ion translocase. *Trends in Biochemical Sciences*, *17*(9), 348–353. https://doi.org/10.1016/0968-0004(92)90313-X.

Schultz, A., & Baltscheffsky, M. (2003). Properties of mutated *Rhodospirillum rubrum* H^+-pyrophosphatase expressed in *Escherichia coli*. *Biochimica et Biophysica Acta—Bioenergetics*, *1607*(2–3), 141–151. https://doi.org/10.1016/j.bbabio.2003.09.003.

Scott, D. A., de Souza, W., Benchimol, M., Zhong, L., Lu, H. G., Moreno, S. N. J., et al. (1998). Presence of a plant-like proton-pumping pyrophosphatase in acidocalcisomes of *Trypanosoma cruzi*. *The Journal of Biological Chemistry*, *273*(34), 22151–22158. https://doi.org/10.1074/jbc.273.34.22151.

Shah, N. R., Wilkinson, C., Harborne, S. P. D., Turku, A., Li, K.-M., Sun, Y.-J., et al. (2017). Insights into the mechanism of membrane pyrophosphatases by combining experiment and computer simulation. *Structural Dynamics*, *4*(32105), 1–12. https://doi.org/10.1063/1.4978038.

Vidilaseris, K., Kellosalo, J., & Goldman, A. (2018). A high-throughput method for orthophosphate determination of thermostable membrane-bound pyrophosphatase activity. *Analytical Methods*, *10*(6), 646–651. https://doi.org/10.1039/C7AY02558K.

Villalba, J. M., Palmgren, M. G., Berberián, G. E., Ferguson, C., & Serrano, R. (1992). Functional expression of plant plasma membrane H^+-ATPase in yeast endoplasmic reticulum. *The Journal of Biological Chemistry*, *267*(17), 12341–12349. Retrieved from http://www.jbc.org/content/267/17/12341.long.

Waterhouse, A. M., Procter, J. B., Martin, D. M. A., Clamp, M., & Barton, G. J. (2009). Jalview version 2—A multiple sequence alignment editor and analysis workbench. *Bioinformatics*, *25*(9), 1189–1191. https://doi.org/10.1093/bioinformatics/btp033.

Yoon, H.-S., Kim, S.-Y., & Kim, I.-S. (2013). Stress response of plant H^+-PPase-expressing transgenic *Escherichia coli* and *Saccharomyces cerevisiae*: A potentially useful mechanism for the development of stress-tolerant organisms. *Journal of Applied Genetics*, *54*(1), 129–133. https://doi.org/10.1007/s13353-012-0117-x.

Zhen, R.-G., Kim, E. J., & Rea, P. A. (1997). Acidic residues necessary for pyrophosphate-energized pumping and inhibition of the vacuolar H^+-pyrophosphatase by *N,N′*-dicyclohexylcarbodiimide. *The Journal of Biological Chemistry*, *272*(35), 22340–22348. Retrieved from http://www.jbc.org/content/272/35/22340.full.pdf.

CHAPTER FIVE

Crystallization of Liganded Phosphatases in the HAD Superfamily

Christine M. Harvey[2], Katherine H. O'Toole[2], Karen N. Allen[1]
Department of Chemistry, Boston University, Boston, MA, United States
[1]Corresponding author: e-mail address: drkallen@bu.edu

Contents

Abstract

Phosphotransferases catalyze reactions on chemically diverse molecules in organisms from all domains of life. The haloalkanoate dehalogenase superfamily (HADSF) is a model system for phosphoryl transfer enzymes as members catalyze phosphoester hydrolase, phosphonate hydrolase, and phosphomutase reactions on sugars, lipids, nucleotides, and peptides. Because these reactions are fundamental to essential metabolic transformations, understanding the mechanism and determinants of substrate specificity in the HADSF is critical. Structure/function relationships in the superfamily have

[2] These authors contributed equally.

Methods in Enzymology, Volume 607
ISSN 0076-6879
https://doi.org/10.1016/bs.mie.2018.06.011

also been leveraged in the development of methodologies for the assignment of enzyme function. Enzyme complexes with substrate, product, and analogs of the ground state or intermediate/transition state can be studied via high-resolution macromolecular crystallography to provide insight to the relative location of residues and ligands, as well as associated enzyme conformational states. This knowledge can aid in inhibitor design for phosphohydrolase reactions and target-specific therapeutics. Here we describe experimental approaches to capture liganded X-ray crystallographic structures of HADSF members. A number of these methods can be employed generally, including other families of phosphohydrolases and enzymes catalyzing phosphoryl transfer.

ABBREVIATIONS

β-PGM β-phosphoglucomutase
Bb *Bordetella bronchiseptica*
Bc *Bacillus cereus*
Bt *Bacteroides thetaiotaomicron*
BT4131 hexose-phosphate phosphatase
Ca *Candida albicans*
cdN cytosolic 5′(3′)-deoxyribonucleotidase
Cn *Cryptococcus neoformans*
$\mathbf{D}_n$ nucleophilic aspartate
$\mathbf{D}_{n+2}$ general acid/base aspartate
Ec *Escherichia coli*
Gal-1-P galactose 1-phosphate
Glc-1-P glucose 1-phosphate
Glc-6-P glucose 6-phosphate
GmhB D-glycero-D-manno-heptose-1,7-bisphosphate phosphatase
HADSF haloalkanoate dehalogenase superfamily
Hi *Haemophilus influenzae*
Hs *Homo sapiens*
KDN 2-keto-3-deoxy-D-glycero-D-galacto-nononic acid
KDN9PP 2-keto-3-deoxy-D-glycero-D-galacto-nononate 9-phosphate phosphatase
KDO 3-deoxy-D-manno-octulosonic acid
KDO8PP 3-deoxy-D-manno-octulosonate 8-phosphate phosphatase
Ll *Lactococcus lactis*
MDP1 magnesium-dependent phosphatase-1
Mj *Methanocaldococcus jannaschii*
Mm *Mus musculus*
Mt *Mycobacteria tuberculosis*
Neu5Ac 5-*N*-acetyl-β-D-neuraminate
Oc *Oryctolagus cuniculus*
PDB Protein Data Bank
PMM1 phosphomannomutase 1
PSP phosphoserine phosphatase
SPP sucrose-6-phosphate phosphatase
Ss *Synechocystis* sp. PCC 6803
St *Salmonella typhimurium*

Suc-6-P sucrose 6-phosphate
T6PP trehalose-6-phosphate phosphatase
T_m thermal transition midpoint

1. INTRODUCTION

Phosphohydrolases are vital to the cellular processes of signal transduction, regulation, and primary and secondary metabolism. Hence, the enzymes that catalyze phosphohydrolase reactions are ubiquitous and catalyze phosphoryl transfer from chemically varied molecules including proteins, nucleotides, carbohydrates, and lipids (Allen & Dunaway-Mariano, 2016). The necessity for Nature to have evolved such catalysts is underscored by the extraordinarily slow rate of uncatalyzed phosphate monoester hydrolysis at $2 \times 10^{-20}\ s^{-1}$ (Lad, Williams, & Wolfenden, 2003). Therefore, understanding the catalytic and chemical mechanisms of phosphoryl transfer is of significant interest. To lend mechanistic insight, a primary approach is the use of X-ray crystallography to visualize the Michaelis complex or its analogs (including product complexes) to provide the active-site residue positions relative to bound substrate. Insights from such "ground-state" complexes are often extended by obtaining intermediate/transition-state analog complexes to visualize interactions integral to catalysis. Using these approaches together with evaluation of steady-state kinetic constants and site-directed mutagenesis provides significant insight into the contribution of enzyme residues to chemical steps, to substrate binding, and to stabilization of the transition state. Additionally, identifying substrate specificity determinants can aid inhibitor design for therapeutics. Herein, we discuss the design and technical aspects of experiments to determine the liganded structures of phosphatases with substrate, substrate mimics, intermediate and transition-state analogs, products, and inhibitors.

Throughout this chapter, the haloalkanoate dehalogenase superfamily (HADSF) will be used as a model system for phosphohydrolases. The HADSF is the largest family of phosphatases consisting of >150,000 members from all domains of life and is comprised mostly of enzymes that perform reactions at phosphorus (99% of annotated members) (Allen & Dunaway-Mariano, 2004; Burroughs, Allen, Dunaway-Mariano, & Aravind, 2006). The many liganded structures deposited in the Protein Data Bank (PDB) (>110 unique liganded structures at the time of this writing)

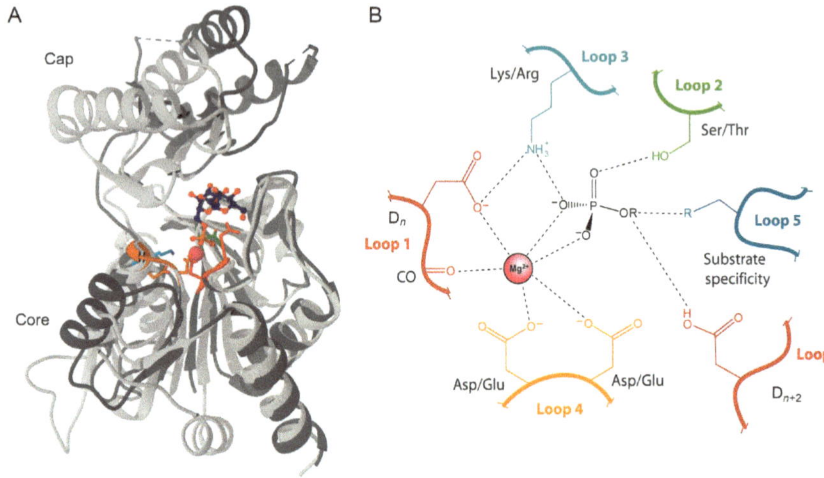

Fig. 1 (A) Ribbon diagram of trehalose-6-phosphate phosphatase in the unliganded, open conformation, PDB ID: 4OFZ (*dark gray*), and in the substrate-bound, closed conformation, PDB ID: 5DX9 (*light gray*). Magnesium is shown as a *pink sphere* and substrate shown as *ball* and *stick*. The active-site loops 1–4 are shown as *sticks* colored as in panel B. (B) The common active site of HADSF members. Note that the substrate specificity loop 5 is a simplification to denote one or more secondary structural elements that provide residues to bind the substrate-leaving group.

make it an ideal superfamily for describing a variety of methods for obtaining enzyme–ligand complexes via crystallography. HADSF members are comprised of a Rossmann-fold containing core domain and an accessory cap domain, which closes onto the core, completing the solvent excluded active site at the interface between cap and core (Fig. 1A) (Burroughs et al., 2006). The major cap types (C0, C1, C2a/C2b) are named by the location of the cap insertion into the Rossmann core domain and by the size and secondary structure topology of the domain. Irrespective of cap type, the active site is found at the interface of the core and cap domain. The cap domain and its dynamics with respect to the core typically determine substrate specificity/promiscuity (Allen & Dunaway-Mariano, 2009; Pandya, Farelli, Dunaway-Mariano, & Allen, 2014).

The mechanism of catalysis in the HADSF phosphatases involves a divalent magnesium ion cofactor, nucleophilic aspartate (D_n), and a general acid/base aspartate located two residues toward the C-terminus in the primary structure (D_{n+2}) (Fig. 1B). The reaction scheme, shown in Fig. 2, proceeds through a trigonal bipyramidal phosphoenzyme intermediate/transition state (Lahiri, 2003; Lu, Dunaway-Mariano, & Allen, 2008; Wang, Kim, Jancarik, Yokota, & Kim, 2001). The substrate undergoes a nucleophilic attack by aspartate, D_n, cleavage of the bond to the substrate-leaving group,

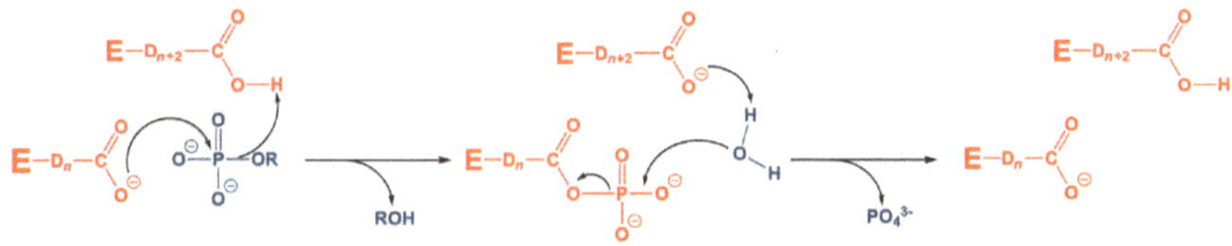

Fig. 2 General mechanism of HADSF members. The phosphoryl transfer occurs in two steps with a stable phosphoaspartyl intermediate.

and subsequent formation of the phosphoaspartyl intermediate. The second half reaction comprises hydrolysis of the phosphoenzyme and regeneration of the nucleophilic aspartate, releasing inorganic phosphate as a second product (Collet, 1998). Each step along the reaction coordinate (or its mimic) can be analyzed structurally via well-designed crystallographic studies. In this chapter, we describe methods to determine the structure of enzyme complexes revealing the ground-state and transition-state geometries using substrates, substrate analogs, products, phosphoryl mimics, and inhibitors (examples discussed listed in Table 1).

Ligands can be introduced to proteins prior to crystal formation, via endogenous ligands carried through purification or exogenously introduced by cocrystallization. Alternatively, ligand addition after protein crystals have formed is accomplished via soaking. It is important to consider the benefits and drawbacks of each method for every protein system as the approaches vary depending on the HADSF member. Cocrystallization is advantageous when the protein requires a significant conformational change for productive ligand binding, which is common among HADSF members. Ligand binding with induced fit often leads to protein stabilization, which has the added benefit of promoting crystallization; however, cocrystallization often requires extensive optimization for each ligand. As soaking requires a preformed crystal, the crystal contacts must allow the protein to adopt a conformer amenable to ligand binding, and the ligand size (usually <500 Da) must allow diffusion through the solvent channels to the binding site. In general, soaking is enabled by the high solvent content in protein crystals (at present 86% of protein X-ray crystallographic structures in the PDB are 35%–65% solvent). Thus, if the crystal lattice can allow a productive conformer, and if ligands are small and soluble, then protein–ligand complexes can be determined via soaking without laborious optimization of the crystallization condition. The soaking method can lead to rapidly obtaining structures of numerous enzyme–ligand complexes.

In both soaking and cocrystallization methods, the concentration of the ligand must be optimized to have >90% of the protein molecules bound as crystallographic electron density is obtained from averaging all

Table 1 Examples of Liganded Structures From the HADSF

Section	Species	Protein	Ligand	Cation	PDB ID	Method	Reference (First Author, Year)
2.1	*Ec*	GmhB	D-Glycero-D-manno-heptose 1,7-bisphosphate	Mg^{2+}	3L8G	Soak	Nguyen (2010)
	Hs	PMM1	D-Mannose 1-phosphate	Mg^{2+}	2FUE	Soak	Silvaggi (2006)
	Mt	T6PP	Trehalose 6-phosphate	Mg^{2+}		Soak	Shan (2016)
2.3	*Mm*	cdN	dGMP	Mg^{2+}	2JAO	Soak	Walldén (2007)
	Mm	cdN	dUMP	Mg^{2+}	2JAR	Soak	Walldén (2007)
	Bc	Phosphonatase	PALD	Mg^{2+}	1SWW	Soak	Zhang (2004)
	Mj	PSP	Phosphoserine	—	1L7P	Cocrystal	Wang (2002)
	Cn	T6PP	Trehalose 6-phosphate	Mg^{2+}	5DX9	Cocrystal	Miao (2016)
2.4	*Hi*	KDO8PP	Sulfate	Co^{2+}	1K1E	Cocrystal	Parsons (2002)
	Mm	MDP-1	Tungstate	Mg^{2+}	1U7P	Soak	Peisach (2004)
	Hs	PSP	—	Ca^{2+}	1NNL	Cocrystal	Peeraer (2003)
	Ss	SPP	Suc-6-P	—	1U2T	Soak	Fieulaine (2005)
2.5	*Ec*, *Bb*	GmhB	Phosphate	Mg^{2+}	3L8F, 3L8H	Soak	Nguyen (2010)
	Mj	PSP	Phosphate	Mg^{2+}	1L7M	Cocrystal	Wang (2002)
3.1	*Ll*	β-PGM	Glucose-1,6-bisphosphorane	Mg^{2+}	1O08, 1O03	Cocrystal	Lahiri (2003)

3.2	*Bt*	BT1666	Sulfate	Mg^{2+}	3R4C	Cocrystal	Lu (2011)
	Bt	BT4131	Sulfate	Mg^{2+}	1YMQ	Cocrystal	Lu (2005)
	Oc	Ca^{2+} ATPase	Beryllium trifluoride	Mg^{2+}	2ZBE	Cocrystal	Toyoshima (2007)
	Oc	Ca^{2+} ATPase	Tetrafluoromagnesate, ADP, thapsigargin	Mg^{2+}	1WPG	Cocrystal	Toyoshima (2004)
	Mj	PSP	Beryllium trifluoride	Mg^{2+}	1J97	Cocrystal	Cho (2001)
	Bt	Pyrophosphatase	Sulfate	—	3QUB, 3QUC	Cocrystal	Huang et al. (2011)
3.3	*Bt*	BT4131	Tungstate	Mg^{2+}	2RB5	Cocrystal	Lu (2008)
	Bt	BT4131	Vanadate	Mg^{2+}	2RBK	Cocrystal	Lu (2008)
	Bt	BT4131 (D10A)	Vanadate	Mg^{2+}	2RAR	Cocrystal	Lu (2008)
	Oc	Ca^{2+} ATPase	AlF_x, ADP	Mg^{2+}	2ZBD	Cocrystal	Toyoshima (2004)
	Bt	KDN9PP	Vanadate, KDN	Mg^{2+}	4HGO	Soak	Daughtry (2013)
	Bt	KDN9PP	Vanadate, Neu5Ac	Mg^{2+}	3E81	Soak	Lu (2009)
	Hi	KDO8PP	Vanadate, KDO	Mg^{2+}	4HGP	Soak	Daughtry (2013)
4.1	*Bc*	Phosphonatase	Vinyl sulfonate	Mg^{2+}	1RQL	Cocrystal	Morais (2004)
	Hs	PSP	2-Amino-3-phosphonopropionic acid	—	1L8L	Cocrystal	Kim (2002)
	Ss	SPP	Cellobiose	Mg^{2+}	2B1R	Soak	Fieulaine (2007)
	Ss	SPP	Maltose	Mg^{2+}	2D2V	Soak	Fieulaine (2007)
	Ss	SPP	Trehalose	Mg^{2+}	2B1Q	Soak	Fieulaine (2007)
	Ll	β-PGM	Gal-1-P	Mg^{2+}	1Z4O	Cocrystal	Tremblay (2005)

molecules exposed to the X-ray source. We suggest a protein:ligand ratio of 1:1.2 to supply excess ligand while preventing nonspecific ligand binding. Nonspecific binding can cause heterogeneity, disrupt crystal contacts and prevent crystallization, or degrade crystal quality and lead to a decrease in resolution. Additionally, considerations for the binding affinity should be made to ensure homogeneity; the ligand concentration should ideally be 5–10 times greater than the dissociation constant (K_D). Practically, the Michaelis constant (K_M), inhibition constant (K_I), and the half-maximal inhibitory concentration (IC_{50}) can be used as estimates for K_D. Both soaking and cocrystallization approaches have proven successful for members of the HADSF. Considerations for experimental design, reagents, and methods are described in this chapter.

2. SUBSTRATE AND PRODUCT COMPLEXES

2.1 Substrate Soaking

Optimally, a crystal structure of the enzyme–substrate complex in the catalytically competent conformation with the residues aligned for catalysis (i.e., the Michaelis complex) is desired. Unfortunately, using standard single-crystal macromolecular X-ray crystallographic approaches, data collection is not rapid enough to capture this complex prior to catalysis. Therefore, catalysis occurs and the product(s) either diffuse out of the active site, leading to an unliganded structure, or remain bound resulting in determination of the enzyme–product complex. However, based on the equilibrium between substrate and product(s), the enzyme conformation, and the rate of turnover *in crystallo*, catalysis may not occur or occur on a timescale slow enough to trap a Michaelis-like complex. Hence, we suggest attempting to soak crystals with the substrate. The resulting data must be closely analyzed to determine which species is bound: the substrate, the product(s), buffers, salts, or precipitants from the crystallization milieu, solvent, or any combination thereof. The population of any one bound species must be greater than 10%–20% of the ensemble (depending on the resolution and ligand motion); otherwise, the corresponding electron density may be weak or unobservable (Rupp, 2010). Notably, the occupancies and B-factors of solvent/ligands are correlated at the resolutions commonly observed for protein complexes (~2.5 Å) (Bhat, 1989). If multiple species are bound, all ligands should be modeled and B-factors and occupancies refined for appropriate interpretation of the data (for an example of refinement of multiple species, see Stiers & Beamer, 2018).

Alternative to soaking with substrate alone, shifting the equilibrium toward substrate by the addition of products can be effective. This approach was applied in determining the complex of D-glycero-D-manno-heptose-1,7-bisphosphate phosphatase (GmhB) from *Escherichia coli* (*Ec*) with D-glycero-D-manno-heptose 1,7-bisphosphate (PDB ID: 3L8G). The crystals were soaked with a 1:1 ratio of substrate to product, shifting the equilibrium toward bound substrate (Nguyen et al., 2010).

An additional aspect in substrate soaking exists, in that flexible enzymes may crystallize in a conformation that is not catalytically competent. In the HADSF, the cap domain typically closes over the core, providing a solvent exclusive environment and maintaining electrostatic control of the active site to promote catalysis (Lahiri, Zhang, Dunaway-Mariano, & Allen, 2002; Lu, Dunaway-Mariano, & Allen, 2005; Wang et al., 2002; Zhang et al., 2005). Thus, if the enzyme crystallizes in the cap-open conformer, soaked substrate may bind to either the core or cap domain alone precluding identification of all binding and catalytic determinants, as is the case for *Mycobacterium tuberculosis* (*Mt*), trehalose-6-phosphate phosphatase (T6PP) (Shan, Min, Liu, Jiang, & Rao, 2016), and *Homo sapiens* (*Hs*) phosphomannomutase 1 (PMM1) (PDB ID: 2FUE) (Silvaggi et al., 2006). As the *Mt*T6PP structure shows interactions of the substrate with only the core domain, interaction with catalytic residues (D_n and D_{n+2}) and a subset of substrate-binding determinants (for the most part, those involved in binding the phosphoryl group) can be identified. In contrast, D-mannose 1-phosphate is bound solely by residues of the cap domain of *Hs*PMM1, and therefore, only interactions with the substrate-leaving group are identified.

In these two examples, the large conformational change necessary for turnover is precluded by crystal contacts. However, it is possible that even in a domain–domain closed conformation, small, but necessary, changes to reach the transition state are prevented by decreased mobility in the crystalline form, causing the catalytic rate of the enzyme to be lower than that of enzyme in solution. For example, the turnover rate (k_{cat}) of aldolase is two times slower *in crystallo* than in solution (Sygusch & Beaudry, 1984). Thus, a short soak (1–5 min) prior to flash freezing could yield a substrate-bound structure for an enzyme that is slow ($k_{cat} < 1\,s^{-1}$) with a natural substrate or a poor substrate ($k_{cat}/K_M < 10^3\,M^{-1}\,s^{-1}$). The success of a quick soak depends on three characteristics of the crystalline enzyme: the viscosity of the mother liquor, the crystal morphology, and the k_{cat} of the enzyme. The viscosity of the mother liquor determines the diffusion rate of the ligand. For example, a system with a viscous mother liquor or soaking solution and a larger ligand will have a slower diffusion rate. The crystal morphology also determines the rate at which a ligand fully diffuses into a crystal. Small, thin crystals (plates, microcrystals, etc.) with high surface-to-volume ratios equilibrate with the substrate faster than large, thick crystals (cubic, pyramidal, etc.). This rate has been estimated in our laboratory empirically by measuring the time for a Coomassie-based or other dye (e.g., Izit Dye (Hampton Research)) to color the crystal. Alternatively, the time for diffusion into the crystal can be calculated using the binding affinity of the ligand for the enzyme and physical and chemical properties of the protein crystal and ligand (Geremia, Campagnolo, Demitri, & Johnson, 2006). Once the crystal is cryocooled, it is assumed that the catalytic reaction is terminated. Experimental trial and error is necessary to optimize the amount of time the substrate is soaked with the crystal.

A structure of the Michaelis-like complex may be obtained by simply soaking a preformed crystal with substrate. But, one must consider the species present at equilibrium, the conformation of enzyme *in crystallo*, and soaking time. Section 2.2 provides experimental procedures for ligand soaking.

2.2 Ligand-Soaking Experiments

2.2.1 Equipment

- Crystal trays (i.e., 24 well Hampton Research VDX plate)
- Siliconized cover slips (Hampton Research)
- Stereo-Microscope (e.g., Olympus SZX12)
- CryoLoops™ (Hampton Research)

2.2.2 Buffers and Reagents

- Pregrown crystals (>0.7 mm on a side tend to be more fragile than smaller crystals)
- Components of crystallization solution
- Ligand stock solution (as concentrated as allowed by solubility)
- Cryoprotectant

2.2.3 Procedure

1. Select crystals for soaking.
2. Create a soaking solution of mother liquor with at least a 5%–20% increase in precipitant (i.e., polyethylene glycol (PEG), ammonium sulfate, etc.) and ligand of interest at a concentration such that a final

molar ratio of 1:1.2 protein:ligand (using the total concentration of protein in the drop) or 10 × K_D is reached (whichever is larger).

3. Carefully transfer crystals from original drop to soaking drop using a loop large enough to allow for collection of multiple crystals (if applicable), seal the well, observe under a microscope for any immediate crystal cracking/decay. Visible signs of crystal decay include crystal dissolution, precipitation, phase separation, and crystal cracking. Occasionally, crystals will appear solid, but upon disturbance, they crack and dissolve. Additionally, crystals may lose diffraction upon soaking.
4. Allow for crystal to equilibrate (soak) in ligand-containing drop prior to cryoprotection and cryocooling for data collection.

2.2.4 Notes and Examples

- The soaking drop size can vary between 2 and 20 μL and is generally larger than the crystallization drop. We prefer a 5 μL soaking drop (increased with respect to a typical 2 μL crystallization drop). One advantage to a larger drop is that it will not dry out as quickly under the microscope during manipulation for crystal mounting.
- The suggested concentration of ligand for soaking is a minimum and greater concentrations may be necessary and should be empirically determined.
- Initial soak time is typically determined as described in Section 2.1 then optimized via trial and error depending on resulting diffraction quality and achievement of the desired bound complex.
- If possible, it is best to introduce the cryoprotectant in the soaking solution to decrease the number of crystal transfers and damage to the crystal. This can be accomplished by adding glycerol, ethylene glycol, or other cryoprotectants to the mother liquor solution (Pflugrath, 2015).
- To shift the equilibrium toward bound substrate Nguyen and colleagues added product to a soaking solution (e.g., the GmhB structure (see Section 2.1)) (Nguyen et al., 2010). In this case the crystal was transferred to a drop of well solution containing an additional 5% PEG3350 and equal concentrations of substrate and product (30 μ*M*).

2.3 Mutation of the Catalytic Residue(s)

Point mutations are utilized in crystallography to decrease surface entropy, add crystal contacts, and introduce residues necessary for experimental phasing (Goldschmidt, Cooper, Derewenda, & Eisenberg, 2007; Lawson et al., 1991; Leahy, Erickson, Aukhil, Joshi, & Hendrickson, 1994).

Mutating an amino acid essential to enzyme catalysis but not to affinity oblates activity but generates a substrate-binding protein. In the HADSF, the D_n is often mutated to obtain substrate-bound crystal structures (Miao et al., 2016; Walldén et al., 2007; Wang et al., 2002; Zhang et al., 2004). Point mutations have varying effects on the production, stability, and crystallization of the resulting protein. Mutating residues on the solvent-exposed surface of the protein may disrupt or promote crystal contacts, while changing amino acids on the protein interior can affect the structural stability. It has been shown that replacing a polar residue with a nonpolar equivalent in the enzyme active site can stabilize the protein (Loladze, Ermolenko, & Makhatadze, 2001) and consequently lead to higher quality crystals and enhanced resolution. As the active site of the HADSF is at the cap–core interface, the residues can be either solvent exposed or solvent excluded. Therefore, mutations at the active site can shift the equilibrium between the open and closed conformers.

Taking into account these considerations, conservative mutation at the active site is recommended; for instance, a mutation of the nucleophilic Asp (D_n) to Asn has most commonly resulted in stabilization of the Michaelis complex for HADSF members (Miao et al., 2016; Walldén et al., 2007; Wang et al., 2002). Even with this conservative approach, results from mutation of the active site can vary. For instance, in the structure determination of *Mus musculus* (*Mm*) cytosolic 5′(3′)-deoxyribonucleosidase (cdN), the nucleophilic Asp (D_n) to Asn mutation had little effect on the stability of the enzyme (PDB IDs: 2JAR, 2JAO) (Walldén et al., 2007). Variant protein was expressed, purified, and crystallized under the same conditions as wild type, and then simply soaked with substrate. However, cocrystallization of the D_n11N variant of *Methanocaldococcus jannaschii* (*Mj*) phosphoserine phosphatase (PSP) with phosphoserine (PDB ID: 1L7P) required optimization of the crystal conditions (Wang et al., 2002). After successfully determining the structure of wild-type *Candida albicans* (*Ca*) T6PP bound to a substrate analog (PDB ID: 5DXI), cocrystallization of the D_n559N variant with substrate yielded no crystals. A structure with substrate bound was instead determined for the D_n704N variant of T6PP from *Cryptococcus neoformans* (*Cn*) (PDB ID: 5DX9) (Miao et al., 2016).

Although Asp (D_n) to Asn mutation has proven successful for HADSF enzymes, a different tactic was taken in the case of *Bacillus cereus* (*Bc*) phosphonatase, where the nucleophilic Asp was mutated to Ala. Whereas the wild-type structure was determined to 3.0 Å resolution (PDB ID: 1FEZ) (Morais et al., 2000), this mutation afforded an unliganded structure with increased resolution (2.3 Å) (PDB ID: 1SWV), as well as a substrate-bound (phosphonoacetaldehyde) structure refined to a resolution of 2.55 Å (PDB ID: 1SWW) (Zhang et al., 2004). It should be noted that the absence of the D_n side chain in this position caused the substrate phosphoryl group to move into the position usually occupied by the aspartyl carboxylate.

2.4 Exchange Metals

In phosphoryl transfer reactions the magnesium ion acts to stabilize both the leaving group of the substrate and the oxygens of the phosphoryl group in the transition state (Herschlag & Jencks, 1989) and indeed, HADSF members that catalyze phosphoryl transfer have an absolute requirement for a divalent magnesium ion cofactor. Thus, when magnesium ion is replaced with another divalent metal, activity is either abolished as is the case with calcium ion or significantly decreased as is the case for cobalt ion and some other period 4 elements (Peeraer et al., 2003). In systems with metal-dependent catalysis, it is possible to exchange the metal to prevent turnover and capture substrate-bound structures. Furthermore, in the HADSF, the catalytic magnesium ion is not typically essential for proper folding and stability, so it may be removed altogether.

One method to allow for calcium or cobalt ion binding in place of magnesium ion is simply to increase the concentration of the divalent metal chloride in the mother liquor solution, as is the case in *Hs*PSP (PDB ID: 1NNL) (Peeraer et al., 2003). Similarly, a structure of *Haemophilus influenzae* (*Hi*) 3-deoxy-D-manno-octulosonate 8-phosphate phosphatase (KDO8PP) was determined with cobalt ion in place of magnesium ion by omitting magnesium from the crystallization buffer and adding 10 m*M* cobalt (II) chloride to the drop solution (PDB ID: 1K1E) (Parsons et al., 2002). Although these metal swaps were not performed to obtain substrate-bound structures, the procedures show a valid means of replacing the catalytic magnesium with an alternative divalent metal. Moreover, the Ca^{2+}-substituted structure of human PSP provides evidence that the loss of activity with calcium ion is due to the differing coordinate bonds between the calcium and magnesium ions. The calcium ion prevents the placement of phosphoserine in a catalytically competent orientation (Peeraer et al., 2003).

Furthermore, substrate-bound structures may be determined by removing the magnesium from the active site by chelation via soaking of the crystal in ethylenediaminetetraacetic acid (EDTA) and substrate. This is exemplified by the structure of *Synechocystis* sp. PCC 6803 (*Ss*) sucrose phosphatase (SPP), wherein the substrate, sucrose 6-phosphate (Suc-6-P), was bound

(PDB ID: 1U2T) in an orientation that is consistent with a Michaelis-like complex but electron density consistent with that of a water molecule was observed in the binding site for the magnesium cofactor. Thus, in the absence of magnesium, it is expected that water binds in the metal-binding site, affording hydrogen bonds in place of coordinate bonds to the phosphoryl group of the substrate. Notably, *Ss*SPP crystals grown under the same condition and soaked in a solution containing Suc-6-P (no EDTA) allows for turnover *in crystallo* of substrate and produces a structure of SPP with sucrose and phosphate bound (PDB ID: 1TJ5) (Fieulaine, Lunn, Borel, & Ferrer, 2005). This confirms that removal of magnesium via chelation by EDTA *in crystallo* may not destroy the crystal but can prohibit turnover, enabling comparison of substrate and product complexes using the same crystal form.

The affinity of the enzyme for its metal cofactor should be considered as it varies substantially in the HADSF. In some members, such as *Bc* phosphonatase, the protein is destabilized and unfolds upon magnesium removal (Zhang et al., 2004). In contrast, *Salmonella typhimurium* (*St*) T6PP has a low affinity for magnesium; soaking with 50 m*M* magnesium chloride prior to cryocooling does not provide full occupancy of the magnesium ion in the active site (unpublished observation). Lastly, magnesium salts prevent the crystallization of *Mm* magnesium-dependent phosphatase-1 (MDP-1). Even soaking crystals in magnesium alone did not allow for cofactor binding. However, *Mm*MDP-1 crystals grown in the absence of magnesium and then soaked in a solution containing both sodium tungstate (as a phosphate analog) and magnesium resulted in full occupancy of the magnesium- and phosphate-binding sites (PDB ID: 1U7P) (Peisach, Selengut, Dunaway-Mariano, & Allen, 2004). The tungstate itself provides one coordinate bond to aid in completion of the magnesium ion coordination sphere, synergizing tungstate and magnesium-ion binding.

The metal cofactor of the HADSF may be removed or replaced by another divalent metal before or after crystallization to generate a non-catalytic enzyme. As the affinity for the magnesium ion is different for each member of the superfamily, perturbation of the ion will have varied effects.

2.5 Phosphate Binding

The location, coordination state, and geometry of phosphate in the active site of phosphatases act as a bridge between substrate-bound and product-bound complexes. Oftentimes, these structures are a result of in-crystal hydrolysis of soaked substrate. In the case of *Bordetella bronchiseptica* (*Bb*) GmhB and *Ec*GmhB, the alternate substrates imidodiphosphate and fructose-1,6-bisphosphate were soaked in crystals for 5 min prior to cryocooling. The ligand density in both structures clearly shows only phosphate bound in the phosphoryl transfer site of the active site (PDB IDs: 3L8F and 3L8H, respectively) (Nguyen et al., 2010).

As phosphate is such a common precipitant in crystallization buffers, it is possible for phosphate to cocrystallize with the enzyme. This is observed in the structure of *Mj*PSP (PDB ID: 1L7M) with phosphate bound (Wang et al., 2002). In this case, phosphate was not added to the protein or soaked into the crystal, but the crystallization conditions included 0.5 *M* sodium phosphate dihydrate. These phosphate-bound structures corroborate the phosphoryl-binding site and provide coordinate bonds to magnesium allowing for identification of the active site. The position of phosphate can also act as a valuable "anchor" for computational positioning of substrates and phosphoryl- or phosphoryl mimic-bearing ligands via superposition.

3. GROUND-STATE, INTERMEDIATE, AND TRANSITION-STATE MIMIC COMPLEXES

Using structural information on the Michaelis complex, transition-state mimics, and intermediates, the overall mechanism of catalysis and determinants of ligand binding can be better understood to guide efforts in rational inhibitor design as therapeutics and as probes in chemical biology.

In phosphatases, this is typically accomplished using phosphate analogs to mimic either the ground state (tetrahedral geometry) or transition state (trigonal bipyramidal geometry) (Fig. 3). Vanadate and tungstate, the most commonly used phosphate analogs, may adopt either the tetrahedral or trigonal bipyramidal geometries, because they are more flexible in the number of coordinate bonds in which they participate (McLauchlan, Peters, Willsky, & Crans, 2015). Orthovanadate (VO_4^{3-}) and orthotungstate (WO_4^{2-}) are tetrahedral and mimic the ground-state geometry. In HADSF members, the ground state can be considered a substrate-bound, product-bound, or phosphoenzyme complex. We include the phosphoenzyme as ground state because of the relative stability, varying between family members but exemplified by the fact that the phosphoenzyme can be isolated (Collet, Gerin, Rider, Veiga-da-Cunha, & Van Schaftingen, 1997). Metavanadate (VO_3^{-}), in the absence of additional coordinate bonds from

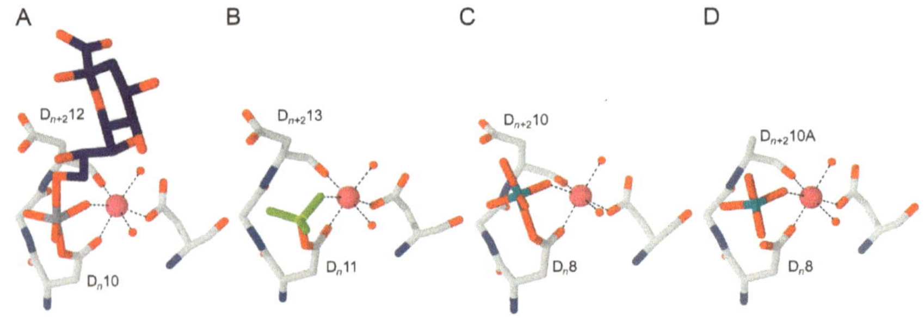

Fig. 3 Active sites of HADSF members KDN9PP, PSP, and BT4131 (PDB IDs: (A) 4HGO; (B) 1J97; (C) 2RB5; (D) 2RAV) shown as *gray sticks* with phosphoryl mimics representing the transition and ground states. Magnesium ion is shown as *pink sphere*, water molecules as *red spheres*, and coordinate bonds to the magnesium ion as *dashed lines*. (A) The trigonal bipyramidal transition-state mimic of the first half reaction. The apical positions of vanadate are occupied by the oxygens of KDN (*navy sticks*) and D_n. (B) BeF_3, forming a coordinate bond with D_n acts as a mimic of the phosphoaspartyl intermediate. (C) A trigonal bipyramidal geometry formed by tungstate with the apical positions occupied by D_n and a water molecule illustrating the second half reaction, hydrolysis of the phosphoaspartyl intermediate. (D) Tetrahedral tungstate-bound structure of the $D_{n+2}10A$ variant mimicking the product-bound complex, wherein D_n is not coordinated to tungstate.

the enzyme or waters, is trigonal planar. Pentacoordinate forms of tungstate and vanadate exhibit trigonal bipyramidal geometries, mimicking the transition state and include coordinate bonds donated by substrate, water, and/or oxygen from the nucleophilic aspartate, D_n (Daughtry et al., 2013; Lu et al., 2008; Lu, Wang, Dunaway-Mariano, & Allen, 2009; Toyoshima, Nomura, & Tsuda, 2004). As pentacoordinated tungstate is less frequently observed in complex with phosphatases (Lu et al., 2008; Peck, Sunden, Andrews, Pande, & Herschlag, 2016), the versatility in coordinate-bond formation makes vanadate most appealing for mimicking transition-state geometry (McLauchlan et al., 2015). Sulfate and some metal fluoride complexes, such as aluminum trifluoride (AlF_3) and beryllium trifluoride (BeF_3^-), mimic the ground-state geometries; therefore, when added together with the leaving group they can simulate enzyme–substrate complexes (Cho et al., 2001; Toyoshima et al., 2004; Toyoshima, Norimatsu, Iwasawa, Tsuda, & Ogawa, 2007). Notably BeF_3^- with one coordinate bond from D_n resembles the phosphoenzyme. Here we describe how the use of phosphate mimics in the presence and absence of the leaving group product can serve as ground-state, intermediate, and transition-state complex analogs for phosphatases in the HADSF.

3.1 Intermediate-Bound Complexes

In the example of *Lactococcus lactis* (Ll) β-phosphoglucomutase (β-PGM), Lahiri was able to successfully observe the pentavalent bisphosphorane intermediate (PDB ID: 1O03 and 1O08). This was accomplished through equilibration of crystals with magnesium and either glucose 1-phosphate (Glc-1-P) or glucose 6-phosphate (Glc-6-P) (Lahiri, 2003). In mutases, isomerases, and racemases (enzymes with a single substrate and single product), equilibration of a crystal or cocrystallization with either substrate or product at concentrations greater than the K_M results in observation of the species that predominates at equilibrium (Lavie, Allen, Petsko, & Ringe, 1994). The structure clearly shows the apical groups belonging to the nucleophilic oxygen of aspartate—D_n8 and the oxygen of the substrate-leaving group (Lahiri, 2003). By gaining insight into intermediate-bound structures, such as the bisphosphorane, one can discern important contributions to catalysis. For instance, the position of the D_{n+2} that forms a hydrogen bond to the bridging oxygen in this structure supported its identification as the general acid–base and highlighted the change in conformation the residues undergo between the intermediate/transition state and unliganded structures. This conformational change is a key component of the mechanism by which the enzyme discriminates between substrate hexose sugar and water, allowing mutase over phosphatase activity (Dai et al., 2009).

3.2 Ground-State Mimic Complexes

Phosphoryl mimics in the absence of additional ligands in the active site resemble the ground state, by taking on a tetrahedral geometry. From these structures, one can identify binding determinants for the phosphate moiety. These are usually tetravalent complexes of any of the phosphoryl mimics discussed in Section 3.

For the HADSF sulfate-bound structures deposited in the PDB at the time of this writing, the sulfate ligand was introduced from the crystallization milieu (for example, PDB IDs: 3R4C, 3QUB, 3QUC, and 1YMQ). These structures allow for identification of the phosphoryl-binding site and the relative placement of catalytic residues.

The fluorophosphoryl mimics, like beryllium fluorides, are exemplified in structures of *Mj*PSP (Cho et al., 2001) and calcium ATPase from *Oryctolagus cuniculus* (*Oc*) (PDB ID: 2ZBE) (Toyoshima et al., 2004, 2007). To identify the necessary ratios to form BeF_3^-, Cho et al. observed BeF_3^- complex formation using $^1H–^{15}N$ FHSQC spectra. Sodium fluoride, beryllium chloride, and magnesium chloride at 54 m*M*, 10.8 m*M*, and 90 m*M*, respectively, were added to *Mj*PSP prior to crystallization. In this *Mj*PSP structure (PDB ID: 1J97), there is an oxygen–beryllium bond (1.55 Å) observed with the oxygen of the nucleophilic aspartate (D_n11) completing the tetrahedral geometry and mimicking the phosphoenzyme (Cho et al., 2001).

The *Oc* calcium ATPase bound to tetrafluoromagnesate (MgF_4^{2-}) (PDB ID: 1WPG) mimics the phosphate-bound product complex with the MgF_4^{2-} species exhibiting tetrahedral geometry (Toyoshima et al., 2004). This geometry affords insight into the conformation of enzyme during the second half reaction prior to product release.

3.3 Transition-State Mimic Complexes

As described in Section 3, the choice of phosphate mimic can dictate whether the ground-state (tetrahedral) or transition-state (trigonal bipyramidal) geometry is captured. Using tungstate or vanadate as the phosphoryl mimic is advantageous in obtaining a transition-state-like complex as they can make more than four coordinate bonds. Structures of bacterial homologs, *Bacteroides thetaiotaomicron* (*Bt*) 2-keto-3-deoxy-D-glycero-D-galacto-nononate 9-phosphate phosphatase (KDN9PP) (PDB ID: 4HGO) and *Hi*KDO8PP (PDB ID: 4HGP), in the presence of vanadate and product 2-keto-3-deoxy-D-glycero-D-galacto-nononic acid (KDN) or 3-deoxy-D-manno-octulosonic acid (KDO), respectively, adopt transition-state-like geometry through vanadium participating in five coordinate bonds. The equatorial positions are occupied by three oxygens coordinated to vanadium, and the carboxylate of the nucleophilic aspartate and the C9O (KDN) or C8O (KDO) of the substrate are in the apical positions, completing the trigonal bipyramidal geometry. This complex mimics the first step of phosphoryl transfer, in which the nucleophilic aspartate attacks the phosphoryl group of the substrate and provides further evidence of an in-line, backside displacement mechanism (Daughtry et al., 2013).

Notably, several crystals of *Bt*KDN9PP grown in a condition containing sodium acetate did not yield liganded structures. In these structures, acetate binds in the subsite of the active site where the carboxylate moiety of the sialic acid-leaving group of the substrate would bind. This problem was circumvented by soaking the crystals in well solution containing the ligands of interest with an increased concentration of PEG (5%) in the absence of acetate. This allows for the ligand of interest, in this case an alternate substrate, 5-*N*-acetyl-β-D-neuraminate (Neu5Ac), and vanadate, to outcompete any bound or remaining acetate, yielding a transition-state complex structure (PDB ID: 3E81) (Lu et al., 2009).

In the case of *Bt* hexose-phosphate phosphatase (BT4131), the trigonal bipyramidal geometry was observed in a structure with tungstate bound (PDB ID: 2RB5) as well as with vanadate bound (PDB ID: 2RBK). Three W–O coordinate bonds are in the equatorial positions, and the apical positions are occupied by the fourth W–O coordinate bond and the carboxylate oxygen of D_n8. Additionally, the apical oxygen is hydrogen bonding with the carboxylate of $D_{n+2}10$. This structure mimics the transition state of phosphoenzyme hydrolysis—the second half reaction of phosphoryl transfer. The same pentacoordinate geometry is assumed by vanadate in the BT4131 active site; these structures support the role of the enzyme active site in stabilizing this geometry along the normal reaction coordinate.

Notably, in a structure of a variant of BT4131 with the $D_{n+2}10$ replaced by alanine, VO_3^- is bound as a tetravalent complex with the D_n8 supplying the fourth coordinate bond (PDB ID: 2RAR). This complex displays distorted tetrahedral geometry in which the bond angles to the nonbridging oxygens are close to 90° (the VO_3^- moiety is close to planar) and the tetravalent D_n8-vanadate adduct is a mimic of the aspartyl phosphate intermediate and not the pentavalent trigonal bipyramidal transition state. This tetrahedral geometry illustrates the importance of the D_{n+2} in stabilizing the trigonal bipyramidal geometry, further supporting the general base mechanism of phosphoryl transfer in the HADSF (Lu et al., 2008).

Toyoshima et al. generated AlF_x, ADP, *Oc* calcium ATPase complexes by cocrystallizing with ADP and adding aluminum (III) chloride and sodium fluoride directly to the mother liquor prior to crystallization (PDB ID: 2ZBD). It was not possible to identify from the electron density whether AlF_3 or AlF_4^{2-} was present in the structure (Toyoshima et al., 2004). However, it is obvious that aluminum forms at least six coordinate bonds: the three (or four) equatorial positions are fulfilled by the fluoride ions, with carboxylate of D_n8 in one of the apical positions and an oxygen from ADP in the second apical position. This displays a trigonal (or tetragonal) bipyramidal geometry resembling the transition state (Toyoshima et al., 2004).

Whether the reaction mechanism is an elimination–addition (dissociative), addition–elimination (associative) or concerted, the intermediate/transition state takes on a trigonal bipyramidal geometry. Any transition-state/intermediate analog should mimic that geometry. Previously, it has been noted that although the complexes of HADSF members with AlF_x

and VO_4^{3-} provide evidence for the trigonal bipyramidal geometry of the transition state, they do not rigorously test the ability of the enzyme active site to play a role in the formation and stabilization of the trigonal bipyramidal complex (Lu et al., 2008). Tungstate provides a better probe as pentavalent trigonal bipyramidal geometry is rare for tungstate in the active site of phosphotransferases (Lima, Klein, & Hendrickson, 1997; Lu et al., 2008; Peck et al., 2016). Notably, in the case of BT4131, comparison of the structure of the enzyme-tungstate complex and the enzyme-vanadate complex reveals the same pentavalent trigonal bipyramidal geometry and stabilizing hydrogen-bond interactions. In BT4131, the ~130-fold lower affinity of tungstate to vanadate is considered to reflect the amount of binding energy to stabilize this higher valency state of tungstate (Lu et al., 2008). But the extent to which any of these complexes mimic the transition state cannot be determined solely by crystallographic structure determination or binding affinity to wild-type enzyme. Instead, the relationship between catalytic activity and affinity for a series of site-directed variants has been used to show mimicry of the transition state. Indeed, such correlations show that vanadate is not a true transition-state analog for protein–tyrosine phosphatase (Deng, Callender, Huang, & Zhang, 2002), and that tungstate is a mimic for the transition-state properties of alkaline phosphatase (Peck et al., 2016).

These complex structures together provide evidence useful for defining the catalytic and chemical mechanism of phosphohydrolase and phosphomutase activity of HADSF members. Coordination of the nucleophilic aspartate and the leaving group to vanadium completes the trigonal bipyramidal geometry and allows visualization of the transition-state analog complex of the first half reaction. Coordination of the nucleophilic aspartate and water to vanadium or tungsten allows visualization of the transition-state analog complex of the second half reaction. Rational inhibitor design should take advantage of the hydrogen bonds formed to the equatorial oxygens and oxygens bridging to the leaving group as these are expected to be optimized in the transition state. Tungstate ions are simply added as tungstic acid or as the sodium salt (Lu et al., 2008; Peisach et al., 2004), whereas procedures for preparation of vanadate solutions are detailed in Section 3.4.

3.4 Activation of Sodium Orthovanadate

The activation of vanadate for use in forming ground-state or transition-state analogs in kinetics is essential to preventing the polymerization into decavanadate polyanion. Sections 3.4.1–3.4.4 describe our method for activation of sodium orthovanadate at pH 10 to be used for crystallization purposes.

3.4.1 Equipment

- pH probe, calibrated
- Hot/stir plate

3.4.2 Buffers and Reagents

- 6 *M* hydrochloric acid
- 10 *M* sodium hydroxide
- Sodium orthovanadate (CAS 13721-39-6)

3.4.3 Procedure

1. Add sodium orthovanadate (3.6782 g) to Millipore water (20.0 mL, half final volume) in a heat-resistant container (e.g., a 100 mL pyrex bottle).
2. Add 6 *M* hydrochloric acid to reach pH 10 in 100–200 μL aliquots (solution will turn yellow with addition of acid due to the formation of decavanadic acid).
3. Cover container with aluminum foil and boil on a hot/stir plate until colorless (~15 min). Then, cool to room temperature.
4. Measure pH and add either 6 *M* hydrochloric acid or 10 *M* sodium hydroxide until the solution reaches pH 10.
5. Repeat steps 3–4 until the pH stabilizes at pH 10.
6. Add Millipore water to achieve the desired final volume (40 mL).

3.4.4 Notes

- Detailed protocol is for making a 40 mL, 500 m*M* sodium orthovanadate stock solution.
- 250–500 m*M* stocks are ideal for minimizing volume change due to the addition of ligand.
- Aliquots of ~1.5 mL in Eppendorf tubes can be stored at −20°C. Samples may withstand 2–3 freeze-thaws. If solution is no longer colorless (begins to turn yellow), then the sample should be discarded.

4. INHIBITORS

4.1 Substrate Mimics as Inhibitors

Another means of obtaining information about the residues comprising the substrate-binding site in phosphatases is to utilize a nonhydrolyzable substrate mimic. Because phosphoryl transfer is not possible, these molecules are likely to compete for the active site, and as such, affinity can be found through measurement of K_I. One option is to use the sulfonate- or sulfate-substituted version of the phospho-substrate. Trehalose 6-sulfate is a 180 μ*M* inhibitor of *St*T6PP (Liu, Dunaway-Mariano, & Mariano, 2017) and was used to mimic the Michaelis complex. This approach was utilized for *Bc* phosphonoacetaldehyde hydrolase, as well (see phosphonoacetaldehyde hydrolase–vinyl sulfonate complex, PDB ID: 1RQL) (Morais et al., 2004). Alternatively, for phosphatases, substituting the substrate phosphate group with a phosphonate group removes the leaving group and prevents turnover. For *Hs*PSP, the inhibitor 2-amino-3-phosphonopropionic acid which lacks the bridging oxygen to phosphorous was used as a substrate mimic with an IC_{50} of 1.4 m*M* (PDB ID: 1L8L) (Kim et al., 2002).

Generally, in the HADSF the leaving group does not bind tightly. However, in some cases inhibitors may resemble only parts of the original substrate. For sugar phosphatases, other sugars can be used as inhibitors. For instance, structures of *Ss*SPP were determined with the inhibitory disaccharides cellobiose, maltose, and trehalose, which have poor affinity with K_I values of 101, 100, and 26 m*M*, respectively (PDB IDs: 2BIR, 2D2V, and 2BIQ) (Fieulaine, Lunn, & Ferrer, 2007). Additionally, α-galactose 1-phosphate (Gal-1-P) acts as a substrate mimic of the phosphorylated enzyme or a 30 μ*M* inhibitor of *Ll*β-PGM. Gal-1-P was cocrystallized with *Ll*β-PGM to reveal structural insight into the overall mechanism of catalysis (PDB ID: 1Z4O) (Tremblay, Zhang, Dai, Dunaway-Mariano, & Allen, 2005).

Fortunately, most mimics of phosphatase substrates will be polar and therefore soluble in water. However, the ultimate goal of many projects is small molecule drug design that requires molecules to permeate the cell membrane. Cell-permeable drugs are nonpolar and rarely charged (Lipinski, Lombardo, Dominy, & Feeney, 2001); thus, they are generally not freely soluble in water. As crystallization requires high concentrations of protein (generally ~300 μ*M*), high concentrations of inhibitors in crystallization buffer are necessary to achieve equimolar solutions. In Section 4.2, we describe methods to test protein and crystal stability in the presence of solvents and inhibitors and strategies to increase stability, if necessary.

4.2 Protein and Crystal Stability in the Presence of Inhibitors/Solvents

4.2.1 Determining Protein Stability

4.2.1.1 Equipment

- White PCR strips or plates
- Real-time thermocycler

4.2.1.2 Buffers and Reagents

- Solvents of interest
- Protein (10 μL of 20 μ*M* per sample)
- SYPRO® Orange Protein Dye (5000 ×, Thermo-Fisher, catalog number S-6650)

4.2.1.3 Procedure

1. Prepare solvent of interest in protein buffer. Example: Prepare 2.5%, 5%, 10%, and 20% dimethyl sulfoxide (DMSO) solutions in buffer containing HEPES pH 7.0 and 50 m*M* NaCl. Add 10 μL of solvent mix to a well in a white PCR plate. Repeat for each solvent and concentration to be tested. Use 10 μL of protein buffer as a control.
2. Add SYPRO® Orange dye to 20 μ*M* protein solution at 1:500 dilution.
3. Add 10 μL protein with SYPRO® Orange dye to each solvent mix.
4. Cover wells with clear plastic.
5. Record the fluorescence at 520 nm while temperature is increased from 20°C to 95°C at 1°C per minute.
6. Analyze data: Raw data will show an increase in fluorescence as the protein unfolds and the SYPRO® Orange dye binds to exposed hydrophobic residues. The inverted derivative of the fluorescence signal will result in peaks corresponding to the thermal transition midpoint (T_m). Select solvents at concentrations that minimally decrease the T_m ($\leq$5°C).

4.2.1.4 Notes

- Centrifuge prepared PCR strip/plate to remove bubbles if necessary.

4.2.2 Optimizing Crystal Stability

4.2.2.1 Equipment

- Crystal trays
- Siliconized cover slips
- Stereo-Microscope

4.2.2.2 Buffers and Reagents

- Pregrown crystal
- Solvent

4.2.2.3 Procedure

1. Follow procedures in Section 2.2.3 for soaking crystals. If ligand is precious, test using solvent.

2. Monitor crystal decay over time at various time points such as 5 min, 30 min, 1 h, 2 h, 6 h, and overnight (see Section 2.2.3 for discussion of decay).
3. If crystals are stable, cryoprotect and cryocool for data collection. If not, try cross-linking (Lusty, 1999) or the use of alternative solvents or mixtures of solvents as described in Ciccone et al. (2015) and Hassell et al. (2006).

4.2.2.4 Notes

- Prepare soaking solutions with varying concentrations of solvent (if material is precious) or inhibitor in solvent. Limiting the solvent content to less than 10% v/v is preferable.
- The stability of protein in solution versus in the crystalline state may differ. In solution, *St*T6PP is stable with up to 20% DMSO, but crystals rapidly decay as assessed macroscopically with >5% DMSO.
- When cross-linking, we use 5 μL of 8% glutaraldehyde on a microbridge in the crystallization well incubated for 5 min. During the first trial, monitor crystals every 5 min for signs of crystal decay.
- Molecular Dimensions sells the CryoSol kit (catalog number MD1-90) of multicomponent solutions for solubilizing hydrophobic ligands and for cryoprotection.

5. SUMMARY AND CONCLUSION

In this chapter, we have described various techniques using single-crystal X-ray crystallography to gain mechanistic insight into the HADSF of phosphatases. Using procedures such as site-directed mutagenesis and metal exchange, we demonstrate how one can capture a substrate-bound structure and decipher the binding determinants for substrate, affording mechanistic understanding and insight for inhibitor design. Phosphoryl mimics, such as vanadate, tungstate, sulfate, aluminum fluoride, and beryllium fluoride, allow for structural characterization of both half reactions in the overall mechanism of catalysis: nucleophilic attack producing the phosphoenzyme intermediate, and hydrolysis to release an inorganic phosphate and regenerate the enzyme. Lastly, characterization of substrate-like inhibitors can provide additional support for hypotheses regarding the binding determinants and mechanism of catalysis.

Furthermore, the methods described can be applied to other phosphatase families or generally to protein–ligand complexes. The methods for soaking substrate into crystals in Sections 2.1 and 2.2 may be useful to any ligand-binding protein. Sections 2.3–2.5 are applicable to other families of phosphatases. Mutagenesis of analogous residues important for catalysis or removal/replacement of the metal cofactor will oblate activity and enable substrate binding. Additionally, cocrystallization with phosphate allows one to pinpoint the active site. In Section 3, the information about phosphate mimics can be leveraged in other phosphoryl transfer enzymes to capture various snapshots of the catalytic mechanism. Finally, the concepts and methods to determine an inhibitor-bound structure in Section 4 can be used for any protein. These methods highlight complications in attaining liganded structures and some approaches that circumvent these problems. The HADSF acts as a testbed for phosphoryl mimics, with particularly strong applications to phosphoryl transferases that proceed through a covalent intermediate.

REFERENCES

Allen, K. N., & Dunaway-Mariano, D. (2004). Phosphoryl group transfer: Evolution of a catalytic scaffold. *Trends in Biochemical Sciences*, *29*(9), 495–503.

Allen, K. N., & Dunaway-Mariano, D. (2009). Markers of fitness in a successful enzyme superfamily. *Current Opinion in Structural Biology*, *19*(6), 658–665.

Allen, K. N., & Dunaway-Mariano, D. (2016). Catalytic scaffolds for phosphoryl group transfer. *Current Opinion in Structural Biology*, *41*, 172–179.

Bhat, T. N. (1989). Correlation between occupancy and temperature factors of solvent molecules in crystal structures of proteins. *Acta Crystallographica, Section A: Foundations of Crystallography*, *45*(1), 145–146.

Burroughs, A. M., Allen, K. N., Dunaway-Mariano, D., & Aravind, L. (2006). Evolutionary genomics of the HAD superfamily: Understanding the structural adaptations and catalytic diversity in a superfamily of phosphoesterases and allied enzymes. *Journal of Molecular Biology*, *361*(5), 1003–1034.

Cho, H., Wang, W., Kim, R., Yokota, H., Damo, S., Kim, S. H., et al. (2001). BeF3- acts as a phosphate analog in proteins phosphorylated on aspartate: Structure of a BeF3- complex with phosphoserine phosphatase. *Proceedings of the National Academy of Sciences of the United States of America*, *98*(15), 8525–8530.

Ciccone, L., Vera, L., Tepshi, L., Rosalia, L., Rossello, A., & Stura, E. A. (2015). Multicomponent mixtures for cryoprotection and ligand solubilization. *Biotechnology Reports*, 7, 120–127.

Collet, J.-F. (1998). A new class of phosphotransferases phosphorylated on an aspartate residue in an amino-terminal DXDX(T/V) motif. *Journal of Biological Chemistry*, *273*(23), 14107–14112.

Collet, J.-F., Gerin, I., Rider, M. H., Veiga-da-Cunha, M., & Van Schaftingen, E. (1997). Human L-3-phosphoserine phosphatase: Sequence, expression and evidence for a phosphoenzyme intermediate. *FEBS Letters*, *408*(3), 281–284.

Dai, J., Finci, L., Zhang, C., Lahiri, S., Zhang, G., Peisach, E., et al. (2009). Analysis of the structural determinants underlying discrimination between substrate and solvent in beta-phosphoglucomutase catalysis. *Biochemistry*, *48*(9), 1984–1995.

Daughtry, K. D., Huang, H., Malashkevich, V., Patskovsky, Y., Liu, W., Ramagopal, U., et al. (2013). Structural basis for the divergence of substrate specificity and biological function within HAD phosphatases in lipopolysaccharide and sialic acid biosynthesis. *Biochemistry*, *52*(32), 5372–5386.

Deng, H., Callender, R., Huang, Z., & Zhang, Z.-Y. (2002). Is the PTPase – vanadate complex a true transition state analogue? *Biochemistry*, *41*(18), 5865–5872.

Fieulaine, S., Lunn, J. E., Borel, F., & Ferrer, J.-L. (2005). The structure of a cyanobacterial sucrose-phosphatase reveals the sugar tongs that release free sucrose in the cell. *The Plant Cell*, *17*(7), 2049–2058.

Fieulaine, S., Lunn, J. E., & Ferrer, J.-L. (2007). Crystal structure of a cyanobacterial sucrose-phosphatase in complex with glucose-containing disaccharides. *Proteins: Structure, Function, and Bioinformatics*, *68*(3), 796–801.

Geremia, S., Campagnolo, M., Demitri, N., & Johnson, L. N. (2006). Simulation of diffusion time of small molecules in protein crystals. *Structure*, *14*(3), 393–400.

Goldschmidt, L., Cooper, D. R., Derewenda, Z. S., & Eisenberg, D. (2007). Toward rational protein crystallization: A web server for the design of crystallizable protein variants. *Protein Science: A Publication of the Protein Society*, *16*(8), 1569–1576.

Hassell, A. M., An, G., Bledsoe, R. K., Bynum, J. M., Carter, H. L., Deng, S. J. J., et al. (2006). Crystallization of protein-ligand complexes. *Acta Crystallographica, Section D: Biological Crystallography*, *63*(1), 72–79.

Herschlag, D., & Jencks, W. P. (1989). Phosphoryl transfer to anionic oxygen nucleophiles. Nature of the transition state and electrostatic repulsion. *Journal of the American Chemical Society*, *111*(19), 7587–7596.

Huang, H., Patskovsky, Y., Toro, R., Farelli, J. D., Pandya, C., Almo, S. C., et al. (2011). Divergence of structure and function in the haloacid dehalogenase enzyme superfamily: *Bacteroides thetaiotaomicron* BT2127 is an inorganic pyrophosphatase. *Biochemistry*, *50*(41), 8937–8949.

Kim, H.-Y., Heo, Y.-S., Kim, J. H., Park, M. H., Moon, J., Kim, E., et al. (2002). Molecular basis for the local conformational rearrangement of human phosphoserine phosphatase. *Journal of Biological Chemistry*, *277*(48), 46651–46658.

Lad, C., Williams, N. H., & Wolfenden, R. (2003). The rate of hydrolysis of phosphomonoester dianions and the exceptional catalytic proficiencies of protein and inositol phosphatases. *Proceedings of the National Academy of Sciences of the United States of America*, *100*(10), 5607–5610.

Lahiri, S. D. (2003). The pentacovalent phosphorus intermediate of a phosphoryl transfer reaction. *Science*, *299*(5615), 2067–2071.

Lahiri, S. D., Zhang, G., Dunaway-Mariano, D., & Allen, K. N. (2002). Caught in the act: The structure of phosphorylated beta-phosphoglucomutase from *Lactococcus lactis*. *Biochemistry*, *41*(26), 8351–8359.

Lavie, A., Allen, K. N., Petsko, G. A., & Ringe, D. (1994). X-ray crystallographic structures of D-xylose isomerase-substrate complexes position the substrate and provide evidence for metal movement during catalysis. *Biochemistry*, *33*(18), 5469–5480.

Lawson, D. M., Artymiuk, P. J., Yewdall, S. J., Smith, J. M. A., Livingstone, J. C., Treffry, A., et al. (1991). Solving the structure of human H ferritin by genetically engineering intermolecular crystal contacts. *Nature*, *349*(6309), 541–544.

Leahy, D. J., Erickson, H. P., Aukhil, I., Joshi, P., & Hendrickson, W. A. (1994). Crystallization of a fragment of human fibronectin: Introduction of methionine by site-directed mutagenesis to allow phasing via selenomethionine. *Proteins: Structure, Function, and Genetics*, *19*(1), 48–54.

Lima, C. D., Klein, M. G., & Hendrickson, W. A. (1997). Structure-based analysis of catalysis and substrate definition in the HIT protein family. *Science*, *278*(5336), 286–290.

Lipinski, C. A., Lombardo, F., Dominy, B. W., & Feeney, P. J. (2001). Experimental and computational approaches to estimate solubility and permeability in drug discovery and development settings. *Advanced Drug Delivery Reviews*, *46*(1–3), 3–26.

Liu, C., Dunaway-Mariano, D., & Mariano, P. S. (2017). Rational design of first generation inhibitors for trehalose 6-phosphate phosphatases. *Tetrahedron*, *73*(10), 1324–1330.

Loladze, V. V., Ermolenko, D. N., & Makhatadze, G. I. (2001). Heat capacity changes upon burial of polar and nonpolar groups in proteins. *Protein Science*, *10*(7), 1343–1352.

Lu, Z., Dunaway-Mariano, D., & Allen, K. N. (2005). HAD superfamily phosphotransferase substrate diversification: Structure and function analysis of HAD subclass IIB sugar phosphatase BT4131. *Biochemistry*, *44*(24), 8684–8696.

Lu, Z., Dunaway-Mariano, D., & Allen, K. N. (2008). The catalytic scaffold of the haloalkanoic acid dehalogenase enzyme superfamily acts as a mold for the trigonal bipyramidal transition state. *Proceedings of the National Academy of Sciences of the United States of America*, *105*(15), 5687–5692.

Lu, Z., Dunaway-Mariano, D., & Allen, K. N. (2011). The X-ray crystallographic structure and specificity profile of HAD superfamily phosphohydrolase BT1666: Comparison of paralogous functions in *B. thetaiotaomicron*. *Proteins*, *79*(11), 3099–3107.

Lu, Z., Wang, L., Dunaway-Mariano, D., & Allen, K. N. (2009). Structure-function analysis of 2-keto-3-deoxy-D-glycero-D-galactonononate-9-phosphate phosphatase defines specificity elements in type C0 haloalkanoate dehalogenase family members. *The Journal of Biological Chemistry*, *284*(2), 1224–1233.

Lusty, C. J. (1999). A gentle vapor-diffusion technique for cross-linking of protein crystals for cryocrystallography. *Journal of Applied Crystallography*, *32*(1), 106–112.

McLauchlan, C. C., Peters, B. J., Willsky, G. R., & Crans, D. C. (2015). Vanadium-phosphatase complexes: Phosphatase inhibitors favor the trigonal bipyramidal transition state geometries. *Coordination Chemistry Reviews*, *301–302*, 163–199.

Miao, Y., Tenor, J. L., Toffaletti, D. L., Washington, E. J., Liu, J., Shadrick, W. R., et al. (2016). Structures of trehalose-6-phosphate phosphatase from pathogenic fungi reveal the mechanisms of substrate recognition and catalysis. *Proceedings of the National Academy of Sciences of the United States of America*, *113*(26), 7148–7153.

Morais, M. C., Zhang, W., Baker, A. S., Zhang, G., Dunaway-Mariano, D., & Allen, K. N. (2000). The crystal structure of *Bacillus cereus* phosphonoacetaldehyde hydrolase: Insight into catalysis of phosphorus bond cleavage and catalytic diversification within the HAD enzyme superfamily. *Biochemistry*, *39*(34), 10385–10396.

Morais, M. C., Zhang, G., Zhang, W., Olsen, D. B., Dunaway-Mariano, D., & Allen, K. N. (2004). X-ray crystallographic and site-directed mutagenesis analysis of the mechanism of schiff-base formation in phosphonoacetaldehyde hydrolase catalysis. *Journal of Biological Chemistry*, *279*(10), 9353–9361.

Nguyen, H. H., Wang, L., Huang, H., Peisach, E., Dunaway-Mariano, D., & Allen, K. N. (2010). Structural determinants of substrate recognition in the HAD superfamily member D-glycero-D-manno-heptose-1,7-bisphosphate phosphatase (GmhB). *Biochemistry*, *49*(6), 1082–1092.

Pandya, C., Farelli, J. D., Dunaway-Mariano, D., & Allen, K. N. (2014). Enzyme promiscuity: Engine of evolutionary innovation. *Journal of Biological Chemistry*, *289*(44), 30229–30236.

Parsons, J. F., Lim, K., Tempczyk, A., Krajewski, W., Eisenstein, E., & Herzberg, O. (2002). From structure to function: YrbI from *Haemophilus influenzae* (HI1679) is a phosphatase. *Proteins*, *46*(4), 393–404.

Peck, A., Sunden, F., Andrews, L. D., Pande, V. S., & Herschlag, D. (2016). Tungstate as a transition state analog for catalysis by alkaline phosphatase. *Journal of Molecular Biology, 428*(13), 2758–2768.

Peeraer, Y., Rabijns, A., Verboven, C., Collet, J.-F., Van Schaftingen, E., & De Ranter, C. (2003). High-resolution structure of human phosphoserine phosphatase in open conformation. *Acta Crystallographica, Section D: Biological Crystallography, 59*(6), 971–977.

Peisach, E., Selengut, J. D., Dunaway-Mariano, D., & Allen, K. N. (2004). X-ray crystal structure of the hypothetical phosphotyrosine phosphatase MDP-1 of the haloacid dehalogenase superfamily. *Biochemistry, 43*(40), 12770–12779.

Pflugrath, J. W. (2015). Practical macromolecular cryocrystallography. *Acta Crystallographica, Section F: Structural Biology Communications, 71*(6), 622–642.

Rupp, B. (2010). *Biomolecular crystallography: Principles, practice, and application to structural biology*. New York, NY: Garland Science.

Shan, S., Min, H., Liu, T., Jiang, D., & Rao, Z. (2016). Structural insight into dephosphorylation by trehalose 6-phosphate phosphatase (OtsB2) from *Mycobacterium tuberculosis*. *The FASEB Journal, 30*(12), 3989–3996.

Silvaggi, N. R., Zhang, C., Lu, Z., Dai, J., Dunaway-Mariano, D., & Allen, K. N. (2006). The X-ray crystal structures of human α-phosphomannomutase 1 reveal the structural basis of congenital disorder of glycosylation type 1a. *Journal of Biological Chemistry, 281*(21), 14918–14926.

Stiers, K. M., & Beamer, L. J. (2018). Assessment and impacts of phosphorylation on protein flexibility of the α-D-phosphohexomutases. *Methods in Enzymology, 607*, 241–268.

Sygusch, J., & Beaudry, D. (1984). Catalytic activity of rabbit skeletal muscle aldolase in the crystalline state. *The Journal of Biological Chemistry, 259*(16), 10222–10227.

Toyoshima, C., Nomura, H., & Tsuda, T. (2004). Lumenal gating mechanism revealed in calcium pump crystal structures with phosphate analogues. *Nature, 432*(7015), 361–368.

Toyoshima, C., Norimatsu, Y., Iwasawa, S., Tsuda, T., & Ogawa, H. (2007). How processing of aspartylphosphate is coupled to lumenal gating of the ion pathway in the calcium pump. *Proceedings of the National Academy of Sciences of the United States of America, 104*(50), 19831–19836.

Tremblay, L. W., Zhang, G., Dai, J., Dunaway-Mariano, D., & Allen, K. N. (2005). Chemical confirmation of a pentavalent phosphorane in complex with β-phosphoglucomutase. *Journal of the American Chemical Society, 127*(15), 5298–5299.

Walldén, K., Rinaldo-Matthis, A., Ruzzenente, B., Rampazzo, C., Bianchi, V., & Nordlund, P. (2007). Crystal structures of human and murine deoxyribonucleotidases: Insights into recognition of substrates and nucleotide analogues. *Biochemistry, 46*(48), 13809–13818.

Wang, W., Cho, H. S., Kim, R., Jancarik, J., Yokota, H., Nguyen, H. H., et al. (2002). Structural characterization of the reaction pathway in phosphoserine phosphatase: Crystallographic "snapshots" of intermediate states. *Journal of Molecular Biology, 319*(2), 421–431.

Wang, W., Kim, R., Jancarik, J., Yokota, H., & Kim, S.-H. (2001). Crystal structure of phosphoserine phosphatase from Methanococcus jannaschii, a hyperthermophile, at 1.8 Å resolution. *Structure, 9*(1), 65–71.

Zhang, G., Dai, J., Wang, L., Dunaway-Mariano, D., Tremblay, L. W., & Allen, K. N. (2005). Catalytic cycling in β-phosphoglucomutase: A kinetic and structural analysis. *Biochemistry, 44*(27), 9404–9416.

Zhang, G., Morais, M. C., Dai, J., Zhang, W., Dunaway-Mariano, D., & Allen, K. N. (2004). Investigation of metal ion binding in phosphonoacetaldehyde hydrolase identifies sequence markers for metal-activated enzymes of the HAD enzyme superfamily. *Biochemistry, 43*(17), 4990–4997.

SECTION III

Substrate Specificity and Enzyme Conformational States

CHAPTER SIX

Structure, Mechanism, and Substrate Profiles of the Trinuclear Metallophosphatases from the Amidohydrolase Superfamily

Swapnil V. Ghodge*, Frank M. Raushel†,1
*Eshelman School of Pharmacy, University of North Carolina at Chapel Hill, Chapel Hill, NC, United States
†Department of Chemistry, Texas A & M University, College Station, TX, United States
[1]Corresponding author: e-mail address: raushel@chem.tamu.edu

Contents

Abstract

The rate of reliable protein function annotation has not kept pace with the rapid advances in genome sequencing technology. This has created a gap between the number of available protein sequences, and an accurate determination of the respective physiological functions. This investigation has attempted to bridge the gap within the confines of members of the polymerase and histidinol phosphatase family of proteins in cog1387 and cog0613, which is related to the amidohydrolase superfamily. The adopted approach relies on using the mechanistic knowledge of a known enzymatic reaction, and discovering functions of closely related homologs using various tools including bioinformatics and rational library screening. The initial enzymatic reaction was that of L-histidinol phosphate phosphatase. Extensive structural, biochemical, and bioinformatic analysis of enzymes capable of hydrolyzing L-histidinol phosphate provided useful insights in predicting substrates and mechanistic studies of related enzymes. This led to the discovery of unprecedented catalytic functions such as a cyclic phosphate dihydrolase that specifically hydrolyzed a cyclic phosphodiester to inorganic phosphate and a vicinal diol; a phosphoesterase that hydrolyzes the 3′-phosphate of 3′,5′-adenosine bisphosphate and similar nucleotides; and the first reported

Methods in Enzymology, Volume 607
ISSN 0076-6879
https://doi.org/10.1016/bs.mie.2018.04.019

5′–3′ exonuclease for 5′-phosphorylated oligonucleotides from *Escherichia coli* and related organisms. This work provides a template for developing sequence–structure–function correlations within a family of enzymes that helps expedite new enzyme function discovery and more accurate annotations in protein databases.

ABBREVIATIONS

AHS amidohydrolase superfamily
BLAST basic local alignment search tool
COG cluster of orthologous groups
DNA deoxyribonucleic acid
E-value expectation value
NCBI National Center for Biotechnology Information
NMR nuclear magnetic resonance
PHP polymerase and histidinol phosphatase
RNA ribonucleic acid

1. INTRODUCTION

Phosphorylated substrates constitute a very important part of the metabolome of any living organism. Consequently, a detailed understanding of the enzymes that metabolize these phosphoesters, with respect to their mechanism and substrate specificity is crucial. RNA/DNA repair and synthesis, signal transduction, phospholipid metabolism, energy storage, transcription control, protein activation/deactivation, cell transformation, virulence, and other biological phenomena rely on phosphoesterases (Widlansky & Taylor, 1999). Numerous synthetic compounds including pesticides, insecticides, nerve agents, and plasticizers have an organophosphate functional group, and engineered phosphatases offer a convenient bioprocessing solution to degrade these compounds under ambient and eco-friendly conditions (Masson & Lushchekina, 2016). Clearly, the study of phosphatases within biological as well as biotechnological context is indispensable.

Phosphomonoesterases have been historically classified into four broad families based on their reaction mechanisms, and more specifically, their respective active site catalytic residue: histidine phosphatases, cysteine phosphatases, serine phosphatases, and metallophosphatases (Van Etten, 1982). The metallophosphatases utilize divalent metal ions to activate the nucleophilic water/hydroxide for a direct attack at the phosphorus center of the substrate being hydrolyzed. Biochemically characterized metallophosphatases that belong to the amidohydrolase superfamily (AHS) of enzymes contain two or three divalent metal ion cofactors within their active site. The AHS was first identified by Holm and Sander (1997), based on the significant resemblance between the structural features of phosphotriesterase (PDB: 1PTA), urease (PDB: 1KRC), and adenosine deaminase (PDB: 2ADA) (Holm & Sander, 1997). Subsequent studies of members of the AHS have revealed important insights into the chemical reaction mechanisms and the breadth of substrate specificity within this superfamily (Seibert & Raushel, 2005). The enzyme active site is located at the C-terminal end of a β-barrel motif in a distorted TIM-barrel fold, in which seven or eight parallel β-strands are flanked by seven or eight α-helices respectively, on the outer faces. In general, the metal-binding residues are located at the C-terminal end of the β-strands, while the loops that link the β-strands and α-helices bear the residues responsible for substrate specificity. Members of the AHS catalyze hydrolysis of amide, ester, halogen, or other functional groups at carbon and phosphorus centers (Holm & Sander, 1997). The AHS has been divided into 24 clusters of orthologous groups (COGs) of proteins (Tatusov, Galperin, Natale, & Koonin, 2000). Of these, seven cogs have been shown to contain phophoesterases (cogs 1735, 0084, 1099, 1831, 4464, 1387, and 0613). This chapter focuses on studies directed toward the mechanistic understanding and new function discovery of enzymes from two closely related cogs: 1387 and 0613. Both cogs consisted of enzymes that belong to the polymerase and histidinol phosphatase (PHP) family, and therefore, shared similarity in terms of their overall structure, active site, and potentially, mechanism and substrate specificity. Enzymes from cog1387 had been annotated as L-histidinol phosphate phosphatase-like proteins, while those from cog0613 were generically annotated as metal-dependent phosphoesterases, in various public databases such as UniProt (Universal Protein Resource) and NCBI (National Center for Biotechnology Information). We initiated our studies with the known enzymatic function of L-histidinol phosphate phosphatase (HPP) from cog1387. We explored the chemical mechanism of phosphate hydrolysis, the optimal conditions for enzymatic catalysis, and the structural details of substrate recognition and orientation at the active site of this enzyme and its orthologs. The results and the experience thus obtained were leveraged to discover unknown functions among other phosphoesterases. This work attempted to systematically bridge the gap in the databases of unannotated protein sequences, and the knowledge of their physiological substrates and enzymatic mechanism of phosphate hydrolysis, and contribute toward the broad objective of developing a sequence-based strategy for annotating enzyme functions correctly.

2. L-HISTIDINOL PHOSPHATE PHOSPHATASE (GHODGE, FEDOROV, ET AL., 2013)

HPP catalyzes the hydrolysis of L-histidinol phosphate to L-histidinol and inorganic phosphate, the penultimate step in the biosynthesis of L-histidine (Scheme 1) (Alifano et al., 1996). Enzymes known to catalyze this reaction belong to two different classes: (a) the haloacid dehalogenase (HAD) superfamily (Rangarajan et al., 2006; Thaller, Schippa, & Rossolini, 1998) and (b) the PHP family (Aravind & Koonin, 1998). Structural studies on enzymes from the PHP family have revealed that these enzymes possess a trinuclear metal center at the active site (Omi et al., 2007; Teplyakov et al., 2003). These metal-binding sites have been denoted as α, β, and γ, respectively. The α and β sites correspond to the conserved metal centers in prototypical binuclear amidohydrolases (Kim et al., 2008; Lai et al., 2004; Thoden, Phillips, Neal, Raushel, & Holden, 2001), while the γ-metal center is unique to PHP family members (Fig. 1). Prior structural studies on HPP enzymes from *Thermus thermophilus* (PDB: 2YXO, 2YZ5, and 2Z4G) and *Listeria monocytogenes* (PDB: 3DCP) indicated that the recombinantly expressed enzymes bound iron in the α- and β-metal sites, and zinc in the γ site (Omi et al., 2007). Our studies predominantly focused on the characterization of HPP from *Lactococcus lactis* (Ghodge, Fedorov, et al., 2013). Recombinant enzyme expressed and purified from *E. coli* contained iron and zinc in a 2:1 ratio. The kinetic parameters at pH 9.0 for this enzyme were determined to be $k_{cat} = 1.9\ s^{-1}$ and $k_{cat}/K_m = 2.6 \times 10^3\ M^{-1}\ s^{-1}$. The presence of iron in the enzyme, which is prone to oxidation, and a relatively low catalytic efficiency prompted us to replace iron with other

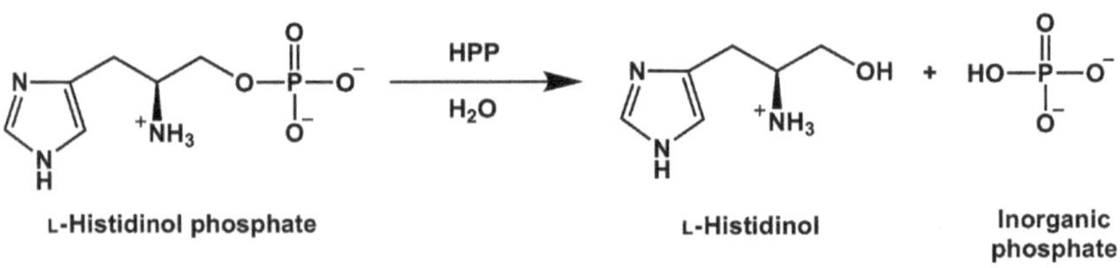

Scheme 1 Reaction catalyzed by L-histidinol phosphate phosphatase.

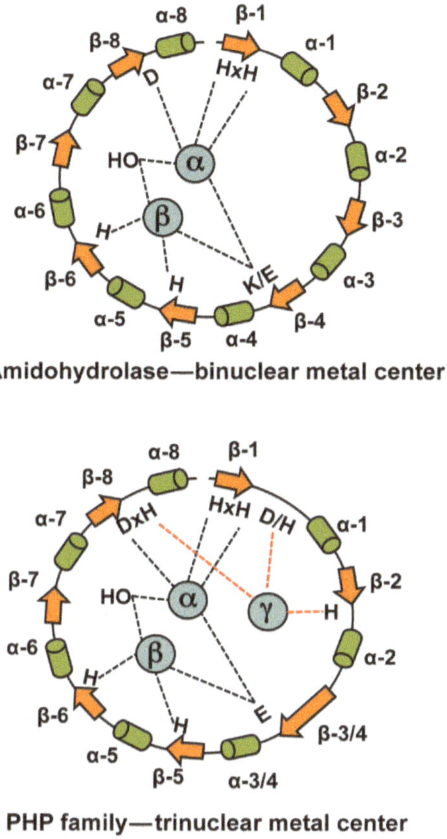

Fig. 1 Comparative schematic depiction of the secondary structural elements and metal ligation scheme of members of the amidohydrolase superfamily and the polymerase and histidinol phosphatase superfamily containing two and three divalent metal ions at the active site, respectively. Metal centers are shown as *blue spheres*, β-strands are shown as *orange arrows*, and α-helices are shown as *green cylinders*.

divalent metal ions. Efforts to add or exchange metal ions in vitro were unsuccessful. The same clone was subsequently expressed under iron-free conditions and in the presence of Zn^{2+} and/or Mn^{2+} ions. This protocol involved addition of the iron chelator 2,2′-bipyridyl when the OD of the expression cultures was between 0.15 and 0.2, addition of Zn^{2+} and Mn^{2+} ions when the OD was approximately 0.4, followed by addition of 0.25 m*M* IPTG to induce protein overexpression. This protocol yielded an HPP enzyme with catalytic activity enhanced by two orders of magnitude as compared to the one expressed in regular LB medium. This optimally active enzyme was used for further biochemical characterization.

X-ray crystallographic studies on the HPP enzyme from *L. lactis* yielded three structures: HPP with sulfate (PDB: 4GC3), HPP with L-histidinol and phosphate (PDB: 3UMU), and a complex of L-histidinol and arsenate (PDB: 4GK8). These provided snapshots of the reaction with a phosphate mimic bound (PDB: 4GC3), products bound (PDB: 3UMU), and a substrate mimic bound (PDB: 4GK8). The enzyme exhibited a distorted $(\beta/\alpha)_7$-barrel structural fold. The trinuclear metal center in the active site is located at the mouth of the barrel, with the α and β sites buried deep in the active site, while the γ site is situated closer to the bulk solvent. All three metal ions were shown to be indispensable for catalysis by systematic point mutations of enzyme residues that ligate the metal ions. All three X-ray structures have a hydroxide molecule bridging the α- and β-metal centers that serves as the nucleophile in the hydrolysis reaction (Martí-Arbona et al., 2005). The substrate binds at the active site such that three out of the four oxygen atoms of the phosphate serve as ligands to the three metal centers (Fig. 2). The fourth oxygen atom was found to form ionic interactions with the guanidino side chains of Arg160 and Arg197. Single-site variants of these arginine residues diminished catalytic efficiency by 1–2 orders of magnitude, indicating that they serve to make the phosphorus center more electrophilic. The leaving oxygen atom ligates to the γ-metal ion. In the L-histidinol arsenate complex

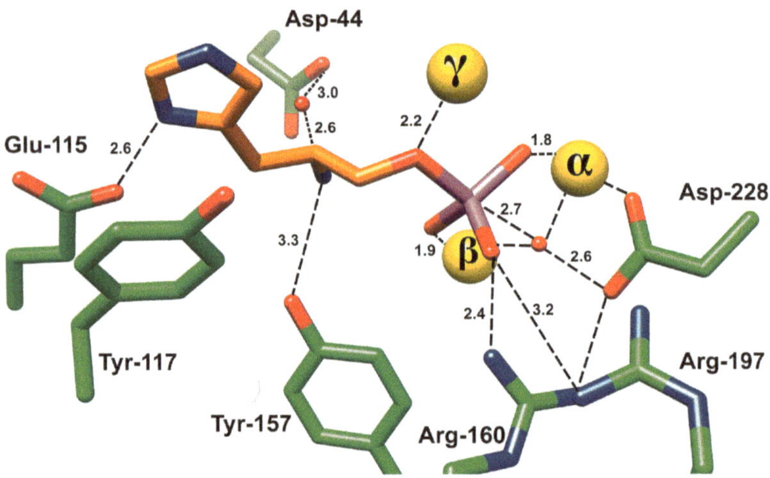

Fig. 2 Complex of L-histidinol and arsenate as a substrate mimic at the active site of L-histidinol phosphate phosphatase from *L. lactis*. Active site metal ions (zinc) are shown as *large yellow spheres*, enzyme residues are shown in *green*, water molecules are shown as *small red spheres*, L-histidinol is in *orange*, and vanadate is shown in *magenta*. Distances between atoms are indicated in angstrom units.

with the HPP enzyme, the bridging hydroxide is 2.7 Å away from the As-atom while the bond angle between the bridging hydroxide, As-atom, and the O-atom from L-histidinol is 166 degrees. The O-atom from L-histidinol is 2.2 Å away from the γ-metal center. This evidence suggested that the α- and β-metal ions served to orient and activate the nucleophile for the hydrolytic chemistry, while the γ-metal ion served as a Lewis acid for the leaving alkoxide/alcohol, in addition to ligating and orienting the substrate for catalysis.

Additional mechanistic insights were obtained by studying the pH-rate profiles, solvent isotope effects, and solvent viscosity effects on the HPP-catalyzed reaction. pH-rate profiles were studied in the pH range between 7 and 10, with enzyme instability and deactivation precluding access to pH values below 7, and above 10 (Fig. 3) (Cook & Cleland, 2007). In the pH-rate profile for k_{cat}, there is a loss of activity at low pH with an apparent p*Ka* of 7.5. This ionization most likely reflects the protonation of the bridging hydroxide to water in the HPP–substrate complex, similar to that observed in the binuclear amidohydrolases (Aubert, Li, & Raushel, 2004; Martí-Arbona et al., 2005). However, the loss of activity in the reaction catalyzed by HPP is not observed in the pH-rate profile for k_{cat}/K_m at low pH values. This observation suggests that the apparent p*Ka* for the protonation of the bridging hydroxide in the free enzyme is <7. When polyanionic substrates ligate to the metal ions within the enzyme active site, the apparent pK_a for protonation of the bridging hydroxide must shift to ~7.5 in the HPP–substrate complex. In the pH-rate profile for k_{cat}/K_m, there is a loss of activity at high pH with an apparent p*Ka* of 9.7. In the active site of HPP, the most likely group that accounts for this ionization is the primary amino group of the substrate. This group may ultimately serve as the general acid during protonation of the leaving group oxygen of the alcohol. Alternatively, P–O bond cleavage may occur with formation of an alkoxide anion intermediate that is initially stabilized by the γ-metal. In the latter mechanism, the loss of activity at high pH in the k_{cat}/K_m profile may be attributed to the nonproductive binding of the substrate where the neutral amino group coordinates to the γ-metal and subsequently decreases the Lewis acidity of this metal ion toward the bridging O atom of the phosphate and/or distorts the coordination geometry of the bound substrate. *N*-formyl L-histidinol phosphate was tested as a substrate analog and the rate of hydrolysis were found to be reduced by more than two orders of magnitude relative to the hydrolysis of L-histidinol phosphate. This observation supported the critical ionization state of the amino group for binding and/or proton transfer.

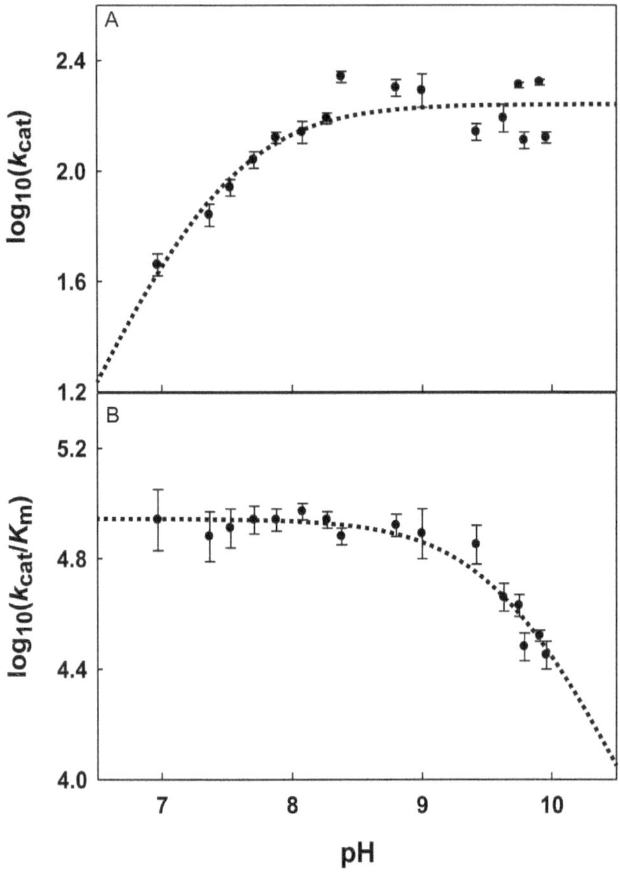

Fig. 3 pH-Rate profile for the hydrolysis of L-histidinol phosphate by HPP from *L. lactis* at 22 ± 1°C. (A) Variation of log k_{cat} vs pH. The *dotted line* represents a fit of the data to the equation: $\log k_{cat} = \log [c/(1 + [H^+]/K_a)]$. (B) Variation of log k_{cat}/K_m vs pH. The *dotted line* represents a fit of the data to the equation: $\log k_{cat}/K_m = \log [c/(1 + K_b/[H^+])]$.

The solvent deuterium isotope effects on either k_{cat} or k_{cat}/K_m are relatively small (1.4 and 1.3, respectively) at pH 8.5, and thus, the proton transfer steps do not appear to limit the rate of the enzyme-catalyzed transformation. An increase in solvent viscosity had relatively minor effects on the kinetic parameters. At pH 8.6, the slope for the k_{cat} profile is 0.1 and that for k_{cat}/K_m is 0.3. Clearly, the enzyme–substrate complex is not sticky. The lack of a solvent isotope effect and the absence of a solvent viscosity effect leave an undefined conformational change as the most likely rate-limiting step in the overall reaction mechanism. The proposed mechanism of enzymatic hydrolysis of L-histidinol phosphate is shown in Scheme 2.

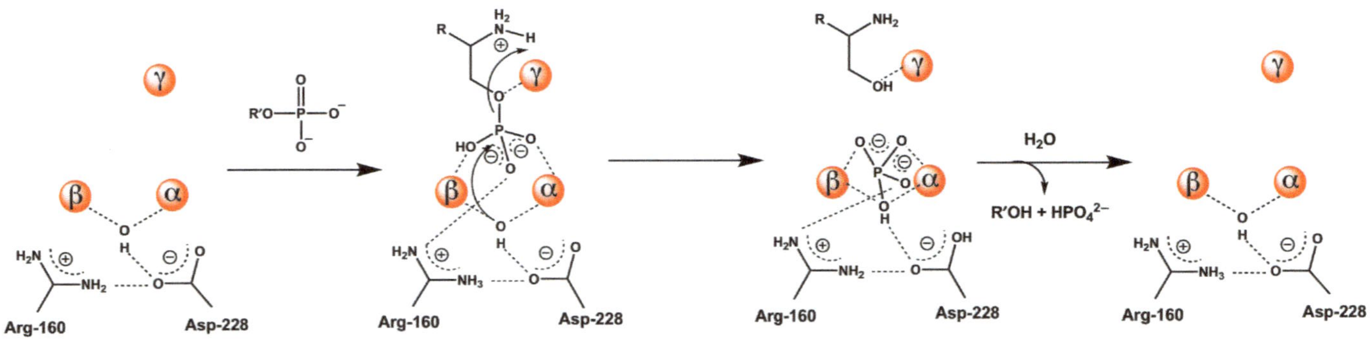

Scheme 2 Proposed enzymatic reaction mechanism of HPP enzyme for the hydrolysis of L-histidinol phosphate.

We verified the ability of eight other enzymes from various organisms that were annotated as HPP to hydrolyze L-histidinol phosphate (Table 1). Primary sequence alignment of these eight enzymes, along with HPP enzymes from *L. lactis*, *T. thermophilus*, and *L. monocytogenes*, helped to identify structural elements that governed the substrate specificity of these enzymes. Representative sequence alignment of six of these HPP enzymes is shown in Fig. 4. The substrate-binding pocket of HPP contains a positively charged pocket of the divalent metal ions and arginine residues (Arg160 and Arg197 from *L. lactis* HPP) that bind the phosphate functional group in the substrate, and a hydrophobic pocket consisting of tyrosine and/or phenylalanine residues (Tyr117 from *L. lactis* HPP) that binds the imidazole ring via π-stacking interactions. Other residues that were found to interact with the substrate in the crystal structure were Glu115 and Tyr157. However, these residues were not conserved in all HPP sequences, and therefore, are not

Table 1 Kinetic Parameters of HPP Enzymes of cog1387 From Different Organisms

Locus Tag	Organism	k_{cat} (s^{-1})	K_m (m*M*)	k_{cat}/K_m ($M^{-1}\ s^{-1}$)	Metal Content per Enzyme Monomer
MCCL_0344	*Macrococcus caseolyticus* JCSC5402	30	1.5	2.0×10^4	Mn (0.9); Fe (0.6); Zn (0.3)
BBR47_00270	*Brevibacillus brevis* NBRC 100599	7.2	2.6	2.8×10^3	Zn (1.6); Mn (0.3)
BCE_1533	*Bacillus cereus* ATCC10987	8.1	1.5	5.6×10^3	Mn (1.4); Zn (0.4)
BcerKBAB4_1335[a]	*Bacillus weihenstephanensis* KBAB4	1.2	4.9	2.5×10^2	Zn (1.5); Fe (1.2)
BcerKBAB4_1335		34	1.3	2.6×10^4	Mn (1.0); Zn (0.6)
BSU29620[a]	*Bacillus subtilis* subsp. subtilis str. 168	0.37	0.10	3.7×10^3	Fe (2.6); Zn (0.4)
BH3206	*Bacillus halodurans* C-125	28	0.64	4.4×10^4	Mn (1.1); Zn (0.3)
GK2799	*Geobacillus kaustophilus* HTA426	7.6	8	9.0×10^2	Mn (1.0); Zn (0.4); Fe (0.3)
SMU_1486c	*Streptococcus mutans* UA159	0.31	0.58	5.3×10^2	Zn (0.4); Fe (0.3)

[a]Enzymes were overexpressed without the addition of 2,2′-bipyridyl in the growth medium.

```
                        β-1
L37351          1 MSLKKLDYHFHSHFSA-----------------------DS-EELPRK--------------------HVTEAIAHGL 34
MCCL_0344       1 ---MKVDFHIHLEEGPYTNNFFNKTITSIDTVKGI-QATHTLDDIERKAQLFNERMEKGDYSEWWLDLYLEMSLQKGL 74
BCE_1533        1 ---MKVDYHIHLEEGPYSIGWLAKINESLQHYEPLKEEKHSMEWLVKIQERLQRRVNEGPFTTKWIDLYLEEALRKGI 75
SMU_1486c       1 ----MRDNHLHTHFSY-----------------------DS-------------------------DASFEDYLTHYD 26
TTHA0331        1 ----MVDSHVHTPLCG-----------------------HA-EGHPEA--------------------YLEEARAKGL 30
LMOh7858_0629   1 ---MKRDGHTHTEFCP-----------------------HGTHDDVEE--------------------MVLKAIELDF 32

                       β-2                                                          β-3/4
L37351         35 EEICFTEH-----R--DFYFPGMDFSL-----------------NLPEYFQEINQLQAEF-KDKIKIKIGLEMGID  85
MCCL_0344      75 KQVGIVDHLYRFQETRNYFLKYMDVSDTDLGCRQREWLNQVMTHKMDDFVTFINSQKEKWDKAGVELKLGIEADYF 150
BCE_1533       76 KEVGIVDHLYRFYEAKEYYEKYVDISDSGLGRLQKEWLDQVRVASLHDFTKAIEEAKERWSKRGVTLKLGIEADYF 151
SMU_1486c      27 GEIVTTEH----YDLSNPYTQQDDV------------------PDYEAYSKEIAELNAKY---GNRIKRGIEIGYY  77
TTHA0331       31 KGVVFTDH----SPMPPW-YDP--------ES------RMRLEALPFYLLALERVRERAQ---DLYVGIGLEADFH  84
LMOh7858_0629  33 DEYSIVEH----APLSSEFMKNTAGDKEAVTT------ASMAMSDLPYYFKKMNHIKKKY-ASDLLIHIGFEVDYL  97

                                  β-5
L37351         86 LRFKSEINQFIDSA--PFDFVIASVHEI-G---DIE-VYDGTEFYLQK-----TKEEAQREYLLACLDVVQ-NFE 147
MCCL_0344     151 IGGEEELKSLLAPY--EFDYIIGSVHFNHGW------GFDNPE-LENKFNE-YDLVKLYTDHFNTVIKAAESGIF 215
BCE_1533      152 IGGEQQLQSLLALG--DFDYVIGSVHFLNGW------GFDNPD-TKEYFKE-HDLYALYDTFFKTVECAVRSELF 216
SMU_1486c      78 QPREADILSFLADK--DYDLKLLSVHHN-G---VN--DYLDDEVAD------MDKETIIQEYLDKLEYAI----G 136
TTHA0331       85 PGTEGFLAQLLRRY--PFDYVIGSVHYL-G---AWPLDHPDHQEEYAW----RDLKEVFRAYFQEVEKAAR-S-- 146
LMOh7858_0629  98 IGYEDFTRDFLNEYGPQTDDGVLSLHFLEGQGGFRSIDFSAEDYNEGIVQFYGGFEQAQLAYLEGVKQSIEADLG 172

                     β-6                                           β-7
L37351        148 --NYNSFGHLDYVAR-YGPYT-DK-SIKFAENREILFEILRALASKEKALEIN-----TRL-FDDPKTEQFYSDL 211
MCCL_0344     216 ----SFIAHLDNL-KVFNYRP------EEALLIPLYEQVAEALAKNDVATEVN-VGLKYRYPV----KEQCPSER 274
BCE_1533      217 ----DIIAHLDNI-KVFNYRL------DENEQISYYKKIARALVETNTATEIN-AGLYYRYPV----REMCPSPL 275
SMU_1486c     137 RVEADVLAHFDYGFRLFDLT----VDELKTHEAQLRRIFHKMIDHNLAFELNS------KSMYLYGHEHLYRYAL 199
TTHA0331      147 -GLFHAIGHLDLPKK-FGH----R--LPEEALLELAEPALRAVAEAGLFLDVNTAGL-RRP-AKEVYPAPALLRR 211
LMOh7858_0629 173 LFKPRRMGHISLCQK-FQQFFGEDTSDFSEEVMEKFRVILALVKKRDYELDFNTAGL-FKPLCGETYPPKKIVTL 245

                           β-8
L37351        212 LINFKRLGGKFITLGTDSHIAKRDWLSIHKARTLIKKAGFHELATFSGMKIDKNKKSIKEKLAAALEHHHHHH 284
MCCL_0344     275 FIKVLSEYDVKFTTSSDSHFPHDIGIYNDEIRNLLKRNGVKSIVTFSKMKREEKDI---KSIIH--------- 335
BCE_1533      276 YLQVLAKHEVPITLSSDAHYPNDLGKYVEENIKTLRNHDIAHLATFTKRVRTMRLL-EEVTISK--------- 338
SMU_1486c     200 SLV-KNLGCHKYTIGSDGHKLEHFRLAFDKIQDILDEYGIEEGEII--------------------------- 244
TTHA0331      212 AREL----GIGLVLGSDAHRPEEVGFAFPEVQALLAGLGFREAYYFVEGSPVAYPL----SRAS--------- 267
LMOh7858_0629 246 ASEL----QIPFVYGSDSHGVQDIGRGYSTYCQK--------------------------------------- 275
```

Fig. 4 Primary sequence alignment of HPPs from various organisms is shown. L37351 is the HPP from *L. lactis* subsp. lactis Il1403 characterized in this study. The organisms from which the other HPPs were derived were *M. caseolyticus* JCSC5402 (MCCL_0344), *B. cereus* ATCC10987 (BCE_1533), *S. mutans* UA159 (SMU_1486c), *T. thermophilus* HB8 (TTHA0331), and *L. monocytogenes* str. 4b h7858 (LMOh7858_0629). The primary sequence of L37351 shown here is the actual sequence of the recombinant protein characterized in this study, while the remaining sequences are from the NCBI database. Metal-binding residues are colored *red*, while the conserved residues interacting with the substrate are colored *blue*. The β-strands observed in the available crystal structures of the three HPP enzymes are highlighted in *gray* and labeled.

critical for substrate recognition. This pattern of conserved residues in the active site of HPP was used to identify additional sequences that likely encode proteins that catalyze the hydrolysis of L-histidinol phosphate. The critical residues include the nine amino acids that coordinate the three divalent cations. In addition to these residues, there must be arginine and lysine residues shortly after the C-terminal ends of β-strands 6 and 7 and a tyrosine or phenylalanine after the lysine or arginine at the end of β-strand 6, and on the loop after β-strand 5. These criteria were used to define the boundaries of HPP activity in those sequences represented in a sequence similarity network (SSN) of protein sequences from cog1387. This exercise clearly delineated authentic HPP enzyme sequences from those of unknown function that had significant sequence similarity to HPP enzymes.

These studies provided a solid foundation for our understanding of the structural elements critical for enzymatic activity of phosphoesterases from the PHP family as well as the biochemical conditions for studying them in vitro. The knowledge and understanding generated by this investigation were leveraged to discover unprecedented enzymatic functions and substrate scope of PHP family members.

3. CYCLIC PHOSPHATE DIHYDROLASE (GHODGE, CUMMINGS, WILLIAMS, & RAUSHEL, 2013)

We turned our attention to enzymes from cog0613. There were three proteins from this cog whose crystal structures were available in the PDB: 3E0F/3O0F, 2ANU, and 3E38 (Fig. 5). However, there was no enzyme whose physiological function was known at the time. Careful bioinformatic analysis revealed that one nonredundant protein sequence from this cog was placed in the operon that encoded genes involved in the metabolism of phosphonates. Several bacteria are capable of utilizing phosphonates as a source of phosphorus under phosphate-deficient conditions (Chen, Ye, Zhu, Wanner, & Walsh, 1990). In *E. coli* and related organisms, the 14-cistron phnCDEFGHIJKLMNOP operon is responsible for the conversion of hydrolytically stable phosphonates into physiologically usable phosphates (Metcalf & Wanner, 1993). Of these, the phnGHIJKLM genes encode the minimal set of proteins needed to cleave the C–P bond, while phnNOP encode protein sequences that catalyze accessory biochemical reactions. The remaining genes, phnCDEF, encode transporters of phosphonate substrates and regulatory elements. The catalytic functions of enzymes PhnGHIJKLM convert unactivated phosphonate substrates into 5-phosphoribosyl-1,2-cyclic phosphate (PRcP) (Kamat, Williams, Dangott, Chakrabarti, & Raushel, 2013; Kamat, Williams, & Raushel, 2011). The enzyme PhnP hydrolyzes the cyclic phosphate of PRcP to ribose 1,5-bisphosphate (1,5-RbP), which is then converted to 5-phosphoribosyl-1-pyrophosphate (PRPP) by the enzymatic action of PhnN (Hove-Jensen, McSorley, & Zechel, 2011; Kamat et al., 2011). PhnO is an accessory enzyme that has been shown to acylate 1-aminoalkylphosphonic acids (Hove-Jensen, McSorley, & Zechel, 2012). Bioinformatic analysis of the C–P lyase pathway in several bacteria revealed that a small subset of bacteria, which included the human intestinal pathogens *Clostridium difficile* and

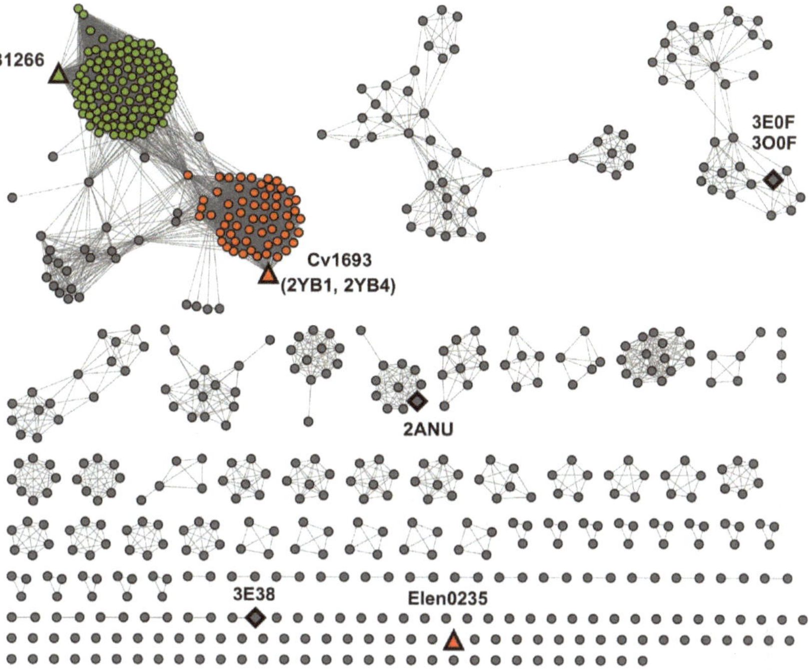

Fig. 5 Sequence similarity network of nonredundant enzyme sequences belonging to cog0613 at a BLAST E-value cutoff of 1×10^{-60}. Each node represents a nonredundant protein sequence, while each edge (*line*) represents pairs of sequences that are more closely related than the arbitrary E-value cutoff (10^{-60}). The crystal structures that were available at the beginning of this work are shown as *rhombus*, and their respective PDB codes are indicated. The enzymes that have been biochemically characterized in this work are shown as *triangles*. Cv1693 is a 3′,5′-nucleotide bisphosphate-3′-phosphatase, Elen0235 is a cyclic phosphate dihydrolase (cPDH), while B1266 is a 5′–3′ exonuclease. Protein sequences that are sufficiently similar to Cv1693 and B1266 in sequence (E-value of $<10^{-70}$, which is approximately equivalent to a sequence identity of >43%) are colored *orange* and *green*, respectively.

Eggerthella lenta, possess a phosphoesterase of unknown function from cog0613 in place of the accessory enzymes PhnNOP from *E. coli* (Fig. 6). This presented an intriguing possibility that the metabolite PRcP, produced as a product of the reaction catalyzed by the radical SAM enzyme PhnJ in these organisms, may have a distinct fate as compared to *E. coli* and related organisms.

We cloned, heterologously expressed and purified using affinity chromatography, the metal-dependent phosphoesterase from *E. lenta* (Elen0235), with a C-terminal 6 × His-tag. We also obtained PhnP from *E. coli* using a similar strategy. The product of the PhnJ enzymatic reaction, PRcP,

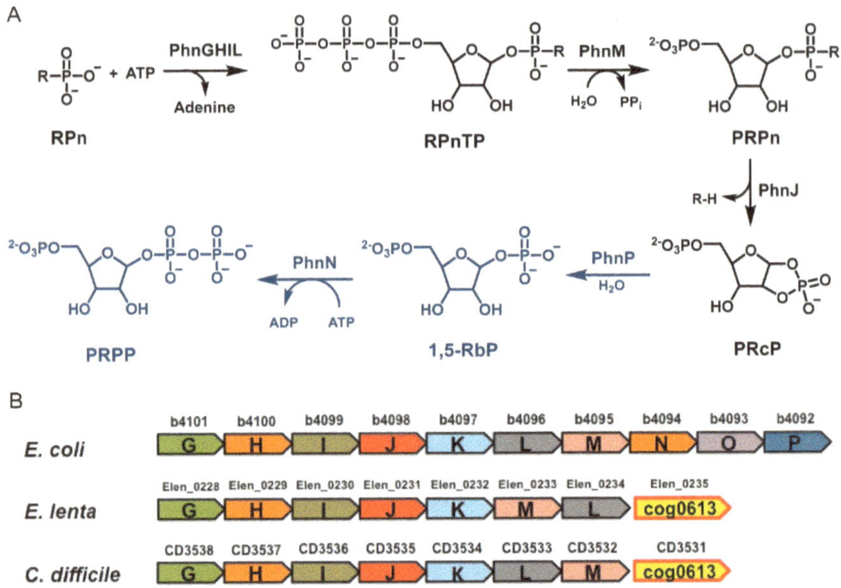

Fig. 6 (A) Scheme showing the intermediates in the metabolism of phosphonates by the C–P lyase pathway in *E. coli* and related organisms. (B) Schematic representation of organization of open reading frames encoding enzymes that constitute the C–P lyase pathway in select organisms.

was chemically synthesized using a modified procedure from the literature (Hove-Jensen et al., 2011). The ^{31}P NMR spectrum of this substrate showed two resonances corresponding to the two phosphorus atoms: the phosphate attached to the hydroxyl at C5 resonated at 4.57 ppm and the 1,2-cyclic phosphate at 19.31 ppm. The corresponding proton-coupled ^{31}P NMR spectrum showed a triplet and a doublet of doublets for the two chemical shifts, respectively. When PRcP was incubated with *E. coli* PhnP, it was hydrolyzed to 1,5-RbP, as determined by the disappearance of the signal at 19.31 ppm corresponding to the cyclic phosphate and the presence of two resonance signals at 4.57 ppm for the phosphate at C5 (triplet in ^{1}H-coupled NMR) and 3.02 ppm for the phosphate at C1 (doublet in ^{1}H-coupled NMR). Incubation of PRcP and *E. lenta* Elen0235 showed complete consumption of the cyclic phosphate species; however, the product mixture showed resonance signals at 4.56 ppm for the phosphate at C5 and 3.20 ppm for P_i. The ^{1}H-coupled spectrum demonstrated that the resonance at 4.56 ppm was a triplet, while the one at 3.20 was a singlet. This led us to the hypothesis that the enzymatic reaction resulted in the formation of ribose-5-phosphate (R5P) and inorganic phosphate (P_i).

These conclusions were confirmed by NMR analysis of product mixtures with independent addition of standard samples of R5P and P_i. These data showed that both PhnP and Elen0235 hydrolyzed PRcP efficiently, albeit to distinct products: 1,5-RbP by PhnP, and R5P and P_i by Elen0235.

The above experiments showed that Elen0235 catalyzed an unprecedented enzymatic reaction of an efficient hydrolysis of a cyclic phosphodiester to inorganic phosphate and a vicinal diol. Since both phosphodiester bonds are being hydrolyzed in this reaction, either 1,5-RbP or 2,5-ribose bisphosphate (2,5-RbP) should function as reaction intermediates. The enzymatic hydrolysis of PRcP by PhnP was used to obtain 1,5-RbP. Acid hydrolysis of PRcP followed by anion exchange chromatography was used to obtain pure 2,5-RbP (Fathi & Jordan, 1988). When 1,5-RbP and 2,5-RbP were incubated with Elen0235 in separate reactions, there was no change in the ^{31}P NMR spectrum of 1,5-RbP, while the reaction mixture with 2,5-RbP showed the formation of two new signals. These two resonance signals coincided with the products of enzymatic hydrolysis of PRcP by Elen0235 and were confirmed to be D-ribose-5-phosphate and inorganic phosphate (P_i). This showed that the enzyme hydrolyzed the cyclic phosphodiester to a phosphomonoester (2,5-RbP), which was further hydrolyzed to inorganic phosphate.

Prior studies on the enzymatic mechanism of hydrolysis by AHS enzymes, including our study on HPPs, demonstrated that the hydrolysis of phosphoesters by this family of enzymes occurs by nucleophilic attack of water at the phosphorus center (Ghodge, Fedorov, et al., 2013). We expected the hydrolysis of 2,5-RbP to R5P and inorganic phosphate to occur by this mechanism. It was not clear, however, whether the hydrolysis of the initial cyclic phosphate occurred by the nucleophilic attack of water at C1 of ribose (PRcP) or the cyclic P-center. This question was addressed by carrying out the enzymatic reaction in ^{18}O-labeled water. Literature reports have shown that for every substitution of ^{18}O for ^{16}O, the ^{31}P-resonance of phosphate esters shifts upfield by about 0.021 ppm and is additive for multiple substitutions (Cohn & Hu, 1978; Villafranca & Raushel, 1980). If the hydroxide attacks at the P-center in both hydrolytic steps, then one should expect three species corresponding to the phosphate product, containing 0, 1, and 2 oxygen-18 labels, and three corresponding resonance signals separated by 0.021 ppm each. The results showed three species corresponding to P_i and the percentage of the phosphate species containing 0, 1, and 2 atoms were consistent with the ratio of ^{16}O and ^{18}O in the water used to hydrolyze the substrate. The results clearly demonstrated that both water molecules used in the overall reaction attack directly at the phosphorus center.

These studies showed that the enzyme hydrolyzed phosphoester bonds by facilitating nucleophilic attack at the P-center. Additionally, the enzyme is capable of distinguishing between the phosphoester linkage at the anomeric carbon of ribose in the cyclic phosphate of PRcP and the phosphomonoester of 1,5-RbP. The enzyme efficiently distinguished between the two phosphomonoesters: 1,5-RbP and 2,5-RbP. We next determined the Michaelis–Menten constants of this enzymatic reaction starting with PRcP and 2,5-RbP, respectively. The apparent kinetic constants for the hydrolysis of PRcP at pH 8.0 and 25°C are as follows: $k_{cat} = 2.0\,s^{-1}$, $K_m = 14\,\mu M$, and $k_{cat}/K_m = 1.4 \times 10^5\,M^{-1}\,s^{-1}$. The kinetic constants for hydrolysis of 2,5-RbP by Elen0235 are as follows: $k_{cat} = 7.4\,s^{-1}$, $K_m = 23\,\mu M$, and $k_{cat}/K_m = 3.3 \times 10^5\,M^{-1}\,s^{-1}$. The rate constants for the enzymatic hydrolysis of PRcP and 2,5-RbP by Elen0235 are thus very similar to one another. To determine whether or not Elen0235 catalyzed the hydrolysis of PRcP processively, 4.0 m*M* PRcP was incubated with 0.5 μ*M* Elen0235 at pH 8.0 until the reaction proceeded to ~50% completion. The reaction was quenched by addition of 0.5 m*M* EDTA, and the reaction mixture was passed through a 10-kDa molecular weight cutoff centrifugal filter to remove the enzyme. The pH of the filtrate was adjusted to 5.8, and the products of the reaction were analyzed by ^{31}P NMR spectroscopy to determine whether or not the intermediate, 2,5-RbP, was released into solution. The products mixture clearly showed the formation of a significant amount of 2,5-RbP as an intermediate. This observation demonstrates that the reaction mechanism is not processive and confirms that 2,5-RbP is a kinetically competent intermediate that is released from the enzyme site as the initial product of PRcP hydrolysis.

This study helped to elucidate the mechanism of hydrolysis of PRcP to R5P and P_i. The enzyme hydrolyzes the cyclic phosphate in PRcP to 2,5-RbP, which in turn is hydrolyzed to R5P and P_i sequentially. Based on the conservation of metal-binding residues between HPP and Elen0235, it is clear that Elen0235 possesses a trinuclear active site. We hypothesize that the α- and β-metal ions activate the bridging hydroxide ligand for nucleophilic attack at the P-center of the cyclic phosphate of PRcP and the phosphomonoester of 2,5-RbP, while the γ-metal ion serves as the Lewis acid for the leaving hydroxyl/alcohol functional group (Scheme 3).

A protein sequence BLAST analysis using Elen0235 primary sequence as bait revealed that a very limited number of organisms possess this gene.

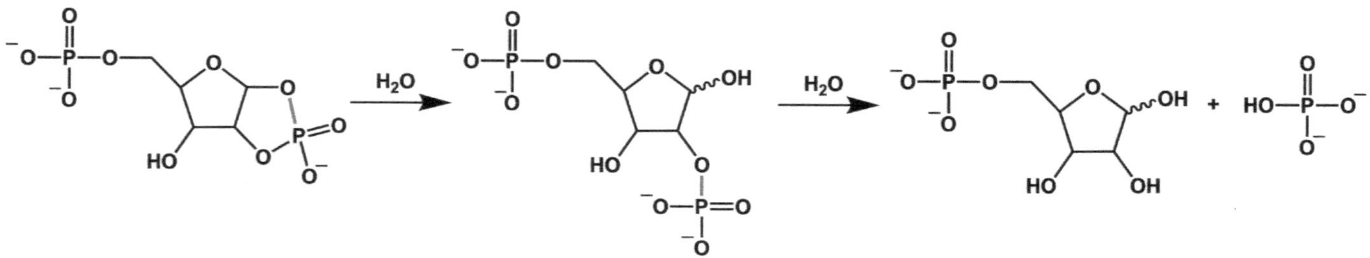

PRcP 2,5-RbP R5P Pi

Scheme 3 Overall reaction catalyzed by cyclic phosphate dihydrolase (cPDH).

The relatively short list of organisms, however, includes some highly virulent human intestinal pathogens like *C. difficile* and *E. lenta*, which are known to resist several antibiotics. This raises the question of whether the conversion of a phosphonate to inorganic phosphate provides any specific metabolic benefit(s) to these organisms. The other product, R5P, may be converted to PRPP by PRPP synthetase that transfers the pyrophosphate group from ATP to R5P, and a copy of which is present in every organism that possesses an ortholog of Elen0235.

4. 3′,5′-ADENOSINE BISPHOSPHATE-3′-PHOSPHATASE (CUMMINGS ET AL., 2014)

After the discovery of the unique cyclic phosphate dihydrolase (cPDH) enzymatic activity in cog0613, we turned our focus to two of the most populated clusters from this cog. The gene product for Cv1693 from *Chromobacterium violaceum* was cloned, heterologously expressed, and purified. The functional annotation of this enzyme began with the determination of its crystal structure. Preliminary attempts at crystallization of this protein yielded thin spindly needles, and their properties did not improve by grid screening protocols. A survey of the PDB showed that the structure of Bad1165 (PDB: 3E0F, 3O0F) had been determined with bound adenosine monophosphate (AMP) in the active site. Fluorescence-based thermal denaturation studies showed that addition of mono- and diphosphorylated nucleotides improved the thermal stability of Cv1693. The addition of AMP, cytidine monophosphate (CMP), or inosine monophosphate (IMP) to the original crystallization conditions of Cv1693 yielded higher quality crystals. Crystals grown in the presence of IMP were the best quality from the point of view of the X-ray diffraction data; however, they had low occupancy of metal ions in the anticipated trinuclear active site of Cv1693 based on a partially refined X-ray crystal structure and ICP-MS analysis. An alternate strategy was adopted in which EDTA was added to remove all metal ions, and divalent manganese was added directly to the protein solution. The final crystal structure contained AMP and phosphate bound within the trinuclear (Mn^{2+}) active site of Cv1693.

The structure of Cv1693 is unique among amidohydrolases in general, and even quite distinct as compared to that of HPP, although they possess the same overall distorted TIM-barrel fold (Fig. 7). In Cv1693, the third β-strand of the barrel, β-strand 3, is significantly longer than the strand usually seen for other members of the AHS. Furthermore, the bridging ligand

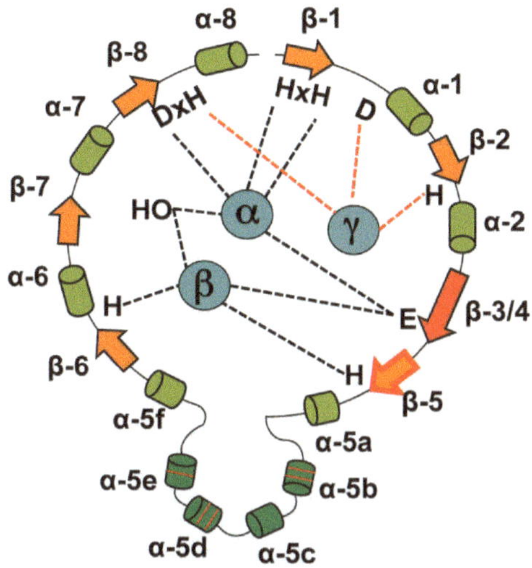

Fig. 7 Schematic representation of the secondary structure of Cv1693. The α-, β-, and γ-metal sites are shown as *blue spheres*. The strands and helices that constitute the distorted TIM- or (β/α)7-barrel are colored *orange* and *light green*, respectively. The long β-strand-3/4 is colored *red*, while the antiparallel β-strand 5, unique to members of cog0613 among AHS members, is highlighted with a *red border*. The α-helices in the insertion element are colored *dark green*, and the positions of the enzyme residues seen to interact with the bound 5′-AMP are shown as *red lines*.

between two of the metals (M_α and M_β) is usually located at the end of β-strand 3 or 4, but in Cv1693, the bridging ligand (Glu64) is positioned at the center of elongated β-strand 3. Therefore, in keeping with the β-strand numbering scheme for other members of the AHS used to designate key active site residues with respect to their β-strand number, we chose to designate β-strand 3 as β-strand 3/4. While most AHS (β/α)-barrels are composed completely of parallel β-strands, the fifth β-strand of Cv1693, β-strand 5, is antiparallel to β-strand 3/4 and β-strand 6. In addition, as compared to HPP, Cv1693 lacks two (β/α)-barrel α-helices (α_3 and α_4) and includes an insertion (amino acids 83–171, α_{5b}–α_{5e}) between β-strand 5 and α_{5f}. One of the helices, α_{5a}, assumes the typical position of α_3, and the other, α_{5a}–α_{5e}, form a four-helix bundle that caps the top of the (β/α)-barrel and contains several residues that contact adenosine 5′-monophosphate. These residues include Arg99, Arg102, Arg134, Thr135, and Phe154.

The active site of Cv1693, with AMP and inorganic phosphate ligands bound, is shown in Fig. 8. Mn_α is coordinated to His7 and His9 at the end of β-strand 1 and Asp248 at the end of β-strand 8. Mn_β is ligated to His75 (β-strand 4) and His191 (end of β-strand 6). The third metal coordinates to Asp14 (β_1–α_1 loop), His39 (β_2–α_2 loop), and His250 (β_8–α_8 loop). The orthophosphate forms a bridge from Mn_α to Mn_β but also from Mn_α to Mn_γ, and is additionally coordinated by the guanidinium group of Arg194 projecting from the β_6–α_6 loop. All of the contacts between Cv1693 and AMP originate from the insertion domain between β-strand 5 and α_{5f}. Arg99 and Arg103 from α_{5b} and Arg134 from α_{5d} form salt bridges to the 5′-phosphate of AMP. In the structure without a nucleotide ligand, this binding pocket was occupied by a sulfate ion. The only contact to the adenine base is a stacking interaction with Phe154 (α_{5e}). The 2′-OH group of the ribose moiety forms a hydrogen bond to Thr135, and the 2′-OH (2.4 Å) and 3′-OH (2.5 Å) groups are coordinated to the third metal (M_γ). Interestingly, the 3′-OH group is only 2.5 Å from an oxygen of the metal-bound orthophosphate.

While determining the substrate specificity of Cv1693, we hypothesized that it is a phosphatase, based on the conservation of metal-binding residues that characterize the active site of HPP, as well as cPDH (Ghodge, Cummings, et al., 2013; Ghodge, Fedorov, et al., 2013). In addition, Bad1165, whose structure was solved with bound AMP, but whose substrate

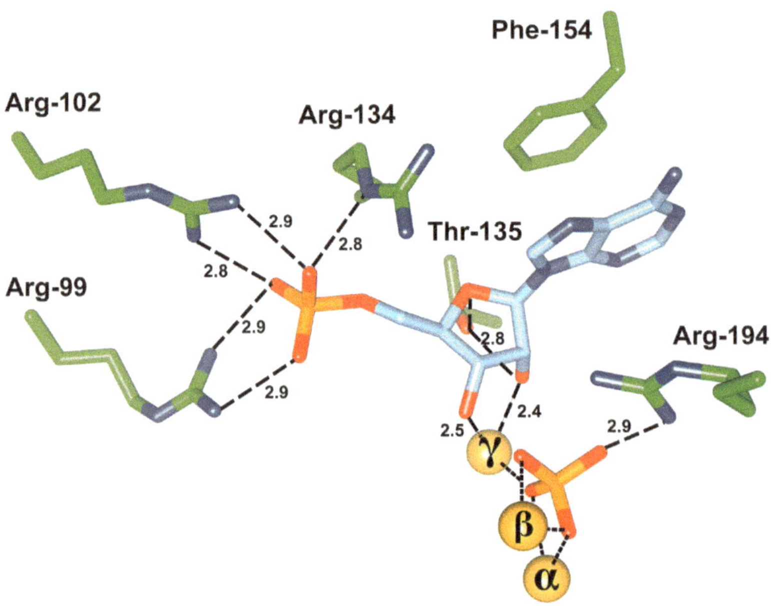

Fig. 8 Interactions of AMP and phosphate with enzyme residues at the active site of Cv1693. Enzyme residues are shown in *green*, active metal ions (manganese) are shown as *yellow spheres*, the AMP backbone is shown in *light blue*, while inorganic phosphate is shown in *yellow* and *orange*. Distances between atoms are indicated in angstrom units.

specificity was unknown, had been shown to catalyze the hydrolysis of *p*-nitrophenyl phosphate, but the hydrolysis of pNp-like substrates has apparently not been investigated (Han et al., 2011). Finally, the proximity of orthophosphate in the active site of Cv1693 to the 3′-OH group of the bound AMP in the ternary complex suggested that this enzyme would likely hydrolyze compounds such as 3′,5′-cAMP, 3′,5′-pAp, 2′,5′-pAp, or 3′-AMP. Testing a small substrate library demonstrated that Cv1693 efficiently hydrolyzes 3′,5′-pAp to orthophosphate and 5′-AMP with k_{cat} and k_{cat}/K_m values of $22\,s^{-1}$ and $1.4 \times 10^6\ M^{-1}\ s^{-1}$, respectively (Scheme 4; Table 2). The enzyme has a rather broad specificity for the identity of the

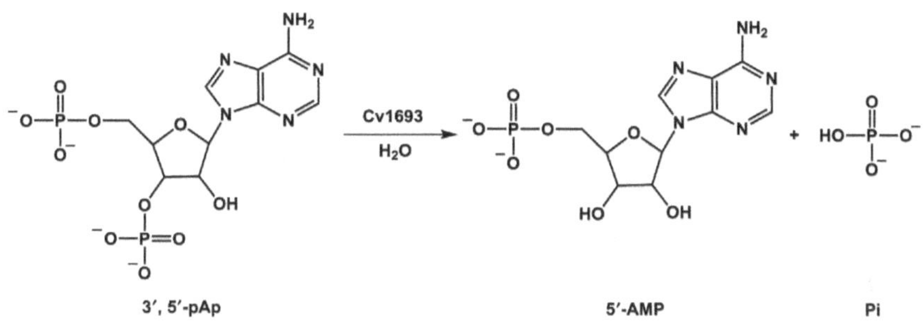

Scheme 4 Reaction catalyzed by 3′,5′-adenosine bisphosphate 3′-phosphatase.

Table 2 Kinetic Parameters of Cv1693 With a Library of Substrates

Compound	k_{cat} (s^{-1})	K_m (μM)	k_{cat}/K_m ($M^{-1}\ s^{-1}$)
pAp	22	16	1.4×10^6
2′-Deoxy-pAp	7.1	10	7.0×10^5
pCp[a]	8.6	14	6.2×10^5
2′-Deoxy-pCp	9.4	26	3.6×10^5
pGp[a]	12	17	7.1×10^5
2′-Deoxy-pGp	4.2	5.2	7.9×10^5
pUp[a]	5.4	8.8	6.1×10^5
2′-Deoxy-pUp	5.2	29	1.8×10^5
pTp	3.3	9.3	3.6×10^5
2′-Deoxy-pIp	4.3	22	1.9×10^5
pApA	0.05	1.6	3.2×10^4
N^6-methyl-3′,5′-pAp	12	5.4	2.1×10^6

[a]These assays were conducted with 1:1 mixtures of the 2′,5′- and 3′,5′-nucleotide bisphosphates.

aromatic base, and the 2′-hydroxyl is relatively unimportant for substrate recognition. Additionally, the enzyme can slowly hydrolyze short oligonucleotides with a terminal 5′-phosphate but will not hydrolyze short oligonucleotides without a terminal 5′-phosphate. The rate of hydrolysis of the dinucleotide pApA is two orders of magnitude lower relative to 3′,5′-pAp. Thus, short oligonucleotides may not be the biological substrates of this enzyme.

These results showed that Cv1693 was capable of efficiently hydrolyzing 3′,5′-adenosine bisphosphate in vitro. We wanted to determine if the enzyme would carry out the physiological function of hydrolyzing 3′,5′-pAp in vivo. There are two biological pathways/reactions in which 3′,5′-pAp is released as a by-product: the transfer of a sulfate from 3′-phosphoadenosine 5′-phosphosulfate (PAPS) and the transfer of the phosphopantetheine group of CoASH to the acyl-carrier protein (Hatzios, Iavarone, & Bertozzi, 2008; Mootz, Finking, & Marahiel, 2001). The CysQ enzyme from *E. coli* and *Mycobacterium tuberculosis*, and the NrnA (YtqI) enzyme from *Bacillus subtilis*, *M. tuberculosis*, *Mycoplasma pneumoniae*, and *Streptococcus mutans* have been assigned the physiological role of hydrolyzing 3′,5′-pAp (Hatzios et al., 2008; Mechold, Fang, Ngo, Ogryzko, & Danchin, 2007; Neuwald et al., 1992; Spielberg, Xiong, Smith, Gu, & York, 1999). CysQ belongs to cog1218 and NrnA belongs to cog0618. We retrieved 964, 830, and 856 protein sequences from cogs 0613, 1218, and 0618, respectively, from a total of 1752 representative bacterial genomes deposited in the MicrobesOnline database (October 2013) (Dehal et al., 2009). SSN diagrams were constructed for each cog independently at a BLAST E-value cutoff of 1×10^{-70}. We then mapped Cv1693, CysQ, and NrnA in the three SSN diagrams, as well as other enzymes reported to hydrolyze 3′,5′-pAp 3′-phosphate from these cogs. All enzyme sequences that were connected to these mapped sequences possessed a sequence identity of $\geq$40% in these diagrams and were then annotated as 3′,5′-pAp 3′-phosphatases. We were able to provisionally annotate 118 enzymes from cog0613, 309 from cog1218, and 283 from cog0618 as pAp phosphatases in this manner. When we compared the list of organisms to which each of these enzymes belonged, we found that only ~6% of the organisms had apparent pAp phosphatases from two of the three cogs discussed here, while the remaining organisms had a pAp phosphatase from just one of the three cogs. There was no instance of an organism possessing pAp phosphatases from all three cogs. Further examination of the phylogenetic classification of each organism revealed that organisms possessing the pAp phosphatase from cog0613 are predominantly betaproteobacteria (~97%), those from cog1218 are largely gammaproteobacteria (~74%), while those from cog0618 are largely bacilli from the phylum Firmicutes (80%). This bioinformatic analysis showed that different classes of organisms have evolved enzymes from different families to catalyze identical molecular function. It also provided strong evidence that Cv1693 and its closely related orthologs from cog0613 performed the function of hydrolyzing 3′,5′-pAp in vivo, since the vast majority of organisms that possess a closely related ortholog of Cv1693 do not possess an enzyme from cog1218 (CysQ) or cog0618 (YtqI or NrnA) that can catalyze the same reaction.

5. RNase AM (GHODGE & RAUSHEL, 2015)

Out of the two most populated clusters from the SSN diagram of cog0613, one contained Cv1693 and related protein sequences, which were shown to efficiently hydrolyze 3′,5′-pAp and similar nucleotides. The other cluster contained a group of protein sequences of unknown function that contained a gene product annotated as TrpH or YciV (now renamed as RNase AM) from *E. coli*. The original annotation nomenclature alluded to the protein being involved in tryptophan metabolism; however, the sequence network analysis and conservation of metal-binding residues from the PHP family indicated that it was more likely to be a phosphatase that hydrolyzes nucleotide substrates similar to those of Cv1693. The purified RNase AM protein, when incubated with 3′,5′-pAp, released phosphate over time. This enzyme was capable of hydrolyzing 3′,5′-pAp to 5′-AMP and phosphate with a k_{cat} of $10\,s^{-1}$ and a k_{cat}/K_m of $1.8 \times 10^5\ M^{-1}\ s^{-1}$. Removal of the 5′-phosphoryl group from 3′,5′-pAp resulted in a 360-fold reduction in k_{cat}/K_m, obtained from the kinetic parameters for the hydrolysis of 3′-AMP. The enzyme is able to hydrolyze the 2′-phosphate from 2′,5′-pAp with a k_{cat}/K_m of 100 $M^{-1}\ s^{-1}$, which is 1000-fold lower than that of 3′,5′-pAp. RNase AM also hydrolyzed other 3′,5′-bisphosphonucleotides, including 2′-deoxy 3′,5′-bisphosphonucleotides, with k_{cat}/K_m in the range of 10^4–10^5 $M^{-1}\ s^{-1}$ (Table 3). The identity of the purine/pyrimidine base in the 3′,5′-bisphosphonucleotide substrates did not have a significant effect on the kinetic parameters of hydrolysis by RNase AM. In general, RNA substrates had catalytic parameters higher than those of DNA substrates of the same length. The substrate specificity of RNase AM resembled that of Cv1693.

Table 3 Kinetic Parameters of RNase AM With a Library of Substrates

Compound	k_{cat} (s^{-1})	K_m (mM)	k_{cat}/K_m (M^{-1} s^{-1})
pAp	9.9	56	1.8×10^5
2′-Deoxy-pAp	5.6	53	1.1×10^5
pCp (2′,5′/3′,5′ mixture)[a]	6.0	86	7.0×10^4
2′-Deoxy-pCp[a]	1.9	114	1.7×10^4
pGp (2′,5′/3′,5′ mixture)[a]	5.5	45	1.2×10^5
2′-Deoxy-pGp[a]	1.7	30	5.8×10^4
pUp (2′,5′/3′,5′ mixture)[a]	4.7	59	8.0×10^4
2′-Deoxy-pUp[a]	2.1	110	1.9×10^4
2′-Deoxy-pTp[a]	1.6	32	4.9×10^4
2′-Deoxy-pIp[a]	3.3	48	6.9×10^4
3′-AMP	0.22	420	5.0×10^2
2′,5′-pAp	ND	ND	100 ± 10
p(Gp)G	0.60	14	4.1×10^4
p(Ap)A	14	160	9.0×10^4
$p(Ap)_2A$	3.8	55	7.0×10^4
$p(Ap)_3A$	3.6	45	7.9×10^4
$p(Ap)_4A$	2.7	59	4.6×10^4
p(dA)dA	2.5	150	1.7×10^4
$p(dA)_2dA$	0.72	67	1.1×10^4
$p(dA)_3dA$	1.20	176	6.8×10^3
$p(dA)_4dA$	0.79	280	2.9×10^3

[a]Compounds assayed using Pi detection assay. Other substrates were assayed using the adenylate kinase-pyruvate kinase-lactate dehydrogenase coupling assay. For more details, see Ghodge and Raushel (2015).

RNase AM was additionally able to hydrolyze oligonucleotides with much enhanced efficiency as compared to Cv1693. The enzyme hydrolyzed p(Ap)A with a k_{cat} of 14 s^{-1} and a k_{cat}/K_m of 9×10^4 M^{-1} s^{-1}, while the corresponding parameters for the hydrolysis of $p(Ap)_4A$ were 2.7 s^{-1} and 4.6×10^4 M^{-1} s^{-1}. The k_{cat} and k_{cat}/K_m for hydrolysis of the corresponding 2′-deoxy substrates p(dAp)dA and $p(dAp)_4dA$ were 2.5 s^{-1} and 1.7×10^4 M^{-1} s^{-1}, and 0.79 s^{-1} and 2.9×10^3 M^{-1} s^{-1}. Overall, the presence of a

2′-hydroxyl group on the oligonucleotide substrates results in a 2- to 15-fold increase in the catalytic efficiency of hydrolysis by RNase AM compared to that of the corresponding 2′-deoxy substrates. The maximum length of a single stranded oligonucleotide tested during this study was a 14-mer, and in each case, RNase AM hydrolyzed the substrate to the constituent 5′-nucleotide monophosphates. The enzyme was able to hydrolyze p(Ap)$_5$ at a rate comparable to that of p(Ap)$_4$A but unable to hydrolyze (Ap)$_4$A at a detectable rate. Substrates must therefore be phosphorylated at the 5′-end of the oligonucleotide, but it makes no difference whether the 3′-end is phosphorylated. Monitoring the progress of a model 6-mer oligonucleotide using anion exchange chromatography revealed the progressive formation and hydrolysis of oligonucleotides of shorter lengths (5-mer, 4-mer, and 3-mer) with corresponding accumulation of 5′-nucleotide monophosphates. This observation implied that the enzyme functions as an exonuclease by hydrolyzing 5′-phosphorylated oligonucleotides in the 5′ → 3′ direction (Scheme 5).

The enzyme is additionally capable of hydrolyzing double-stranded oligonucleotides (Fig. 9). However, the rate of hydrolysis of double-stranded DNA is approximately 10-fold lower than that of corresponding single-stranded DNA. The enzyme is able to hydrolyze an overhang at the 5′-terminus of a double-stranded DNA with a terminal 5′-phosphate, at a rate comparable to that of the corresponding single-stranded DNA. Double-stranded DNA with a 3′-extension on the reverse strand was hydrolyzed at a rate twofold lower than that of completely paired double-strand, and a rate 20-fold lower than that of the single-stranded DNA.

The best substrate for RNase AM according to this study in terms of the catalytic parameters was 3′,5′-pAp. However, as described in the previous section, *E. coli* possesses CysQ, which has been ascribed the physiological role of hydrolysis of 3′,5′-pAp by extensive biochemical studies (Neuwald et al., 1992). RNase AM, therefore, may be involved in hydrolyzing short

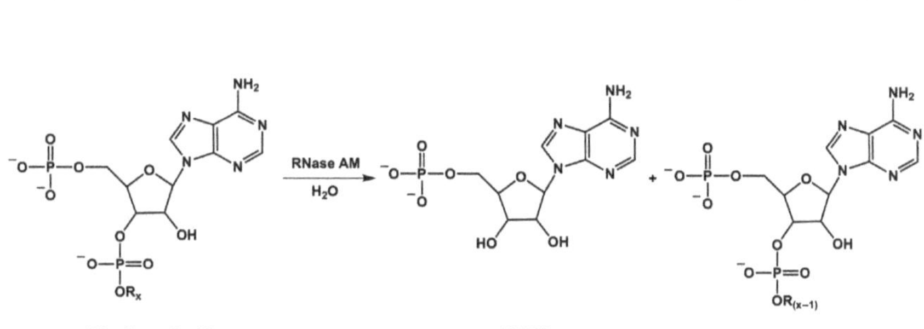

Scheme 5 Reaction catalyzed by RNase AM.

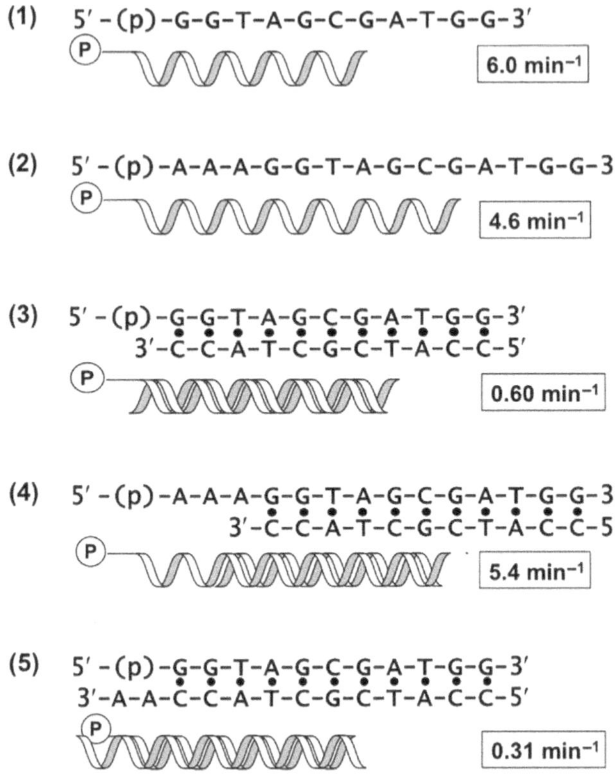

Fig. 9 DNA oligonucleotides tested as substrates with RNase AM, and their respective specific rate constants of hydrolysis in adjacent *boxes*.

oligonucleotides. Exonucleases that act in the $5' \rightarrow 3'$ direction have been identified in several organisms from diverse phylogenetic backgrounds. However, *E. coli* was not known to possess such a ribonuclease until the discovery of RNase AM. Degradation of larger oligonucleotides by the action of other nucleases gives rise to formation of oligonucleotides that may be substrates for RNase AM. Further cell and/or chemical biology studies will help probe the exact physiological role of RNase AM.

In summary, the studies described here demonstrate an effective strategy for annotating previously unknown enzyme functions using a combination of biochemistry, biophysics, and bioinformatics. This work commenced with a focus on understanding the enzyme sequence–structure–function relationship for a known enzymatic reaction. This was accomplished by obtaining crystal structures of the enzyme with substrate/product analogs bound at the active site, an extensive array of biochemical experiments, and extending the characterization to a statistically significant number of diverse orthologs. These efforts helped elucidate the enzymatic mechanism of a trinuclear phosphatase utilizing d-block metal ions at the active site. The knowledge thus generated was leveraged to discover the cognate substrates of a closely related set of enzymes with no previously known functions. The first of these functions discovered was an enzyme, cPDH, which hydrolyzed a cyclic phosphodiester sequentially to inorganic phosphate and vicinal diol, and is found in difficult-to-treat human gut pathogens like *E. lenta* and *C. difficile*. The second function discovered revealed a phosphatase from betaproteobacteria that hydrolyzed the 3′-phosphate from 3′,5′-bisphosphoadenosine, a by-product from sulfate as well as coenzyme A metabolism. The third enzyme function discovered in this work was that of RNase AM, the first 5′–3′ exonuclease reported from the extensively studied *E. coli* and related organisms. The discovery of cPDH and RNase AM has provided two rare specimens among phosphatases, which can hydrolyze phosphomonoesters and phosphodiesters with comparable efficiency, and yet, strict substrate specificity. Further structural biology studies on these enzymes will help inform efforts for de novo protein design and engineering of phosphatases for commercial applications. Replication of this overall strategy for other families of enzymes, as is underway in several cases (Akiva et al., 2014; Gerlt, Babbitt, Jacobson, & Almo, 2012; Huang et al., 2015; Mashiyama et al., 2014) will help accelerate the rate of reliable protein function annotations.

REFERENCES

Akiva, E., Brown, S., Almonacid, D. E., Barber, A. E., II, Custer, A. F., Hicks, M. A., et al. (2014). The structure-function linkage dadatbase. *Nucleic Acids Research*, *42*(D1), D521–D530.

Alifano, P., Fani, R., Lio, P., Lazcano, A., Bazzicalupo, M., Carlomagno, M. S., et al. (1996). Histidine biosynthetic pathway and genes: Structure, regulation, and evolution. *Microbiological Reviews*, *60*, 44–69.

Aravind, L., & Koonin, E. V. (1998). Phosphoesterase domains associated with DNA polymerases of diverse origins. *Nucleic Acids Research*, *26*, 3746–3752.

Aubert, S. D., Li, Y., & Raushel, F. M. (2004). Mechanism for the hydrolysis of organophosphates by bacterial phosphotriesterase. *Biochemistry*, *43*, 5707–5715.

Chen, C. M., Ye, Q. Z., Zhu, Z. M., Wanner, B. L., & Walsh, C. T. (1990). Molecular biology of carbon-phosphorus bond cleavage. Cloning and sequencing of the phn (psiD) genes involved in alkylphosphonate uptake and C-P lyase activity in *Escherichia coli* B. *The Journal of Biological Chemistry*, *265*, 4461–4471.

Cohn, M., & Hu, A. (1978). Isotopic (^{18}O) shift in ^{31}P nuclear magnetic resonance applied to a study of enzyme-catalyzed phosphate—phosphate exchange and phosphate (oxygen)—water exchange reactions. *Proceedings of the National Academy of Sciences of the United States of America*, *75*, 200–203.

Cook, P. F., & Cleland, W. W. (2007). Enzyme kinetics and mechanism. In R. L. Rogers (Ed.), New York: Garland Science Publishing, pp. 325–357.

Cummings, J. A., Vetting, M., Ghodge, S. V., Xu, C., Hillerich, B., Seidel, R. D., et al. (2014). Prospecting for unannotated enzymes: Discovery of a 3′,5′-nucleotide bisphosphate phosphatase within the amidohydrolase superfamily. *Biochemistry, 53*, 591–600.

Dehal, P. S., Joachimiak, M. P., Price, M. N., Bates, J. T., Baumohl, J. K., Chivian, D., et al. (2009). MicrobesOnline: An integrated portal for comparative and functional genomics. *Nucleic Acids Research, 38*, 396–400.

Fathi, R., & Jordan, F. (1988). Certain novel ribofuranosyl phosphates derived from 5-phospho-alpha-D-ribofuranosyl-1-pyrophosphate: Synthesis, structure, and alkaline hydrolytic reactivities. *Journal of Organic Chemistry, 53*, 1997–2001.

Gerlt, J. A., Babbitt, P. C., Jacobson, M. P., & Almo, S. C. (2012). Divergent evolution in the enolase superfamily: Strategies for assigning functions. *The Journal of Biological Chemistry, 287*, 29–34.

Ghodge, S. V., Cummings, J. A., Williams, H. J., & Raushel, F. M. (2013). Discovery of a cyclic phosphodiesterase that catalyzes the sequential hydrolysis of both ester bonds to phosphorus. *Journal of the American Chemical Society, 135*, 16360–16363.

Ghodge, S. V., Fedorov, A. A., Fedorov, E. V., Hillerich, B., Sidel, R., Almo, S. C., et al. (2013). Structural and mechanistic characterization of L-histidinol phosphate phosphatase from the polymerase and histidinol phosphatase family of proteins. *Biochemistry, 52*, 1101–1112.

Ghodge, S. V., & Raushel, F. M. (2015). Discovery of a previously unrecognized ribonuclease from *Escherichia coli* that hydrolyzes 5′-phosphorylated fragments of RNA. *Biochemistry, 2015*(54), 2911–2918.

Han, G. W., Ko, J., Farr, C. L., Deller, M. C., Xu, Q., Chiu, H. J., et al. (2011). Crystal structure of a metal-dependent phosphoesterase (YP_910028.1) from *Bifidobacterium adolescentis*: Computational prediction and experimental validation of phosphoesterase activity. *Proteins: Structure, Function, and Bioinformatics, 79*, 2146–2160.

Hatzios, S. K., Iavarone, A. T., & Bertozzi, C. R. (2008). Rv2131c from *Mycobacterium tuberculosis* is a CysQ 3′-phosphoadenosine-5′-phosphatase. *Biochemistry, 47*, 5823–5831.

Holm, L., & Sander, C. (1997). An evolutionary treasure: Unification of a broad set of amidohydrolases related to urease. *Proteins: Structure, Function, and Genetics, 28*, 72–82.

Hove-Jensen, B., McSorley, F. R., & Zechel, D. L. (2011). Physiological role of phnP-specified phosphoribosyl cyclic phosphodiesterase in catabolism of organophosphonic acids by the carbon-phosphorus lyase pathway. *Journal of the American Chemical Society, 133*, 3617–3624.

Hove-Jensen, B., McSorley, F. R., & Zechel, D. L. (2012). Catabolism and detoxification of 1-aminoalkylphosphonic acids: N-acetylation by the phnO gene product. *PLoS One*, 7, e46416.

Huang, H., Pandya, C., Liu, C., Al-Obaidi, N. F., Wang, M., Zheng, L., et al. (2015). Panoramic view of superfamily of phosphatases through substrate profiling. *Proceedings of the National Academy of Sciences of the United States of America, 112*, E1974–E1983.

Kamat, S. S., Williams, H. J., Dangott, L. J., Chakrabarti, M., & Raushel, F. M. (2013). The catalytic mechanism for aerobic formation of methane by bacteria. *Nature, 497*, 132–136.

Kamat, S. S., Williams, H. J., & Raushel, F. M. (2011). Intermediates in the transformation of phosphonates to phosphate by bacteria. *Nature, 495*, 570–573.

Kim, J., Tsai, P. C., Chen, S. L., Himo, F., Almo, S. C., & Raushel, F. M. (2008). Structure of diethyl phosphate bound to the binuclear metal center of phosphotriesterase. *Biochemistry, 47*, 9497–9504.

Lai, W. L., Chou, L. Y., Ting, C. Y., Kirby, R., Tsai, Y. C., Wang, A. H., et al. (2004). The functional role of the binuclear metal center in D-aminoacylase: One-metal activation and second-metal attenuation. *The Journal of Biological Chemistry*, *279*, 13962–13967.

Martí-Arbona, R., Fresquet, V., Thoden, J. B., Davis, M. L., Holden, H. M., & Raushel, F. M. (2005). Mechanism of the reaction catalyzed by isoaspartyl dipeptidase from Escherichia coli. *Biochemistry*, *44*, 7115–7124.

Mashiyama, S. T., Malabanan, M. M., Akiva, E., Bhosle, R., Branch, M. C., Hillerich, B., et al. (2014). Large scale determination of sequence, structure and function relationships in cytosolic glutathione transferases across the biosphere. *PLoS Biology*, *12*, e1001843.

Masson, P., & Lushchekina, S. V. (2016). Emergence of catalytic bioscavengers against organophosphorus agents. *Chemico-Biological Interactions*, *259*, 319–326.

Mechold, U., Fang, G., Ngo, S., Ogryzko, V., & Danchin, A. (2007). YtqI from Bacillus subtilis has both oligoribonuclease and pAp-phosphatase activity. *Nucleic Acids Research*, *35*, 4552–4561.

Metcalf, W. W., & Wanner, B. L. (1993). Evidence for a fourteen-gene, phnC to phnP locus for phosphonate metabolism in *Escherichia coli*. *Gene*, *129*, 27–32.

Mootz, H. D., Finking, R., & Marahiel, M. A. (2001). 4′-Phosphopantetheine transfer in primary and secondary metabolism of *Bacillus subtilis*. *The Journal of Biological Chemistry*, *276*, 37289–37298.

Neuwald, A. F., Krishnan, B. R., Brikun, I., Kulakausas, S., Suziedelis, K., Tomcsanyi, T., et al. (1992). CysQ, a gene needed for cysteine synthesis in *Escherichia coli* K-12 only during aerobic growth. *Journal of Bacteriology*, *174*, 415–425.

Omi, R., Goto, M., Miyahara, I., Manzoku, M., Ebihara, A., & Hirotsu, K. (2007). Crystal structure of monofunctional histidinol phosphate phosphatase from *Thermus thermophilus* HB8. *Biochemistry*, *46*, 12618–12627.

Rangarajan, E. S., Proteau, A., Wagner, J., Hung, M.-N., Matte, A., & Cygler, M. (2006). Structural snapshots of *Escherichia coli* histidinol phosphate phosphatase along the reaction pathway. *The Journal of Biological Chemistry*, *281*, 37930–37941.

Seibert, C. M., & Raushel, F. M. (2005). Structural and catalytic diversity within the amidohydrolase superfamily. *Biochemistry*, *44*, 6383–6391.

Spielberg, B. D., Xiong, J.-P., Smith, J. J., Gu, R. F., & York, J. D. (1999). Cloning and characterization of a mammalian lithium-sensitive bisphosphate 3′-nucleotidase inhibited by inositol 1,4-bisphosphate. *The Journal of Biological Chemistry*, *274*, 13619–13628.

Tatusov, R. L., Galperin, M. Y., Natale, D. A., & Koonin, E. V. (2000). The COG database: A tool for genome-scale analysis of protein functions and evolution. *Nucleic Acids Research*, *28*, 33–36.

Teplyakov, A., Obmolova, G., Khil, P. P., Howard, A. J., Camerini-Otero, R. D., & Gilliland, G. L. (2003). Crystal structure of the *Escherichia coli* YcdX protein reveals a trinuclear zinc active site. *Proteins: Structure, Function, and Genetics*, *51*, 315–318.

Thaller, M. C., Schippa, S., & Rossolini, G. M. (1998). Conserved sequence motifs among bacterial, eukaryotic, and archaeal phosphatases that define a new phosphohydrolase superfamily. *Protein Science*, 7, 1647–1652.

Thoden, J. B., Phillips, G. N., Jr., Neal, T. M., Raushel, F. M., & Holden, H. M. (2001). Molecular structure of dihydroorotase: A paradigm for catalysis through the use of a binuclear metal center. *Biochemistry*, *40*, 6989–6997.

Van Etten, R. L. (1982). Human prostatic acid phosphatase: A histidine phosphatase. *Annals of the New York Academy of Sciences*, *390*, 27–51.

Villafranca, J. J., & Raushel, F. M. (1980). Biophysical applications of NMR to phosphoryl transfer enzymes and metal nuclei of metalloproteins. *Annual Review of Biophysics and Bioengineering*, *9*, 363–392.

Widlansky, T. S., & Taylor, W. (1999). Chemistry and enzymology of phosphatases. *Comprehensive Natural Products Chemistry*, *5*, 139–162.

FURTHER READING

He, S. M., Wathier, M., Podzelinska, K., Wong, M., McSorley, F. R., Asfaw, A., et al. (2011). Structure and mechanism of PhnP, a phosphodiesterase of the carbon-phosphorus lyase pathway. *Biochemistry*, *50*, 8603–8615.

Hui, M. P., Foley, P. L., & Belasco, J. G. (2014). Messenger RNA degradation in bacterial cells. *Annual Review of Genetics*, *48*, 537–559.

Postic, G., Danchin, A., & Mechold, U. (2012). Characterization of NrnA homologs from *Mycobacterium tuberculosis* and *Mycoplasma pneumoniae*. *RNA*, *18*, 155–165.

CHAPTER SEVEN

Affinities and Comparisons of Enzyme States by Principal Component Analysis of NMR Spectra, Automated using TREND Software

Jia Xu[2], Steven R. Van Doren[1]
Biochemistry Department, University of Missouri, Columbia, MO, United States
[1]Corresponding author: e-mail address: vandorens@missouri.edu

Contents

[2] Current address: Magnetic Resonance Research Facility, University of Iowa, Iowa City, IA, United States.

Methods in Enzymology, Volume 607
ISSN 0076-6879
https://doi.org/10.1016/bs.mie.2018.05.016

Abstract

The new TREND NMR software package makes significant new insights into enzymes and other molecules easily accessible from collections of unassigned NMR spectra. TREND NMR uses unsupervised multivariate statistics to automate extraction of reaction courses for fitting, including binding isotherms from titrations detected by NMR spectra. The package also makes comparisons and groupings of NMR-detected enzyme states straightforward, by using principal component analysis (PCA). Such comparisons are illustrated for human protein tyrosine phosphatase 1B variants and inhibitor complexes. The "unfold" PCA-based comparisons of these protein phosphatase samples in one to three statistical dimensions are consistent with the recent structural characterizations of the samples, suggesting the relevance of quick assessment by PCA implemented semiautomatically in TREND NMR. The software is free for academic use. Step-by-step protocols are provided for measuring affinities and comparing molecular states using TREND NMR.

ABBREVIATIONS

BMRB	Biological Magnetic Resonance Data Bank
CONCISE	COordiNated ChemIcal Shifts bEhavior
FHA	forkhead associated
FID	free induction decay
GUI	graphical user interface
HMQC	heteronuclear multiple quantum coherence
HSQC	heteronuclear single quantum coherence
KAPP	kinase-associated protein phosphatase
NMR	nuclear magnetic resonance
NOESY	nuclear Overhauser effect and exchange spectroscopy
PC	principal component
PCA	principal component analysis
PMM/PGM	phosphomannomutase/phosphoglucomutase
PTP1B	protein tyrosine phosphatase 1B
RMS	root mean squared
ROI	region of interest
SD	standard deviation
SVD	singular value decomposition
TOCSY	total correlation spectroscopy
TREND	track equilibrium or nonequilibrium shifts in data
TROSY	transverse relaxation-optimized spectroscopy
WT	wild type

1. INTRODUCTION

Biophysical investigations of enzymes include studies of conformational equilibria and ligand binding. ^{15}N-resolved NMR spectra of protein backbones are exquisitely sensitive to ligand binding and conformational equilibria (Selvaratnam, Chowdhury, VanSchouwen, & Melacini, 2011; Williamson, 2013). By applying unsupervised machine-learning software to unassigned NMR spectra, it is now convenient to generate binding isotherms (Xu & Van Doren, 2016a) and affinities, 1D ordering of enzyme states (e.g., open–closed continuums) (Cembran, Kim, Gao, & Veglia, 2014), and conformational groupings of covalent and/or noncovalent states of enzymes in two or three dimensions (Sakurai & Goto, 2007; Xu, Sarma, Wei, Beamer, & Van Doren, 2017). Because the calculations bypass the laborious process of assigning NMR peaks, obtaining ligand affinities and comparisons of enzyme states should now be easily accessible to many more investigators. The aforementioned software employs principal component analysis (PCA) or independent component analysis enhanced by preprocessing steps (Xu & Van Doren, 2016a, 2017). The algorithm obviates the assignment of NMR peaks, which consumes much time in traditional NMR studies. This paves the way for novel biophysical insights into enzyme samples shortly after preparing them and collecting routine NMR spectra of them. One-dimensional ordering of enzyme conformational equilibria uses their progression of NMR peak positions and is designated COordiNated ChemIcal Shifts bEhavior (CONCISE) (Cembran et al., 2014). The NMR *spectra* of multiple enzyme conditions can also be compared in two or three dimensions of principal components (PCs) (Xu et al., 2017), as pointed out later. PCA has been used for years to compare and group 1D NMR spectra in metabolomics (Lindon, Holmes, & Nicholson, 2001; Trygg, Holmes, & Lundstedt, 2007). More recently, PCA has been used to monitor or group 2D NMR spectra in protein characterizations (Arbogast, Delaglio, Schiel, & Marino, 2017; Egner et al., 2018; Majumder, DeMott, Burz, & Shekhtman, 2014; Sakurai & Goto, 2007). We report herein the new TREND NMR suite of software for conducting this range of applications conveniently. We use phosphatases and a phosphohexomutase as examples to demonstrate the capabilities of the TREND NMR suite.

2. INSIGHTS AVAILABLE SEMIAUTOMATICALLY USING TREND NMR SOFTWARE

2.1 PCA Comparison of Multiple Enzyme Samples

PCA has long been used to explore relationships and distinctions among *multiple samples*, each with a complex data profile, such as chemometrics studies of spectra (Adams, 2004; Trygg et al., 2007). PCA is commonly calculated using singular value decomposition (SVD). Two of the applications of PCA to NMR spectra of enzymes illustrated below fit the traditional strategy of comparing samples. The most conventional of these PCA applications is to calculate a scores plot or biplot of PC1, PC2, and optionally PC3. NMRpipe has provided such a function for NMR spectra (Delaglio, 2012; Delaglio et al., 1995). Below we point out the biplot capability of the TREND NMR suite. The clever alternative strategy of comparing the amide NMR spectra of several enzyme states, known as CONCISE, was illustrated on the open-to-closed continuum of conformational equilibria of cAMP-dependent protein kinase A and complexes it forms with phospholamban (Cembran et al., 2014; Kim et al., 2015; Srivastava et al., 2014). The CONCISE approach calculates PC1 for each shifting NMR peak and compares the ensemble of these PC1 values for each enzyme state included in the comparison. The relative position of the average PC1 of each state indicates the position along the conformational equilibrium. A narrow peak implies the collective response of the conformation of the enzyme, whereas a broad peak implies lack of coordination of the conformational ensemble (Cembran et al., 2014). CONCISE results are also available for a bacterial phosphomannomutase/phosphoglucomutase (PMM/PGM) (Xu et al., 2017).

2.2 PCA Tracking of Changes to a Single Enzyme Sample

More recently and less often, however, PCA has been used to track changes to a *single sample*, monitored by a series of complex NMR spectra. Examples have included monitoring the time evolution of reactions (Jaumot, Marchan, Gargallo, Grandas, & Tauler, 2004; Majumder et al., 2014), the dependence of protein-folding equilibria upon pH (Sakurai & Goto, 2007) or denaturant (Konuma, Lee, Goto, & Sakurai, 2013), or protein associations with ligands (Xu & Van Doren, 2016a, 2017). The TREND NMR suite is suitable for resolving concurrent processes in these and all kinds of series of equilibrium or nonequilibrium measurements of a single sample. TREND NMR is applied similarly to these diverse types of measurements. What distinguishes the application of TREND NMR to studies of ligand binding, chemical transformation, or folding is primarily the choice of equation for fitting a curve (in TRENDanalysis) to the progress of the reaction.

3. CAPABILITIES OF TREND NMR SUITE WITH EXAMPLES

We present only a part of the wide array of potential applications of TREND NMR to single samples undergoing change, beginning with applications to ligand-binding titrations with challenging NMR behaviors.

We proceed in Section 3.2 to a more typical usage of PCA for grouping. The example compares multiple states of protein tyrosine phosphatase 1B (PTP1B). Then, in Section 3.3, we illustrate the recent CONCISE comparison of conformational equilibria of PTP1B.

3.1 Binding Isotherms and Affinities Directly From Unassigned NMR Spectra

Proteins are regularly titrated with ligands in NMR spectrometers. The convention has been to attempt to fit binding isotherms to NMR peaks only if they have large enough linear changes in peak position, a behavior known as fast exchange (Ding et al., 2007; Furukawa, Konuma, Yanaka, & Sugase, 2016; Williamson, 2013; Yokogawa et al., 2012; Ziarek, Peterson, Lytle, & Volkman, 2011). In fast exchange, NMR peaks at intermediate stages of a titration lie at a weighted average of the peak positions of the free and bound states (Palmer, Kroenke, & Loria, 2001; Williamson, 2013). We use examples with slow exchange or intermediate exchange, previously regarded as unfit for plotting a binding isotherm, to point out that such a limitation should no longer exist when applying an appropriate PCA approach directly to the NMR spectra of the titration (Xu & Van Doren, 2016a).

3.1.1 Titration With NMR Peaks That Disappear and Reappear (Slow Exchange)

In titrations in higher affinity regimes, NMR peaks of the free state can disappear and be replaced by distinct peaks of the bound state elsewhere, without peaks at intermediate positions. A binding isotherm cannot be represented by plotting only the two points of peak positions of free and bound states. This behavior is termed slow exchange due to the dissociation rate constant being much slower than the peak separation (in units of Hz) (Palmer et al., 2001).

Slow-exchange behavior was evident upon xylose 1-phosphate titration of the crowded ^{15}N transverse relaxation-optimized spectroscopy (TROSY) NMR spectra of a 470-residue PMM/PGM from *Pseudomonas* (Xu & Van Doren, 2016a). That is, multiple peaks disappeared and reappeared elsewhere with this inhibitor present (Fig. 1A). Application of TREND NMR directly to the 11 spectra of the titration provided a binding isotherm (Fig. 1B) in agreement with a binding isotherm constructed by manual analysis of peak heights (Xu & Van Doren, 2016a). The automatically computed binding isotherm was then fitted using the TRENDanalysis program to an apparent K_D of $28.6 \pm 11.6\,\mu M$, which agrees within uncertainties with the labor-intensive manual measurement of the K_D at $23 \pm 6\,\mu M$.

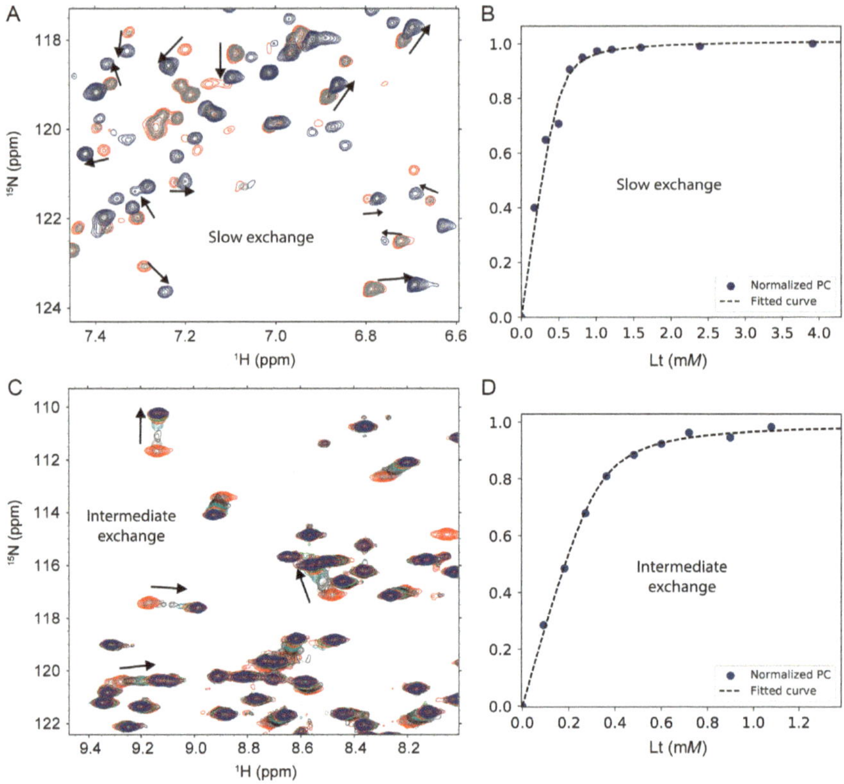

Fig. 1 Reliable binding isotherms and affinities are available semiautomatically in challenging exchange regimes in NMR spectra when using the TREND NMR suite. Panel (A) shows application to spectra with abrupt jumps of NMR peak position (slow-exchange regime) upon titration of a phosphohexomutase with xylose 1-phosphate, an inhibitor. Panel (B) shows the binding isotherm calculated automatically by TREND NMR and fitted to a K_D and plotted with TRENDanalysis. Panel (C) shows a titration with a mixture of both linear peak shifts (fast exchange) and nonlinear shifts with peak broadening (intermediate exchange). A phosphoThr peptide was titrated into the phosphopeptide-binding FHA domain of *Arabidopsis* kinase-associated protein phosphatase and monitored by ^{15}N HSQC spectra across the range of concentrations in the titration. Panel (D) shows the binding isotherms automatically computed by TREND NMR using Pareto scaling to assist with the nonlinearity of peak shifts present.

3.1.2 Titration With Nonlinear Shifts and Broadening of Peaks (Intermediate Exchange)

Intermediate exchange is a common regime in which plotting of NMR peak positions (chemical shifts) or their perturbations by a titration can skew the appearance of binding isotherms and fitting of apparent dissociation constants by up to two orders of magnitude (Williamson, 2013). The sigmoidal distortion of such chemical shift perturbations in the middle of titrations occurs when the dissociation rate constant is of similar magnitude to the spectral separation (in units of Hz) between the NMR peak positions of the free and bound states. The chemical shifts in the middle of the ligand-binding transition then deviate from a weighted average of the free and bound peak positions and undergo line broadening (Palmer et al., 2001; Williamson, 2013).

When some peaks undergo intermediate exchange, other peaks with smaller or larger peak separations between free and bound states often experience fast or slow exchange in the same titration. In a phosphoThr peptide titration of the forkhead-associated (FHA) domain of kinase-associated protein phosphatase (KAPP) from *Arabidopsis*, some peaks displayed intermediate exchange and others fast exchange (Fig. 1C). Applying TREND NMR to the 11 amide NMR spectra of this titration nonetheless provides an orthodox binding isotherm when choosing Pareto scaling in preprocessing (Fig. 1D). The K_D of $20 \pm 3\,\mu M$ measured by titration calorimetry (Ding et al., 2007) agrees well with the apparent K_D of $19 \pm 1\,\mu M$ fitted by TRENDanalysis to these spectra by the procedure described in Section 5.3, despite the problematic, nonlinear peak shifts.

3.2 Quick Comparison of Enzyme NMR Spectra and Conformational States

Multiple covalent and noncovalent (ligand-bound) modifications of human protein phosphatase 1B (PTP1B) were recently characterized carefully (Choy et al., 2017). Although there is no substitute for the authors' elegant and exhaustive structural and dynamic studies, some insights can be obtained quickly from the NMR spectra by PCA without any conventional assignments of the peaks of the spectra (Collins, Oregioni, Robertson, Kelly, & Ramos, 2015; Egner et al., 2018; Xu et al., 2017). For five free states of PTP1B (wild type (WT) and variant) and three inhibitor complexes of PTP1B, we analyzed the NMR peak lists that the authors deposited in the BMRB (Choy et al., 2017), along with the previously deposited WT NMR peak list (Krishnan et al., 2015). Since data consistency and preprocessing strongly influence PCA outcomes, the following steps were included in the preprocessing by TREND NMR: Each peak list was adjusted in referencing to match the recent WT peak list (BMRB ID: 26814). NMR peaks not present in all lists were deleted. Each list was mean centered, which is routine in multivariate statistical studies. PCA biplots were then generated and the results clustered using TRENDanalysis. Clustering used the *K*-means nearest neighbor algorithm (*K*-means). The biplot of PC1 and PC2 suggests three clusters: free states of PTP1B (WT or variant), the TCS401 competitive inhibitor complexes, and the CPT157633 competitive inhibitor complex (Fig. 2A). The inhibitor complexes being distinct are consistent with crystal structures showing that binding of the TCS401 inhibitor closes the WPD loop at the active site and rigidifies helix α7 at the remote allosteric site (Choy et al., 2017). Addition of PC3 to the biplot separates PTP1B (L192A) as distinctive in terms of both PC3 and PC1 (Fig. 2B). This agrees with the significant spectral and structural adjustments accompanying the internal L192A lesion (Choy et al., 2017). Overall, the inhibitor complexes and free states are distinguished—especially by PC1, but also by PC2.

The mutated PTP1B enzyme constructs appear to be resolved into separate clusters when applying the software instead to spectra simulated from the peak lists (Fig. 2C). The simulated spectra contain all the amide NMR peaks the authors assigned and deposited (Choy et al., 2017; Krishnan et al., 2015), and not only the peaks of the residues common to all lists analyzed for Fig. 2A and B. We simulated the nine amide NMR spectra of PTP1B samples corresponding to the BMRB depositions (listed in the legend to Fig. 2) using the spectral simulation feature of Sparky (Goddard & Kneller, 2000). Despite the high complexity of the overlay of the simulated spectra, PCA reduces the complexity tremendously by describing it by a few PCs. Clustering using *K*-means in TRENDanalysis suggests five clusters in biplots of PC1 and PC2 (Fig. 2C). This groups both WT constructs together and the inhibitor complexes together, but separates the site-directed variants into three groups in the PCA biplot. This suggests that T178A and Y152A/Y153A substitutions have related spectral and conformational effects that differ from the effects of the other mutations (Fig. 2B and C). In agreement with this, a recent structural investigation revealed that the side chains of Thr178 and Tyr152 that are in close contact belong to the allosteric pathway that connects to helix α7, and affect one another's NMR peak positions when mutated (Choy et al., 2017). The separation of L192A and N193A variant enzymes (Fig. 2C) seems consistent with L192A having the greater impact in decreasing catalytic activity, altering the crystal structure, and introducing NMR chemical shift perturbations (Choy et al., 2017). Thus, PCA-based statistical inferences from the NMR spectra appear to be validated by the recent careful structural investigation (Choy et al., 2017).

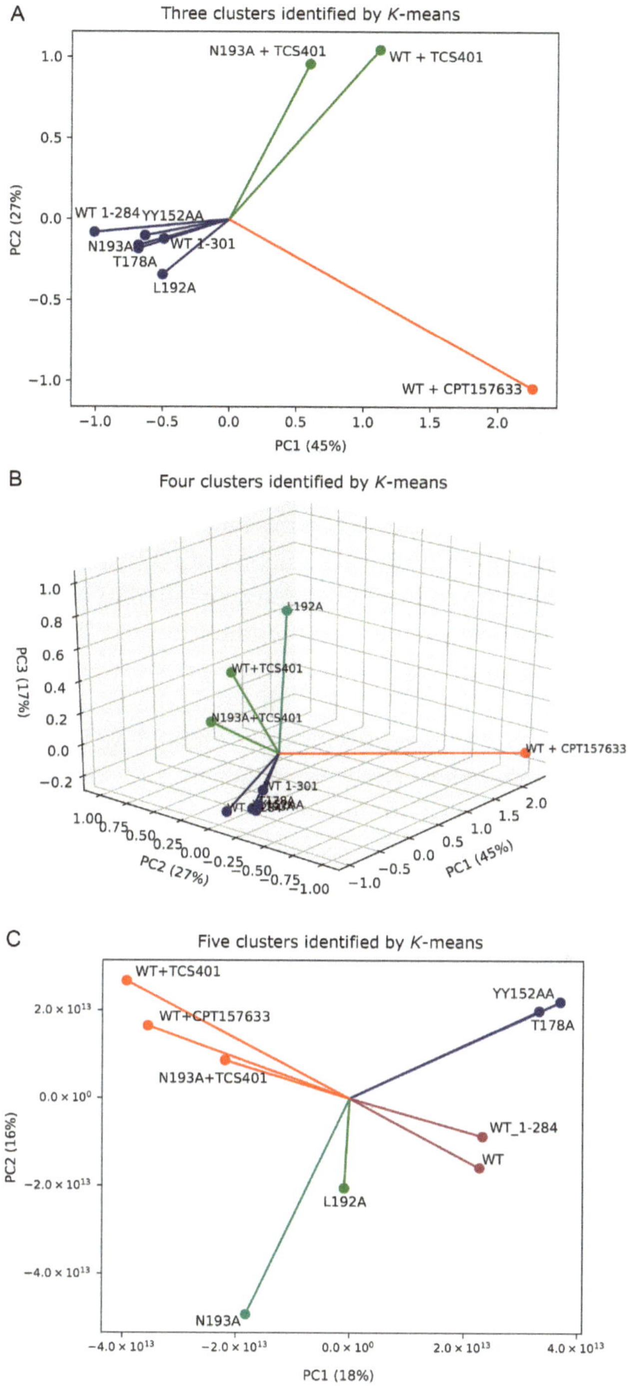

Fig. 2 See legend on next page.

3.3 1D Conformational Ordering of Enzyme Conformational Equilibria

Ligand binding, posttranslational modifications like phosphorylation, and mutations can each affect how much the states of an enzyme are populated. To recognize the shifts among such states, statistical analysis of NMR peak shifts was proposed and named CONCISE (Cembran et al., 2014). It proved informative for comparing positioning in the conformational landscape, especially between open and closed states, of various complexes and covalent states of both cAMP-dependent protein kinase A (Cembran et al., 2014) and the bacterial phosphohexomutase PMM/PGM (Xu et al., 2017).

We have incorporated the CONCISE algorithm into the new TREND NMR and TRENDanalysis programs and have applied it to the aforementioned PTP1B spectra. Five of the nine states of PTP1B studied by NMR (Choy et al., 2017; Krishnan et al., 2015) are compared by CONCISE in Fig. 3. Comparing them initially using the full set of amide peaks suggested that extremes of the conformational equilibria of PTP1B occur in the truncated free state (the 284-residue WT construct plotted in blue in Fig. 3) and the CPT157633 inhibitor complex (301 residues; purple). These two samples could range from the most open to the most rigid, closed states. The CONCISE comparison was then refined by using only the linearly shifted NMR peaks conforming to the pattern with the truncated WT unliganded state and CPT157633 complex lying at opposite ends of the ranges of NMR peak shifts. The TCS401 inhibitor complexes occupy the positions nearer the CPT157633 inhibitor complex on the closed end of the range (Fig. 3). As mentioned, the TCS401 inhibitor closes the WPD loop that is more open when the active site is vacant (Choy et al., 2017). The 284-residue

Fig. 2 Protein tyrosine phosphatase 1B (PTP1B) samples are resolved into groups of covalently or noncovalently modified samples by applying unfold PCA with TREND NMR and clustering with TRENDanalysis. The amide NMR peak lists are available from the BMRB database (Choy et al., 2017; Krishnan et al., 2014). (A) PC1, PC2, and *K*-means divide the PTP1B samples into a cluster for the free-state (both wild type (WT) and variants) and separate clusters for the two competitive inhibitors indicated. (B) PC1–PC3 and *K*-means also divide the samples into these three clusters plus the distinctive PTP1B (L192A). (C) Application of unfold PCA directly to amide NMR spectra simulated from the peak lists resolves the variant enzymes into three groups using PC1, PC2, and *K*-means clustering. The peak lists used have BMRB accession codes of 19223 (WT residues 1–301), 26814 (WT residues 1–284), 26832 (Y152A, Y153A), 26835 (T178A), 26836 (L192A), 26837 (N193A), 26838 (WT+TCS401), 26847 (N193+TCS401), and 25375 (WT+CPT-157633). The PTB1B comprised residues 1–301, except where noted. The peak lists were referenced to wild-type list 26814 using the method of Cembran et al. (2014).

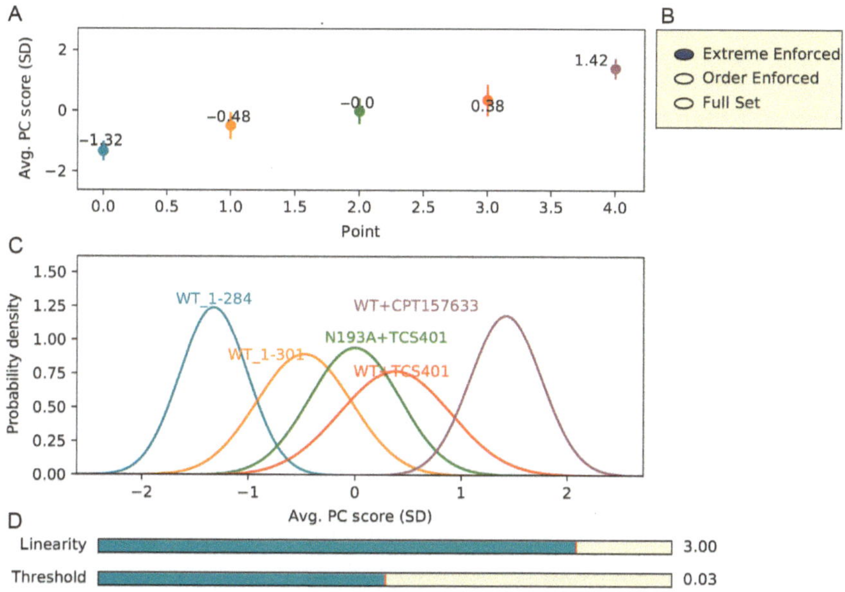

Fig. 3 CONCISE ordering of PTP1B samples into a continuum of conformational equilibria, as calculated and plotted by TREND NMR. For clarity, only five of the NMR peak lists available for PTP1B were used (from *left* to *right*: BMRB IDs 26814, 19223, 26847, 26838, and 25375). The free state of the truncated wild type (residues 1–284) and the CPT157633 inhibitor complex defines the extremes of the NMR peak shifts. (A) The average "PC score" of NMR peak shifts is the average PC1 for the ensemble of NMR peak shifts, expressed as the number of standard deviations (SDs) of PC1. (B) Checkboxes for three choices of NMR peaks to retain for CONCISE are shown. The "Extreme Enforced" option was chosen for this calculation. Extreme Enforced selects the subset of residues following the pattern that the peak of the open (unliganded) state lies at one extreme position and the peak of the most closed complex lies at the other extreme (Cembran et al., 2014). By default extremes are the first and last files in the input files. Other files can be chosen as starting and ending states (i.e., extremes) in TRENDanalysis. The "Order Enforced" option selects only the subset of residues with NMR peaks that shift in the order specified by the order of the file names of peak lists. "Full Set" uses all peaks in the assessment. (C) The conformational populations are represented by replotting the average PC score as a Gaussian distribution with the SD shown. The linearity parameter is defined as the ratio of the SD of PC1 to that of PC2, i.e., SD_{PC1}/SD_{PC2} (Cembran et al., 2014). (D) The interactive slider bars are used to set the minimum linearity and size of the peak shifts. These are displayed with the defaults of linearity (SD_{PC1}/SD_{PC2}) at 3.0 and threshold for peak shifts at 0.03 ppm.

WT construct has helix α7 deleted from the allosteric effector site, but without altering the crystal structure (Choy et al., 2017). Nonetheless, this truncated construct lying at the "open" extreme in the CONCISE analysis of the NMR spectra (Fig. 3) suggests that deletion of helix α7 introduces a perceptible opening of the enzyme structure in solution.

4. NMR DATA

TREND NMR reads NMR spectra and peak lists in a variety of formats. TREND NMR can also utilize raw NMR data (free induction decays or FIDs). Accurate analysis of FIDs from titrations is feasible in the fast- or slow-exchange regimes, but is not recommended when intermediate exchange is present (Xu & Van Doren, 2016a).

4.1 Considerations and Forms of NMR Data

Consistency in the measurements is the chief requirement for PCA to succeed. This is the easiest to achieve in the form of lists of NMR peaks because they free the user from uniformity of data collection. Peak lists and spectral segments (bins) have been the main format of NMR data analyzed by PCA historically (Lindon et al., 2001; Sakurai & Goto, 2007). It remains important that the chemical shifts of the peaks listed be referenced consistently. The free state of the enzyme makes a good choice of reference state for peak lists.

Analyzing the NMR spectra directly adds convenience and broadens the range of NMR spectral behaviors that can be analyzed accurately (Xu & Van Doren, 2016a), as illustrated for ligand binding in Section 3.1 (Fig. 1). To ensure successful PCA-based comparisons and extractions of reaction courses and binding isotherms, investigators should endeavor to collect the NMR spectra uniformly. That is, they should use the same spectral width, gain, signal averaging, carrier frequency, and other basic NMR acquisition parameters, not to mention the preferred practice of measuring the spectra at comparable sample concentrations. Artifacts from the water line in one or more spectrum can compromise an entire PCA-based analysis. In order to minimize interference by artifacts from the water, the user may avoid affected spectral regions by zooming in on the region of interest (ROI) in TREND NMR. When analyzing FIDs, the user may subtract the water signal by choosing a solvent filter to remove the low frequency signal.

4.2 Choices of Format

In its **spectra** tab, TREND NMR automatically detects the formats of NMR spectra from Bruker Topspin and Agilent (Varian) VnmrJ directories, or from NMRpipe, Sparky, and JCAMP-DX files. In its **fid** tab, TREND NMR automatically detects the format of FIDs from Bruker and Agilent in directories, or from NMRpipe, Siemens RDA, and JCAMP-DX files (manual selection of format is necessary when using the original TREND software; Xu & Van Doren, 2017). TREND NMR reads peak lists in Sparky and NMR-STAR formats in its **peaklist** mode. Multiple pieces of NMR software can export peak lists in Sparky and NMR-STAR formats. TREND NMR can retrieve NMR-STAR files archived at the Biological Magnetic Resonance Data Bank (BMRB) (Ulrich et al., 2008) when the accession codes are listed in a text file. The user may specify the data fields and atom types (chemical shifts) to use, as described in Section 5.4.

5. LAYOUT AND USAGE OF TREND NMR SOFTWARE

5.1 Platforms and Installation

The TREND NMR package is available for Windows, recent versions of MacOS, and a recent version of Ubuntu and CentOS/RedHat Linux operating systems. Prospective users may either register for free academic licensing or request a commercial demo version from the TREND website (Xu & Van Doren, 2016c). The simple installation of the downloaded file is described in the online manual (Xu & Van Doren, 2016b).

5.2 TREND NMR and TRENDanalysis Functions

Unfold PCA (Jaumot et al., 2004) is the core algorithm of TREND NMR and its TRENDmain predecessor. The unfold PCA is implemented as a combination of preprocessing steps and SVD (Xu & Van Doren, 2016a, 2017) (Fig. 4). The preprocessing steps are foundational to the success of the TREND family of software in obtaining binding isotherms and reaction progress curves directly from NMR spectra experiencing all combinations of chemical exchange regimes. The data projections by PCA provide simple, information-rich summaries, regardless of the type of reaction monitored by NMR. When monitoring a chemical reaction, whether at equilibrium (such as a titration with a partner) or via a time-dependent reaction progress curve, the PC1 summary of the main course of change by the reaction is a key interest. PC1 can be fitted in TRENDanalysis to an equation that the user selects or enters (Fig. 4).

Much more commonly and traditionally, PCA is used instead to compare and categorize measurements. TRENDanalysis prepares such comparisons as "biplots" of two or three PCs. Normally, it is the best to use the largest PCs, i.e., PC1 through PC3, because they capture the majority of the differences (statistical variances) among the measurements. We propose

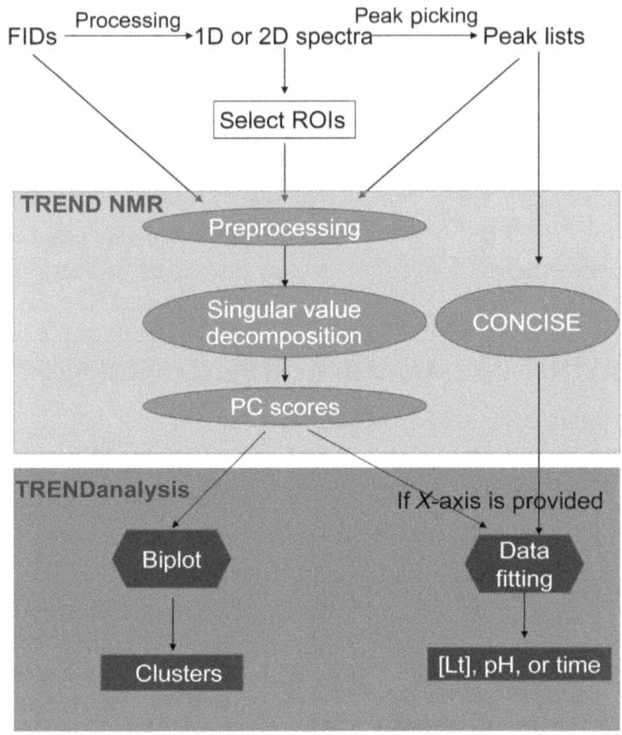

Fig. 4 The functions and data flow of TREND NMR and TRENDanalysis. [Lt] refers to the total ligand concentration.

the use of biplots from PCA of NMR peak lists or spectra to compare enzyme states (Xu et al., 2017) (Fig. 2). These can then be divided into groups manually or semiautomatically using clustering algorithms. Clustering approaches and their properties and limitations have been reviewed (Ronan, Qi, & Naegle, 2016). TRENDanalysis offers clustering algorithms available in Python (Pedregosa et al., 2011), each requiring a parameter to be set. Finding a nearest neighbor is the strategy of centroid-based *K*-means and density-based DBSCAN (Hinneburg & Keim, 1999). Agglomerative clustering is hierarchical (Bouguettaya, Yu, Liu, Zhou, & Song, 2015). Affinity propagation seeks "exemplars" for the clustering (Frey & Dueck, 2007). Mean-shift develops clusters from statistical modes in the data (Yizong, 1995).

TREND NMR and TRENDanalysis both implement the CONCISE strategy of comparing enzymes using their NMR peak lists. CONCISE simplifies the comparison by focusing on PC1, but elaborates the comparison by calculating a distribution of the PC1 values of individual NMR peaks affected by enzyme conditions (Cembran et al., 2014).

5.3 Procedure for Measuring Affinity From a Titration

We illustrate an effective approach to the otherwise difficult case of the titration shown in Fig. 1C and D, which is complicated by nonlinear peaks shifts and line broadening of intermediate exchange. Overcoming these complications require applying the algorithm directly to the NMR spectra and a good choice of preprocessing steps (Xu & Van Doren, 2016a). It is necessary that the several NMR spectra of the titration be collected uniformly with a consistent set of parameters. TREND NMR is used to calculate the binding isotherm. TRENDanalysis is used to fit the isotherm to an equation to obtain the dissociation constant. The specific steps in using the software are given next.

1. *Essential*: Invoke *trendnmr.exe*. If the NMR spectra are in the third-party file format of NMRpipe, Sparky, or JCAMP-DX, go under **file** and browse to the location of the spectra and select them (Fig. 5A). If the spectra were processed in the Bruker or Agilent vendor's software, go instead under **dir** and browse to the several directories containing the spectra and select the directories. In either case, the software automatically recognizes these file types.
2. The prefix of the output files created can be left at the default of "example" or changed.
3. *Recommended*: **Scaling** can normally be set at its default value of "auto" for autoscaling (van den Berg, Hoefsloot, Westerhuis, Smilde, & van der Werf, 2006) of the rows and **threshold** set at "auto" as well. However, in cases with intermediate exchange present, choosing "pareto" scaling (van den Berg et al., 2006) suppresses sigmoidal distortion of the binding isotherm (Xu & Van Doren, 2016a). The peak broadening observed in intermediate exchange motivates keeping peak data closer to the base plane of noise. Consequently, **threshold** is set to "manual" in this case and the **thresholdnumber** is decreased to "2," meaning twofold higher than the noise level.
4. *Essential when fitting K_D*: Prepare a text file containing the total ligand concentration at each point in the titration. Under **xaxis**, browse to and select this file (Fig. 5A). Specify the units of these concentrations under **xunit**.
5. When the NMR spectra are clean, no zooming to select a region is necessary. However, if spectral artifacts from the water line or other sources are present, zooming to avoid these artifacts is highly recommended. The desired spectral region can be specified in points under **xlim** and **ylim**. Or the region may be selected interactively by turning **ROI** "on."

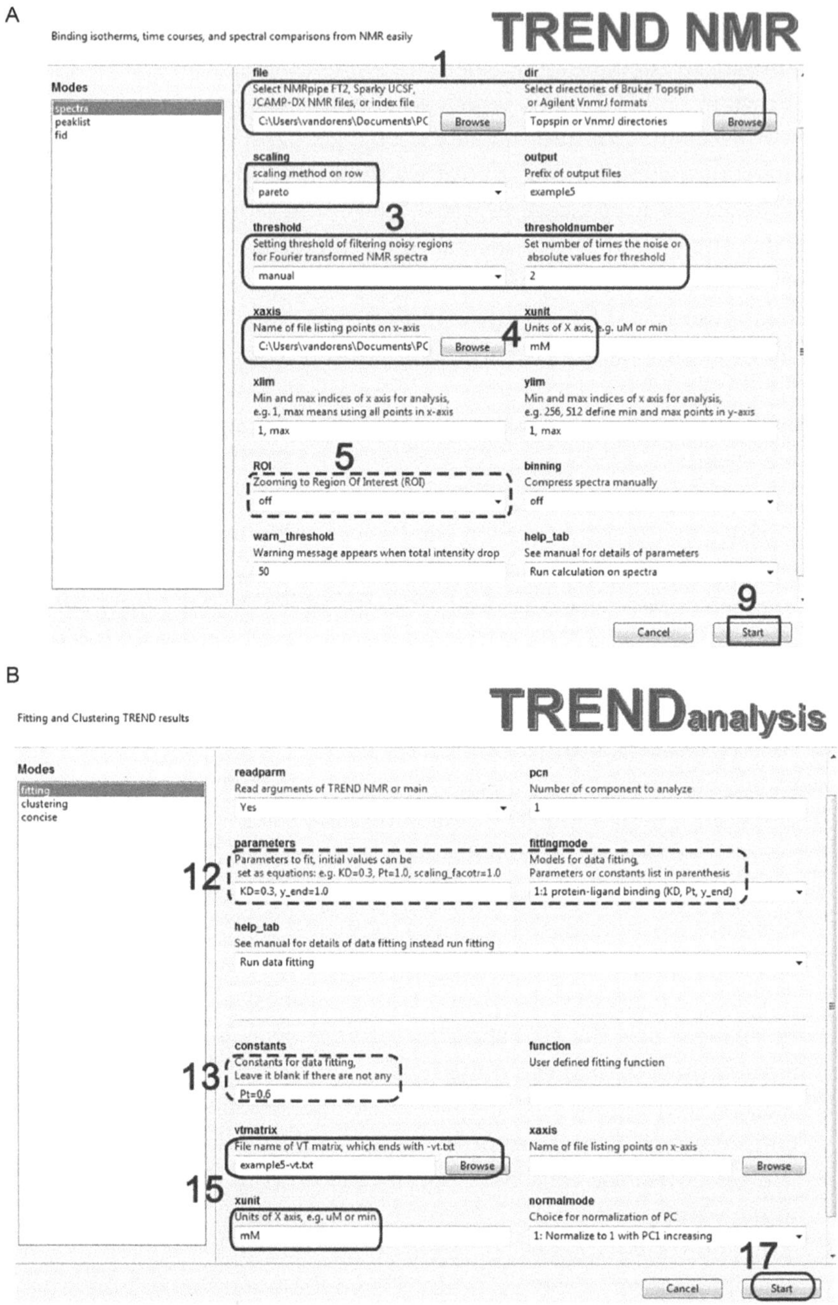

Fig. 5 Usage of the user interfaces when measuring affinity from an NMR-detected titration. Panel (A) illustrates the spectra tab of TREND NMR used to calculate binding isotherms from a series of NMR spectra from a titration. Panel (B) shows the TRENDanalysis fitting mode that fits a K_D to the binding isotherm. Essential steps are boxed with *solid lines* and other important steps with *dashed lines*. The numbers correspond to the steps in the protocol.

In that case, once the spectrum appears, drag a box to define the region and close the window. Also close the window showing the expanded spectral region.

6. If the spectra contain so many points that calculations are slow, turn **binning** "on" to decrease the number of points employed. This launches a window with interactive slider bars. Click on the **Grouped pts** bar on the left end to pool 2–8 points together for calculations on the lower resolution spectrum. Closing the interactive window launches the PCA calculation.
7. If concerned about a drop in peak heights during the titration, e.g., in cases of ligand-induced aggregation or precipitation, **warn_threshold** can be set to a percentage of the initial height to trigger a warning message. The default threshold is 50%.
8. Under **help_tab**, changing to "help" attempts to open the online manual page.
9. Click **start** to launch these operations.
10. Invoke *trendanalysis.exe* for fitting the binding isotherm (or other reaction course). Choose the **fitting** mode (Fig. 5B) and leave **readparm** set to "Yes" in order to read the key files and parameters calculated and written by *trendnmr.exe* (in steps 1–9).
11. Under **pcn**, leave the default at "1," indicating PC1 is to be fitted. Two or more binding reactions could motivate fitting of PC2, chosen by setting to "2."
12. *Important*: The default choices under **parameters** and **fittingmode** specify fitting to 1:1 binding (without correcting for dilution of the protein by the ligand, which is acceptable for the examples of concentrated ligands in Fig. 1). For a more complex binding equation, the user should scroll down under **fittingmode** and select "user defined function" and type in the expression under **function**. Examples of equation syntax can be found within the software installation directory in the path TRENDpro_example\Example_Fitting_Config_Files\. Update **parameters** to match the custom equation.
13. *Important*: The fitted K_D will be more accurate if the protein concentration is known accurately and is fixed at the known concentration under **constants**. Calibration of the protein assay using amino acid analysis is recommended for accuracy. Fixing the "Pt" parameter necessitates removing it from the list of parameters to be fitted under **parameters**.
14. Under **help_tab**, select "help" if desired to open the online manual page. Reset to "run data fitting" in order to perform curve fitting.

15. *Essential*: If an **output** file prefix was entered in TREND NMR, under **vtmatrix** browse to that "prefix-vt.txt" file (Fig. 5B). (The "vt" in the file name refers to the transposed V matrix calculated by SVD.) Under **xunit**, specify the units of the file of ligand concentrations used in TREND NMR.
16. Provided the file of ligand concentrations was specified in TREND NMR, **xaxis** can be left blank. For fitting of ordinary binding isotherms, the **normalmode** default of 1 is the desired choice.
17. Press **start** and wait for the plot with fitted curve to appear. Use the file icon to save the plot if desired. Close the plot window in order for the calculation to be completed. The fitted equation and parameters are then listed in both the HTML web browser window and the graphical user interface (GUI) window.

5.4 Procedures for Statistical NMR Comparisons of Enzyme States

1. *Essential*: Prepare a text file listing the file names (one per line) of each NMR peak list from each protein state to be compared. List first the peak list that will serve as the reference point for any adjustment of referencing of the peak lists. The NMR peak lists may be in Sparky format (Goddard & Kneller, 2000) or NMR-STAR format (Hall, 1991; Spadaccini & Hall, 2012). Alternatively, the file can be a list of BMRB accession codes for direct download from the BMRB (Ulrich et al., 2008). Invoke *trendnmr.exe* and select the **peaklist** tab at left side of the window (Fig. 6A). Under **file**, browse through the directory structure to the text file containing the NMR peak lists and open it.
2. CONCISE analysis of peak lists does not seem sensitive to scaling. Nonetheless, because centering is normally performed before PCA, choosing "noscaling: centered but not scaled" under both **scaling** and **columnscaling** is appropriate.
3. Regarding **output**, the user may retain default prefix of "example" or enter a more meaningful prefix.
4. The **xaxis** and **xunit** fields can be ignored.
5. *Important if not amide peaks*: Under **sparky_fields_to_use_and_scale**, the default of "N=0.2, H=1" is appropriate for selecting and scaling the correct fields from amide peak lists (Fig. 6A). (These peak lists are generated from widely used ^{15}N HSQC and TROSY NMR spectra.) For peak lists from ^{13}C–^{1}H correlation spectra such as heteronuclear single quantum coherence (HSQC), heteronuclear multiple quantum coherence (HMQC), and methyl TROSY, a similar choice of

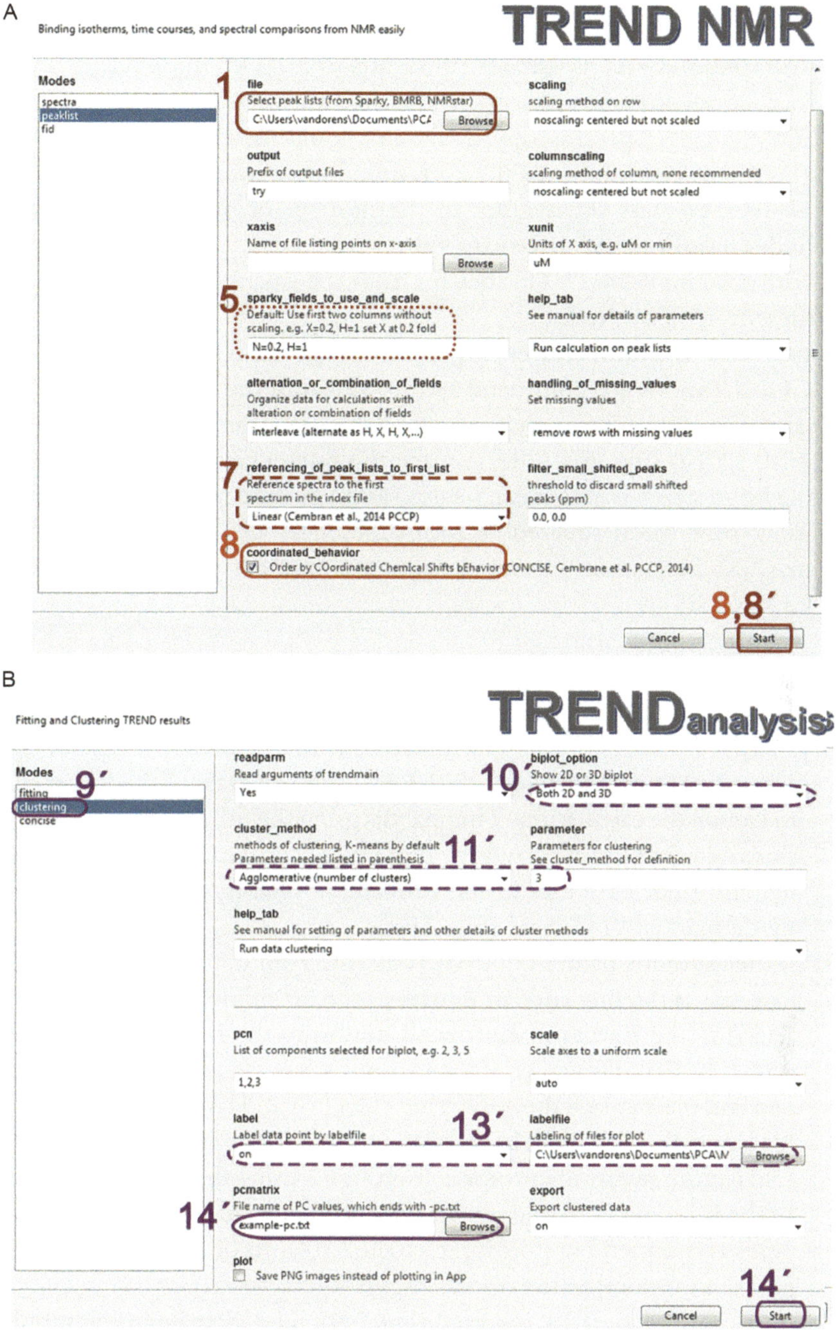

Fig. 6 Usage of the interfaces for comparing enzyme states based on multivariate statistics of NMR spectra. Key steps are each highlighted with a *rounded rectangle* (with *solid line* for essential steps) and *numbered* by the step in the procedure in the text. (A) The peaklist tab in TREND NMR is selected. *Orange rectangles* mark steps specific to the CONCISE strategy. (B) The clustering tab in TRENDanalysis is selected. *Purple rounded rectangles* are for steps specific to 2D/3D resolution of enzyme states.

"C=0.25, H=1" is suitable. Peak lists from 2D homonuclear ^1H–^1H total correlation spectroscopy (TOCSY) or nuclear Overhauser effect and exchange spectroscopy (NOESY) could instead be set to "H=1, H=1" to select and scale the corresponding entries from each peak list.

6. NMR peak lists often differ in the number of peaks. Consequently, under **handling_of_missing_values**, use the default of "remove rows with missing values." This deletes from the calculation those peaks not present in all of the lists.
7. *Important*: Uniform referencing of NMR spectra is essential when using CONCISE. We recommend setting **referencing_of_peak_lists_to_first_list** to the linear style of referencing recommended by Cembran et al. (2014) (Fig. 6A). Please place the reference spectrum first in the file list of step 1 where it can be used as the reference point. Referencing using root mean squared (RMS) differences between peak positions may also be chosen but can decrease the number of peaks retained by CONCISE.

5.4.1 Steps Specific to CONCISE Ordering of States

8. *For CONCISE*: Under **coordinated_behavior**, turn on CONCISE analysis by selecting the checkbox next to "Order by COordiNated ChemIcal Shifts bEhavior" (Fig. 6A). Click on **start** at the lower right to launch the calculation. The analysis using the full set of amide peaks is shown in the plotting window. Note which enzyme states appear at the left and right extremes in the Probability Density Plot in this default mode marked "Full Set" (Fig. 3B).
9. If the list of file names of NMR peak lists from step 1 already lists these extrema as the first and last entries in the list of file names, proceed by selecting the **Extreme Enforced** option in the interactive plot window (Fig. 3B). However, if the extreme states are not already listed first and last, please place these extreme cases first and last in a revision of the file listing the peak lists. In order to proceed to the **Extreme Enforced** calculation, restart the protocol from step 1 using this reordered file list. In the Extreme Enforced mode, the default linearity (SD_{PC1}/SD_{PC2}) of 3.0 is recommended (Cembran et al., 2014). In order to include more peaks in the comparison, use the upper interactive slider bar to decrease the degree of linearity required (Fig. 3D). The increase in the number of peaks (residues) retained in the calculation can be checked by looking in the TREND NMR GUI window.
10. In the plot window, the display of the panels may be adjusted by clicking a button at the bottom with an arrow and dragging the panel.

Save the plot using the rightmost icon (symbolizing a file on a disk) at the bottom of the plot window. The .pdf, .png, and .svg formats are convenient to display.

11. TRENDanalysis can be used to continue and adjust the calculations under the CONCISE tab. For example, the first and last extreme states may be set and the plot can be labeled using a file listing the labels for the peaks.

5.4.2 Alternate Steps for Further Resolution of Enzyme States

8′. Click on **start** at the lower right of the TREND NMR GUI window to launch the calculation.

9′. *For 2D/3D resolution*: Invoke *trendanalysis.exe* and select the **clustering** mode. Use the default setting of "Yes" under **readparm**.

10′. *Important*: Under **biplot_option** choose "2D" or "Both 2D and 3D."

11′. *Recommended*: Under **cluster_method**, choosing either "*K*-means" or "Agglomerative" algorithms for clustering appears to be reliable. They both require the user to decide the number of clusters to find. The default setting of **parameter** to "3," meaning find three clusters, is suitable for the initial trial. Revise this number as needed on subsequent trials.

12′. Leave **pcn** set to the defaults of "1, 2, 3" referring to PC1 to PC3. Leave **scale** set to "auto." Leave **export** set to "on."

13′. To provide a custom set of labels, turn **label** "on." Under **labelfile** browse to a text file listing the labels to use.

14′. *Essential*: Make sure that the file under **pcmatrix** has the prefix used in the TREND NMR calculation. Press **start**.

15′. On the 2D biplot the user can drag text labels for points to better positions (this is not currently feasible in the 3D biplot). Rotate the 3D biplot to a preferred angle of view. Save the plot by clicking the file icon. Close the plot window; this returns control to the TRENDanalysis window.

6. CONCLUSIONS

Both the original TREND (Xu & Van Doren, 2017) and new TREND NMR software packages make the power of semiautomatic extraction of reaction courses from NMR spectra freely available to the academic community. Because NMR peak assignments are not involved, the approach should be feasible for many studies. The new TREND NMR package goes beyond the original TREND software by introducing the following functions: (A) automatic recognition of multiple formats of

NMR data files, (B) fitting of equations and parameters to a variety of reaction courses, including the binding isotherms mentioned earlier, (C) 2D and 3D PCA biplots generated from NMR peak lists and spectra, (D) statistical clustering of spectral states of enzymes and other molecules, and (E) the CONCISE strategy of ordering the conformational landscape of enzymes. Consequently, TREND NMR represents a powerful and convenient enabling technology for a wide range of investigators who can obtain NMR spectra of their molecules of interest. Although this chapter has emphasized applications to enzymes, these functions should also be applicable to a great diversity of molecules and their reactions studied by NMR.

ACKNOWLEDGMENTS

We are grateful to the National Science Foundation for support (MCB-1409898). We recognize Alessandro Cembran, Gianluigi Veglia, and coworkers for pioneering the CONCISE approach. We thank the groups of Wolfgang Peti and Rebecca Page for depositing the NMR peak lists of PTP1B. We appreciate Lesa Beamer and her group for collaboration on phosphohexomutases.

REFERENCES

Adams, M. J. (2004). *Chemometrics in analytical spectroscopy*. Cambridge: Royal Society of Chemistry.

Arbogast, L. W., Delaglio, F., Schiel, J. E., & Marino, J. P. (2017). Multivariate analysis of two-dimensional (1)H, (13)C methyl NMR spectra of monoclonal antibody therapeutics to facilitate assessment of higher order structure. *Analytical Chemistry*, *89*(21), 11839–11845. https://doi.org/10.1021/acs.analchem.7b03571.

Bouguettaya, A., Yu, Q., Liu, X. M., Zhou, X. M., & Song, A. (2015). Efficient agglomerative hierarchical clustering. *Expert Systems with Applications*, *42*(5), 2785–2797. https://doi.org/10.1016/j.eswa.2014.09.054.

Cembran, A., Kim, J., Gao, J., & Veglia, G. (2014). NMR mapping of protein conformational landscapes using coordinated behavior of chemical shifts upon ligand binding. *Physical Chemistry Chemical Physics*, *16*(14), 6508–6518. https://doi.org/10.1039/c4cp00110a.

Choy, M. S., Li, Y., Machado, L., Kunze, M. B. A., Connors, C. R., Wei, X., et al. (2017). Conformational rigidity and protein dynamics at distinct timescales regulate PTP1B activity and allostery. *Molecular Cell*, *65*(4), 644–658. e645. https://doi.org/10.1016/j.molcel.2017.01.014.

Collins, K. M., Oregioni, A., Robertson, L. E., Kelly, G., & Ramos, A. (2015). Protein–RNA specificity by high-throughput principal component analysis of NMR spectra. *Nucleic Acids Research*, *43*(6), e41. https://doi.org/10.1093/nar/gku1372.

Delaglio, F. (2012). *NMRPipe*. Retrieved from http://www.nmrscience.com/screen.html.

Delaglio, F., Grzesiek, S., Vuister, G. W., Zhu, G., Pfeifer, J., & Bax, A. (1995). NMRPipe: A multidimensional spectral processing system based on UNIX pipes. *Journal of Biomolecular NMR*, *6*(3), 277–293. https://doi.org/10.1007/BF00197809.

Ding, Z., Wang, H., Liang, X., Morris, E. R., Gallazzi, F., Pandit, S., et al. (2007). Phosphoprotein and phosphopeptide interactions with the FHA domain from Arabidopsis kinase-associated protein phosphatase. *Biochemistry*, *46*(10), 2684–2696. https://doi.org/10.1021/bi061763n.

Egner, J. M., Jensen, D. R., Olp, M. D., Kennedy, N. W., Volkman, B. F., Peterson, F. C., et al. (2018). Development and validation of 2D difference intensity analysis for chemical library screening by protein-detected NMR. *ChemBioChem*, *19*(5), 448–458. https://doi.org/10.1002/cbic.201700386.

Frey, B. J., & Dueck, D. (2007). Clustering by passing messages between data points. *Science*, *315*(5814), 972–976. https://doi.org/10.1126/science.1136800.

Furukawa, A., Konuma, T., Yanaka, S., & Sugase, K. (2016). Quantitative analysis of protein–ligand interactions by NMR. *Progress in Nuclear Magnetic Resonance Spectroscopy*, *96*, 47–57. https://doi.org/10.1016/j.pnmrs.2016.02.002.

Goddard, T. D., & Kneller, D. G. (2000). *SPARKY*. San Francisco: University of California, San Francisco.

Hall, S. R. (1991). The star file—A new format for electronic data transfer and archiving. *Journal of Chemical Information and Computer Sciences*, *31*(2), 326–333. https://doi.org/10.1021/ci00002a020.

Hinneburg, A., & Keim, D. A. (1999). *Optimal grid-clustering: Towards breaking the curse of dimensionality in high-dimensional clustering. Proceedings of the 25th International Conference on Very Large Databases* (pp. 506–517).

Jaumot, J., Marchan, V., Gargallo, R., Grandas, A., & Tauler, R. (2004). Multivariate curve resolution applied to the analysis and resolution of two-dimensional [^{1}H,^{15}N] NMR reaction spectra. *Analytical Chemistry*, *76*(23), 7094–7101. https://doi.org/10.1021/ac049509t.

Kim, J., Masterson, L. R., Cembran, A., Verardi, R., Shi, L., Gao, J., et al. (2015). Dysfunctional conformational dynamics of protein kinase A induced by a lethal mutant of phospholamban hinder phosphorylation. *Proceedings of the National Academy of Sciences of the United States of America*, *112*(12), 3716–3721. https://doi.org/10.1073/pnas.1502299112.

Konuma, T., Lee, Y. H., Goto, Y., & Sakurai, K. (2013). Principal component analysis of chemical shift perturbation data of a multiple-ligand-binding system for elucidation of respective binding mechanism. *Proteins*, *81*(1), 107–118. https://doi.org/10.1002/prot.24166.

Krishnan, N., Koveal, D., Miller, D. H., Xue, B., Akshinthala, S. D., Kragelj, J., et al. (2014). Targeting the disordered C terminus of PTP1B with an allosteric inhibitor. *Nature Chemical Biology*, *10*(7), 558–566. https://doi.org/10.1038/nchembio.1528.

Krishnan, N., Krishnan, K., Connors, C. R., Choy, M. S., Page, R., Peti, W., et al. (2015). PTP1B inhibition suggests a therapeutic strategy for Rett syndrome. *Journal of Clinical Investigation*, *125*(8), 3163–3177. https://doi.org/10.1172/JCI80323.

Lindon, J. C., Holmes, E., & Nicholson, J. K. (2001). Pattern recognition methods and applications in biomedical magnetic resonance. *Progress in Nuclear Magnetic Resonance Spectroscopy*, *39*(1), 1–40.

Majumder, S., DeMott, C. M., Burz, D. S., & Shekhtman, A. (2014). Using singular value decomposition to characterize protein–protein interactions by in-cell NMR spectroscopy. *ChemBioChem*, *15*(7), 929–933. https://doi.org/10.1002/cbic.201400030.

Palmer, A. G., 3rd, Kroenke, C. D., & Loria, J. P. (2001). Nuclear magnetic resonance methods for quantifying microsecond-to-millisecond motions in biological macromolecules. *Methods in Enzymology*, *339*, 204–238.

Pedregosa, F., Varoquaux, G., Gramfort, A., Michel, V., Thirion, B., Grisel, O., et al. (2011). Scikit-learn: Machine learning in python. *Journal of Machine Learning Research*, *12*, 2825–2830.

Ronan, T., Qi, Z., & Naegle, K. M. (2016). Avoiding common pitfalls when clustering biological data. *Science Signaling*, *9*(432), re6. https://doi.org/10.1126/scisignal.aad1932.

Sakurai, K., & Goto, Y. (2007). Principal component analysis of the pH-dependent conformational transitions of bovine beta-lactoglobulin monitored by heteronuclear NMR. *Proceedings of the National Academy of Sciences of the United States of America*, *104*(39), 15346–15351. https://doi.org/10.1073/pnas.0702112104.

Selvaratnam, R., Chowdhury, S., VanSchouwen, B., & Melacini, G. (2011). Mapping allostery through the covariance analysis of NMR chemical shifts. *Proceedings of the National Academy of Sciences of the United States of America*, *108*(15), 6133–6138. https://doi.org/10.1073/pnas.1017311108.

Spadaccini, N., & Hall, S. R. (2012). Extensions to the STAR file syntax. *Journal of Chemical Information and Modeling*, *52*(8), 1901–1906. https://doi.org/10.1021/ci300074v.

Srivastava, A. K., McDonald, L. R., Cembran, A., Kim, J., Masterson, L. R., McClendon, C. L., et al. (2014). Synchronous opening and closing motions are essential for cAMP-dependent protein kinase A signaling. *Structure*, *22*(12), 1735–1743. https://doi.org/10.1016/j.str.2014.09.010.

Trygg, J., Holmes, E., & Lundstedt, T. (2007). Chemometrics in metabonomics. *Journal of Proteome Research*, *6*(2), 469–479. https://doi.org/10.1021/pr060594q.

Ulrich, E. L., Akutsu, H., Doreleijers, J. F., Harano, Y., Ioannidis, Y. E., Lin, J., et al. (2008). Biological Magnetic Resonance Data Bank. *Nucleic Acids Research*, *36*(Database issue), D402–D408. https://doi.org/10.1093/nar/gkm957.

van den Berg, R. A., Hoefsloot, H. C., Westerhuis, J. A., Smilde, A. K., & van der Werf, M. J. (2006). Centering, scaling, and transformations: Improving the biological information content of metabolomics data. *BMC Genomics*, 7, 142. https://doi.org/10.1186/1471-2164-7-142.

Williamson, M. P. (2013). Using chemical shift perturbation to characterise ligand binding. *Progress in Nuclear Magnetic Resonance Spectroscopy*, *73*, 1–16. https://doi.org/10.1016/j.pnmrs.2013.02.001.

Xu, J., Sarma, A. V. S., Wei, Y., Beamer, L. J., & Van Doren, S. R. (2017). Multiple ligand-bound states of a phosphohexomutase revealed by principal component analysis of NMR peak shifts. *Scientific Reports*, 7(1), 5343. https://doi.org/10.1038/s41598-017-05557-w.

Xu, J., & Van Doren, S. R. (2016a). Binding isotherms and time courses readily from magnetic resonance. *Analytical Chemistry*, *88*(16), 8172–8178. https://doi.org/10.1021/acs.analchem.6b01918.

Xu, J., & Van Doren, S. R. (2016b). *TREND manual*. Retrieved from https://trendmizzou.gitbooks.io/trend-manual/content/.

Xu, J., & Van Doren, S. R. (2016c). *TREND: TRack equilibrium or non-equilibrium shifts in data*. Retrieved from http://biochem.missouri.edu/trend.

Xu, J., & Van Doren, S. R. (2017). Tracking equilibrium and nonequilibrium shifts in data with TREND. *Biophysical Journal*, *112*(2), 224–233. https://doi.org/10.1016/j.bpj.2016.12.018.

Yizong, C. (1995). Mean shift, mode seeking, and clustering. *IEEE Transactions on Pattern Analysis and Machine Intelligence*, *17*(8), 790–799. https://doi.org/10.1109/34.400568.

Yokogawa, M., Kobashigawa, Y., Yoshida, N., Ogura, K., Harada, K., & Inagaki, F. (2012). NMR analyses of the interaction between the FYVE domain of early endosome antigen 1 (EEA1) and phosphoinositide embedded in a lipid bilayer. *Journal of Biological Chemistry*, *287*(42), 34936–34945. https://doi.org/10.1074/jbc.M112.398255.

Ziarek, J. J., Peterson, F. C., Lytle, B. L., & Volkman, B. F. (2011). Chapter ten—Binding site identification and structure determination of protein–ligand complexes by NMR: A semiautomated approach. In L. C. Kuo (Ed.), *Methods in enzymology: Vol. 493*. (pp. 241–275). Academic Press.

CHAPTER EIGHT

Assessment and Impacts of Phosphorylation on Protein Flexibility of the α-D-Phosphohexomutases

Kyle M. Stiers, Lesa J. Beamer[1]
Biochemistry Department, University of Missouri, Columbia, MO, United States
[1]Corresponding author: e-mail address: beamerl@missouri.edu

Contents

Abstract

Enzymes in the α-D-phosphohexomutase (PHM) superfamily catalyze a multistep reaction, entailing two successive phosphoryl transfers. Key to this reaction is a conserved phosphoserine in the active site, which serves alternately as a phosphoryl donor and acceptor during the catalytic cycle. In addition to its role in the enzyme mechanism, the phosphorylation state of the catalytic phosphoserine has recently been found to have widespread effects on the structural flexibility of enzymes in this superfamily. These effects must be carefully accounted for when assessing other perturbations to these

Methods in Enzymology, Volume 607
ISSN 0076-6879
https://doi.org/10.1016/bs.mie.2018.04.003

enzymes, such as mutations or ligand binding. In this chapter, we focus on methods for assessing and modulating the phosphorylation state of the catalytic serine, as well as straightforward ways to probe the impacts of this modification on protein structure/flexibility. This knowledge is essential for producing homogeneous and stable samples of these proteins for biophysical studies. The methods described herein should be widely applicable to enzymes across the PHM superfamily and may also be useful in characterizing the effects of posttranslational modifications on other proteins.

ABBREVIATIONS

G16P glucose 1,6-bisphosphate
HDXMS hydrogen–deuterium exchange by mass spectrometry
MOPS 3-(*N*-morpholino)propanesulfonic acid
NMR nuclear magnetic resonance
PGM1 phosphoglucomutase 1
PHMs α-D-phosphohexomutases
PMM/PGM phosphomannomutase/phosphoglucomutase
PMSF phenylmethylsulfonyl fluoride
P-Ser phosphoserine

1. INTRODUCTION

The α-D-phosphohexomutases (PHMs) comprise a ubiquitous enzyme superfamily, found in all kingdoms of life (Stiers, Muenks, & Beamer, 2017). Enzymes in this superfamily catalyze the isomerization of different sugar phosphates, with subgroups defined according to their preferences for the sugar moiety (e.g., glucose, mannose, glucosamine, and *N*-acetylglucosamine). The diversity of these phosphosugar substrates/products is consistent with the varied roles of the PHMs in metabolic and carbohydrate biosynthetic pathways. While the biological roles of the PHMs are well established, only in recent years have several proteins in the superfamily emerged as model systems for biophysical study. These included investigations of the molecular basis of diverse substrate recognition, the role of conformational change in enzyme mechanism, and the association of networks of coevolving residues with protein dynamics (Lee, Mick, Furdui, & Beamer, 2012; Luebbering et al., 2012; Regni, Naught, Tipton, & Beamer, 2004; Regni, Schramm, & Beamer, 2006; Schramm, Mehra-Chaudhary, Furdui, & Beamer, 2008). Of particular interest herein is the recently revealed impact of the phosphorylation state of the catalytic phosphoserine (P-Ser) on the structural flexibility of these enzymes

(Lee, Villar, Artigues, & Beamer, 2014; Stiers, Xu, et al., 2017; Xu, Lee, Beamer, & Van Doren, 2015).

The PHMs typically catalyze a reversible phosphoryl transfer from the 1- to the 6-position of their sugar substrates (Stiers, Muenks, et al., 2017). This reaction relies on a conserved P-Ser in the active site that participates in phosphoryl transfer. During the multistep reaction, the P-Ser first donates its phosphoryl group to substrate (e.g., glucose 1-phosphate), creating a bisphosphorylated sugar intermediate (Fig. 1). Following a 180-degree reorientation of the intermediate in the active site, the serine then accepts the alternate phosphoryl group, creating product and regenerating the active form of the enzyme. Thus, the protein cycles between phosphorylated and unphosphorylated states at different points in its multistep reaction. This mechanism, and the associated critical importance of the P-Ser, was established many decades ago by William Ray and coworkers in their pioneering studies of rabbit phosphoglucomutase (Ray, Mildvan, & Long, 1973).

In addition to its direct role in phosphoryl transfer, the phosphorylation state of the catalytic serine was recently found to have multiple, indirect effects on the structural flexibility of the PHMs. Detailed biophysical studies of *Pseudomonas aeruginosa* phosphomannomutase/phosphoglucomutase (PMM/PGM), including hydrogen–deuterium exchange by mass spectrometry (HDXMS) and nuclear magnetic resonance (NMR), showed that phosphorylated and unphosphorylated versions of the enzyme differ significantly in

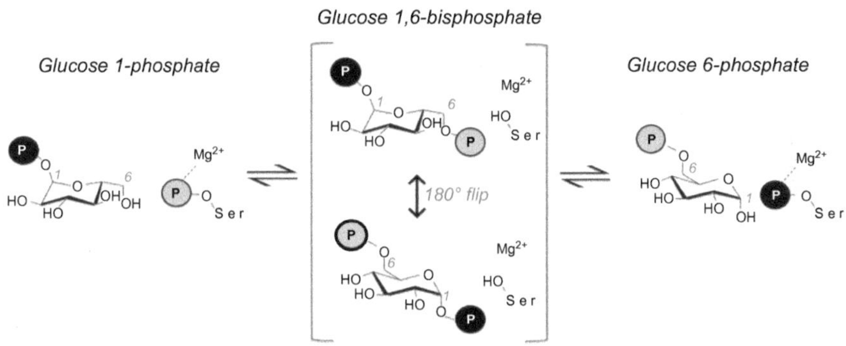

Fig. 1 The multistep enzyme mechanism of the PHMs illustrated for the conversion of glucose 1-phosphate to glucose 6-phosphate. The reaction initiates with transfer of the enzymic phosphoryl group (*gray*) to substrate, to form a bisphosphorylated sugar intermediate. Reorientation of glucose 1,6-bisphosphate (*black arrow*) places the 1-phosphate group (*black*) near the catalytic serine, positioning it for the second phosphoryl transfer back to enzyme. The product is now phosphorylated at the 6-position with the phosphoryl group (*gray*) derived from the enzyme.

their flexibility, with more than 100 of the 450 residues in the enzyme affected despite the modest chemical difference of a single phosphorylation site (Lee, Villar, et al., 2014; Xu et al., 2015). The correlation between the unphosphorylated enzyme and increased structural flexibility was intriguing, as this is the state of the protein that coincides with the required 180-degree reorientation of the intermediate, the most unusual feature of the PHM mechanism. This observation suggested that increased structural flexibility of the protein might facilitate the release of high-affinity interactions between the enzyme and intermediate, and thereby encourage its reorientation (Lee, Villar, et al., 2014). Evidence is mounting for analogous phosphorylation-dependent differences in protein flexibility in other members of the superfamily (Lee, Stiers, Kain, & Beamer, 2014; Stiers, Xu, et al., 2017), suggesting that this may be a conserved feature of the unique catalytic mechanism of the PHMs.

In this chapter, we describe approaches for determining and modifying the phosphorylation level of the catalytic P-Ser, as well as for assessing phosphorylation-dependent effects on protein flexibility for this enzyme superfamily. Our goal is to enable the preparation of homogeneous and stable samples suitable for biophysical studies, including NMR, X-ray crystallography, and HDXMS. Given the widespread evolutionary distribution of the PHMs and their importance in metabolism, bacterial pathogenesis, and human disease, these enzymes have been oddly underrepresented in such studies. This situation, however, may be starting to change based on recent structural depositions in the Protein Data Bank (Stiers, Muenks, et al., 2017). The influence of phosphorylation must also be considered in the interpretation of the effects of mutations on these proteins, including those involved in human inherited disease (Lee, Stiers, et al., 2014). Without establishing competence of the catalytic P-Ser for phosphorylation, and defining the phosphorylation level of the P-Ser in mutant enzymes, the effects of mutations can be easily confused with those due to alterations in phosphorylation state.

2. ASSESSMENT AND MODULATION OF THE PHOSPHORYLATION STATE OF THE CATALYTIC P-Ser

In the preparation of homogeneous PHM samples for biophysical analysis, two factors must be considered. First, a reliable method for assessing the phosphorylation state of the catalytic P-Ser must be available, so the effects of experimental conditions can be determined. Second, approaches for modifying the level of phosphorylation are needed, in order to produce the enzyme state desired for analysis. Possible ways of addressing these two issues are found in the following sections.

2.1 Electrospray Ionization Mass Spectrometry

Modern mass spectrometric techniques are well suited to detecting covalent modifications of protein structure, including phosphorylation, glycosylation, and other posttranslational modifications (Schweppe, Haydon, Lewis, Resing, & Ahn, 2003). Our method of choice for characterizing the phosphorylation of the catalytic serine in the PHMs is electrospray ionization mass spectrometry (ESI-MS). This technique has sufficient resolution to permit analysis of intact protein samples, avoiding the need for proteolytic digestion and peptide identification. Although the 80-Da mass difference between phosphorylated and unphosphorylated protein is small relative to the size of these proteins (~50–60 kDa), this difference is clearly discernible in high-quality spectra obtained from purified protein samples (Fig. 2). The relative fractions of phosphorylated and unphosphorylated protein can then be calculated based on the ratio of the individual peak heights to the sum of the two. Given availability of a mass spectrometry facility with appropriate instrumentation and expert staff, the experiment is fast and sample requirements are minimal (~20 μg of protein).

2.1.1 Equipment

- ESI-MS capability for mass range of ~50 kDa. We use an Agilent ESI Q-TOF MS with a coupled Agilent 1200 series nano-flow liquid chromatography system and a C8 reverse-phase column
- Software for analysis of ESI-MS data capable of deconvoluting highly charged protein ions observed in the spectra

2.1.2 Buffers and Reagents

- Purified PHM (>50 μ*M*) in compatible buffer, such as 50 m*M* MOPS, pH 7.4, and 1 m*M* $MgCl_2$. Avoid high salt (<100 m*M* prior to dilution, <10 m*M* as analyzed), detergents (preferably none), and glycerol (<20% prior to dilution, <2% as analyzed)
- 1% formic acid

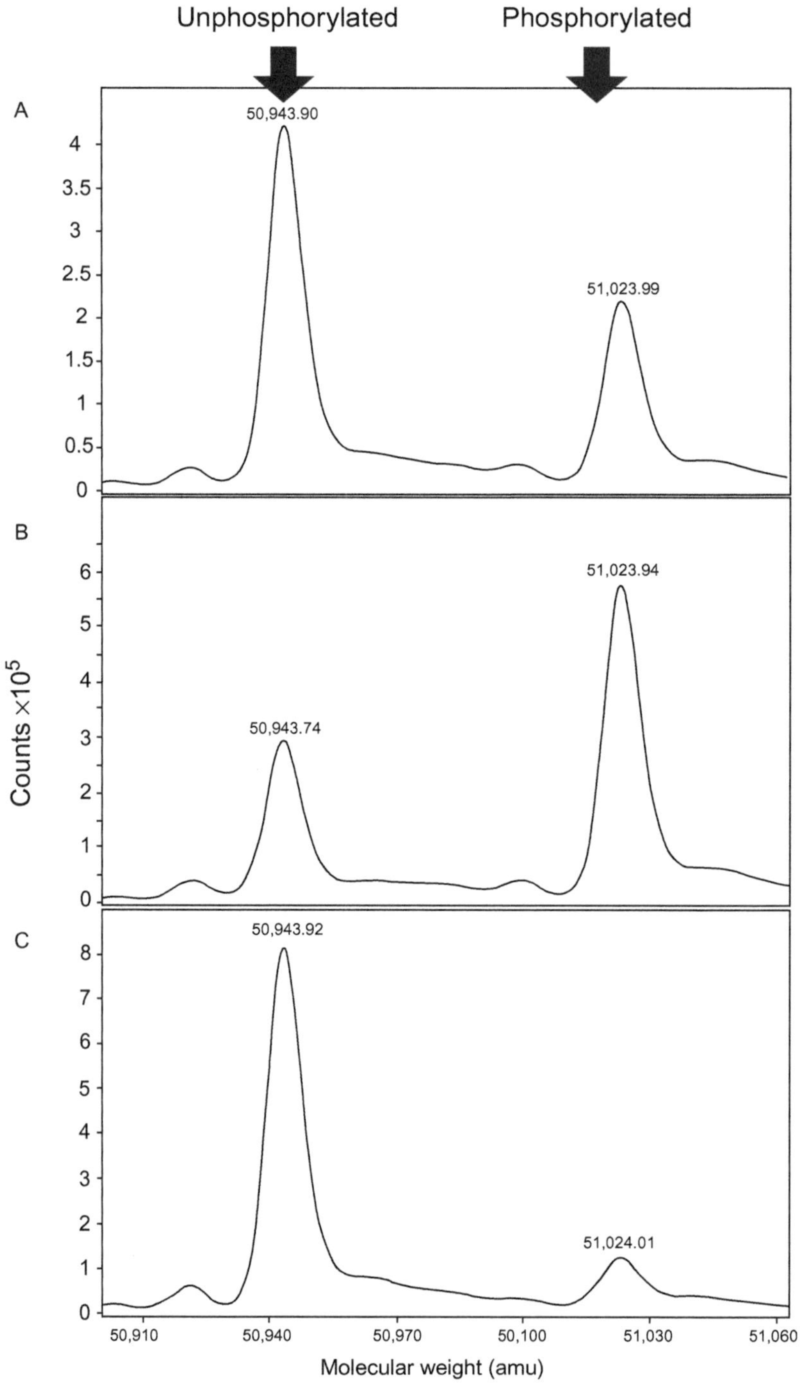

Fig. 2 ESIMS for assessment of the phosphorylation state of the catalytic serine of PHMs. Deconvoluted spectra of *Xanthomonas citri* PGM as purified after heterologous expression in *E. coli* (A), or after treatment with a molar excess of G16P (B), or glucosamine 1-phosphate (C). The measured molecular masses of 51,024 and 50,944 Da for these peaks correspond to phosphorylated and unphosphorylated enzyme (calculated masses 51,023.43 and 50,943.43, respectively).

2.1.3 Procedure

1. Prepare PHM sample (~25 μg total; 5 μL of >50 μ*M*) in a compatible buffer. Thaw sample on ice when ready for analysis.
2. Dilute sample to 1 pmol/μL in 1% formic acid. Vortex at the lowest setting for 10 s and centrifuge (refrigerated) at maximum speed for 5 min.
3. Aliquot into autosampler vial and place in autosampler cooled to 4–7°C.
4. Begin liquid chromatography–mass spectrometry using the following parameters: mass range 295–2500 *m/z*, 0.63 spectra/s, acetonitrile gradient from 3% to 90%. Total run time is approximately 18.5 min per sample. The gradient can be altered to best suit sample complexity, e.g., run time as short as 9 min on the Agilent 1200 series work for highly pure samples (>95% target protein).
5. Deconvolute (using maximum entropy algorithm) consecutive charge state peaks for corresponding mass range of samples with assistance of expert. See also Note 6 below.
6. In the deconvoluted spectra, use the calculated average molecular mass of the protein (based on amino acid sequence) to identify peaks for the phosphorylated and unphosphorylated protein species.
7. Calculate the ratio of phosphorylated and unphosphorylated protein by dividing the individual peak heights by the sum of the two peak heights.

2.1.4 Notes

1. PHM samples awaiting ESI-MS analysis should be frozen, to avoid changes in the phosphorylation level of the P-Ser over time (see below).
2. Typically, we characterize the proteins by ESI-MS immediately after purification, as well as following treatment with reagents to promote phosphorylation of the catalytic serine or to remove this phosphorylation (see Section 2.2).
3. Be aware that some enzymes in the superfamily may have phosphorylation sites in addition to the catalytic serine, due to posttranslational modification (Gururaj, Barnes, Vadlamudi, & Kumar, 2004; Lee, Furdui, & Beamer, 2017). This can lead to significant peak complexity making assignment of individual peaks more difficult. However, with highly pure protein, it is still possible to examine hyperphosphorylated intact proteins (Taylor et al., 2016).
4. While alternative methods for quantification of relative phosphorylation exist, such as various commercial stains for use with SDS gels, we have not verified their efficacy for detecting the P-Ser modification of the PHMs.

5. As these ESI-MS experiments are done with intact (full-length) protein samples, the mass spectrometric analysis is not affected by signal suppression that is common in phosphoprotein digests (Pan et al., 2013).
6. Deconvolution settings may need to be tuned for specific cases. We used the following parameters: mass, 45–65 kDa; mass step, 0.1 Da; *m/z* range restricted, 700–1200; baseline subtraction factor, 7; peak signal, noise ratio of 30; 25% peak height for mass averaging; minimum charge state required for averaging, 6; and iterations, 40.

2.2 Reagents and Variables to Promote/Remove Modification of the P-Ser

In most cases, phosphorylation of the catalytic serine is efficiently achieved via incubation of the PHMs with their respective reaction intermediates: sugar 1,6-bisphosphates, such as glucose 1,6-bisphosphate (G16P), mannose 1,6-bisphosphate, etc. Generation of phosphoenzyme occurs spontaneously when the unphosphorylated protein is mixed with an appropriate bisphosphosugar, under conditions consistent with enzyme activity (e.g., in presence of Mg^{2+}). This produces phosphoenzyme and a monophosphosugar. For kinetic studies, it is routine to include G16P in the assay mixture to ensure that the P-Ser is fully phosphorylated and that the protein is maximally active. The G16P serves to phosphorylate any enzyme that is initially unphosphorylated, and also regenerates the catalytic P-Ser should the phosphoryl group be lost during the time course of the assay. [For the two PHMs studied in detail, this was found to occur approximately once every 15–20 catalytic cycles (Naught & Tipton, 2005; Ray & Roscelli, 1964).]

Kinetic studies are generally short in time frame (minutes) and performed at low concentrations and moderate temperatures (e.g., 25°C). Biophysical studies, in contrast, may require fully homogeneous samples that, ideally, are stable for days/weeks, sometimes under extreme conditions (high temperatures and sample concentrations). Thus, additional effort is required to obtain and maintain homogeneous samples for these methods. To date, most of the PHMs tested have purified with their catalytic serine as a mixture of phosphorylated and unphosphorylated forms. As noted earlier, treatment with a bisphosphosugar is most effective at fully phosphorylating the P-Ser. [Treatment with high concentrations of a monophosphosugar may also suffice (Naught & Tipton, 2001).] With the exception of G16P, however, the other bisphosphosugar "activators" (e.g., mannose 1,6-bisphosphate or glucosamine 1,6-bisphosphate) are not readily available. Fortunately, for most PHMs, it appears that G16P is an effective substitute, presumably because most enzymes in the superfamily have at least some residual PGM activity (Stiers, Muenks, et al., 2017). A potential alternative to G16P is fructose 1,6-bisphosphate (Lee et al., 2017; Namboori & Graham, 2008), which is readily available and low in price relative to G16P.

Production of unphosphorylated enzyme is less straightforward, unless the protein happens to purify in this state (Stiers, Xu, et al., 2017). However, a fortuitous observation identified glucosamine 1-phosphate as an effective "dephosphorylating agent" for *P. aeruginosa* PMM/PGM (Sarma et al., 2012), which was subsequently shown to be effective with human PGM1 as well (Lee, Stiers, et al., 2014). [The phosphorylation level of the P-Ser of *P. aeruginosa* PMM/PGM is reduced from 60% to 10% after 4h at 4°C in the presence of a fourfold molar excess of glucosamine 1-phosphate (Lee et al., 2017).] The mechanism of action is not fully understood, but presumably takes place either by transfer of the enzymic phosphoryl group to glucosamine 1-phosphate (without a subsequent phosphoryl transfer) or by binding of glucosamine 1-phosphate facilitating the hydrolysis of the phosphoryl group. However, glucosamine 1-phosphate appears to have no effect on the phosphorylation level of the P-Ser of the phosphoglucosamine mutases (Lee et al., 2017). Enzymes in this subgroup of the superfamily, which preferentially use glucosamine 1- and 6-phosphate as substrates, will require the development of other methods for the production of unphosphorylated enzyme.

Later, we describe a typical protocol for producing either fully phosphorylated (>90%) or unphosphorylated (<10%) enzyme. Although other variables, such as temperature (Lee et al., 2017), also effect the phosphorylation level of the P-Ser over time and might also be of utility in this regard, we have found use of these reagents to be more reliable, reproducible, and gentler in their effects on the protein. Effectiveness should be confirmed, however, when working on novel members of the PHM superfamily, as substantial variations in response are observed between proteins.

2.2.1 Equipment

- Dialysis cups, cassettes, or tubing with appropriate MW cutoff (~30kDa)
- Nanodrop or other UV/Vis spectrophotometer (for determining protein concentration)

2.2.2 Buffers and Reagents

- Purified PHM in appropriate buffer such as 50m*M* MOPS, pH 7.4, with 1m*M* $MgCl_2$
- Dialysis buffer such as 50m*M* MOPS, pH 7.4, with 1m*M* $MgCl_2$

- Agents to promote phosphorylation: G16P or fructose 1,6-bisphosphate
- Agents to promote loss of phosphoryl group from the P-Ser: glucosamine 1-phosphate

2.2.3 Procedure

1. Mix purified PHM with desired reagent (see Section 2.2.2) at 6- to 10-fold molar excess of protein, e.g., 50 μ*M* PHM with 500 μ*M* G16P.
2. Incubate at 4°C overnight or longer if time course desired.
3. Dialyze protein sample to remove reagents (not necessary for ESI-MS); concentrate protein if necessary.
4. Prepare sample for ESI-MS or other applications.
5. Determine % phosphorylation as described in Section 2.1.3.

2.2.4 Notes

1. Phosphorylation is typically rapid (i.e., minutes); an overnight incubation is just for convenience.
2. Typically dilute protein concentrations are used (~50 μ*M*), to reduce the possibility of aggregation, especially of unphosphorylated protein. When it occurs, aggregation typically becomes apparent as a white precipitate that appears over time during dialysis. While most PHMs studied to date are quite stable and homogeneous, use of dynamic light scattering or gel filtration chromatography to assess monodispersity should be considered, especially when working with novel enzymes in the family.
3. The effect of temperature was investigated in detail with *P. aeruginosa* PMM/PGM, where phosphorylation of the P-Ser was reduced by 50% after 1 day at 37°C, but stable for many months in samples stored at −80°C (Lee et al., 2017). Higher temperatures generally accelerate spontaneous loss of the phosphoryl group, but may promote protein aggregation.
4. To determine a time course of phosphorylation/dephosphorylation, take aliquots of sample at desired time points (Step 2 above) and store frozen until ready for analysis.
5. The ratio of phosphorylating/dephosphorylating reagents to enzyme can be increased if needed to produce homogeneous samples. Longer incubation times with glucosamine 1-phosphate may also increase homogeneity.
6. Another option for removal of the enzymic phosphate group is successive additions of substrate followed by dialysis (Ma & Ray, 1980). However, in our experience, this protocol is less effective than treatment with glucosamine 1-phosphate and would require optimization.

3. STRAIGHTFORWARD ASSAYS TO CHARACTERIZE DIFFERENCES BETWEEN THE PHOSPHORYLATED AND UNPHOSPHORYLATED ENZYME STATES

The initial detailed studies showing differences in protein flexibility between the phosphorylated and unphosphorylated states of PHMs used rigorous biophysical methods, including HDXMS and NMR. Such studies can be quite labor- and sample-intensive, particularly for proteins of ~500 residues, and require a significant investment of time and resources. Fortunately, it was subsequently found that several classical biochemical assays are also sensitive to phosphorylation-dependent differences in protein flexibility in these enzymes. Later, we outline several that have been used successfully with multiple members of the PHM superfamily.

3.1 Limited Proteolysis

Partial proteolytic digestion can provide information on protein structure and flexibility, with cuts preferentially occurring at loops and other flexible regions (Fontana et al., 2004). Such preferential cleavage is easily accomplished by adding low concentrations of nonspecific proteases to the sample of interest and running a time course of digestion (Fig. 3). Related methods, such as pulse proteolysis (Park & Marqusee, 2005) that combine limited proteolysis with addition of a denaturant, may also be useful. Below is a limited proteolysis protocol used for *P. aeruginosa* PMM/PGM (Xu et al., 2015).

3.1.1 Equipment

- Standard SDS-PAGE equipment including 12% or 14% SDS-PAGE gels

3.1.2 Buffers and Reagents

- Purified PHM (1.0–1.5 mg/mL) in 50 m*M* MOPS, pH 7.4, and 1 m*M* $MgCl_2$
- Proteinase K from *Tritirachium album* (Sigma-Aldrich)—5 mg/mL stock dissolved in 50 m*M* Tris, pH 8.0, and 1 m*M* $CaCl_2$
- Phenylmethylsulfonyl fluoride (PMSF)—50 m*M* stock in H_2O
- SDS-PAGE sample buffer, running buffer, and stain

3.1.3 Procedure

1. Use the 5 mg/mL stock of proteinase K to prepare a diluted stock at 0.1 mg/mL (a volume of 50 μL is sufficient).

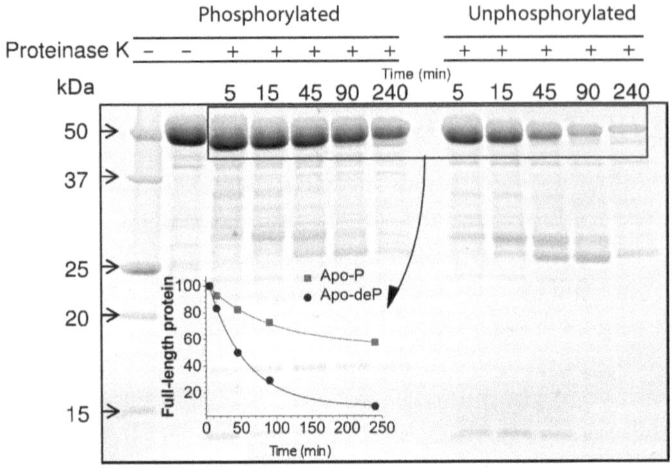

Fig. 3 Use of limited proteolysis to characterize the phosphorylated and unphosphorylated states of *P. aeruginosa* PMM/PGM. A time course of digestion with proteinase K shows the more rapid and extensive digestion of unphosphorylated enzyme (lanes 9–13). Inset highlights the differences in rate of decay of the major band corresponding to the full-length protein (~50 kDa). *Reprinted from Xu, J., Lee, Y., Beamer, L. J., & Van Doren, S. R. (2015). Phosphorylation in the catalytic cleft stabilizes and attracts domains of a phosphohexomutase.* Biophysical Journal, 108, *325–327, with permission from Elsevier.*

2. Mix 95 μL of the PHM with 5 μL of the diluted proteinase K stock and incubate at room temperature. This is the reaction mixture, which is sufficient for ~7 points on a time course. Adjust volume accordingly for additional points.
3. As a control, prepare a second PHM sample (~30 μL total) with buffer added in place of proteinase K.
4. At desired time-points (e.g., 5, 15, 45, 90, 150, and 240 min), remove a 14-μL aliquot from the reaction mixture and terminate the digestion with the addition of 1 μL of 50 m*M* PMSF (final concentration 3 m*M*). Vortex briefly to mix.
5. Add 15 μL of 2 × SDS sample buffer to each of the aliquots from Step 4, flash freeze, and store at −20°C until ready for SDS-PAGE.
6. Load control and digested protein aliquots on a 12% or 14% SDS-PAGE gel. Run gel and stain.

3.1.4 Notes

1. The reaction mixture above has ~300:1 ratio (w/w) of PHM to protease. Both this ratio and the length of digestion may need to be optimized depending on the rate and extent of proteolysis observed on the gel.

2. Different proteases (e.g., thermolysin, trypsin, etc.) can also be selected to fine-tune the assay, depending on experimental needs.
3. Limited proteolysis may also be useful to assess protection due to binding of ligands or the effects of disease-related mutants in these enzymes (Lee, Stiers, et al., 2014).
4. If desired, semiquantitative analysis of the digestion can be done by densitometry to determine the rate of decay of the primary band's intensity (Fig. 3).

3.2 Binding of the Fluorescent Dye ANS

Binding of 8-anilino-1-naphthalenesulfonic acid (ANS) has long been used to probe the conformational flexibility of proteins because its fluorescence in solution is greatly enhanced upon binding to proteins (Hawe, Sutter, & Jiskoot, 2008; Ray & Balaram, 1999). In the case of *P. aeruginosa* PMM/PGM, ANS shows a significant increase in fluorescence emission when mixed with unphosphorylated enzyme relative to phosphorylated enzyme (Xu et al., 2015) (Fig. 4). Similar effects were observed with human PGM1 and phosphoglucosamine mutase from *Bacillus anthracis* (Lee, Stiers, et al., 2014; Stiers, Xu, et al., 2017). This suggests that phosphorylation decreases the access of ANS to crevices within the folded structure of the enzyme, consistent with an overall decrease in protein flexibility due to the P-Ser modification. The protocol used for *P. aeruginosa* PMM/PGM is detailed later (Xu et al., 2015).

3.2.1 *Equipment*

- Microplate fluorimeter
- 96-well plates suitable for fluorescence measurements

3.2.2 *Buffers and Reagents*

- Purified phosphorylated and unphosphorylated PHM at ~0.5 mg/mL in 50 m*M* MOPS, pH 7.4, and 1 m*M* $MgCl_2$
- ANS—1.0 *M* stock prepared in 1.0 *M* NaOH

3.2.3 *Procedure*

1. Use a 1.0 *M* stock of ANS to prepare a 50 m*M* stock in 50 m*M* MOPS, pH 7.4. A volume of ~25 μL is sufficient for 12 reactions.
2. Aliquot 200 μL of the 0.5 mg/mL PHM into 12 microcentrifuge tubes or directly into 96-well plate as desired.
3. Add 2 μL of 50 m*M* ANS to each sample.

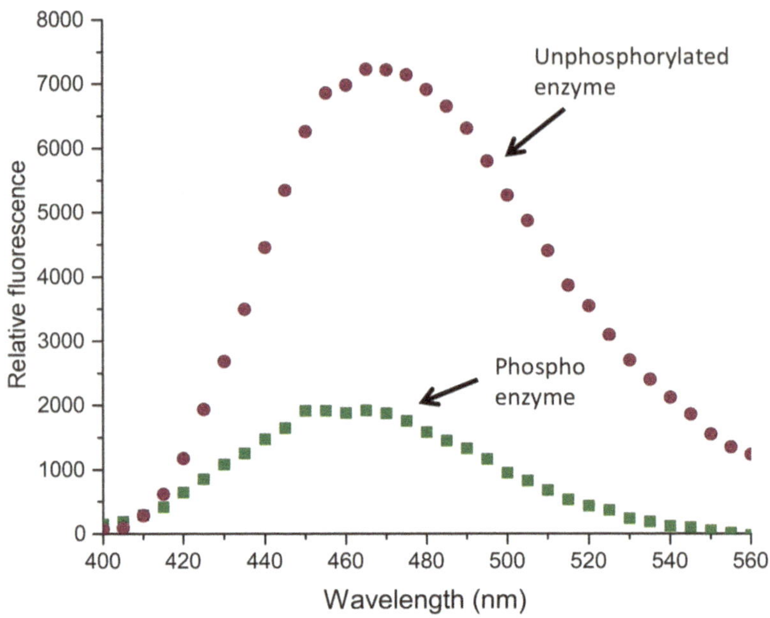

Fig. 4 Binding of the fluorescent probe ANS to the phosphorylated and unphosphorylated states of *P. aeruginosa* PMM/PGM. The increased fluorescence of ANS in the presence of unphosphorylated enzyme is consistent with greater access of the dye to hydrophobic crevices within the structure and with greater protein flexibility. *Reprinted from Xu, J., Lee, Y., Beamer, L. J., & Van Doren, S. R. (2015). Phosphorylation in the catalytic cleft stabilizes and attracts domains of a phosphohexomutase.* Biophysical Journal, 108, *325–327, with permission from Elsevier.*

4. For a control, prepare an additional sample with 2 μL of 50 m*M* ANS in buffer (no PHM) in the same volume.
5. Incubate samples at 25°C for 1 h; transfer into 96-well plate if needed.
6. Measure fluorescence of samples (excitation wavelength = 365 nm; for emission, scan from 400 to 560 nm). Subtract the background emission of free ANS in solution (Step 4) from the spectrum of each protein sample.

3.2.4 Notes

1. For reproducibility, the concentrations of protein and ANS in each tube need to be precisely matched. Take care in pipetting. Making measurements in duplicate or triplicate may also be helpful.
2. Relative to Fig. 4, smaller differences in ANS-binding were observed when comparing the phosphorylated and unphosphorylated states of

B. anthracis phosphoglucosamine mutase (Stiers, Xu, et al., 2017), although the trend was the same in that the unphosphorylated enzyme shows greater fluorescence.

3. This method can also be used to assess the effects of disease-associated mutations on protein flexibility (Lee, Stiers, et al., 2014).

3.3 Chemical Denaturation

Changes in protein stability are commonly monitored with methods such as thermal shift assays or thermal melts by circular dichroism. However, the temperature dependence of the P-Ser modification in the PHMs precludes the use of such methods (Fig. 5A). Chemical denaturation, such as that performed with urea or guanidine, has been used successfully to characterize differences between the phosphorylated and unphosphorylated states of *P. aeruginosa* PMM/PGM (Xu et al., 2015) (Fig. 5B). In this system, intrinsic fluorescence measurements show that a 3.0 *M* concentration of urea is needed for 50% denaturation (C_m) of the phosphoenzyme, suggesting that it has greater stability than the unphosphorylated state, which has a C_m of 2.6 *M*. C_m values are compared because *P. aeruginosa* PMM/PGM, like most proteins in the PHM superfamily, cannot be refolded after denaturation, indicating a lack of the reversibility needed to obtain the difference in the Gibbs free energy (ΔG) of folding.

3.3.1 Equipment

- Microplate fluorimeter
- 96-well plates suitable for fluorescence measurements (e.g., Corning #3720)
- Software for curve fitting, e.g., Origin Pro (OriginLab, Northampton, MA)

3.3.2 Buffers and Reagents

- A stock of purified PHM at >10 μ*M* in 50 m*M* MOPS, pH 7.4, 0.1 m*M* DTT
- Urea stock at 10.0 *M*

3.3.3 Procedure

1. Prepare a series of samples at a constant concentration of protein, but with varying concentrations of urea (e.g., from 0.0 to 8.0 *M* in steps of 0.25 *M*). We used PHM at 1.3 μ*M* (in buffer above) in a total final volume of 150 μL. [Samples can be prepared in tubes or directly in 96-well plate as desired.] Incubate for 12 h at 25°C.

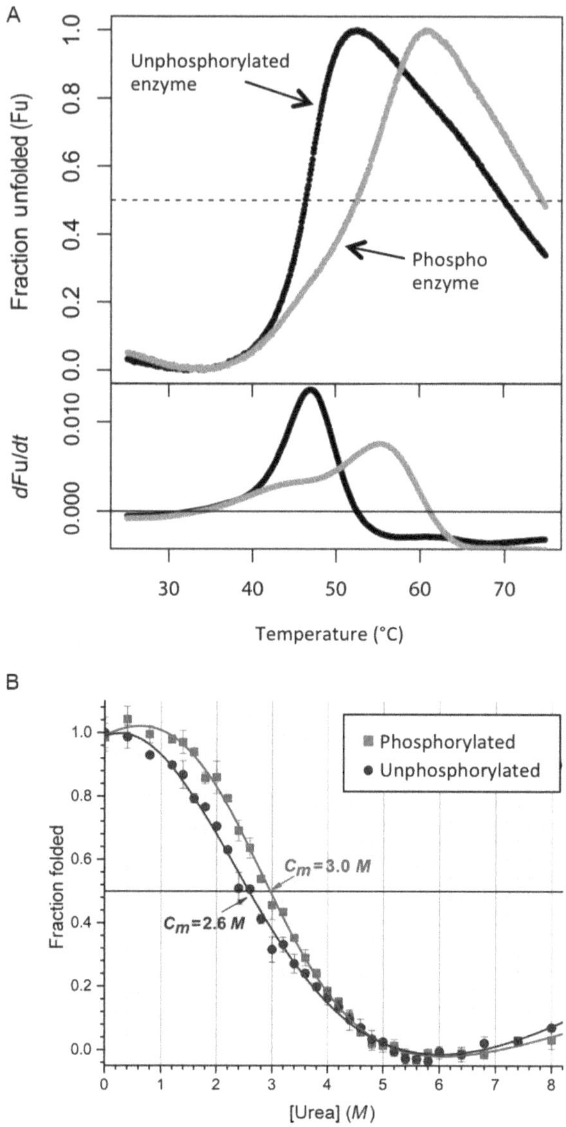

Fig. 5 Thermal and chemical denaturation of phosphorylated and unphosphorylated PHMs. (A) Thermal denaturation of PHMs is complicated by the temperature dependence of the P-Ser modification. A thermal shift assay (Huynh & Partch, 2015) using samples of human PGM1. Unphosphorylated enzyme (*black*) displays a typical, monophasic melting curve. In contrast, phosphoenzyme (*gray*) shows a broadened and biphasic melting curve, which is particularly evident in the first derivative (*lower panel*). The complex behavior of the phosphoenzyme appears to result from the loss of the P-Ser modification during the time course of heating. (B) Urea unfolding of the phosphorylated and unphosphorylated states of *P. aeruginosa* PMM/PGM monitored by intrinsic fluorescence. A comparison of the midpoint of the curves (C_m) indicates that phosphoenzyme requires a higher concentration of denaturant to reach 50% unfolded. *Panel (B): Reprinted from Xu, J., Lee, Y., Beamer, L. J., & Van Doren, S. R. (2015). Phosphorylation in the catalytic cleft stabilizes and attracts domains of a phosphohexomutase.* Biophysical Journal, 108, *325–327, with permission from Elsevier.*

2. Prepare matching blanks for each urea concentration used in Step 1 (no PHM).
3. Transfer all samples into a 96-well plate and measure intrinsic tryptophan fluorescence (excitation at 280 nm and detection at 320 nm). For each protein sample, subtract the corresponding spectrum of urea (samples from Step 2).
4. Use curve-fitting software to determine the midpoint of the denaturation curves to obtain C_m.

3.3.4 Notes

1. Be sure to seal the 96-well plate to prevent evaporation.
2. When working with urea stocks at high concentration (e.g., 10 *M*) the solution can be heated in 40°C water bath to dissolve any crystallized urea.
3. Measurements in duplicate/triplicate are encouraged to allow determination of errors.
4. The detection wavelength (~320 nm) should be optimized for each protein by scanning its emission spectra from 300 to 400 nm (excitation at 280 nm) and identifying the wavelength with maximum emission intensity.
5. To normalize fluorescence signal (RFUs) to the fraction of folded protein (0.0–1.0) the following equation can be used: Fraction folded $= \frac{RFU_i - RFU_{min}}{RFU_{control} - RFU_{min}}$ where RFU_i is each fluorescence signal, $RFU_{control}$ is the fluorescence at 0.0 *M* urea, and RFU_{min} is the minimum fluorescence over all urea concentrations for a sample. The normalized data can then be fit to a generic (or custom) logistic function to determine midpoint, such as in Fulcher and Van Doren (2011).

3.4 NMR Spectra of Isotopically Labeled Protein

Various types of NMR experiments, such as TROSY spectra (Cai, Huang, Yang, Craigie, & Clore, 2016; Studier, 2005), can provide insight into protein folding/flexibility in solution, even in the absence of peak assignments (i.e., knowing which residue is associated with each peak in the spectrum). Measurement of such spectra for large proteins requires production of isotopically labeled protein samples, either ^{15}N or $^{2}H/^{15}N$ depending on spectral quality. Due to their generally high-expression levels and solubility in recombinant expression systems, production of isotopically labeled versions of the PHMs has been quite feasible, following established protocols (Sarma et al., 2012; Stiers, Xu, et al., 2017). In the PHMs studied to date, deuteration has been necessary to achieve sufficient quality spectra to assess differences between the phosphorylated and unphosphorylated states of these enzymes (Lescop, Schanda, & Brutscher, 2007).

We have collected BEST-TROSY spectra (Studier, 2005) to assess potential differences between the phosphorylated and unphosphorylated states of *P. aeruginosa* PMM/PGM and *B. anthracis* phosphoglucosamine mutase (Fig. 6). Below is a condensed protocol for the production of isotopically labeled samples from a 0.5-L culture of PG media (Xu et al., 2015) prepared with ammonium-^{15}N chloride and D_2O. This was sufficient to produce ~20 mg of labeled protein for these two enzymes. With the assistance of an expert spectroscopist, a BEST-TROSY spectrum can usually be collected within an hour, allowing qualitative comparisons of phosphorylated and unphosphorylated samples of PHMs.

3.4.1 Equipment

- Standard protein expression/purification equipment including temperature-controlled shakers and spectrophotometer for measuring optical density of cell cultures
- Bruker Advance 800 MHz spectrometer with TCI cryoprobe or similar
- Shighemi or other NMR tube

3.4.2 Buffers and Reagents

- Glycerol stock of *E. coli* strain expressing PHM of interest
- PG media supplemented with appropriate antibiotic for bacterial strain and expression system
- 1 *M* stock of isopropyl β-D-1-thiogalactopyranoside for inducing protein expression
- Buffer A: 50 m*M* MOPS, pH 7.4, 1 m*M* $MgCl_2$, and 1 m*M* DTT
- Reagents for isotopic labeling: ammonium-^{15}N chloride, 99% D_2O. These are available from several vendors including Cambridge Isotope Laboratories and Isotec (Sigma-Aldrich). Obtain a quote for best pricing.

3.4.3 Procedure

1. Using sterile technique, take scrapes from the frozen glycerol stock and inoculate two culture tubes each containing 2.5 mL of PG media with antibiotic. Shake at 37°C overnight or until cloudy.

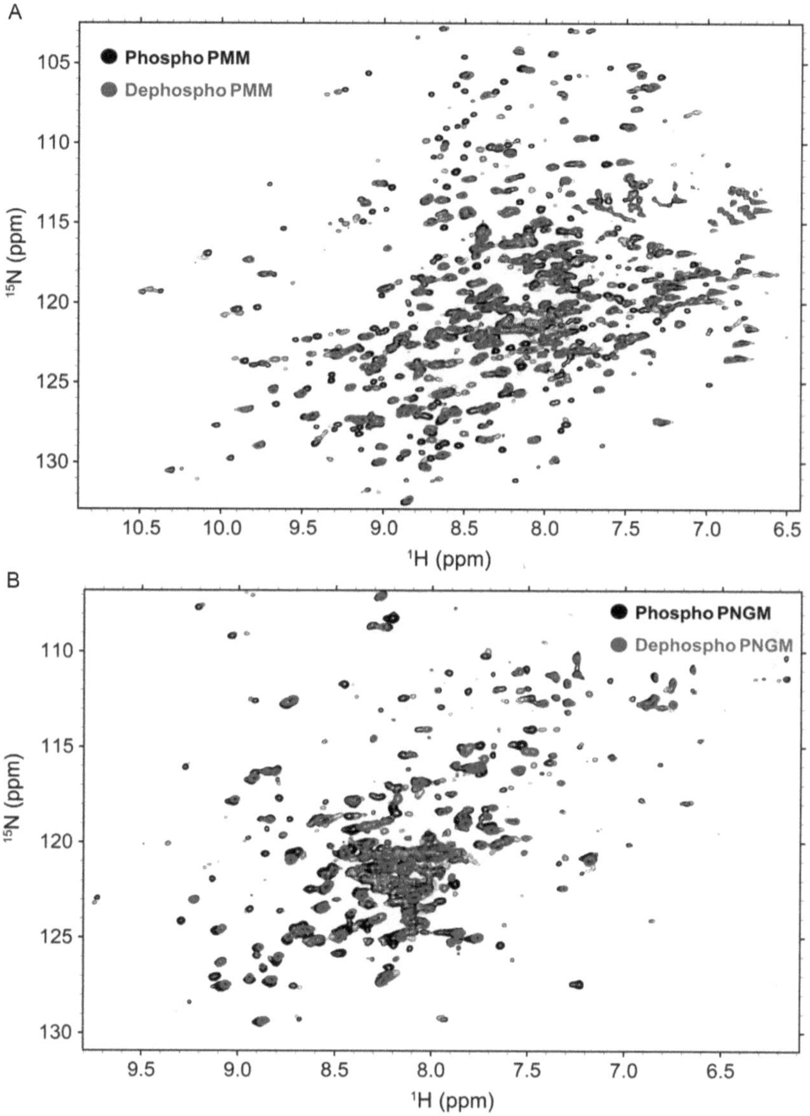

Fig. 6 BEST-TROSY spectra of enzymes from two different subgroups of the PHM superfamily. A comparison of spectra measured on an 800 MHz magnet from the phosphorylated (*black*) and unphosphorylated (*gray*) states of (A) *P. aeruginosa* PMM/PGM (Xu & Van Doren, 2017) and (B) *B. anthracis* phosphoglucoasmine mutase (Stiers, Xu, et al., 2017). Qualitative differences can be seen within each pair of spectra, with the phosphorylated samples showing both an increased number of and an increased sharpness of peaks, consistent with increased compactness and reduced structural fluctuations for this enzyme state.

2. Use the 2.5 mL cultures to inoculate two flasks each containing 250 mL PG media (500 mL total) with antibiotic. Shake at 37°C until cultures reach an optical density of ~0.8 (usually 4–8 h). Induce as appropriate (e.g., 0.4 m*M* IPTG) and grow overnight at 18°C.
3. Harvest cells by centrifugation, flash freeze pellets in liquid nitrogen, and store at −80°C until ready for purification. Purify labeled protein using standard protocols.
4. Divide the sample of purified protein and treat to either phosphorylate or remove phosphorylation of the catalytic serine (see Section 2.2.3). Dialyze overnight in Buffer A to remove small molecule reagents.
5. Determine the concentration of phosphorylated and unphosphorylated protein samples. Concentrate/dilute as needed so that the samples are equivalent, to allow qualitative comparisons of their spectra.
6. Carefully load protein samples into Shighemi or other NMR tube. Avoid bubbles and frothing of sample. Collect BEST-TROSY spectra with help from expert spectroscopist.

3.4.4 Notes

1. For the PHMs tested so far, yields of labeled proteins have been generally similar to those obtained from cultures grown in rich media, although growth times are often much longer.
2. The major expense of labeling is the cost of deuterium (currently >$400/L) so smaller cultures may be grown if protein yield from the recombinant system is sufficient.
3. It is important to optimize the conditions for producing phosphorylated or unphosphorylated samples (see Section 2) for the protein of interest prior to making labeled samples, since the nonuniformity of isotopic labeling complicates interpretation of mass spectra.
4. Avoid freezing of samples prior to NMR measurements if possible, as spectra are very sensitive to aggregation. Inclusion of dithiothreitol or other reductant in the sample buffer may help prevent oxidative changes to protein samples.
5. We recommend NMR data collection at 25–35°C. Higher temperature sharpens peaks in the spectra by decreasing solvent viscosity and increasing the rate of rotational diffusion, which is helpful given the typical 50–60 kDa size of the PHM protomer. However, higher temperatures may also accelerate loss of the P-Ser modification and lead to protein precipitation (see Section 3.3). Lower temperatures may avoid these issues but will likely require longer collection times.

Both protein aggregation and the lifetime of the phosphorylation should be tested with unlabeled protein to assess the effects of the experimental conditions.

6. A series of TROSY spectra taken over time may be used to determine the time course of dephosphorylation of the sample in the NMR tube. Semiautomatic fitting of such data with the software package TREND (Xu & Van Doren, 2016) has been described.

4. CRYSTALLOGRAPHIC REFINEMENT OF PARTIAL OCCUPANCY P-Ser

An essential aspect of three-dimensional structural analyses of the active site of the PHMs is an appropriate modeling of the catalytic P-Ser during crystallographic refinement. In various crystal structures, this residue has been observed as fully phosphorylated (e.g., PDB ID 1K35), fully unphosphorylated (e.g., PDB ID 3PDK), or a mixture of both (e.g., PDB ID 5VBI). Correct modeling of this residue is essential for understanding its critical role in the enzyme mechanism. It is also important for studies that rely on the crystallographic coordinates, such as molecular dynamics and electrostatic calculations (Bras, Fernandes, Ramos, & Schwartz, 2018; Stiers, Xu, et al., 2017; Xu et al., 2015).

Moreover, incorrect modeling (e.g., P-Ser in unphosphorylated enzyme) will lead to spurious difference electron density peaks for the phosphate group (especially on the phosphorus atom) and can produce artifacts such as high B-factors (Fig. 7A and B). Such artifacts can lead to incorrect analyses and mistaken conclusions in later studies.

The most complicated case for crystallographic refinement is a mixture of phosphorylated and unphosphorylated forms of the catalytic serine. Due to constraints of the refinement programs [e.g., refinement with Phenix is discussed here (Adams et al., 2010)], it is not possible to model the residue as both a serine and P-Ser (PDB residue name SEP) as this results in two different residue types at the same residue position. To avoid this, the residue must be modeled as two conformers of P-Ser (denoted as ASEP and BSEP in the coordinate file), the latter of which has the terminal phosphoryl group deleted (atoms O1, O2, O3, and P of BSEP). As an added complication, it is common to see water molecules in the structure of the unphosphorylated enzyme that occupy positions near those of the phosphoryl group oxygens in the phospho-conformer (ASEP). Hence, the occupancy of these waters

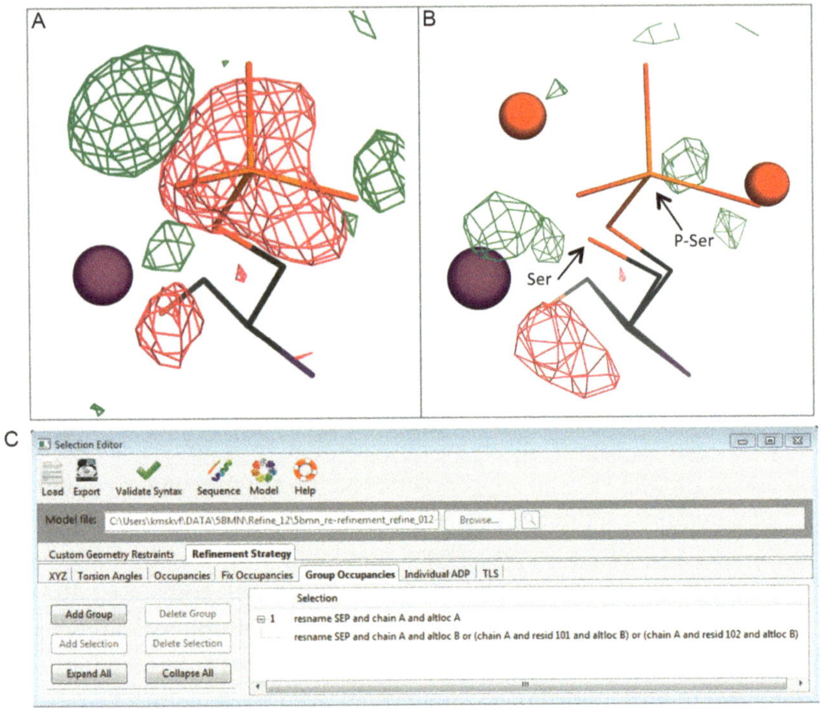

Fig. 7 Modeling of the catalytic serine in the crystal structure of PGM from *Xanthomonas citri* (PDB ID 5BMN) (Goto et al., 2016). (A) An F_o–F_c electron density map contoured at $\pm 3.0\sigma$ in the vicinity of Ser119 as calculated from the deposited coordinates and structure factors. The residue is modeled as P-Ser with an occupancy of 1.0, but the large peak of negative density (*red*) on the phosphorus atom indicates an occupancy <1.0. The nearby metal ion is shown as *purple sphere*. (B) The same residue, following refinement as a mixture of P-Ser and Ser, using the ASEP and BSEP conformers as described in text. The refined occupancy of the P-Ser conformer is 0.37, and the negative density on the phosphorus atom has disappeared. The B-factor of the phosphorus atom has also decreased from 46.6 to 21.9 Å^2. Two water molecules (*red spheres*) coupled to the BSEP conformer (Ser) are found in positions similar or near to the oxygens of the P-Ser. (C) A screenshot of the Phenix GUI indicating how to setup the mixed P-Ser/Ser refinement.

must be explicitly coupled with the BSEP conformer (i.e., by designating the waters as BHOH) to avoid atomic clashes during refinement. The procedure below accomplishes this, with the occupancies of ASEP and the BSEP/BHOH group allowed to vary during refinement, but with their sum constrained to 1.0. This protocol assumes as a starting point that two conformers of SEP (as above) are present in the PDB file and that waters have been built into appropriate density for the BSEP conformer.

4.1 Equipment

- Suitable computer with Phenix and Coot (Emsley & Cowtan, 2004) software packages installed

4.2 Required Files

- PDB coordinate file for protein of interest
- Corresponding structure factor file (.mtz format) for refinement in Phenix

4.3 Procedure

1. Identify the residue numbers of the waters to couple with the BSEP conformer and change their residue names in the PDB file to "BHOH." This can be done by directly editing the PDB file or via the "Residue Info" window in the "Measures" tab of Coot.
2. Create multioccupancy selections in phenix.refine: under Refinement Strategy, change tab to "Group Occupancies" and select "Add Group" (Fig. 7C).
3. In the first selection box, enter: resname SEP and chain A and altloc A. This is all the information needed for the phosphorylated ASEP conformer.
4. Click "Add Selection" (do NOT add a new group) for the BSEP conformer and waters (e.g., residue numbers 101 and 102 in chain S). In second selection box, enter: resname SEP and chain A and altloc B or (chain S and resid 101 and altloc B) or (chain S and resid 102 and altloc B). This couples the truncated BSEP conformer with the waters in chain S that occupy positions near atoms O1, O2, or O3 of the ASEP conformer.
5. Run refinement with occupancy refinement on and check electron-density maps. Rebuild the model with Coot as necessary.

4.4 Notes

1. If two protein chains are present, create a second group and enter SEP selections as above but with chain identifier B.
2. "Altloc A" defines a conformer (e.g., ASEP), for a multiconformer residue.
3. Use of Boolean "or" in second selection box links BSEP with the residues that follow (i.e., two waters). This refinement strategy can be generalized for any refinement of atoms occupying the same space, such as alternate side chain conformations coupled to waters.

4. Starting occupancies for BSEP atoms O1, O2, O3, and P as well as the coupled waters (BHOH oxygens) should be set to less than 1.0 in the PDB file, so that their occupancy will vary during refinement.
5. Be sure to turn on the "refine occupancy" option in Phenix.refine (Configure → Refinement Settings → Refinement Strategy → Occupancies).

5. SUMMARY AND CONCLUSION

To ensure reproducibility and clear interpretation of experimental results, biochemical and biophysical characterizations of enzymes in the PHM superfamily must take into account the phosphorylation state of the catalytic P-Ser. The methods described herein have proven useful for enzymes in distinct subgroups of the superfamily, which have varying substrate preferences, different oligomeric states, and overall limited sequence identity (~25%). Despite these differences, the impacts of phosphorylation on the structural flexibility of these proteins are similar, suggesting this is a widespread phenomenon in the superfamily. These methods may also provide insights into the effects of other types of posttranslational modifications on protein structure, which may be longer in range and more prevalent than currently appreciated (Chaffey et al., 2017; Elbaum & Zondlo, 2014; Xin & Radivojac, 2012).

ACKNOWLEDGMENTS

Studies of PHMs in the Beamer Laboratory have been supported by the National Science Foundation, including Grants MCB-0918389 and MCB-1409898. K.M.S. was supported by NIH T32 training Grant GM008396-26 and American Heart Association predoctoral fellowship 17PRE33400210. We thank Steven Van Doren, Brian Mooney, and Jia Xu for critical reading of the manuscript and assistance with figure preparation.

REFERENCES

Adams, P. D., Afonine, P. V., Bunkóczi, G., Chen, V. B., Davis, I. W., Echols, N., et al. (2010). PHENIX: A comprehensive Python-based system for macromolecular structure solution. *Acta Crystallographica Section D: Biological Crystallography*, *66*(2), 213–221. https://doi.org/10.1107/S0907444909052925.

Bras, N., Fernandes, P., Ramos, M., & Schwartz, S. (2018). Mechanistic insights on human phosphoglucomutase revealed by transition path sampling and molecular dynamics calculations. *Chemistry*, *24*, 1978–1987. https://doi.org/10.1002/chem.201705090.

Cai, M., Huang, Y., Yang, R., Craigie, R., & Clore, G. M. (2016). A simple and robust protocol for high-yield expression of perdeuterated proteins in *Escherichia coli* grown in shaker flasks. *Journal of Biomolecular NMR*, *66*(2), 85–91. https://doi.org/10.1007/s10858-016-0052-y.

Chaffey, P. K., Guan, X., Chen, C., Ruan, Y., Wang, X., Tran, A. H., et al. (2017). Structural insight into the stabilizing effect of O-glycosylation. *Biochemistry, 56*(23), 2897–2906. https://doi.org/10.1021/acs.biochem.7b00195.

Elbaum, M. B., & Zondlo, N. J. (2014). OGlcNAcylation and phosphorylation have similar structural effects in a-helices: Post-translational modifications as inducible start and stop signals in a-helices, with greater structural effects on threonine modification. *Biochemistry, 53*(14), 2242–2260. https://doi.org/10.1021/bi500117c.

Emsley, P., & Cowtan, K. (2004). Coot: Model-building tools for molecular graphics. *Acta Crystallographica Section D: Biological Crystallography, 60*(12), 2126–2132. https://doi.org/10.1107/S0907444904019158.

Fontana, A., De Laureto, P. P., Spolaore, B., Frare, E., Picotti, P., & Zambonin, M. (2004). Probing protein structure by limited proteolysis. *Acta Biochimica Polonica, 51*(2), 299–321. http://doi.org/035001299.

Fulcher, Y. G., & Van Doren, S. R. (2011). Remote exosites of the catalytic domain of matrix metalloproteinase-12 enhance elastin degradation. *Biochemistry, 50*(44), 9488–9499. https://doi.org/10.1021/bi2009807.

Goto, L. S., Vessoni Alexandrino, A., Malvessi Pereira, C., Silva Martins, C., D'Muniz Pereira, H., Brandão-Neto, J., et al. (2016). Structural and functional characterization of the phosphoglucomutase from Xanthomonas citri subsp. citri. *Biochimica et Biophysica Acta (BBA) - Proteins and Proteomics, 1864*(12), 1658–1666.

Gururaj, A., Barnes, C. J., Vadlamudi, R. K., & Kumar, R. (2004). Regulation of phosphoglucomutase 1 phosphorylation and activity by a signaling kinase. *Oncogene, 23*(49), 8118–8127. https://doi.org/10.1038/sj.onc.1207969.

Hawe, A., Sutter, M., & Jiskoot, W. (2008). Extrinsic fluorescent dyes as tools for protein characterization. *Pharmaceutical Research, 25*(7), 1487–1499. https://doi.org/10.1007/s11095-007-9516-9.

Huynh, K., & Partch, C. L. (2015). Analysis of protein stability and ligand interactions by thermal shift assay. *Current Protocols in Protein Science, 2015,* 19.26.1-19.26.14. https://doi.org/10.1002/0471140864.ps1926s79.

Lee, Y., Furdui, C. M., & Beamer, L. J. (2017). Data on the phosphorylation state of the catalytic serine of enzymes in the α-D-phosphohexomutase superfamily. *Data in Brief, 10,* 398–405.

Lee, Y., Mick, J., Furdui, C., & Beamer, L. J. (2012). A coevolutionary residue network at the site of a functionally important conformational change in a phosphohexomutase enzyme family. *PLoS ONE,* 7(6), e38114.

Lee, Y., Stiers, K. M., Kain, B. N., & Beamer, L. J. (2014). Compromised catalysis and potential folding defects in in vitro studies of missense mutants associated with hereditary phosphoglucomutase 1 deficiency. *Journal of Biological Chemistry, 289*(46), 32010–32019.

Lee, Y., Villar, M. T., Artigues, A., & Beamer, L. J. (2014). Promotion of enzyme flexibility by dephosphorylation and coupling to the catalytic mechanism of a phosphohexomutase. *Journal of Biological Chemistry, 289*(8), 4674–4682.

Lescop, E., Schanda, P., & Brutscher, B. (2007). A set of BEST triple-resonance experiments for time-optimized protein resonance assignment. *Journal of Magnetic Resonance, 187*(1), 163–169. https://doi.org/10.1016/j.jmr.2007.04.002.

Luebbering, E. K., Mick, J., Singh, R. K., Tanner, J. J., Mehra-Chaudhary, R., & Beamer, L. J. (2012). Conservation of functionally important global motions in an enzyme superfamily across varying quaternary structures. *Journal of Molecular Biology, 423*(5), 831–846. Retrieved from https://doi.org/10.1016/j.jmb.2012.08.013.

Ma, C., & Ray, W. J. (1980). Structural comparisons among the central complexes in the phosphoglucomutase system by means of spectral techniques. *Biochemistry, 19,* 751–759.

Namboori, S. C., & Graham, D. E. (2008). Acetamido sugar biosynthesis in the euryarchaea. *Journal of Bacteriology, 190*(8), 2987–2996. https://doi.org/10.1128/JB.01970-07.

Naught, L., & Tipton, P. (2001). Kinetic mechanism and pH dependence of the kinetic parameters of *Pseudomonas aeruginosa* phosphomannomutase/phosphoglucomutase. *Archives of Biochemistry and Biophysics*, *396*(1), 111–118.

Naught, L. E., & Tipton, P. A. (2005). Formation and reorientation of glucose 1,6-bisphosphate in the PMM/PGM reaction: Transient-state kinetic studies. *Biochemistry*, *44*(18), 6831–6836.

Pan, N., Liu, P., Cui, W., Tang, B., Shi, J., & Chen, H. (2013). Highly efficient ionization of phosphopeptides at low pH by desorption electrospray ionization mass spectrometry. *The Analyst*, *138*(5), 1321. https://doi.org/10.1039/c3an36737a.

Park, C., & Marqusee, S. (2005). Pulse proteolysis: A simple method for quantitative determination of protein stability and ligand binding. *Nature Methods*, *2*(3), 207–212. https://doi.org/10.1038/nmeth740.

Ray, S. S., & Balaram, P. (1999). 1-Anilino-8-naphthalene-sulfonate (ANS) binding to proteins investigated by electrospray ionization mass spectrometry: Correlation of gas-phase dye binding to population of molten globule states in solution. *The Journal of Physical Chemistry B*, *103*(34), 7068–7072. https://doi.org/10.1021/jp990801s.

Ray, W. J., Jr., & Roscelli, G. A. (1964). A kinetic study of the phosphoglucomutase pathway. *The Journal of Biological Chemistry*, *239*(4), 1228–1236.

Ray, W. J., Mildvan, A. S., & Long, J. W. (1973). Arrangement of the phosphate- and metal-binding subsites of phosphoglucomutase. Intersubsite relationships by means of inhibition patterns. *Biochemistry*, *12*(19), 3724–3732. https://doi.org/10.1021/bi00743a023.

Regni, C., Naught, L., Tipton, P. A., & Beamer, L. J. (2004). Structural basis of diverse substrate recognition by the enzyme PMM/PGM from *P. aeruginosa*. *Structure*, *12*(1), 55–63.

Regni, C., Schramm, A. M., & Beamer, L. J. (2006). The reaction of phosphohexomutase from Pseudomonas aeruginosa: Structural insights into a simple processive enzyme. *Journal of Biological Chemistry*, *281*(22), 15564–15571.

Sarma, A. V. S., Anbanandam, A., Kelm, A., Mehra-Chaudhary, R., Wei, Y., Qin, P., et al. (2012). Solution NMR of a 463-residue phosphohexomutase: Domain 4 mobility, substates, and phosphoryl transfer defect. *Biochemistry*, *51*(3), 807–819.

Schramm, A. M., Mehra-Chaudhary, R., Furdui, C. M., & Beamer, L. J. (2008). Backbone flexibility, conformational change, and catalysis in a phosphohexomutase from Pseudomonas aeruginosa. *Biochemistry*, *47*(35), 9154–9162. https://doi.org/10.1021/bi8005219.

Schweppe, R. E., Haydon, C. E., Lewis, T. S., Resing, K. A., & Ahn, N. G. (2003). The characterization of protein post-translational modifications by mass spectrometry. *Accounts of Chemical Research*, *36*(6), 453–461.

Stiers, K. M., Muenks, A. G., & Beamer, L. J. (2017). Biology, mechanism, and structure of enzymes in the α-D-phosphohexomutase superfamily. *Advances in Protein Chemistry and Structural Biology*, *109*, 265–304.

Stiers, K. M., Xu, J., Lee, Y., Addison, Z. R., Van Doren, S. R., & Beamer, L. J. (2017). Phosphorylation-dependent effects on the structural flexibility of phosphoglucosamine mutase from *Bacillus anthracis*. *ACS Omega*, *2*(11), 8445–8452.

Studier, F. W. (2005). Protein production by auto-induction in high-density shaking cultures. *Protein Expression and Purification*, *41*(1), 207–234. https://doi.org/10.1016/j.pep.2005.01.016.

Taylor, I., Wang, Y., Seitz, K., Baer, J., Bennewitz, S., Mooney, B. P., et al. (2016). Analysis of phosphorylation of the receptor-like protein kinase HAESA during Arabidopsis floral abscission. *PLoS ONE*, *11*(1), e0147203. https://doi.org/10.1371/journal.pone.0147203.

Xin, F., & Radivojac, P. (2012). Post-translational modifications induce significant yet not extreme changes to protein structure. *Bioinformatics*, *28*(22), 2905–2913. https://doi.org/10.1093/bioinformatics/bts541.

Xu, J., Lee, Y., Beamer, L. J., & Van Doren, S. R. (2015). Phosphorylation in the catalytic cleft stabilizes and attracts domains of a phosphohexomutase. *Biophysical Journal*, *108*(2), 325–337.

Xu, J., & Van Doren, S. R. (2016). Binding isotherms and time courses readily from magnetic resonance. *Analytical Chemistry*, *88*(16), 8172–8178. https://doi.org/10.1021/acs.analchem.6b01918.

Xu, J., & Van Doren, S. R. (2017). Tracking equilibrium and nonequilibrium shifts in data with TREND. *Biophysical Journal*, *112*(2), 224–233. https://doi.org/10.1016/j.bpj.2016.12.018.

CHAPTER NINE

Chemical Tools for Studying the Impact of *cis/trans* Prolyl Isomerization on Signaling: A Case Study on RNA Polymerase II Phosphatase Activity and Specificity

Nathaniel Tate Burkholder*, Brenda Medellin*, Seema Irani†, Wendy Matthews*, Scott A. Showalter‡,§, Yan Jessie Zhang*,¶,1

*Department of Molecular Biosciences, The University of Texas at Austin, Austin, TX, United States
†Department of Chemical Engineering, The University of Texas at Austin, Austin, TX, United States
‡Department of Chemistry, Pennsylvania State University, University Park, PA, United States
§Department of Biochemistry and Molecular Biology, Pennsylvania State University, University Park, PA, United States
¶Institute for Cellular and Molecular Biology, The University of Texas at Austin, Austin, TX, United States
[1]Corresponding author: e-mail address: jzhang@cm.utexas.edu

Contents

Abstract

Proline isomerization is ubiquitous in proteins and is important for regulating important processes such as folding, recognition, and enzymatic activity. In humans, peptidyl-prolyl isomerase *cis–trans* isomerase NIMA interacting 1 (Pin1) is responsible

Methods in Enzymology, Volume 607
ISSN 0076-6879
https://doi.org/10.1016/bs.mie.2018.04.020

for mediating fast conversion between *cis*- and *trans*-conformations of serine/threonine–proline (S/T–P) motifs in a large number of cellular pathways, many of which are involved in normal development as well as progression of several cancers and diseases. One of the major processes that Pin1 regulates is phosphatase activity against the RNA polymerase II C-terminal domain (RNAPII CTD). However, molecular tools capable of distinguishing the effects of proline conformation on phosphatase function have been lacking. A key tool that allows us to understand isomeric specificity of proteins toward their substrates is the usage of proline mimicking isosteres that are locked to prevent *cis/trans*-proline conversion. These locked isosteres can be incorporated into standard peptide synthesis and then used in replacement of native substrates in various experimental techniques such as kinetic and thermodynamic assays as well as X-ray crystallography. We will describe the application of these chemical tools in detail using CTD phosphatases as an example. We will also discuss alternative methods for analyzing the effect of proline conformation such as ^{13}C NMR and the biological implications of proline isomeric specificity of proteins. The chemical and analytical tools presented in this chapter are widely applicable and should help elucidate many questions on the role of proline isomerization in biology.

ABBREVIATIONS

IDP intrinsically disordered protein
MGPD malachite green phosphate detection method
NMR nuclear magnetic resonance
pS_5 phosphorylated serine 5
Pin1 peptidyl-prolyl isomerase *cis–trans* isomerase NIMA interacting 1
PPIase peptidyl-prolyl *cis/trans* isomerase
PTM posttranslational modification
RNAPII CTD RNA polymerase II C-terminal domain
SP motif serine–proline motif

1. INTRODUCTION

Among the 20 natural amino acids, proline is the only residue with a significant faction (5%–20% depending on context) found in isomeric *cis*-conformations (Fig. 1A). The capacity for interconversion between *cis*- and *trans*-conformations at peptidyl-prolyl bonds can promote energetically favorable folding patterns in proteins or even act as regulatory switches for protein recognition (Lu, Liou, & Zhou, 2002). However, the presence of *cis*-proline is generally low and isomerization of peptidyl-prolyl bonds is inherently slow (Wedemeyer, Welker, & Scheraga, 2002). In order to regulate *cis/trans*-proline isomerization, organisms have evolved a large

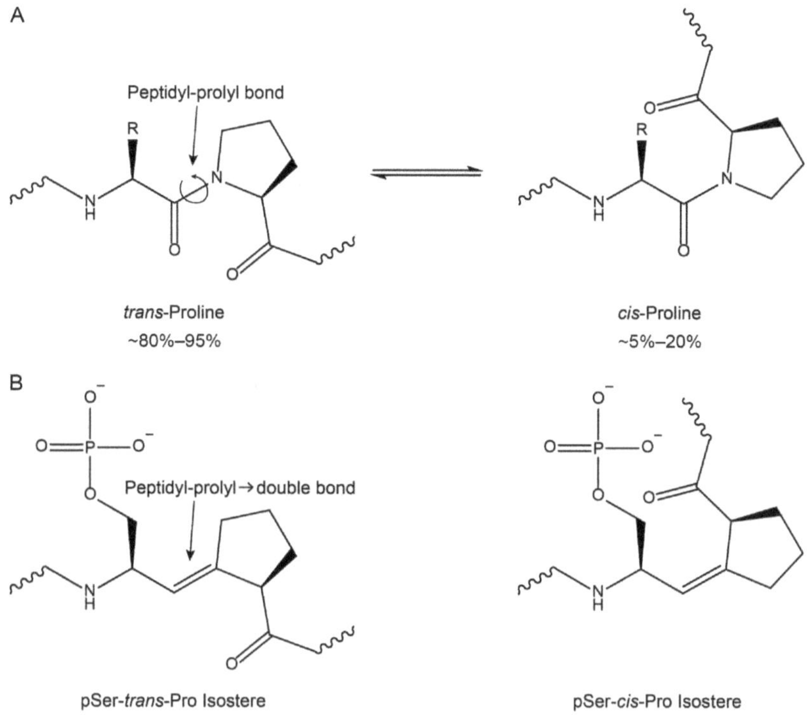

Fig. 1 Isomers of dipeptide motifs. (A) General structures of *cis*- and *trans*-proline peptides. Isomerization occurs around the peptidyl-prolyl bond between proline and the preceding residue. The *trans*-proline conformation is more abundant that the *cis*-conformation (~5%–20%) in nature. (B) Core structures of the phosphorylated SP *cis*- and *trans*-isosteres. Installing a double bond between the serine and proline residues locks the isostere in the *cis*- or *trans*-conformation.

family of peptidyl-prolyl *cis/trans* isomerases (PPIases) that increase the rate of interconversion between *cis*- and *trans*-proline residues to mediate the equilibration of the population of *cis/trans*-prolines (Shaw, 2002). An important PPIase in humans is called PPIase NIMA-interacting 1 (Pin1), which targets phosphorylated Ser/Thr–Pro (S/T–P) motifs that are enriched in phosphorylation-mediated signaling pathways (Yaffe et al., 1997). Regulation of S/T–P motifs by Pin1 can alter the enzymatic activity, protein stability, and cellular localization of its substrates (Liou, Zhou, & Lu, 2011; Lu, Finn, Lee, & Nicholson, 2007; Wulf, Finn, Suizu, & Lu, 2005; Zhou & Lu, 2016). Studying the roles and mechanisms of *cis/trans*-proline-mediated regulation is critical for understanding key signaling pathways as well as elucidating the underlying mechanisms of certain cancers and diseases (Liou, Zhou, & Lu, 2011; Lu & Zhou, 2007). However, it is methodologically difficult to distinguish *cis*- and *trans*-isomers due to their identical chemical composition and molecular weight. Thus, innovative strategies are needed to detect proline conformation in various biological contexts.

One of the essential processes that proline isomerization regulates is transcription mediated by RNA polymerase II (RNAPII), the main enzyme responsible for generation of messenger RNAs as well as small nuclear RNAs in eukaryotic cells (Hanes, 2014). In order to efficiently coordinate transcription of vastly complex genomes, RNAPII has evolved a highly repetitive C-terminal domain (CTD) consisting of consensus heptad repeats $Y_1S_2P_3T_4S_5P_6S_7$ that undergo a variety of posttranslational modifications (PTMs). Various PTMs on the CTD of RNAPII specifically and temporally recruit different transcriptional proteins to execute their appropriate functions at different stages of transcription (Eick & Geyer, 2013). The combination of writing, reading, and erasing of posttranslational marks on the RNAPII CTD is collectively referred to as the "CTD code," which is essential for proper cycling through each stage of general transcription (Buratowski, 2003; Corden, 2013; Eick & Geyer, 2013; Jeronimo, Bataille, & Robert, 2013; Mayfield, Burkholder, & Zhang, 2016). Of the five residues within the heptad repeats that can be phosphorylated, the two SP motifs ($Y\underline{\mathbf{S_2P_3}}T\underline{\mathbf{S_5P_6}}S$) are the major sites of dynamic phosphorylation/dephosphorylation required in each round of transcription (Eick & Geyer, 2013; Harlen & Churchman, 2017; Mayfield et al., 2016; Schüller et al., 2014). The presence of SP motifs in the CTD raises the possibility that *cis/trans* prolyl isomerization plays a role in CTD recognition. Indeed, temperature sensitive mutants of Ess1 (Pin1 homolog in *Saccharomyces cerevisiae*) have a slow growth or temperature sensitive phenotype along with transcriptional defects in elongation and termination (Wu et al., 2000). The termination defects may be in part related to an important regulator of termination called Nrd1 which prefers pS_5-*cis*-P_6 as its substrate (Kubicek et al., 2012). Studying how *cis/trans* prolyl isomerization impacts the specificity of CTD-binding proteins, as well as how altered recognition impacts transcriptional processing, is crucial for deciphering the CTD code.

In addition to altering protein–protein recognition, another mechanism for *cis/trans*-proline isomerization affecting the CTD code is through altering the enzymatic activity of CTD modification enzymes. A number of phosphatases have been found to dephosphorylate RNAPII CTD, some of which play critical roles in transcription. Ssu72 is one such CTD phosphatase which has a high specificity for phosphorylated S_5 of CTD heptad repeats (Krishnamurthy, He, Reyes-Reyes, Moore, & Hampsey, 2004) and has been implicated in both 3′ end processing and transcriptional termination (Steinmetz & Brow, 2003). Interestingly, the crystal structure of the complex between Ssu72 and phosphorylated S_5 (pS_5) CTD substrate revealed that the pS_5P_6 is selectively captured in the *cis*-conformation (Luo et al., 2013; Werner-Allen et al., 2011; Xiang et al., 2010; Fig. 2A) compared to another CTD phosphatase that captured P_6 in the *trans*-conformation (Zhang et al., 2006; Fig. 2B). For crystal structures showing substrate bound in the *trans*-conformation, this does not preclude the possibility of bound substrate being in the *cis*-conformation in some of the crystal units. However, since the *trans*-form substrate is more abundant in these structures the signal from bound *cis*-form substrates will be averaged out. The structural evidence for Ssu72 recognition of pS_5-*cis*-P_6 is especially convincing then, as the averaging effect resulted in the proline being observed only in the *cis*-conformation (Xiang et al., 2010). Furthermore, the availability of *cis* proline is critical for efficient dephosphorylation of pS_5 substrate by Ssu72 as shown through addition of Pin1 (Mayfield et al., 2015; Zhang et al., 2012). However, to fully elucidate the dynamics of Ssu72 recognition of *cis* and *trans* proline conformations, we and others have developed specialized chemical tools and methods to decipher the effects of one conformation vs the other.

While *cis*/*trans*-proline isomerization plays important biological roles in the function of many proteins, this process is difficult to study because tools such as DNA sequencing or mass spectrometry are not able to detect the differences in *cis*- and *trans*-proline within proteins. On the other hand,

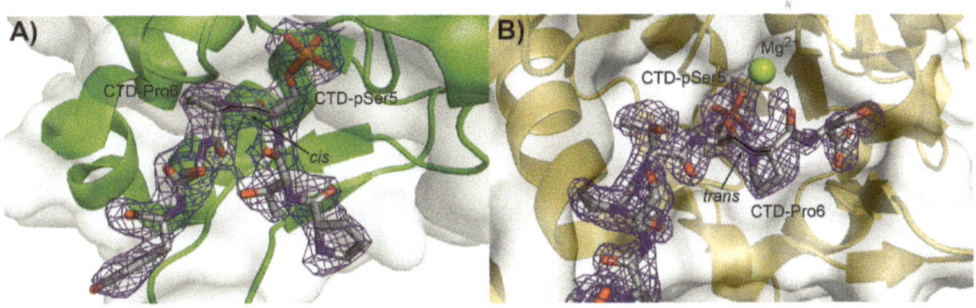

Fig. 2 Crystal structures of Ssu72 and Scp1 bound to native pS_5-P_6 CTD peptides. (A) Ssu72 (*green*) bound to the native pS_5 CTD peptide (*gray*) with P_6 in the *cis*-conformation (PDB: 3O2Q). (B) Scp1 (*goldenrod*) bound to native pS_5 CTD peptide with P_6 in the *trans*-conformation (PDB: 2GHT). The Van der Waals exterior of each protein is represented as a transparent *gray* surface. The 2Fo-Fc maps representing electron density are contoured to 1σ and illustrated as *blue meshes* around the CTD substrates.

structural biology methods such as X-ray crystallography and nuclear magnetic resonance (NMR) can provide detailed information on the interaction between the target protein's isomeric specificity and the ligand containing the proline residue. However, both of these techniques are highly dependent on the applicability of the methodology for each target protein and substrate. If the protein does not crystallize or does not incorporate the peptide ligand easily, then X-ray crystallography will be not suitable. NMR is a powerful method for detecting atomic detailed structure that can distinguish *cis*- and *trans*-prolines, but this technique is incompatible with larger proteins and complexes. In this chapter, we introduce the design of peptidomimetic isosteres that are locked in specific isomeric conformations that can be incorporated into any ligand peptide sequence for testing isomeric specificity of a target protein (Fig. 1B). As an example, we show how we generated peptidomimetic compounds with the sequence of the heptad repeats of RNAPII CTD and used them to understand how prolyl isomerization affects the activity and specificity of CTD phosphatases. In this chapter, we describe in detail how the locked proline peptidomimetics were designed and how these compounds were used as tools to determine the isomeric preference of CTD phosphatases. In essence, we used these locked peptidomimetics in place of native substrate in kinetic assays to determine the rates of dephosphorylation as well as in X-ray crystallography to show the binding modes of substrates with different isomeric conformations. We also compared the crystal structures of CTD phosphatases bound to locked peptidomimetics with structures bound to native peptides, showing that the locked isosteres accurately mimic natural substrates. Complementary studies using ^{13}C NMR to analyze isomeric states of prolines in CTD are also discussed briefly. Finally, how Pin1-mediated regulation of CTD phosphatase substrate availability can lead to differential transcription resulting in alternative cellular outcomes will be considered. While we will focus attention toward the effect of *cis/trans*-proline isomerization on CTD phosphatase activity and specificity, analogous strategies can be adapted for studying any protein whose function is regulated by the proline conformation.

2. USING *cis/trans*-LOCKED SERINE–PROLINE ISOSTERES TO STUDY ACTIVITY AND SPECIFICITY OF CTD PHOSPHATASES

Enzymatic reactions against protein substrates can often be biochemically characterized by producing a piece of the substrate protein sequence and using it in in vitro kinetic assays. However, native proline-containing polypeptides cannot be used to assess the effect of *cis*/*trans*-proline isomerization because a mixture of *cis*- and *trans*-isomers of prolines will coexist and autoconvert during the measurement. To prevent interconversion and obtain a homogenous population of specific isomers, we designed peptidomimetic compounds using unnatural proline homologues (Wang et al., 2003). In such peptidomimetics, the proline residues are replaced by locked isostere moieties in which the peptide bond connecting proline to the preceding residue is replaced by a carbon double bond either in the *cis*- or *trans*-configuration. Unlike peptidyl-prolyl bonds, a carbon double bond cannot be easily rotated preventing interconversion between *cis*- and *trans*-prolines. Combining the aspect of isomeric locking with efficient synthesis and isolation strategies for one or the other isomer essentially guarantees that the final product will consist of a stable and homogenous species of a specific isomer configuration. Thus, these compounds are invaluable tools for studying the effect of *cis*/*trans*-proline isomerization on protein function.

2.1 Design and Synthesis of *cis/trans*-Locked Serine–Proline Isosteres

When designing *cis*- and *trans*-locked proline peptidomimetics, we aim to alter the chemical nature of the peptidyl-prolyl bond to prevent free rotation but minimize structural alterations to the rest of the peptide. Therefore, we kept the primary sequence of the substrate peptides the same as the natural protein sequence while replacing the peptidyl-prolyl bond with a carbon–carbon double bond. In order to accommodate a double bond between the proline and its preceding amino acid, the prolyl nitrogen was substituted with a carbon and the carbonyl oxygen of the preceding residue was removed (Wang et al., 2003). Using the carbon–carbon double bond to lock the isostere has several advantages: (1) the geometry of a carbon–carbon double bond is similar to a peptide bond, but the p-orbitals of the pi-bond between double bonded carbons must remain parallel to each other and therefore cannot be rotated (resulting in a planar locked configuration); (2) carbon–carbon double bonds are more stable than carbon–nitrogen double bonds in aqueous solution, increasing their versatility in experimental design; and (3) carbon–carbon double bonded peptidomimetics are less likely to be degraded by proteases that target peptidyl-prolyl bonds. Loss of the carboxyl group and adoption of a planar conformation in the dipeptide isosteres are unlikely to significantly change conformation of the entire peptide. Overall, modifying peptidyl-prolyl dipeptides with carbon–carbon double bonds poses minimal structural disturbances to the peptide chains while securing proline conformation.

Locked XP (X for any amino acid) isosteres can be chemically synthesized and then incorporated during solid-phase peptide synthesis (Hart, Sabat, & Etzkorn, 1998). Locked proline isosteres are not yet commercially available, but there are relatively efficient synthesis strategies for their production. The key aspect in the chemical synthesis of these isosteres is obtaining sufficient quantities of product with the correct stereochemistry in the double-bond formation step. Two chemical synthesis strategies have been successfully applied to generate stereospecific XP-locked isosteres. The first scheme uses careful application of organocopper-mediated anti-S_N2' reactions (Otaka et al., 2002; Sasaki et al., 2006). This strategy is particularly effective with the synthesis of the *trans*-isostere, in which an SP-like oxazolidinone (**4a**) is modified with a silylmethyl group in an organocopper-mediated anti-S_N2' attack to predominantly form the *trans*-alkene precursor **5a** (Fig. 3; Otaka et al., 2002). However, syntheses based on organocopper chemistry have very low yield in the production of L-form X-*cis*-P isosteres which mimic naturally occurring amino acids (Otaka et al., 2002; Sasaki et al., 2006, Table 1). Instead, rearrangement chemistry can be used to position a double bond in the desired peptidyl-prolyl position to efficiently synthesize SP analogs in either *cis*- or *trans*-configuration (Wang et al., 2003). Synthesis of either isostere begins by making a dipeptide-like precursor through condensation of a Weinreb amide (**2b** and **1c**) with cyclopentyllithium derived from 1-iodocyclopentene (**3b**) (Figs. 4 and 5). The key stereoselective step for the S-*cis*-P is a Still–Wittig rearrangement of the ether functional group and the cyclopentene to form ~53% of the *cis*-isomer (Fig. 4, **6b** → **7b**-*cis*). Yield for the *trans*-isomer in this synthesis is poor (~28%), so instead an Ireland–Claisen rearrangement between an ester functional group and the double bond of a cyclopentene can be performed to acquire ~52% of the *trans*-isomer (Fig. 5, **4c** → **5c**; Wang et al., 2003). Synthesis of the S-*trans*-P isosteres using either method takes eight steps with overall yields of about 17% using organocopper chemistry vs 13% for Ireland–Claisen rearrangement (Table 1). On the other hand, S-*cis*-P can be synthesized with Still–Wittig rearrangement with many fewer steps and higher overall yield (10 steps, 20% yield) than with organocopper chemistry (Table 1).

The synthesis strategies described in this section can be generally performed for any Xaa-Pro sequence and produce abundant amounts of *cis*- or *trans*-isomer with high purity. Although these isosteres are not yet commercially available, increased usage will hopefully encourage industrial

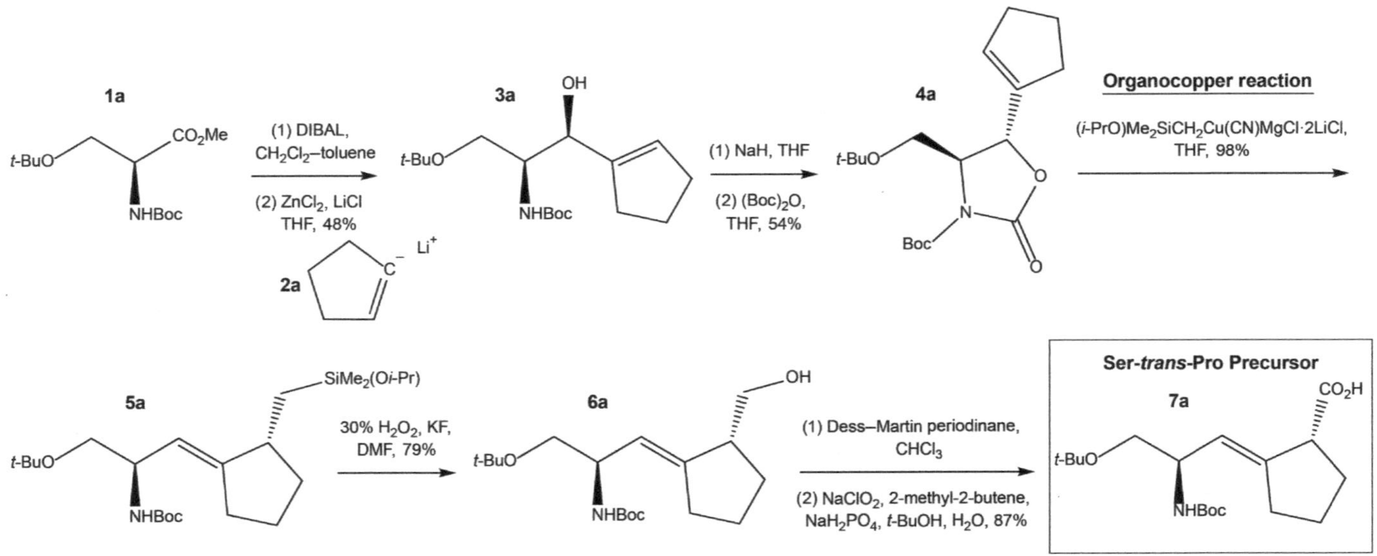

Fig. 3 Synthesis of the S-*trans*-P isostere via organocopper-mediated chemistry. A protected serine derivative (**1a**) is first reacted with cyclopentenyl-lithium (**2a**) to generate an SP-like precursor. Cyclization and protection of **3a** produces ~54% of the *trans*-oxazolidinone derivative (**4a**), which can then be treated with an silyl–organocopper complex ((*i*-PrO)$Me_2SiCH_2Cu(CN)MgCl\cdot 2LiCl$) to form ~98% of the anti-S_N2' product (**5a**). Conversion of the silyl group to a carboxyl group results in the final S-*trans*-P isostere precursor (**7a**), which can then have the *tert*-butyl blocking group removed to form serine.

Table 1 Strategies for Synthesizing *cis/trans* Proline Dipeptide Isosteres

Synthesis Strategy	Product	Number of Steps	Overall Yield (%)
Organocopper[a]	S-*trans*-P	8	17
Organocopper[b]	A-*cis*-P	>20	<11
Ireland–Claisen[c]	S-*trans*-P	8	13
Still–Wittig[c]	S-*cis*-P	10	20

[a]Otaka et al., 2002.
[b]Sasaki et al., 2006.
[c]Wang et al., 2003.

chemical providers to generate locked dipeptide mimetic compounds for scientific use. Once the dipeptide mimetic precursors are synthesized and purified, they can be incorporated into routine solid-phase peptide synthesis as a unit as if they were a single amino acid. Phosphate groups can be added at this stage to the target serine/threonine dephosphorylation site. The polypeptides (incorporating X-*E* or X-C=C-P isosteres) are then purified through chromatography and used as substrates or ligands in characterizing the isomeric specificity of proteins of interest. For our studies, we used peptides consisting of RNAPII CTD sequences with phosphorylated S-*cis*/*trans*-P-locked moieties to determine the effect of SP motif conformation on CTD phosphatase function (Table 2).

2.2 Malachite Green Kinetic Analysis of CTD Phosphatase Activity Against *cis/trans*-Locked Isosteres

In order to study RNAPII CTD recognition, we generated two sets of peptide mimetics with either SP motif of the CTD replaced by locked, dipeptide isosteres (phosphoryl-S-(*E*, *Z*)C=CH)-P] (Table 2). These peptidomimetic compounds containing locked pSP isosteres can thus be used in place of natural substrates in kinetic or thermodynamic measurements to investigate whether CTD-binding proteins prefer *cis*- or *trans*-SP-motifs. Here, we used CTD peptidomimetics containing proline-locked isosteres to investigate proline isomeric specificity of different CTD phosphatases including: (1) Fcp1, which dephosphorylates pS_2 during general transcription (Ghosh, Shuman, & Lima, 2008; Hausmann & Shuman, 2002; Kobor et al., 1999); (2) Ssu72, the phosphatase responsible for pS_5 dephosphorylation during general transcription (Krishnamurthy et al., 2004; Xiang et al., 2010); and (3) Scp1, which inhibits transcription of neuronal genes and dephosphorylates pS_5 (Yeo et al., 2005; Yeo, Lin, Dahmus, & Gill, 2003; Zhang et al., 2006).

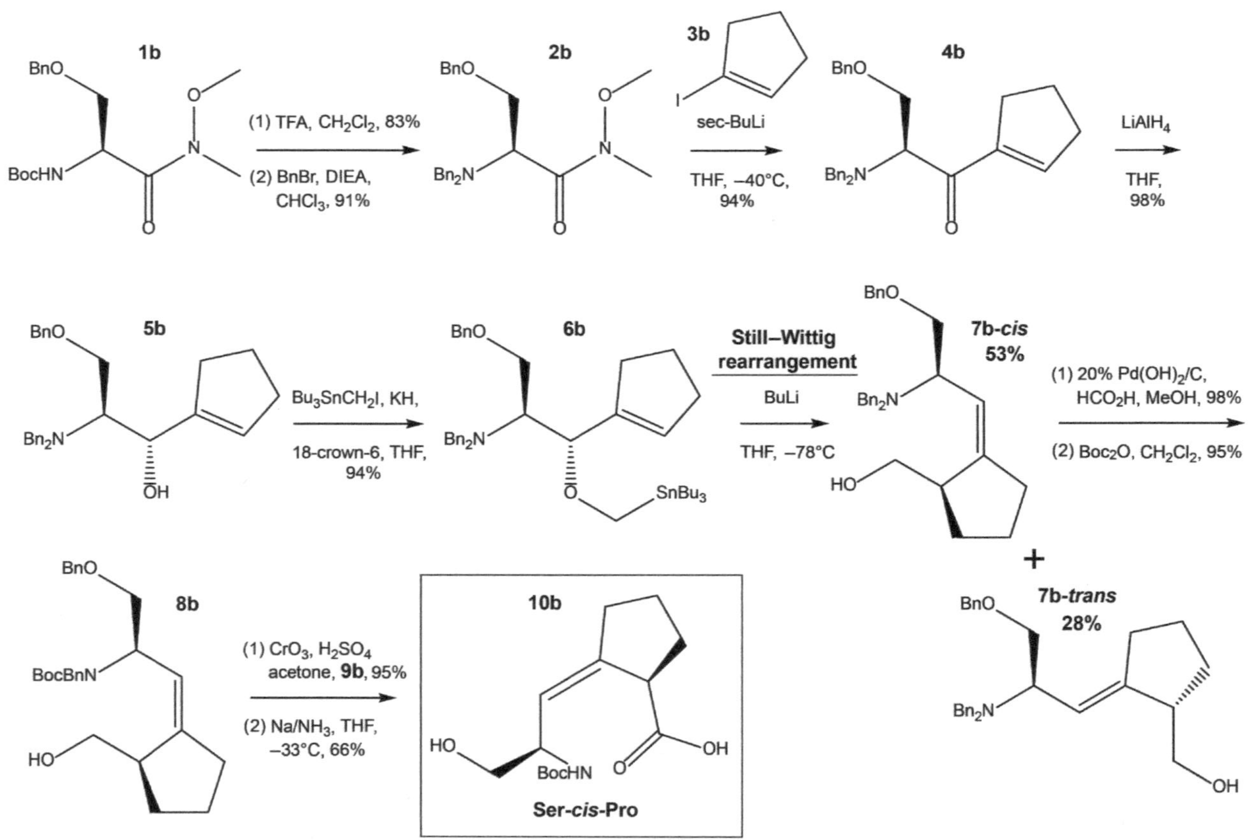

Fig. 4 Synthesis of the S-*cis*-P isostere via Still–Wittig rearrangement. A protected serine derivative (**2b**) is reacted with 1-iodo-cyclopentene (**3b**) in the second synthesis step to generate an SP-like precursor. The Still–Wittig rearrangement of **6b** results in two major isomeric products: ~28% in the *trans*-conformation (**7b**-*trans*) and ~53% in the *cis*-conformation (**7b**-*cis*) which is used in the subsequent reaction. Product **9b** is the acid form of **8b** which is converted to the final S-*cis*-P isostere (**10b**).

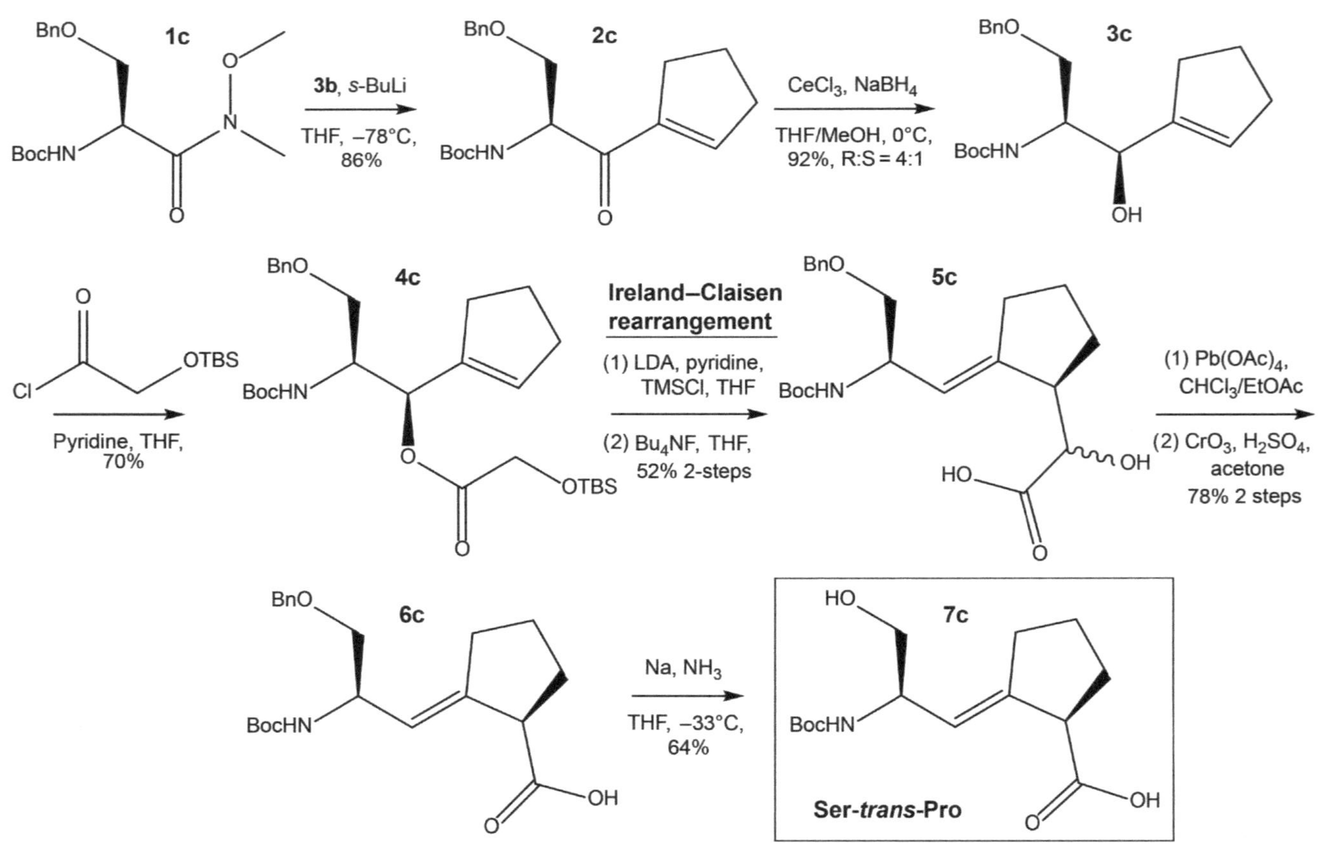

Fig. 5 Synthesis of the S-*trans*-P isostere via Ireland–Claisen rearrangement. 1-Iodo-cyclopentene (**3b**) is used as a reactant in the first synthesis step to generate a proline-like precursor. In the third step of synthesis (**3c** → **4c**), *tert*-butyldimethylsilyloxyacetyl chloride is used to generate the Ireland–Claisen ester precursor. The Ireland–Claisen rearrangement of **4c** is carried out in two steps with activation of TMSCl followed by removal of the TBS group with *tert*-butylammonium fluoride (Bu_4NF) in order to generate ~52% of the *trans*-like precursor (**5c**). Subsequent oxidation and deblocking steps result in the final S-*trans*-P isostere (**7c**).

Table 2 Synthetic Peptide Sequences Used in Malachite Green Kinetics and X-Ray Crystallography

Peptide	Sequence
Native pS_5	Ac-S-P-Y-S-P-T-S(PO_3H_2)-P-S-Y-S-NH_2
trans-locked pS_5	Ac-S-P-Y-S-P-T-S(PO_3H_2)ψ[(*E*)C=CH)]-P-S-Y-S-NH_2
cis-locked pS_5	Ac-S-P-Y-S-P-T-S(PO_3H_2)ψ[(*Z*)C=CH)]-P-S-Y-S-NH_2
Native pS_2	Ac-S-P-S-Y-S(PO_3H_2)-P-T-S-P-S-Y-S-NH_2
trans-locked pS_2	Ac-S-P-S-Y-S(PO_3H_2)ψ[(*E*)C=CH)]-P-T-S-P-S-Y-S-NH_2
cis-locked pS_2	Ac-S-P-S-Y-S(PO_3H_2)ψ[(*Z*)C=CH)]-P-T-S-P-S-Y-S-NH_2

The symbol *E* represents *trans* and the symbol *Z* represents *cis*-conformations. The symbol ψ represents the psi bond of the SP-like isostere.

The specific activity of phosphatases is often measured by the malachite green phosphate detection method (MGPD) using short phosphorylated peptides as substrates. MGPD is a colorimetric assay based on a complex formation between malachite green molybdate (BIOMOL® Green, Enzo Life Sciences) and free orthophosphate, the absorbance of which can be easily detected at wavelengths of 620–650 nm. The overall procedure for MGPD is listed below:

Step 1: Phosphate Standard curve determination (see note 1)

1. Allow an aliquot of Malachite Green Reagent to warm up to room temperature (see note 2).
2. Perform triplicate serial dilutions of a phosphate standard (see note 3) (usually Malachite Green Reagents will be sold along with a phosphate standard). For a 96-well clear microplate scale assay, perform eight serial dilutions in which the last tube should be solvent only acting as a blank.
3. Using a multichannel pipette, take 20 μL of each dilution and add to a clean, clear microplate.
4. Using a multichannel pipette, add 40 μL of room temperature Malachite Green Reagent (as an extra precaution, filtered tips may be used to prevent phosphate contamination).
5. Cap microplate or seal with tape. Incubate at room temperature for 15–20 min to allow green color development.
6. Measure absorbance at 620 nm (precipitation should not be observed as it indicates oversaturation of the solution and will yield inaccurate readings).

7. Average triplicate absorbance measurements and subtract the blank from each reading. Plot absorbance at 620 nm on the y-axis and [phosphate] (M) on the x-axis. Fit to a linear equation and evaluate goodness of fit.

Notes

1. Ideally, the standard curve will be performed using the protein buffer as solvent as this will provide a more comparable standard to the enzymatic reaction. For different buffers and/or old reagent, it is recommended to repeat the standard curve.
2. Malachite Green Reagent is light sensitive so using dark opaque tubes or wrapping tubes in foil is recommended.
3. Malachite Green Reagent should be highly sensitive to phosphate in the nanomolar range; however, carefully inspecting the linearity of standard curve is advised to ensure that the lower limit of phosphate concentration can be reasonably detected.

Step 2: Enzyme activity assay: Phosphatase activity against peptidomimetic substrate

1. Allow an aliquot of Malachite Green Reagent to warm up to room temperature.
2. Prepare 10 × concentrated working stocks of buffer, substrate, and enzyme.
3. Perform a serial dilution of the 10 × substrate at a concentration that when diluted 10-fold results in a maximal rate of phosphate release for the target phosphatase[%]. Leave the last dilution without enzyme as a control for background phosphate signal.
4. Mix buffer, water, and enzyme in a total volume of 36 μL each (excluding substrate) in an eight-strip tube (3 × reactions and 1 × control without enzyme for background phosphate from each substrate dilution).
5. Prepare a 96-well clear microplate with 40 μL of Malachite Green Reagent for quenching each reaction. Seal the plate with plate to prevent contamination.
6. Start reactions by multichannel pipetting 4 μL of 10 × substrate serial dilutions to each reaction and control strip tube. Incubate reactions at an appropriate temperature and time for the target enzyme for as long as the reactions remain in the steady state (see note). Cap strip tubes to prevent evaporation.
7. Quench reactions by adding 20 μL from each strip tube to 40 μL of Malachite Green Reagent in the microplate.

8. Reseal the microplate with tape to avoid contamination and incubate at room temperature for 10–20 min for green color development. During this time the P_i + MG complex will be formed.
9. Read the absorbance in the plate reader at 620 nm. Using the standard curve, convert absorbance to concentration of phosphate in order to calculate the rate of phosphate release for each phosphatase reaction.

Notes

Determine the linear range of the steady state and the concentration of substrate to result in maximal rate of dephosphorylation for the target enzyme by performing a time course with constant concentrations of substrate.

We used the highly sensitive malachite green assay in combination with proline-locked CTD peptides to determine the rates of CTD phosphatases against *cis*- and *trans*-specific substrates (Mayfield et al., 2015). We determined the specificity constants for each enzyme and substrate by plotting the substrate concentration against the initial rate and fitting to the classical steady-state Michaelis–Menten kinetic equation. Using MGPD, we found that Scp1 has an almost 10-fold preference for the pS_5-*trans*-P_6 substrate with a k_{cat}/K_m of $323 \pm 22 \times 10^3\,M^{-1}\,s^{-1}$ compared to the *cis*-proline substrate with a k_{cat}/K_m of only $38.0 \pm 1.3 \times 10^3\,M^{-1}\,s^{-1}$ (Mayfield et al., 2015). Similarly, Fcp1 can dephosphorylate pS_2-P_3 of the CTD in either conformation, but it has a twofold preference for the *trans*-locked peptidomimic compared to the *cis*-proline substrate (Mayfield et al., 2015). In contrast, Ssu72 had virtually no activity against the pS_5-*trans*-P_6 peptide and very low activity against the native peptide. Ssu72 did have robust activity against the *cis*-peptide with a k_{cat}/K_m of $5.24 \pm 0.08 \times 10^3\,M^{-1}\,s^{-1}$ (Mayfield et al., 2015), which supports the crystallographic data suggesting that Ssu72 is CTD phosphatase is specific for *cis*-proline in substrates. Combining the malachite green assay with our proline-locked peptides yielded accurate and quantitative results revealing the *cis*/*trans*-proline specificity of several CTD phosphatases as well as their activity toward each distinct isomeric substrate.

While the malachite green assay is highly sensitive, there are issues that must be addressed for successful application of this assay on studying phosphatase activity. Like all kinetic assays, the enzymatic conditions (e.g., enzyme concentration, temperature, etc.) need to be optimized in order to obtain the best signal/noise and sensitivity. The reaction time is a particular important factor to consider for this assay because the color development buffer tends to precipitate when it is saturated with the orthophosphate complex, leaving a narrow window for detection. Additionally, one must take into account that certain additives may interfere with the assay by either showing a strong absorbance at 620 nm or interfering with molybdate–phosphate complex formation. For example, commonly used detergents such as Triton X-100, SDS, Tween-20, and NP-40 can adversely affect signal readings at 620 nm (Carter & Karl, 1985). Divalent cations such as Zn^{2+} and Ca^{2+} can also affect signal by forming salts with phosphates which frequently precipitate due to their low solubility. Finally, the high sensitivity of the malachite green assay makes it prone to high noise interference from any source of free inorganic phosphate. While it is difficult to completely eliminate phosphate contamination, negative controls lacking enzyme and/or phosphorylated substrate should be used to subtract unwanted noise from activity curves.

2.3 X-Ray Crystallography of CTD Phosphatases Bound to *cis/trans*-Locked Peptides

After kinetically quantifying the proline conformation preference of several CTD phosphatases, we obtained the binding mode of locked peptidomimic substrates using X-ray crystallography. X-ray crystallography has unexpectedly been an invaluable tool for studying the substrate specificity of CTD phosphatases (Xiang et al., 2010; Zhang et al., 2006). It can directly visualize the bound substrate conformation and identify protein residues involved in substrate recognition. However, since X-ray crystallography captures the overall signal from many repeating units of protein molecules, signal from any minor conformations that exist in the crystal may not be great enough to be observed in the occupancy-weighted average. Structure determination of CTD phosphatases bound to locked peptidomimetics on the other hand allowed us to directly visualize binding modes of less favorable but still biologically relevant substrate proline-conformations.

Since locked peptidomimetics are structurally analogous to native peptides, the procedure for determining complex structures by X-ray crystallography is identical to structural determination of proteins in complex with natural peptide ligands. If the target protein turns over the substrate through enzymatic activity, a mutant variant of the target protein must be used in order to capture the binding of substrate in the active site without releasing the product. Protein crystallization is generally an empirical process often requiring much trial and error for each different protein. Müller's recent review provides a good general procedure of the protein–ligand complex crystallization process if you are interested in a more detailed overview (Müller, 2017). To obtain protein–ligand complexes within target protein crystals, soaking is often a good strategy to start with. We begin by testing a range of ligand concentrations and soaking times with good quality crystals while watching for signs of crystal deterioration upon complex formation. Stably soaked crystals can then be cryocooled and data can be collected to obtain the structures of protein–ligand complex. Ligand binding may cause substantial conformational changes and disrupt crystal packing, so alternatively cocrystallization of the ligand and protein can be attempted. However, there is no guarantee that cocrystallization will work in the unliganded crystallization conditions thus requiring rescreening with ligand included. Incorporating ligand into protein crystals depends on both the protein and substrate, but obtaining complex structures with locked peptides should be no different than for native peptides due to the highly similarity in structure between peptidomimetic and natural peptides.

For soaking experiments with Ssu72, we began by crystallizing inactive variants of *Drosophila* Ssu72 (C13D/D144N)–symplekin complexed to substrate peptides. Symplekin is a scaffold protein bound to Ssu72 during transcription, which has been shown to stabilize Ssu72 (Luo et al., 2013; Xiang et al., 2010). We have shown that the association between Ssu72 and symplekin binding does not compromise the enzymatic activity but significantly improves the diffraction limits of crystals (Luo et al., 2013). Preformed crystals of Ssu72–symplekin were transferred to crystal trays containing a serial dilution of locked peptidomimetic compounds (highest concentration at 2 m*M*) in mother liquor (Mayfield et al., 2015). During soaking, Ssu72–symplekin crystals appeared to be stable and showed no sign of visible deterioration. The crystals were harvested and cryoprotected after incubation at room temperature overnight. In cases where protein crystals lose diffraction after compound soaking, several strategies should be attempted to obtain optimal diffraction of the complex crystals: (1) reducing the ligand concentration; (2) reducing soak time (even soaking less than a minute may successfully incorporate the ligand); and (3) including additives that may help improve the stability of a protein crystal upon ligand complexation (Hassell et al., 2007). If the data resolution is good but the ligand density is weak, increasing the ligand concentration and soaking times may improve complex formation. Obtaining optimal protein–ligand complex structures via crystal soaking requires a delicate balance between ligand binding and crystal maintenance.

Guided by these strategies, we obtained the complex structure of Ssu72–symplekin bound to the *cis*-proline isostere compound at high resolution (2.95 Å; Fig. 6A) but not of the *trans*-proline peptidomimic (Mayfield et al., 2015). Comparison of the structures of Ssu72–symplekin bound to the pS_5-*cis*-P_6 isostere or native pS_5P_6 revealed near identical binding modes (Fig. 6B and C), demonstrating that the *cis*-locked isostere is a great mimic for *cis*-proline. Careful inspection of the substrate-bound structure of Ssu72 provides a plausible reason for its specificity toward *cis*-proline. Effective substrate recognition by Ssu72 requires phosphoserine occupancy at its phosphate-binding site as well as the flanking proline residue in an adjacent hydrophobic pocket (Fig. 6D). P_6 of the CTD fits tightly in the hydrophobic pocket consisting of residues L_{45}, P_{46}, L_{82}, and M_{85} of Ssu72, but only the *cis*-isomer seems to be compatible with the cavity shape. When we tried to model a *trans*-proline into the substrate, unfavorable steric clashes occurred with active site residues L_{82} and M_{85}. Overall, the structure of Ssu72–symplekin bound to the *cis*-proline peptidomimic replicates the native substrate-binding mode very well, solidifying the hypothesis that Ssu72 is a *cis*-proline-dependent phosphatase.

Unlike Ssu72, our kinetic characterization Scp1 indicated that while it heavily favors the *trans*-form of CTD substrates it can still efficiently dephosphorylate either isomer (Mayfield et al., 2015). With significantly lower activity against *cis*-proline substrates that occur much less frequently in nature, it is not surprising that the structure of Scp1 in complex with native CTD peptides only shows the *trans*-proline configuration (Zhang et al., 2006). To compare the binding modes of Scp1 with CTD substrates of different isomeric configurations, we obtained the complex structures of Scp1 bound to *cis*- or *trans*-locked isostere peptidomimetic compounds (Mayfield et al., 2015). We were able to obtain diffracting crystals of Scp1 D96N (inactive variant) after soaking with 1 m*M* of either peptidomimic for 2 h at room temperature (crystals diffract to 2.20 Å for the complex with *cis*- and 2.36 Å for the complex with *trans*-locked compounds). The structure of Scp1 in complex with *trans*-locked peptidomimetic compounds echoed the conformation of Scp1 bound to the natural CTD peptide, confirming again that locked isosteres accurately mimic native peptidyl-prolyl conformations (Fig. 7A and B). The comparison between the structures containing the two locked isosteres shows that there are conserved molecular interactions that are essential for Scp1 to recognize its substrates (Fig. 7C and D). One of the major sites of substrate recognition for Scp1 is its hydrophobic pocket, which appears to tightly

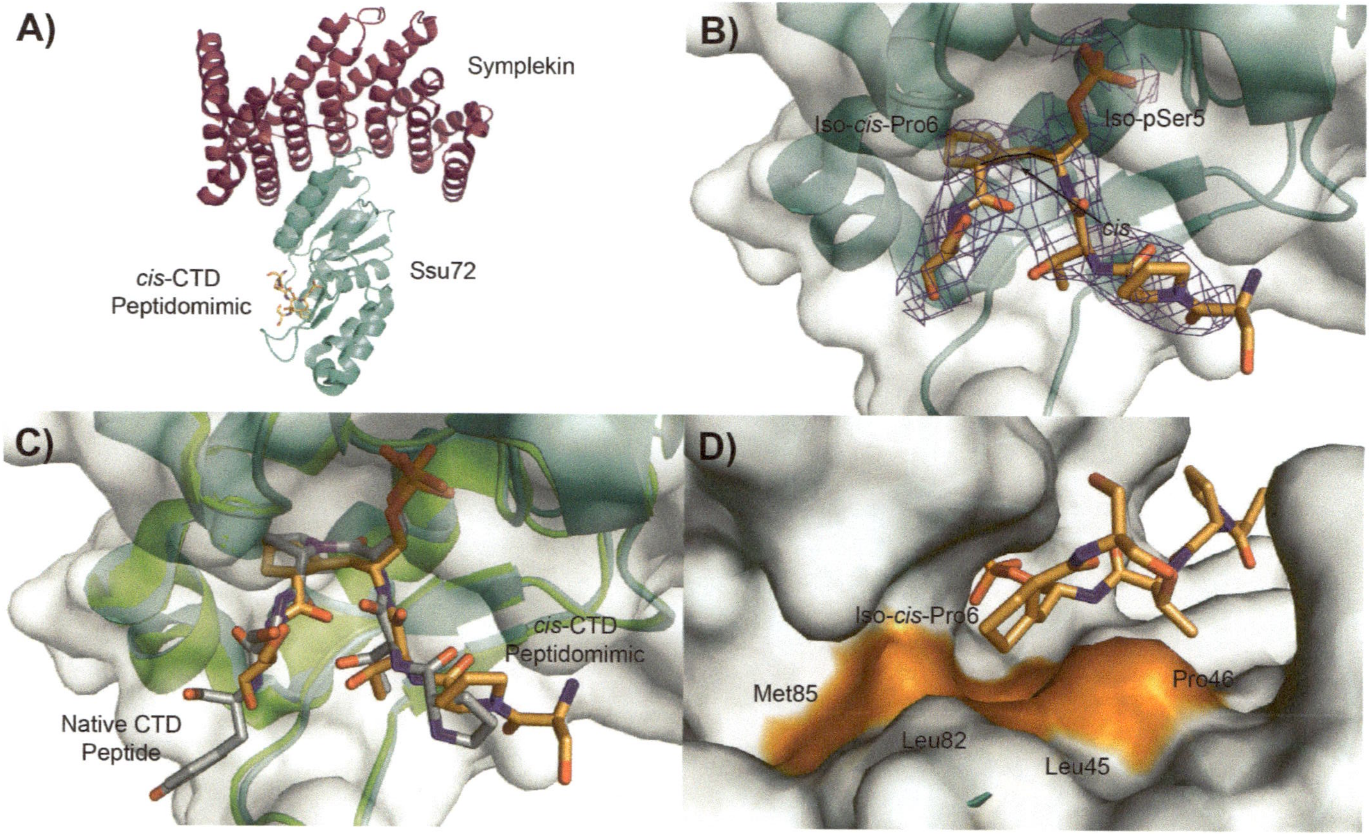

Fig. 6 Crystal structures of Ssu72–symplekin bound to pS_5-*cis*-P_6 peptidomimetic or native pS_5 peptide. (A) Overall structure of Ssu72 (*light blue*) complexed to symplekin (*purple*) and pS_5-*cis*-P_6 peptidomimic (sand; PDB: 4YGX). (B) Structure of *cis*-locked peptidomimic bound at the Ssu72 active site. The Van der Waals exterior of Ssu72 is represented as a transparent *gray* surface. The 2Fo-Fc map representing electron density is contoured to 1σ and illustrated as a *blue mesh* around the CTD peptidomimic. (C) Overlay of Ssu72 bound to pS_5-*cis*-P_6 peptidomimic and Ssu72 bound to native pS_5 peptide (*green* and *gray*, respectively; PDB: 3O2Q). (D) Hydrophobic pocket of Ssu72 consisting of residues L_{45}, P_{46}, L_{82}, and M_{85} (*orange*) that contribute specificity for *cis*-P_6 substrates.

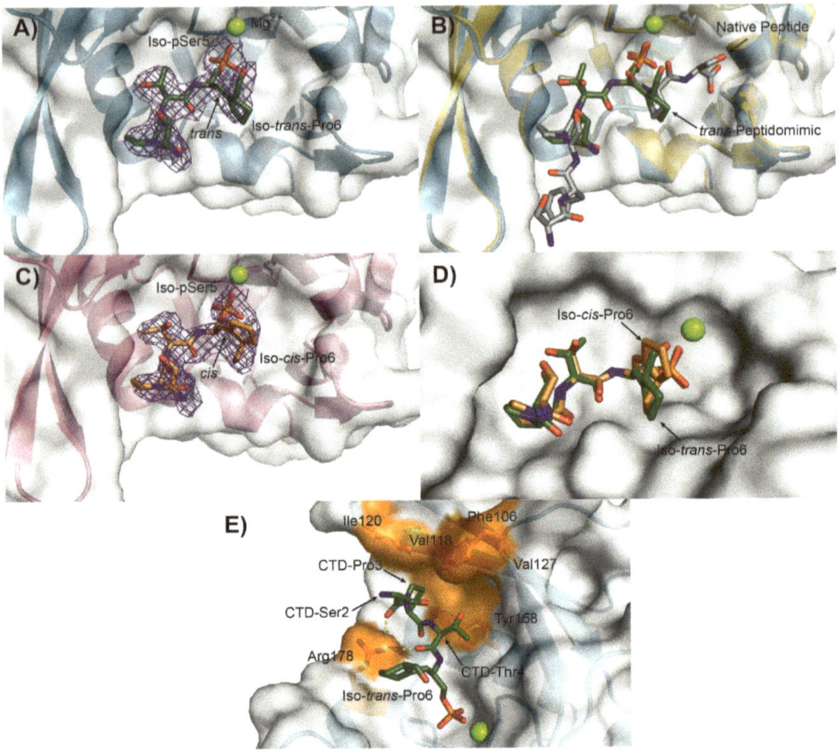

Fig. 7 Crystal structures of Scp1 bound to *cis/trans*-locked pS_5 CTD peptidomimetics or native pS_5 peptide. (A) Structure of Scp1 D96N (*cerulean*) bound to pS_5-*trans*-P_6 peptidomimic (*dark green*, PDB: 4YGY). (B) Overlay of Scp1 bound to pS_5-*trans*-P_6 peptidomimic and Scp1 bound to native pS_5 peptide (*yellow* and *gray*, respectively; PDB: 2GHT). (C) Structure of Scp1 D96N (*violet*) bound to pS_5-*cis*-P_6 peptidomimic (sand, PDB: 4YH1). The Van der Waals exterior of Scp1 is represented as a transparent *gray* surface. The 2Fo-Fc maps representing electron density are contoured to 1σ and illustrated as *blue meshes* around the CTD peptidomimetics. (D) Overlay of the Scp1 structures bound to either *trans*- or *cis*-locked peptidomimetics. (E) Scp1-binding pocket interactions stabilizing complexation formation with peptidomimic substrates. Residues F_{106}, V_{118}, I_{120}, and V_{127} form a hydrophobic pocket around CTD-P_3. Y_{158} interacts with the side chain of CTD-T_4 while R_{178} can form hydrogen bounds with the backbone carboxyl groups of CTD-S_2/T_4.

interact with residues T_4 and P_6 of both CTD phosphomimetics (Fig. 7E). This conserved interaction demonstrates how Scp1 can still efficiently dephosphorylate substrates with P_6 in either isomeric conformation. However, unlike in Ssu72 where P_6 strongly contributes to substrate recognition by tightly binding in a deep pocket, Pro6 likely contributes little to Scp1 substrate recognition as it is situated at the rim of the active site with very few discernible interactions. In particular, locked *cis*-P_6 is much more solvent exposed than locked *trans*-Pro6 (Fig. 7D), which likely explains why Scp1 prefers *trans*- over *cis*-proline substrates. The availability of peptidomimetics with *cis*-locked isosteres allows us to directly visualize the binding mode of a less favored substrate conformation in our target protein that X-ray crystallography would normally not be able to detect.

3. DETERMINATION OF PHOSPHATASE PROLINE CONFORMATION SPECIFICITY USING ^{13}C NMR

A powerful method to detect subtle structural changes in biological molecules is protein NMR, which is embraced by scientists who try to study changes in protein conformations due to its ability to accommodate and even quantify those structural dynamics. However, an obstacle to observing the proline isomeric configuration in a biological sample is the lack of an amide proton on these residues, which often impairs necessary chemical shift assignments near prolines through proton-detected NMR. In any case, the absence of an amide proton prevents the direct detection of signal from the proline peptide bond for the purpose of assigning the isomerization state, although readout of the proline Cβ and Cγ chemical shifts via the adjacent residue is possible if the amide chemical shifts of that residue have been assigned. Using these methods, structural information can also be derived on the bound state of ligand prolines, as was reported for Nrd1 binding of *cis*-P_6 in native pS_5 CTD peptides (Kubicek et al., 2012). Recently, advances in ^{13}C NMR methodology and instrumentation have provided an opportunity to gain straightforward information on the backbone conformation of proteins, especially in the study of intrinsically disorder proteins (IDPs; Brutscher et al., 2015). One key aspect of ^{13}C NMR is that it is not limited by pH as in proton NMR. Further limiting the application of traditional proton NMR in studying IDPs, the lack of unique chemical environments and rapid conformational averaging cause the amide proton chemical shift dispersion for these systems to collapse to a roughly 1 ppm window, thus foiling higher resolutions typically achievable with modern spectrometers. Simultaneously, the mostly exposed backbone protons of these proteins often rapidly exchange with solvent which causes loss in signal. ^{13}C NMR improves resonance dispersion of IDPs and provides structural information of prolines which lack amide protons (Gibbs &

Showalter, 2015), thus leading to many novel discoveries concerning IDPs (Gibbs et al., 2017; Lawrence & Showalter, 2012). ^{13}C NMR can also be used in combination with the other forms of NMR to obtain more comprehensive and clear assignment of resonances of a protein–ligand complex. A particularly applicable review covers strategies for overcoming the technical challenges of employing ^{13}C NMR for studying IDPs (Bastidas, Gibbs, Sahu, & Showalter, 2015).

For the investigation of RNAPII CTD, the application of ^{13}C NMR provides a very useful complement to the peptidomimic approaches described in Sections 2.1–2.3. In each heptad of the CTD, there are generally two critical SP motifs where reversible phosphorylation occurs during transcription, but the resonances associated with S_2P_3 and S_5P_6 are very difficult to distinguish using traditional protein NMR methodologies. By including ^{13}C NMR in the study of the CTD, information on the overall conformation as well as the relative population of *cis*- and *trans*-conformations for these prolines can be determined (Gibbs et al., 2017). The yeast and human CTD sequences are highly repetitive making it difficult to sequence-specifically assign residues, even when using ^{13}C NMR, but the *Drosophila* CTD has more divergent heptad sequences making it an ideal subject for NMR (Fig. 8). Intriguingly, developmentally complex organisms including all mammals and *Drosophila* have evolved multiple CTD heptads in which the S_7 position has been substituted with an asparagine residue. In the context of the *Drosophila* CTD sequence, $pS_5P_6N_7$ heptads were found to possess significantly increased *cis*-P_6 populations in solution using ^{13}C NMR (Gibbs et al., 2017). Formation of an additional intramolecular hydrogen bond between N_7 and T_4 in solution likely helps stabilize the *cis*-conformation (Fig. 9), which should make these heptads favorable substrates for Ssu72. Indeed, kinetic data using pS_5 CTD peptides with N at the seventh position showed that they are better substrates for Ssu72 dephosphorylation compared to consensus CTD peptides (Gibbs et al., 2017). Collectively, peptidomimic chemistry allows for unambiguous demonstration that Ssu72 requires *cis*-proline in its substrates, while ^{13}C NMR clearly demonstrates that certain conserved sequence variations of the CTD heptad dramatically enhance the equilibrium population of pS_5-*cis*-P_6 available in solution. Altogether, the advances in ^{13}C direct-detection NMR and its application on CTD structure and function support the role of this technique as an invaluable tool for studying IDPs.

	H. sapiens	*D. melanogaster*	*S. cerevisiae*
1	YSPTSPA	SPSVSPS	FGVSSPG
2	YEPRSPGG	YSPTSPN	FSPTSPT
3	YTPQSPS	YTASSPGG	YSPTSPA
4	YSPTSPS	ASP---N	YSPTSPS
5	YSPTSPS	YSPSSPN	YSPTSPS
6	YSPTSPN	YSPTSPL	YSPTSPS
7	YSPTSPS	YA--SPR	YSPTSPS
8	YSPTSPS	YASTTPN	YSPTSPS
9	YSPTSPS	FNPQSTG	YSPTSPS
10	YSPTSPS	YSPSSSG	YSPTSPS
11	YSPTSPS	YSPTSPV	YSPTSPS
12	YSPTSPS	YSPTV-Q	YSPTSPS
13	YSPTSPS	FQ-SSPS	YSPTSPS
14	YSPTSPS	FAGSGSNI	YSPTSPS
15	YSPTSPS	YSPGN-A	YSPTSPS
16	YSPTSPS	YSPSSPN	YSPTSPS
17	YSPTSPS	YSPNSPS	YSPTSPS
18	YSPTSPA	YSPTSPS	YSPTSPA
19	YSPTSPS	YSPSSPS	YSPTSPS
20	YSPTSPS	YSPTSPC	YSPTSPS
21	YSPTSPS	YSPTSPS	YSPTSPS
22	YSPTSPN	YSPTSPN	YSPTSPS
23	YSPTSPN	YTPVTPS	YSPTSPN
24	YTPTSPS	YSPTSPN	YSPTSPS
25	YSPTSPS	YS-ASPQ	YSPTSPG
26	YSPTSPN	YSPASPA	YSPGSPA
27	YTPTSPN	YSQTGVK	YSPKQDEQKHNENENSR
28	YSPTSPS	YSPTSPT	
29	YSPTSPS	YSPPSPS	
30	YSPTSPS	YDGSPGSPQ	
31	YSPSSPR	YTPGSPQ	
32	YTPQSPT	YSPASPK	
33	YSPSSPS	YSPTSPL	
34	YSPSSPS	YSPSSPQ	
35	YSPTSPK	HSPSN-Q	
36	YTPTSPS	YSPTGST	
37	YSPSSPE	YSATSPR	
38	YTPTSPK	YSPNMSI	
39	YSPTSPK	YSPSSPK	
40	YSPTSPK	YSPTSPS	
41	YSPTSPT	YTPTARN	
42	YSPTTPK	YSPTSPM	
43	YSPTSPT	YSPTAPSH	
44	YSPTSPV	YSPTSPA	
45	YTPTSPK	YSPSSPTFEESED	
46	YSPTSPT		
47	YSPTSPK		
48	YSPTSPT		
49	YSPTSPKGST		
50	YSPTSPG		
51	YSPTSPT		
52	YSLTSPAISPDDSDEEN		

Fig. 8 Comparison of complete RNAPII CTD sequences from *Homo sapiens*, *Drosophila melanogaster*, and *Saccharomyces cerevisiae*. Each heptad within the CTDs has been numbered and aligned to highlight residue variations from the consensus $Y_1S_2P_3T_4S_5P_6S_7$ sequence. Residue variations were colored as follows: *orange* for aromatic residues (F, Y, and W), *green* for hydrophobic residues (e.g., A and V), *light blue* for polar uncharged residues (e.g., S and T), *purple* for basic residues (H, K, and R), and *red* for acidic residues (D and E).

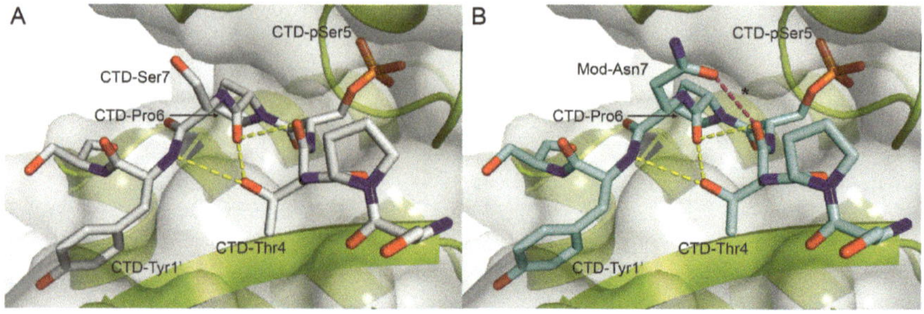

Fig. 9 Structural comparison of Ssu72 (*light green*) bound to native CTD peptide and model of S7N peptide. (A) The *cis*-conformation of P_6 in the native peptide (*white*) is stabilized by three intramolecular hydrogen bonds between T_4, S_7, and Y_1' (*apostrophe* denotes residue from the next heptad; PDB: 3O2Q). (B) Modeling N in place of S_7 in the native peptide (*yellow*) forms an additional hydrogen bond between the N_7 side chain and T_4 carboxyl group that may stabilize the *cis*-conformation of P_6. The Van der Waals exterior of Ssu72 is represented as a transparent *gray surface*.

4. SELECTIVE REGULATION OF PHOSPHATASE ACTIVITY BY PIN1 PROLYL ISOMERIZATION

The critical question regarding protein specificity for proline isomeric states of its substrate is how this special property can rewire downstream signaling pathways. Differences in proline isomeric specificity of regulatory proteins can provide unique opportunities for differential regulation of signaling pathways. For proteins that can easily accommodate either proline conformation, the availability of *cis*- or *trans*-proline substrate should not significantly affect the whole pathway. On the other hand, if the protein only recognizes one isomeric proline substrate, it is likely that equilibrium interconversion between *cis*- and *trans*-proline is too slow to replenish the consumed isomeric form of substrate thus becoming the rate-limiting step of the biological pathway. This effect should be more prominent in proteins specific for *cis*-proline, since this conformation is the minority species in nature thus making it easier to be depleted. Under such a scenario, mechanisms that accelerate reequilibration of the *cis/trans* isomer ratio will increase the amount of apparent enzymatic activity. Pin1 in eukaryotes (or Ess1 in *S. cerevisiae*) catalyzes the conversion between *cis*- and *trans*-proline which can help more quickly equilibrate the population (Hanes, 2014; Lu et al., 2007).

One example of how Pin1 can affect enzymatic function is when coincubated with Ssu72 there is a three- to fourfold increase in pS_5 CTD dephosphorylation compared to control reactions when the phosphate turnover was measured using the malachite green assay (Mayfield et al., 2015; Xiang et al., 2010). This increase in apparent Ssu72 activity is dependent on accelerated CTD proline isomerization since coincubation with an activity-deficient variant of Pin1 does not produce the same effect (Werner-Allen et al., 2011). In contrast, the activity of Scp1 and Fcp1 is not affected by Pin1 in vitro (Mayfield et al., 2015), consistent with their general activity against either *cis*- or *trans*-SP motifs.

To show whether the specific enhancement of Ssu72 activity by Pin1 is physiologically relevant, we tested if the phosphatase activity of Ssu72 is linked to Pin1 activity in cells (Mayfield et al., 2015). We performed lipofectamine (Thermo Fisher Scientific) transfections of HEK293T cells with a pLKO.1 plasmid containing shPin1 along with packaging and envelope plasmids to produce viral particles containing the shRNA coding region. HeLa cells were infected with either shPin1 or control viral particles using 4 μg/mL of hexadimethrine bromide and selected for stable cell lines using 2 μg/mL of puromycin over 48 h. These cells could then be probed with phospho-specific CTD antibodies (Eick & Geyer, 2013) to examine whether Pin1 function is related to CTD phosphorylation levels.

Knockdown of Pin1 expression in HeLa cells resulted in the accumulation of pS_5, likely due to the reduction in apparent Ssu72 activity. In contrast, the levels of pS_2 were not affected by the changes of Pin1 protein level, which was expected since the pS_2 phosphatase Fcp1 does not depend on the presence of *cis*-proline substrate for physiological function (Mayfield et al., 2015). These results strongly suggest that Pin1 specifically regulates the removal of pS_5 of the CTD through controlling substrate availability for Ssu72. It has been established that the phosphatase activity of Ssu72 is necessary for accurate termination of transcription (Dichtl et al., 2002; Ganem et al., 2003; Steinmetz & Brow, 2003). This corresponds with mutations in Ess1, the yeast homolog of Pin1, resulting in transcriptional read-through that was functionally linked to Ssu72 activity (Krishnamurthy et al., 2004). Our studies on the *cis*-proline specificity of Ssu72 provide a molecular explanation for the transcriptional defects in these various in vivo systems (Fig. 10).

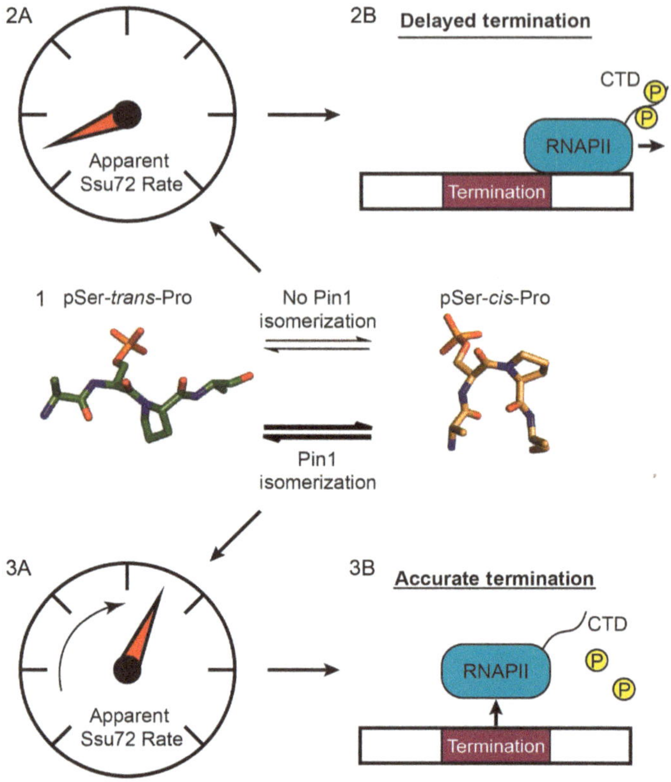

Fig. 10 Model of Pin1 effect on Ssu72-dependent dephosphorylation of RNAPII CTD. (1) Rate of isomerization of phosphorylated SP motifs with or without Pin1. The *thin arrows* without Pin1 represent the slow rate of normal proline isomerization, while the *bold arrows* represent the accelerated rate of isomerization with Pin1. (2A) In cells with low Pin1 activity, Ssu72 activity will become limited as it consumes its preferred pS_5-*cis*-P_6 substrate and the relatively slow rate of isomerization limits further dephosphorylation at this site. (2B) Therefore, disruption in Pin1 and subsequently Ssu72 causes transcriptionally active RNAPII to remain hyperphosphorylated and read through termination sites. (3A) With normally functioning Pin1 catalyzing the *cis/trans* isomerization, Ssu72 rate will no longer be limited by a lack of pS_5-*cis*-P_6 substrate as *trans*-substrate can be converted to the *cis*-conformation in the appropriate time frame. (3B) Proper functioning of Pin1 and Ssu72 results in accurate transcriptional termination and thus better coordination of transcriptional cycling.

5. SUMMARY AND CONCLUSION

In this chapter, we have described in detail a method using conformation-locked proline peptidomimetics to determine the activity and specificity of proteins for *cis*- and *trans*-proline substrates. Using CTD phosphatases as an example, we showed how such compounds can be used to determine without ambiguity the substrate proline conformation preference of a protein. We accomplished this using a highly sensitive colorimetric assay to quantitatively assess enzymatic activity against either proline isomer as well as captured complex structures with their preferred isomeric peptidomimetics using X-ray crystallography. These structures showed that such proline isostere compounds are very good mimics of natural prolines; and in one case, we were able to directly observe the recognition mode of a substrate with a less favored peptidyl-proline bond conformation. This would not have been possible using X-ray crystallography analysis with natural substrates which typically obviates alternative binding modes. We have also briefly discussed alternative tools such as ^{13}C NMR that can provide new information on the dynamics of proline isomerization on enzymatic function. In terms of the biological relevance of proline conformation specificity, proteins with different isomeric preferences can be differentially regulated and achieve alternative cellular outcomes. Regulating the function of proline conformation sensitive proteins with prolyl isomerases like Pin1 thus appears to be critical for normal cellular function.

ACKNOWLEDGMENTS

This work is made possible by Grants from the National Institutes of Health (R01 GM104896 to Y.J.Z.) and Welch Foundation (F-1778 to Y.J.Z.).

REFERENCES

Bastidas, M., Gibbs, E. B., Sahu, D., & Showalter, S. A. (2015). A primer for carbon-detected NMR applications to intrinsically disordered proteins in solution. *Concepts in Magnetic Resonance Part A*, *44*(1), 54–66.

Brutscher, B., Felli, I. C., Gil-Caballero, S., Hošek, T., Kümmerle, R., Piai, A., et al. (2015). NMR methods for the study of instrinsically disordered proteins structure, dynamics, and interactions: General overview and practical guidelines. *Advances in Experimental Medicine & Biology*, *870*, 49–122.

Buratowski, S. (2003). The CTD code. *Nature Structural Biology*, *10*(9), 679–680.

Carter, S. G., & Karl, D. W. (1985). Inorganic phosphate assay with malachite green: An improvement and evaluation. *Journal of Biochemical and Biophysical Methods*, 7(1), 7–13.

Corden, J. L. (2013). RNA polymerase II C-terminal domain: Tethering transcription to transcript and template. *Chemical Reviews*, *113*(11), 8423–8455.

Dichtl, B., Blank, D., Ohnacker, M., Friedlein, A., Roeder, D., Langen, H., et al. (2002). A role for SSU72 in balancing RNA polymerase II transcription elongation and termination. *Molecular Cell*, *10*(5), 1139–1150.

Eick, D., & Geyer, M. (2013). The RNA polymerase II carboxy-terminal domain (CTD) code. *Chemical Reviews*, *113*(11), 8456–8490.

Ganem, C., Devaux, F., Torchet, C., Jacq, C., Quevillon-Cheruel, S., Labesse, G., et al. (2003). Ssu72 is a phosphatase essential for transcription termination of snoRNAs and specific mRNAs in yeast. *The EMBO Journal*, *22*(7), 1588–1598.

Ghosh, A., Shuman, S., & Lima, C. D. (2008). The structure of Fcp1, an essential RNA polymerase II CTD phosphatase. *Molecular Cell*, *32*(4), 478–490.

Gibbs, E. B., Lu, F., Portz, B., Fisher, M. J., Medellin, B. P., Laremore, T. N., et al. (2017). Phosphorylation induces sequence-specific conformational switches in the RNA polymerase II C-terminal domain. *Nature Communications*, *8*(15233), 1–11.

Gibbs, E. B., & Showalter, S. A. (2015). Quantitative biophysical characterization of intrinsically disordered proteins. *Biochemistry*, *54*(6), 1314–1326.

Hanes, S. D. (2014). The Ess1 prolyl isomerase: Traffic cop of the RNA polymerase II transcription cycle. *Biochimica et Biophysica Acta*, *1839*(4), 316–333.

Harlen, K. M., & Churchman, L. S. (2017). The code and beyond: Transcription regulation by the RNA polymerase II carboxy-terminal domain. *Nature Reviews. Molecular Cell Biology*, *18*(4), 263–273.

Hart, S. A., Sabat, M., & Etzkorn, F. A. (1998). Enantio- and regioselective synthesis of a (*Z*)-alkene *cis*-proline mimic. *The Journal of Organic Chemistry*, *63*(22), 7580–7581.

Hassell, A. M., An, G., Bledsoe, R. K., Bynum, J. M., Carter, H. L., 3rd, Deng, S. J., et al. (2007). Crystallization of protein-ligand complexes. *Acta Crystallographica. Section D, Biological Crystallography*, *63*(Pt. 1), 72–79.

Hausmann, S., & Shuman, S. (2002). Characterization of the CTD phosphatase Fcp1 from fission yeast preferential dephosphorylation of serine 2 versus serine 5. *The Journal of Biological Chemistry*, *277*(24), 21213–21220.

Jeronimo, C., Bataille, A. R., & Robert, F. (2013). The writers, readers, and functions of the RNA polymerase II C-terminal domain code. *Chemical Reviews*, *113*(11), 8491–8522.

Kobor, M. S., Archambault, J., Lester, W., Holstege, F. C., Gileadi, O., Jansma, D. B., et al. (1999). An unusual eukaryotic protein phosphatase required for transcription by RNA polymerase II and CTD dephosphorylation in *S. cerevisiae*. *Molecular Cell*, *4*(1), 55–62.

Krishnamurthy, S., He, X., Reyes-Reyes, M., Moore, C., & Hampsey, M. (2004). Ssu72 is an RNA polymerase II CTD phosphatase. *Molecular Cell*, *14*(3), 387–394.

Kubicek, K., Cerna, H., Holub, P., Pasulka, J., Hrossova, D., Loehr, F., et al. (2012). Serine phosphorylation and proline isomerization in RNAP II CTD control recruitment of Nrd1. *Genes & Development*, *26*(17), 1891–1896.

Lawrence, C. W., & Showalter, S. A. (2012). Carbon-detected 15N NMR spin relaxation of an intrinsically disordered protein: FCP1 dynamics unbound and in complex with RAP74. *Journal of Physical Chemistry Letters*, *3*(10), 1409–1413.

Liou, Y. C., Zhou, X. Z., & Lu, K. P. (2011). Prolyl isomerase Pin1 as a molecular switch to determine the fate of phosphoproteins. *Trends in Biochemical Sciences*, *36*(10), 501–514.

Lu, K. P., Finn, G., Lee, T. H., & Nicholson, L. K. (2007). Prolyl *cis-trans* isomerization as a molecular timer. *Nature Chemical Biology*, *3*(10), 619–629.

Lu, K. P., Liou, Y. C., & Zhou, X. Z. (2002). Pinning down proline-directed phosphorylation signaling. *Trends in Cell Biology*, *12*(4), 164–172.

Lu, K. P., & Zhou, X. Z. (2007). Pinning down proline-directed phosphorylation signaling. *Trends in Cell Biology*, *12*(4), 164–172.

Luo, Y., Yogesha, S. D., Cannon, J. R., Yan, W., Ellington, A. D., Brodbelt, J. S., et al. (2013). Novel modifications on C-terminal domain of RNA polymerase II can fine-tune the phosphatase activity of Ssu72. *ACS Chemical Biology*, *8*(9), 2042–2052.

Mayfield, J. E., Burkholder, N. T., & Zhang, Y. J. (2016). Dephosphorylating eukaryotic RNA polymerase II. *Biochimica et Biophysica Acta*, *1864*(4), 372–387.

Mayfield, J. E., Fan, S., Wei, S., Zhang, M., Li, B., Ellington, A. D., et al. (2015). Chemical tools to decipher regulation of phosphatases by proline isomerization on eukaryotic RNA polymerase II. *ACS Chemical Biology*, *10*(10), 2405–2414.

Müller, I. (2017). Guidelines for the successful generation of protein-ligand complex crystals. *Acta crystallographica. Section D, Structural biology*, *73*(Pt. 2), 79–92.

Otaka, A., Katagiri, F., Kinoshita, T., Odagaki, Y., Oishi, S., Tamamura, H., et al. (2002). Regio- and stereoselective synthesis of (*E*)-alkene *trans*-Xaa-pro dipeptide mimetics

utilizing organocopper-mediated *Anti*-SN2' reactions. *The Journal of Organic Chemistry*, *67*(17), 6152–6161.
Sasaki, Y., Niida, A., Tsuji, T., Shigenaga, A., Fujii, N., & Otaka, A. (2006). Stereoselective synthesis of (*Z*)-alkene-containing proline dipeptide mimetics. *The Journal of Organic Chemistry*, *71*(13), 4969–4979.
Schüller, R., Forné, I., Straub, T., Schreieck, A., Texier, Y., Shah, N., et al. (2014). Heptad-specific phosphorylation of RNA polymerase II CTD. *Molecular Cell*, *61*(2), 305–314.
Shaw, P. E. (2002). Peptidyl-prolyl isomerases: A new twist to transcription. *EMBO Reports*, *3*(6), 521–526.
Steinmetz, E. J., & Brow, D. A. (2003). Ssu72 protein mediates both poly(A)-coupled and poly(A)-independent termination of RNA polymerase II transcription. *Molecular and Cellular Biology*, *23*(18), 6339–6349.
Wang, X. J., Hart, S. A., Xu, B., Mason, M. D., Goodell, J. R., & Etzkorn, F. (2003). Serine-*cis*-proline and serine-*trans*-proline isosteres: Steroselective synthesis of (*Z*)- and (*E*)-alkene mimics by still-Wittig and Ireland-Claisen rearrangements. *The Journal of Organic Chemistry*, *68*(6), 2343–2349.
Wedemeyer, W. J., Welker, E., & Scheraga, H. A. (2002). Proline *cis-trans* isomerization and protein folding. *Biochemistry*, *41*(50), 14637–14644.
Werner-Allen, J. W., Lee, C. J., Liu, P., Nicely, N. I., Wang, S., Greenleaf, A. L., et al. (2011). *cis*-Proline-mediated Ser(P)5 dephosphorylation by the RNA polymerase II C-terminal domain phosphatase Ssu72. *The Journal of Biological Chemistry*, *286*(7), 5717–5726.
Wu, X., Wilcox, C. B., Devasahayam, G., Hackett, R. L., Arevalo-Rodriguez, M., Cardenas, M. E., et al. (2000). The Ess1 prolyl isomerase is linked to chromatin remodeling complexes and the general transcription machinery. *The EMBO Journal*, *19*(14), 3727–3738.
Wulf, G., Finn, G., Suizu, F., & Lu, K. P. (2005). Phosphorylation-specific prolyl isomerization: Is there an underlying theme? *Nature Cell Biology*, 7(5), 435–441.
Xiang, K., Nagaike, T., Xiang, S., Kilic, T., Beh, M. M., Manley, J. L., et al. (2010). Crystal structure of the human symplekin-Ssu72-CTD phosphopeptide complex. *Nature*, *467*(7316), 729–733.
Yaffe, M. B., Schutkowski, M., Shen, M., Zhou, X. Z., Stukenberg, P. T., Rahfeld, J. U., et al. (1997). Sequence-specific and phosphorylation-dependent proline isomerization: A potential mitotic regulatory mechanism. *Science*, *278*(5345), 1957–1960.
Yeo, M., Lee, S. K., Lee, B., Ruiz, E. C., Pfaff, S. L., & Gill, G. N. (2005). Small CTD phosphatases function in silencing neuronal gene expression. *Science*, *307*(5709), 596–600.
Yeo, M., Lin, P. S., Dahmus, M. E., & Gill, G. N. (2003). A novel RNA polymerase II C-terminal domain phosphatase that preferentially dephosphorylates serine 5. *The Journal of Biological Chemistry*, *278*(28), 26078–26085.
Zhang, Y., Kim, Y., Genoud, N., Gao, J., Kelly, J. W., Pfaff, S. L., et al. (2006). Determinants for dephosphorylation of the RNA polymerase II C-terminal domain by Scp1. *Molecular Cell*, *24*(5), 759–770.
Zhang, M., Wang, X. J., Chen, X., Bowman, M. E., Luo, Y., Noel, J. P., et al. (2012). Structural and kinetic analysis of prolyl-isomerization/phosphorylation cross-talk in the CTD code. *ACS Chemical Biology*, 7(8), 1462–1470.
Zhou, X. Z., & Lu, K. P. (2016). The isomerase PIN1 controls numerous cancer-driving pathways and is a unique drug target. *Nature Reviews. Cancer*, *16*(7), 463–478.

FURTHER READING

Mayfield, J. E., Robinson, M. R., Cotham, V. C., Irani, S., Matthews, W. L., Ram, A., et al. (2017). Mapping the phosphorylation pattern of *Drosophila melanogaster* RNA polymerase II carboxyl-terminal domain using ultraviolet photodissociation mass spectrometry. *Biochimica et Biophysica Acta*, *1839*(4), 316–333.

SECTION IV

Delineating the Action of Phosphatases in Signaling and Regulation

CHAPTER TEN

Quantitative Analysis of Intracellular Response Regulator Phosphatase Activity of Histidine Kinases

Rong Gao*,†, Ann M. Stock*,†,1

*Center for Advanced Biotechnology and Medicine, Rutgers Biomedical and Health Sciences, Piscataway, NJ, United States

†Department of Biochemistry and Molecular Biology, Rutgers—Robert Wood Johnson Medical School, Piscataway, NJ, United States

[1]Corresponding author: e-mail address: stock@cabm.rutgers.edu

Contents

Abstract

Quantitation of two-component protein activities is becoming increasingly important to understand the general design principles for this widely distributed prokaryotic signaling pathway. In many two-component systems (TCSs), phosphatase activity of the sensor histidine kinase (HK) plays a major role in controlling the system output and resetting the system to the prestimulus state. Quantitation of the phosphatase activity is often carried out in vitro, usually with truncated proteins that may not recapitulate the intact HK in the cellular environment. This chapter outlines a method for characterizing the intracellular phosphatase activity by investigating the TCS deactivation dynamics upon stimulus withdrawal. Two experimental approaches,

Methods in Enzymology, Volume 607
ISSN 0076-6879
https://doi.org/10.1016/bs.mie.2018.04.004

the direct Phos-tag gel analysis and the indirect reporter assay, are described for measuring the TCS deactivation dynamics in cell. Combined with a mathematic model, the experimentally determined kinetics can lead to proper evaluation of the intracellular phosphatase activity.

ABBREVIATIONS

CFP cyan fluorescent protein
Ery erythromycin
Fluo. fluorescence
FP fluorescent protein
GFP green fluorescent protein
HK histidine kinase
OD optical density
PAGE polyacrylamide gel electrophoresis
RR response regulator
RR~P phosphorylated response regulator
SDS sodium dodecyl sulfate
TCS two-component system
WT wild type

1. INTRODUCTION

Two-component systems (TCSs) involving phosphotransfer between histidine protein kinases and response regulator (RR) proteins are a prevalent strategy for bacterial signal transduction. One important, although not universal, mechanism for regulating the phosphorylation level of the RR, and thus system output, is a RR phosphatase activity of the histidine protein kinase (Gao & Stock, 2009; Huynh & Stewart, 2011). Archetypical TCSs consist of a transmembrane sensor histidine kinase (HK) and a RR that elicits the output response (Fig. 1A). The HK catalyzes autophosphorylation of a specific histidine residue, generating a high-energy phosphoryl group that is poised for subsequent transfer. Phosphotransfer from the histidine residue in the HK to an aspartate residue in the RR occurs within a complex, with both proteins contributing to catalysis. The phosphorylation level of the RR determines system output.

Two potential fates of the phosphoryl group of the RR are common to most all TCSs. The phosphoaspartate of the RR can be hydrolyzed by an autodephosphorylation activity of the RR or alternatively, the phosphoryl group can be transferred back to the histidine residue of the HK and from

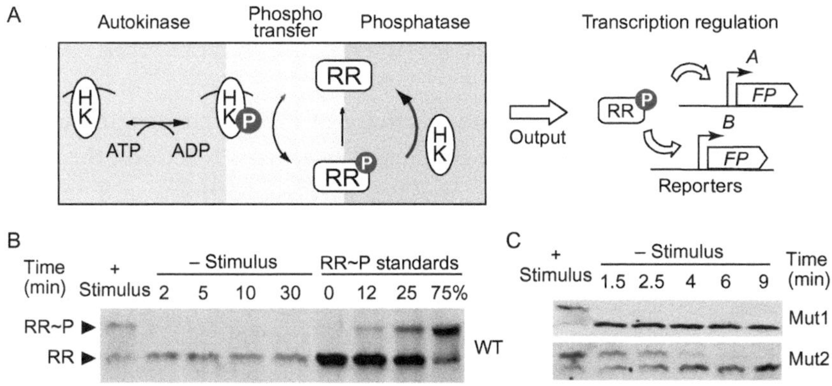

Fig. 1 Measurement of intracellular RR dephosphorylation kinetics. (A) Multiple enzyme activities of two-component proteins. The RR~P level is determined by the autokinase, phosphotransferase, and phosphatase activities of the HK. Promoters of RR-regulated genes can be fused with a fluorescent protein (FP) gene to report RR phosphorylation levels. (B and C) RR dephosphorylation kinetics analyzed by measuring RR~P levels with Phos-tag gels. After removing the stimulus by adding 1 m*M* Pi, PhoB dephosphorylation kinetics were measured in wild type (WT) shown in (B), and phosphatase-diminished mutants $PhoB^{F20D}$ (Mut1) and $PhoR^{T217M}$ (Mut2) shown in (C). Approximately 0.3 OD·mL of cells were harvested at indicated times and cell pellets were lysed by trituration in 55 μL 1 × BugBuster reagent. After addition of 18 μL 4 × SDS loading buffer for denaturation, all samples were immediately flash frozen and later analyzed using Phos-tag gels (10% acrylamide, 25 μ*M* Phos-tag acrylamide) and western blots. To prepare RR~P standards, pure PhoB protein (5 μ*M*) was phosphorylated in the presence of 50 m*M* phosphoramidate for 1.5 h in 50 m*M* Tris, pH 7.4, 100 m*M* NaCl, 5 m*M* $MgCl_2$, followed by denaturation in SDS sample loading buffer and mixing with known amounts of unphosphorylated PhoB.

there to ADP to regenerate ATP in a reversal of the phosphotransfer and autophosphorylation reactions. A third fate exists in the many systems that contain HKs with specific RR phosphatase activity. These HKs hydrolyze the phosphoaspartate of their cognate RRs without transfer to the histidine and regeneration of ATP. In these systems, the phosphatase activity of the HK plays an essential role in determining the level of phosphorylated response regulator and system output. Stimuli control the activities of the HK by stabilizing conformational states that favor either catalysis of autophosphorylation and phosphotransfer or the opposing RR dephosphorylation. The phosphatase activity of an HK is important not only for stimuli-mediated responses but also for maintaining low basal levels of phosphorylated RR in the unstimulated state, counterbalancing phosphorylation of the RR via the cognate HK as well as cross-talk from noncognate HKs and/or intracellular small molecule phosphodonors such as acetyl phosphate (Laub & Goulian, 2007; Siryaporn & Goulian, 2008; Wayne, Li, Kazmierczak, Tsui, & Winkler, 2012).

Quantitation of the phosphatase activity of HKs is challenging because of the intrinsic complexity of the integrated enzymatic reactions, the lability of the phosphoaspartate residue of RRs, and the technical difficulties of biochemical analyses of transmembrane proteins. To circumvent the latter, most quantitation of phosphatase activity of transmembrane HKs using in vitro biochemical analyses has been conducted with cytoplasmic fragments. Such studies have limitations. While the truncated, soluble constructs retain autophosphorylation, phosphotransfer, and RR phosphatase activities, their levels relative to those of the intact proteins are unknown. The importance of HK sensing and transmembrane domains in regulating HK activities raises questions about the signaling state(s) represented by isolated domains. Perhaps due to the absence of the regulatory domains or the natural membrane environment that may be required to maintain the phosphatase signaling state, phosphatase activities measured in vitro using cytoplasmic fragments have been found to be substantially less than those estimated for intact HKs in cellular studies (Gao & Stock, 2017). Furthermore, cellular studies have the advantage of allowing comparison of enzymatic activities in the presence and absence of stimuli if environmental conditions that lead to activation are known, even for systems in which specific ligands or physical parameters have not been identified, or in which sensing is mediated by auxiliary proteins.

One strategy for assessing phosphatase activity is the measurement of system output as an indirect report of RR phosphorylation. For the majority of TCSs that regulate gene expression via RR transcription factors, this can be achieved using reporter gene assays. While this method allows real-time analyses in intact cells, some experimental precautions are necessary. Different operons in a single regulon can show substantially different expression profiles (Gao & Stock, 2015; Zwir et al., 2014), thus it is important that a reporter gene system be constructed with a promoter for which activity has been shown to correlate with levels of RR phosphorylation. It is also important to recognize that reporter output is delayed by the time required for transcription, translation, and maturation of the reporter protein and because of the long lifetime of the reporter protein, measurements of reporter output are cumulative. Furthermore, as for any cellular assay, measurement of reporter proteins is subject to interference by other cellular components.

A second, more direct strategy is the quantitation of RR phosphorylation in cell extracts using the phosphoryl chelator Phos-tag (Barbieri & Stock, 2008). This method relies on gentle lysis of cells to preserve the labile phosphoaspartate; separation of phosphorylated and unphosphorylated forms of the RR by electrophoresis in SDS Phos-tag acrylamide gels; and subsequent detection of the RR bands by immunoblotting. However, the rates of dephosphorylation of RRs catalyzed by HKs upon stimulus withdrawal are typically too rapid for kinetic characterization by this method. Nonetheless, this method can be used to set upper limits for the half-lives of RR dephosphorylation and validate the dephosphorylation kinetics measured by reporter assays. More accurate estimates of phosphatase activity can be obtained by combining such experimental data from wild type (WT) and phosphatase-diminished mutant HKs with modeling to establish boundaries for the kinetic parameters of the phosphatase activity of a wild-type HK.

This chapter will focus on methods for analysis of the phosphatase activity of HKs in cells using Phos-tag analyses, reporter gene assays, and a combined approach of experimental analysis of RR phosphorylation with modeling. These methods were developed to characterize phosphatase activity in the *Escherichia coli* PhoR–PhoB TCS that regulates expression of genes involved in phosphorus assimilation in response to phosphate limitation. However, the approach is generalizable and should be broadly applicable to characterization of intracellular phosphatase activity in other bacterial TCSs.

2. PHOS-TAG GEL ANALYSES OF RR DEPHOSPHORYLATION

Phosphatase activity of HKs plays a major role in resetting the system to the prestimulus state once the stimulus is removed. Characterization of phosphatase activity in cells relies on measurements of RR dephosphorylation kinetics upon stimulus removal. At the heart of such studies on TCS deactivation dynamics is the experimental determination of intracellular RR phosphorylation (RR~P) levels. SDS-PAGE gels containing the phosphoryl chelator Phos-tag (Wako Chemicals) covalently linked to acrylamide allows separation of phosphorylated and unphosphorylated forms of RR from cell lysates, and the subsequent direct quantitation of RR phosphorylation in complex mixtures using immunoblots (Fig. 1B and C). Detailed procedures of Phos-tag SDS-PAGE have been well documented

(Kinoshita, Kinoshita-Kikuta, & Koike, 2009; Kinoshita, Kinoshita-Kikuta, Takiyama, & Koike, 2006) and customized for measuring RR phosphorylation (Barbieri & Stock, 2008; Boulanger, Chen, Hinton, & Stibitz, 2013; Gao & Stock, 2013; Wayne et al., 2012). Thus, the step-to-step protocol of Phos-tag gel analyses is not described in detail here. Instead, practical experimental considerations will be discussed to ensure reliable measurements of RR phosphorylation levels.

2.1 Experimental Considerations

Acyl phosphate on the conserved aspartate residue of an RR is labile under many conditions. Base, strong acid, and heat can lead to significant hydrolysis of phosphoaspartate. Preserving the labile acyl phosphate throughout an experiment is key to the success of measuring phosphorylated RR levels in cell lysates. Common lysis protocols involving sample heating or sonication can cause substantial loss of RR phosphorylation (Barbieri & Stock, 2008). A mild formic acid treatment has been initially used for lysing cells and preparing RR~P samples (Barbieri & Stock, 2008; Boulanger et al., 2013) but it requires an additional sample neutralization step that can increase salt concentration and raise the possibility of electrophoretic distortion in Phos-tag gels.

Detergent-based methods using reagents specialized for bacterial lysis (Gao & Stock, 2013; Richards et al., 2015), such as BugBuster (EMD Millipore) or B-Per (ThermoFisher), provide a gentle and rapid protocol for preparing cell lysate samples. Nuclease or DNase can be added in the detergent-containing lysis buffer to reduce sample viscosity for optimal Phos-tag gel results. Trituration (pipetting up and down) for 15–30 s is usually sufficient for lysing 0.3–0.5 OD · mL (OD600 × volume in mL) of *E. coli* cells. The entire lysis step including bacterial cell harvesting can be accomplished in 1–2 min, which sets a limit of sampling time intervals for kinetic studies. Alternative methods using lysing beads have also proved successful (Park & Groisman, 2014; Wayne et al., 2012). All these methods may need further optimization for individual TCSs in different bacteria species and the central principle for optimization is to minimize both time and the exposure of the RR~P to various hydrolyzing conditions.

Phosphorylated RRs may inevitably lose some of their phosphoryl groups due to the intrinsic lability of acyl phosphate, particularly during electrophoresis when heat is generated. RR~P standards with different fractions of phosphorylated RR are often run along with lysate samples on the same gel (Fig. 1B) to calibrate for the loss of phosphorylation.

Because the chemical stability of phosphoaspartate should be similar in all denatured RRs, it is not essential that the RR~P used for standards be the same RR being experimentally analyzed, however, the use of different RRs complicates detection on a single immunoblot. To generate RR~P standards, purified RR proteins are first phosphorylated, ideally to completion, or at least to a known extent, by excess small molecule phosphodonors, such as phosphoramidate or acetyl phosphate, and subsequently mixed with known amounts of unphosphorylated RR. The resulting standards are often added to lysates of the bacterial cells whose RR gene is deleted to maintain gel-running conditions identical to those of experimental samples. Quantification of these standards on immunoblots is calibrated with the known RR~P fractions to generate a standard curve that ensures accurate quantitation of RR phosphorylation levels.

2.2 Limitations

Direct measurement of RR~P levels by Phos-tag gel analyses relies on harvesting of sufficient bacterial cells and generation of cell lysates, both of which take time and could be laborious if a large number of samples needs to be collected. As discussed earlier, sample preparation takes at least a couple of minutes even with a rapid lysis protocol. However, some HKs have strong phosphatase activity and dephosphorylation of RR~P occurs extremely quickly (Gao & Stock, 2017). For example, upon stimulus removal, PhoB~P is completely dephosphorylated within 2 min, even before completion of processing of the first sample (Fig. 1B). Rapid RR dephosphorylation was still observed in phosphatase-diminished mutants (Fig. 1C). Such rapid dephosphorylation leaves little time to collect enough samples for kinetic analyses but dephosphorylation half-lives can be estimated for these proteins, WT ($\leq$1 min), Mut1 ($\leq$1 min), and Mut2 (2.5–4 min). Moreover, the Phos-tag gel method is inherently low throughput. The number of sample wells in each Phos-tag gel limits the total number of samples that can be analyzed simultaneously. The necessity of including RR~P standards further reduces the sample numbers for kinetic characterization.

3. TRANSCRIPTION REPORTER ASSAYS

Transcription reporters have been routinely used to infer RR phosphorylation levels in cells. A wide variety of genes encoding different reporter proteins, such as β-galactosidase, luciferase, and fluorescent proteins (FPs), can be fused with RR-regulated promoters to report the system output.

Green fluorescent protein (GFP) and its color variants represent one of the most versatile reporters that allow high-throughput, real-time, continuous, and even single-cell measurement of gene expression. Thus, we will use FP reporter as an example to describe necessary considerations for characterization of TCS deactivation dynamics. Upon stimulus removal, temporal dynamics of fluorescence output can be tracked with any fluorescence detector, such as a fluorometer or microplate reader, to report the kinetics of RR dephosphorylation. As an indirect reporter of RR~P, FP reporters have some limitations associated with the complexity of transcription regulation as well as physicochemical properties of FPs. Some of these issues may be specific to FP reporters but common principles are shared with other reporter systems for establishing proper controls and improving the experimental design.

3.1 Time Derivatives of Reporter Dynamics

TCS activity or RR phosphorylation status is frequently indicated by overall fluorescence output. However, the overall fluorescence value at a particular time is the result of all FPs expressed prior to that time point and determined by the cumulative output of promoter activity. In contrast, the first derivative of fluorescence with respect to time, i.e., the rate of fluorescence change, reflects the real-time promoter activity. The time derivative of fluorescence is also influenced by the FP decay rate. FPs are usually very stable and the half-life of GFP is longer than 1 day in *E. coli* (Andersen et al., 1998). At the timescale of activation or deactivation of the exemplified PhoR–PhoB system (1 – 2 h), FP decay is negligible, thus the time derivative of fluorescence is directly used to represent promoter activity and track RR~P dynamics (Fig. 2A and B). For TCSs functioning at a longer time scale (days) or in different organisms, the FP degradation rate may require reevaluation and need to be considered when calculating promoter activity from fluorescence derivatives.

Reporter dynamics constitute discrete data measured at different time points. The first derivative needs to be computed. Considering that the data may contain unequal time intervals, the following equation can be used to calculate the fluorescence derivative by differentiating the second-order Lagrange interpolating polynomial,

$$f'(t_i) = \frac{t_i - t_{i+1}}{(t_{i-1} - t_i)(t_{i-1} - t_{i+1})} f(t_{i-1}) + \frac{2t_i - t_{i-1} - t_{i+1}}{(t_i - t_{i-1})(t_i - t_{i+1})} f(t_i) + \frac{t_i - t_{i-1}}{(t_{i+1} - t_{i-1})(t_{i+1} - t_i)} f(t_{i+1}).$$

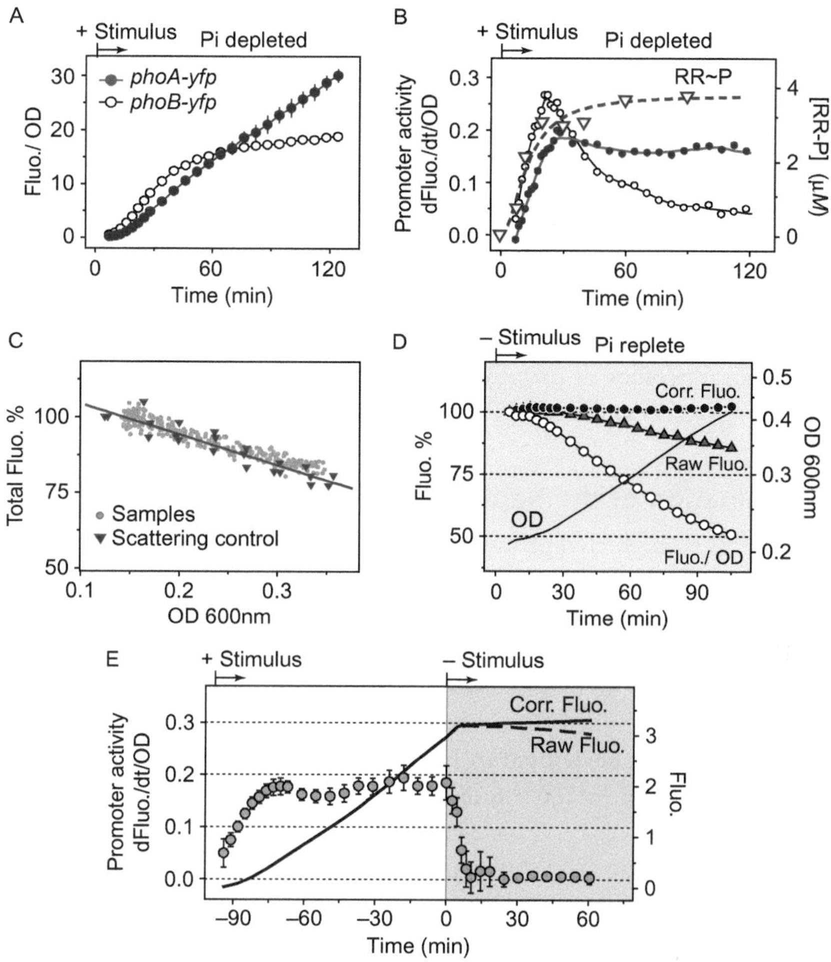

Fig. 2 Data analyses of reporter fluorescence dynamics. (A and B) Evaluation of reporters with time derivatives of fluorescence. Activation dynamics of the *phoA-yfp* and *phoB-yfp* reporters were measured in a fluorescence microplate reader with constant shaking. Eleven replicates for each reporter were assayed simultaneously. Upon stimulation by Pi-depleted conditions, cellular fluorescence (A) increased with distinct dynamics for two reporters. *Solid and open circles* represent data for the *phoA* and *phoB* YFP reporters. Normalized first derivatives of fluorescence (B), (dFluo./dt)/OD, reflect the real-time promoter activities. Dynamics of *phoB* promoter activity suggest an additional transcription inhibition mechanism and are not consistent with RR phosphorylation kinetics measured with Phos-tag gels (*open triangles*). (C) Calibration of fluorescence scattering by bacterial turbidity. Pure yellow fluorescent protein (YFP) solution was diluted to give similar fluorescence readings as the activated *phoA-yfp* reporter. *Dark triangles* represent fluorescence of identical amounts of pure YFP mixed with different amounts (OD) of nonfluorescent bacteria. An empirically fitted linear correlation (*solid line*) was used to correct for the fluorescence of tested samples (*gray dots*).
(Continued)

$f(t_i)$ represents fluorescence of the entire culture at the ith time point. The first derivatives are normalized to the culture optical density (OD) for calculation of the promoter activity (dFluo./dt)/OD. The numerical differentiation process will amplify any noise in the data. When fluorescence readings are particularly noisy, it may be necessary to smooth the time series data by adjacent-averaging of the fluorescence reading with measurements at the preceding and subsequent time points, or by a more sophisticated smoothing algorithm. To boost the signal-to-noise ratio, the FP reporter assay is best performed in multiwell microplates with a large number of replicate samples. Mean fluorescence derivatives from these samples illustrate the temporal profiles of promoter activity, which can be used to evaluate different reporters and define RR~P kinetics.

3.2 Evaluation of the Reporter Promoter

Not every RR-regulated promoter is a suitable reporter for RR phosphorylation kinetics. Transcription of specific promoters may be subject to additional regulation by other transcription factors, small RNAs, or global physiological effects. Temporal profiles of promoter activity can be compared with RR~P dynamics to evaluate whether a specific reporter can recapitulate intracellular RR phosphorylation kinetics. This is best demonstrated by the example shown in Fig. 2A and B. Both *phoA* and *phoB* promoters are activated by the PhoR–PhoB system in response to Pi-depleted conditions. Fluorescence of the two YFP reporters increased monotonically after stimulation (Fig. 2A), but with different temporal patterns in fluorescence derivatives (Fig. 2B). Promoter activity of the *phoB-yfp* reporter peaked at ~20 min and gradually decreased afterward. The *phoA-yfp* reporter reached a steady level of promoter activity after the initial increase, recapitulating the RR phosphorylation kinetics measured by Phos-tag gels. The *phoA* promoter is clearly more suitable to report RR~P levels than the *phoB* promoter. In addition, the steady phase of promoter activity also marks the

Fig. 2—Cont'd (D and E) Correction of the fluorescence scattering. WT cells carrying the *phoA-yfp* reporter were stimulated in a Pi-depleted condition for ~90 min. For the deactivation dynamics shown in (D), stimulus was removed by resuspension of bacteria in Pi replete media. Raw fluorescence (*triangles*) decreased along with the increase in OD (*solid line*). After the scattering correction, corrected fluorescence (*filled circles*) appears constant, while the OD-normalized fluorescence (*open circles*) decreases due to growth-dilution. High temporal resolution upon stimulus removal is achieved by direct addition of Pi to stimulated cultures (E). Time zero is defined as the time of Pi addition. OD-normalized fluorescence derivatives (*gray circles*) depict the deactivation dynamics.

suitable timing for removing stimulus and initiation of TCS deactivation. A steady promoter activity of *phoA-yfp* ensures minimal interference from additional inhibitory mechanisms, such as that apparent in the *phoB* reporter, which would complicate assessment of the decrease in reporter activity expected after stimulus removal.

3.3 Correction of Reporter Fluorescence

Cellular fluorescence signals can be affected by factors not associated with FPs, such as the autofluorescence emitted by other cellular metabolites, which require appropriate corrections before further differentiation analyses. Various methods have been used to correct for bacterial autofluorescence (de Jong, Ranquet, Ropers, Pinel, & Geiselmann, 2010; Zaslaver et al., 2006). Briefly, cells without the corresponding reporter are assayed as the control. Autofluorescence of the control cultures is either directly subtracted or growth-calibrated and then subtracted from fluorescence of bacteria carrying the reporter.

One often-overlooked factor in FP reporter analyses is the absorbance or scattering of the excitation and/or emission light by bacterial suspensions. When the reporter assay involves significant growth of bacterial cultures, the increased bacterial turbidity will reduce the intensity of excitation and/or emission light by light-scattering, leading to reduced fluorescence values. This light-scattering effect may be corrected with a calibration curve that establishes the relation between fluorescence and culture turbidity (Fig. 2C). Identical amounts of either pure FP or fluorescent cells can be mixed with different amounts of nonfluorescent cells to generate an empirical correlation between measured OD and fluorescence. The empirical curve is then used to calibrate the fluorescence reading of the sample of interest.

Fig. 2D and E shows how the scattering correction affects interpretation of fluorescence reporter results measured for deactivation of the PhoR–PhoB system. Once the stimulus was removed and Pi became replete, cells resumed fast growth and raw fluorescence of the entire culture decreased gradually along with the increase in OD. If no correction were made, it would lead to a negative fluorescence derivative and suggest the loss or degradation of FPs. This hypothetical "loss" will potentially interfere with interpretation of the deactivation kinetics. When the scattering effect is corrected with the calibration curve shown in Fig. 2C, fluorescence of the entire culture appeared constant during the majority of the time course, suggesting a rapid shut-off of the promoter activity and a constant total level of the stable FP. FP per cell, or the OD-normalized fluorescence, decreased only due to the growth dilution. The deactivation dynamics were monitored with high temporal resolution in Fig. 2E. Upon stimulus removal, corrected fluorescence continues to rise for ~10 min and the first derivative of fluorescence rapidly decreased with a half-life of ~4 min. The corrected fluorescence allows precise evaluation of the fluorescence derivative. The light-scattering effect is not unique to fluorescent reporters. Any reporter involving optical measurements of light signals, such as luciferase, may also require similar calibration controls to evaluate the influence of bacterial turbidity on reporter output.

3.4 Measuring the Maturation Time of FP

The time needed for the maturation of FP, i.e., formation of the fluorophor, may be of concern for analyzing reporter kinetics. The time required for FP maturation can be ignored only when it is much more rapid than the examined signaling event, otherwise, it needs to be accounted for. Unfortunately, there is not a standard value for FP maturation to assess its impact on the observed fluorescence dynamics. FP maturation time depends on the specific FP variant, the culture conditions, the organism, and even the strain in which FP is expressed (Balleza, Kim, & Cluzel, 2018; Hebisch, Knebel, Landsberg, Frey, & Leisner, 2013). It is necessary to experimentally measure the FP maturation time for the particular reporter used.

FP maturation time in bacteria cells can be determined by measuring the fluorescence increase after addition of a translation inhibitor (Fig. 3A). The FP reporter is activated until the fluorescence derivative reaches a steady state. Protein synthesis is then inhibited by translation inhibitors, such as the antibiotics chloramphenicol or erythromycin (Ery). To ensure complete inhibition, the inhibitor concentration is usually much higher than the minimal inhibitory concentration documented for these antibiotics. It is assumed that the inhibition occurs instantaneously and the subsequent increase in fluorescence results from maturation of FPs that are expressed prior to translation inhibition. The increase in fluorescence can be fitted with a first-order kinetic model to derive the maturation half-time. The maturation half-life of cyan fluorescent protein (CFP) is calculated to be 8.0 ± 0.9 min for the data shown in Fig. 3A, very close to the time scale of PhoR–PhoB deactivation. When reporter dynamics of FP maturation and TCS deactivation are measured in parallel (Fig. 3B), the almost identical temporal profiles of the two suggest that deactivation of the PhoR–PhoB system

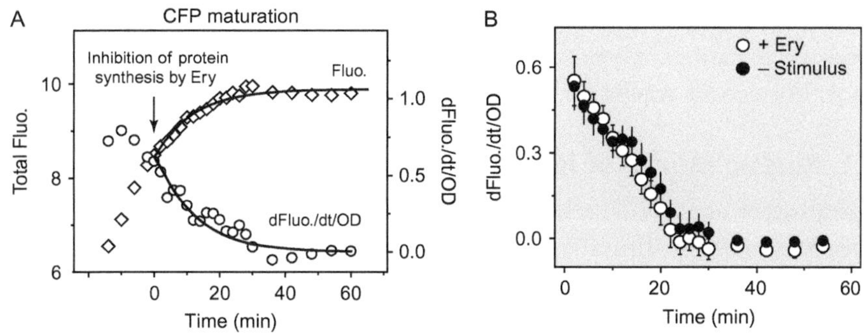

Fig. 3 Measurement of FP maturation time. (A) CFP maturation kinetics. The *phoA-cfp* reporter was activated by Pi-starvation and protein synthesis was inhibited by 250 μg/mL of erythromycin, as indicated by the *arrow*. Fluorescence and fluorescence derivatives are shown in *diamonds and circles*, respectively. *Solid lines* indicate the exponentially fitted line for fluorescence and the corresponding derivatives. (B) Comparison of the deactivation dynamics (*solid circles*) with CFP maturation kinetics (*open circles*).

is extremely fast and the relatively slow FP maturation kinetics dominate the promoter activity dynamics. Effects of FP maturation on reporter dynamics must be considered for understanding the deactivation kinetics and characterization of the PhoR phosphatase activity.

4. MODELING THE RR DEPHOSPHORYLATION KINETICS

As discussed earlier, RR dephosphorylation kinetics can be either directly measured by Phos-tag analyses or indirectly inferred from reporter dynamics. Both methods have limitations that prevent direct quantitation of the phosphatase activity using traditional enzyme kinetic analyses. Due to the sample preparation time, the number of kinetic samples that can be collected for Phos-tag analyses is often limited, especially for HKs with fast dephosphorylation kinetics. Unlike in vitro biochemical assays in which the enzyme concentration can be lowered to slow down the reaction for sufficient sampling, measurement of RR~P dynamics in cells is usually restricted by inherent expression levels of HK and RR. Thus, complete kinetic characterization is not always achievable with limited numbers of data points but the Phos-tag method can provide an estimation of half-lives of RR dephosphorylation. On the other hand, the indirect FP reporter is influenced by transcription and FP maturation rates, thus the measured promoter activity dynamics cannot be directly used for enzyme kinetics. Understanding the reporter dynamics requires a kinetic model that accounts for these non-phosphorylation reactions.

4.1 Kinetic Model of Reporter Dynamics

Phosphatase activity is believed to be the main effector for RR dephosphorylation upon stimulus withdrawal. Therefore, other activities, such as autodephosphorylation of RR~P, reverse phosphotransfer from RR~P to HK, and phosphorylation of RR by HK or nonspecific sources are not considered in order to minimize the number of parameters that must be estimated. Three reactions are modeled to derive the reporter dynamics (Fig. 4A). Dephosphorylation by phosphatase is modeled with a reversible binding equilibrium between RR~P and HK with the on rate k_1 and the dissociation constant K_d, followed by the subsequent dephosphorylation with the catalytic rate constant k_p. Reaction rates are as follows:

$$-d[\mathrm{RR{\sim}P}]/dt = k_1[\mathrm{RR{\sim}P}][\mathrm{HK}] - k_1K_d[\mathrm{HK\cdot RR{\sim}P}] \tag{1}$$

$$-d[\mathrm{HK\cdot RR{\sim}P}]/dt = k_p[\mathrm{HK\cdot RR{\sim}P}] + k_1K_d[\mathrm{HK\cdot RR{\sim}P}] \tag{2}$$

The rate change due to growth dilution is not considered for the PhoR–PhoB system because the growth rate is low with a doubling time near 100 min, much greater than the time scale of dephosphorylation and FP maturation. For TCSs with slow dephosphorylation kinetics or organisms with high growth rates, growth dilution may not be neglected. A Hill function with the DNA-binding affinity K_{DNA} and the maximal expression rate V_m is used to describe RR-activated expression of the nascent FP that has yet to mature to a fluorescent form. Maturation of FP to the fluorescent form FP* follows first-order kinetics with the rate constant k_{mat}, while the scaling factor α denotes the relation between the measured fluorescence values and FP* concentrations.

$$d[\mathrm{FP}]/dt = V_m[\mathrm{RR{\sim}P}]^n/\left([\mathrm{RR{\sim}P}]^n + K_{DNA}^n\right) - k_{mat}[\mathrm{FP}] \tag{3}$$

$$d[\mathrm{FP}^*]/dt = k_{mat}[\mathrm{FP}] \tag{4}$$

$$d[\mathrm{fluo.}]/dt = \alpha * d[\mathrm{FP}^*]/dt \tag{5}$$

The ultimate goal of modeling reporter dynamics is to quantitate the phosphatase constant k_p and K_d. Other kinetic parameters and initial concentrations of each species need to be properly estimated to fit the observed fluorescence dynamics. Total concentrations of HK and RR can be measured with quantitative western blots using purified proteins as standards (Gao & Stock, 2013). The initial concentration of RR~P before stimulus

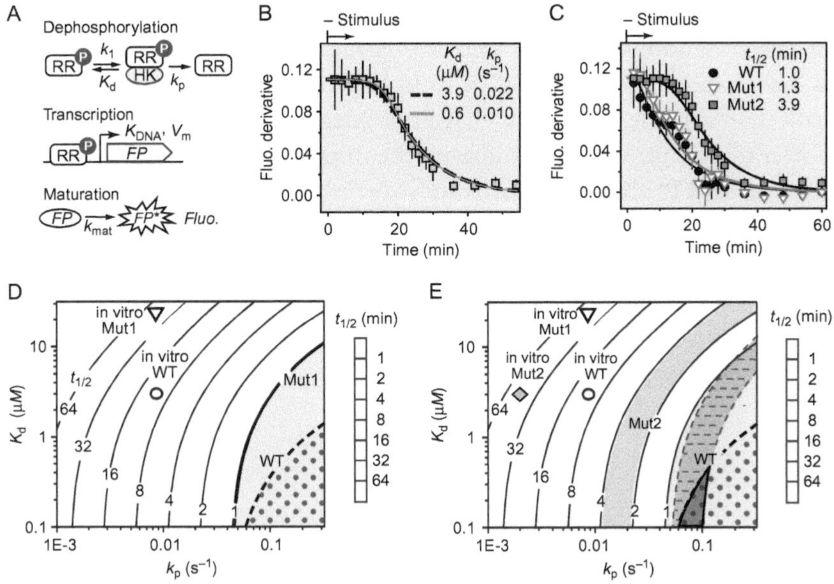

Fig. 4 Modeling the RR dephosphorylation kinetics. (A) Schematic diagram of reactions used for modeling. (B) Simulation of reporter dynamics with different values of K_d and k_p. Derivatives of total CFP fluorescence (*open squares*) are calculated from deactivation dynamics of the *phoA-cfp* reporter in the phosphatase-diminished mutant Mut2 ($PhoR^{T217M}$). Error bars indicate standard deviation from 11 individual wells. *Dashed and solid lines* represent the simulated curves with indicated values of K_d and k_p. Values for the rest of the parameters and initial species concentrations are as following: $[RR{\sim}P]_0$, 6.3 μ*M*; [HK], 0.8 μ*M*; k_{mat}, 0.00144 s^{-1}; K_{DNA}, 0.6 μ*M*; k_1, 0.5 $\mu M^{-1} s^{-1}$; V_m, 0.0019 μ*M*/s; α, 1; $[FP]_0$, 1.28 μ*M*. The first four parameters are experimentally determined as described in Section 4.1, while k_1 is assigned a high value for a fast binding equilibrium. The value of α does not affect the fluorescence derivative because both the initial value $[FP]_0$ and the changing rate can be converted to values relative to $1/\alpha$ and it cancels out the scaling factor α when *d*[Fluo.]/*dt* is computed with Eq. 5. The value α is arbitrarily set at 1 and the corresponding V_m and $[FP]_0$ are calculated as described in Section 4.1. (C) Simulation of reporter dynamics with first-order kinetics of dephosphorylation. Exponential decay, instead of the typical Michaelis–Menten kinetics, is used to approximate the RR dephosphorylation kinetics. Dynamics of fluorescence derivatives (*symbols*) were fitted (*lines*) to determine the dephosphorylation half-lives. (D and E) Contour plots of dephosphorylation half-lives derived from different values of K_d and k_p. *Symbols* indicate parameter values derived from in vitro characterization of truncated HK proteins. *Shaded areas* represent parameter spaces for individual mutants based on their dephosphorylation half-lives in cells. *Dotted and dashed areas* are derived from the relative difference of K_d and k_p between the WT and mutants. Cross-section of the two (*dotted areas* in *dark gray*) highlights the boundaries for WT phosphatase activity.

withdrawal can be calculated from the total RR level and the RR~P fraction determined by Phos-tag analyses. A high value close to the diffusion rate limit can be initially assigned to k_1 because the binding equilibrium is usually fast and the fast rate k_1 has little effect on dephosphorylation. For reporter transcription and maturation, the initial concentration of the immature form of FP protein is not readily available, however, a relative value can be derived based on the steady-state value prior to TCS deactivation. As described in Section 3.2, stimulus is preferably removed during a steady phase of fluorescence derivative, suggesting a steady value for immature FP and a value of zero for $d[\mathrm{FP}]/dt$. Based on Eqs. (3)–(5), the following equations can be derived to assign the initial values:

$$[\mathrm{FP}]_0 = \frac{d[\mathrm{fluo.}]/dt|_0}{\alpha \cdot k_{\mathrm{mat}}}$$

$$V_{\mathrm{m}} = \frac{d[\mathrm{fluo.}]/dt|_0}{\alpha} \cdot \frac{[\mathrm{RR{\sim}P}]_0^n + K_{\mathrm{DNA}}^n}{[\mathrm{RR{\sim}P}]_0^n}$$

in which $d[\mathrm{fluo.}]/dt|_0$ indicates the initial value of the fluorescence derivative. The maturation constant k_{mat} can be measured as described in Section 3.4, while the DNA-binding affinity K_{DNA} and the Hill coefficient n can be approximated by values determined in vitro.

With all the initial conditions and corresponding parameters defined, the model can be simulated with the Simbiology tool of MATLAB® and the phosphatase parameters can be estimated to fit the observed fluorescence dynamics. However, different pairs of K_{d} and k_{p} values can yield very similar time courses of fluorescence derivatives (Fig. 4B). This is likely due to the much higher value of RR~P concentration than the DNA binding constant K_{DNA} for the modeled PhoB protein, which saturates the binding site, maximizes the transcription rate, and masks any early-stage kinetic difference in RR~P concentration caused by different K_{d} and k_{p} values. Accurate estimation of parameters may not be possible with a single set of kinetic data. Global fitting with multiple data sets at different TCS expression levels or RR~P levels is likely to allow more robust fitting and better estimation of the phosphatase activity.

Alternatively, the dephosphorylation kinetics can be approximated by an exponential decay with a single parameter, the half-life $t_{1/2}$. The fluorescence reporter dynamics can be fit to determine the value of $t_{1/2}$. As shown in Fig. 4C, PhoB~P dephosphorylation half-lives derived from FP reporter assays compare well with the half-lives measured by Phos-tag analyses. When combined with in vitro characterization of the phosphatase activity, half-lives of a few phosphatase mutants can be used to define the boundaries for the kinetic parameters of the phosphatase activity in cell.

4.2 Quantitative Estimation of the Phosphatase Activity

In order to use half-lives to estimate the phosphatase activity, half-lives of RR dephosphorylation need to be calculated for different values of K_d and k_p. The same phosphatase model and same initial concentrations as those discussed in Section 4.1 can be used to derive the half-lives of RR dephosphorylation (Fig. 4D). For HKs with strong phosphatase activity, it is typical that only an upper limit of half-life can be experimentally estimated because Phos-tag analyses cannot capture the fast dephosphorylation kinetics and FP reporter dynamics are dominated by FP maturation kinetics. For the PhoR–PhoB system, both the WT and a phosphatase-diminished mutant Mut1, which exhibits lower phosphatase activity that WT, showed fast dephosphorylation kinetics with an estimated upper limit of half-life at ~1 min. A large parameter space (shaded area in Fig. 4D) can give half-lives shorter than 1 min.

The relative difference between kinetic parameters of a WT protein and a phosphatase mutant can be used to establish boundaries for the WT parameter space. Mut1 has been shown to have a K_d value eightfold higher than that of WT by in vitro kinetic analyses (Gao & Stock, 2013). Even though half-lives measured in vitro differ significantly from those determined in cells, the same eightfold difference is assumed for intracellular K_d values between Mut1 and WT. Mut1 has an upper limit of half-life of ~1 min. Lowering the K_d by eightfold establishes a limit for the WT phosphatase parameters (dotted area in Fig. 4D and E). Characterization of a second mutant can further refine the quantitative estimation. Mut2 has a k_p value 4.5-fold lower than that of WT. The intracellular half-life between 2 and 4 min sets the boundaries for WT parameters given a 4.5-fold difference in k_p (Fig. 4E, dashed area shaded in gray). The cross-section of the two areas defined by the two mutants (Fig. 4E, dotted area in dark gray) represents the possible parameter values for WT, suggesting a strong phosphatase activity with a much larger k_p and a smaller K_d than those measured in vitro. Caution should be exercised in such analyses because a single mutation may have complex effects on multiple aspects of TCS activities. The assumption that in vitro differences in parameter values can be readily translated into intracellular parameter differences may not be true for every protein.

5. SUMMARY

Quantitation of HK activities with in vitro methods is often limited by the reconstituted system or the biochemical environment, neither of which mirror the native system under cellular conditions. Analyzing TCS deactivation kinetics in cells using a combined approach provides a quantitative assessment of intracellular phosphatase activity of HKs. For the exemplified PhoR–PhoB system, this approach reveals a strong intracellular phosphatase activity that is necessary to suppress nonspecific phosphorylation in the absence of stimulus (Gao & Stock, 2017). Underestimation of the phosphatase activity by in vitro analyses may be attributed to the lack of regulation from the sensing and transmembrane domains that are truncated. Due to the wide usage of truncated proteins for in vitro characterization of HKs, such underestimation may be common in TCS studies. We hope the approach outlined in this chapter can be extended to other systems for evaluation of the phosphatase activity in the cellular environment. Because TCSs exhibit great diversity in regulatory details, such as the signaling time scale, cellular expression levels, and cell growth rates, various assumptions and considerations discussed earlier need to be thoroughly examined to customize for an individual system.

REFERENCES

Andersen, J. B., Sternberg, C., Poulsen, L. K., Bjorn, S. P., Givskov, M., & Molin, S. (1998). New unstable variants of green fluorescent protein for studies of transient gene expression in bacteria. *Applied and Environmental Microbiology*, *64*(6), 2240–2246.

Balleza, E., Kim, J. M., & Cluzel, P. (2018). Systematic characterization of maturation time of fluorescent proteins in living cells. *Nature Methods*, *15*(1), 47–51.

Barbieri, C. M., & Stock, A. M. (2008). Universally applicable methods for monitoring response regulator aspartate phosphorylation both *in vitro* and *in vivo* using Phos-tag-based reagents. *Analytical Biochemistry*, *376*(1), 73–82.

Boulanger, A., Chen, Q., Hinton, D. M., & Stibitz, S. (2013). In vivo phosphorylation dynamics of the *Bordetella pertussis* virulence-controlling response regulator BvgA. *Molecular Microbiology*, *88*(1), 156–172.

de Jong, H., Ranquet, C., Ropers, D., Pinel, C., & Geiselmann, J. (2010). Experimental and computational validation of models of fluorescent and luminescent reporter genes in bacteria. *BMC Systems Biology*, *4*, 55.

Gao, R., & Stock, A. M. (2009). Biological insights from structures of two-component proteins. *Annual Review of Microbiology*, *63*, 133–154.

Gao, R., & Stock, A. M. (2013). Probing kinase and phosphatase activities of two-component systems in vivo with concentration-dependent phosphorylation profiling. *Proceedings of the National Academy of Sciences of the United States of America*, *110*(2), 672–677.

Gao, R., & Stock, A. M. (2015). Temporal hierarchy of gene expression mediated by transcription factor binding affinity and activation dynamics. *MBio*, *6*(3), e00686-00615.

Gao, R., & Stock, A. M. (2017). Quantitative kinetic analyses of shutting off a two-component system. *MBio*, *8*(3), e00412-00417.

Hebisch, E., Knebel, J., Landsberg, J., Frey, E., & Leisner, M. (2013). High variation of fluorescence protein maturation times in closely related *Escherichia coli* strains. *PLoS One*, *8*(10), e75991.

Huynh, T. N., & Stewart, V. (2011). Negative control in two-component signal transduction by transmitter phosphatase activity. *Molecular Microbiology*, *82*(2), 275–286.

Kinoshita, E., Kinoshita-Kikuta, E., & Koike, T. (2009). Separation and detection of large phosphoproteins using Phos-tag SDS-PAGE. *Nature Protocols*, *4*(10), 1513–1521.

Kinoshita, E., Kinoshita-Kikuta, E., Takiyama, K., & Koike, T. (2006). Phosphate-binding tag, a new tool to visualize phosphorylated proteins. *Molecular and Cellular Proteomics*, *5*(4), 749–757.

Laub, M. T., & Goulian, M. (2007). Specificity in two-component signal transduction pathways. *Annual Review of Genetics*, *41*(1), 121–145.

Park, S. Y., & Groisman, E. A. (2014). Signal-specific temporal response by the Salmonella PhoP/PhoQ regulatory system. *Molecular Microbiology*, *91*(1), 135–144.

Richards, C. L., Lawrence, K. A., Su, H., Yang, Y., Yang, X. F., Dulebohn, D. P., et al. (2015). Acetyl-phosphate is not a global regulatory bridge between virulence and central metabolism in *Borrelia burgdorferi*. *PLoS One*, *10*(12), e0144472.

Siryaporn, A., & Goulian, M. (2008). Cross-talk suppression between the CpxA-CpxR and EnvZ-OmpR two-component systems in *E. coli*. *Molecular Microbiology*, *70*(2), 494–506.

Wayne, K. J., Li, S., Kazmierczak, K. M., Tsui, H. C., & Winkler, M. E. (2012). Involvement of WalK (VicK) phosphatase activity in setting WalR (VicR) response regulator phosphorylation level and limiting cross-talk in *Streptococcus pneumoniae* D39 cells. *Molecular Microbiology*, *86*(3), 645–660.

Zaslaver, A., Bren, A., Ronen, M., Itzkovitz, S., Kikoin, I., Shavit, S., et al. (2006). A comprehensive library of fluorescent transcriptional reporters for *Escherichia coli*. *Nature Methods*, *3*(8), 623–628.

Zwir, I., Yeo, W. S., Shin, D., Latifi, T., Huang, H., & Groisman, E. A. (2014). Bacterial nucleoid-associated protein uncouples transcription levels from transcription timing. *MBio*, *5*(5), e01485-01414.

CHAPTER ELEVEN

Measuring the Activities of Two-Component Regulatory System Phosphatases

Robert B. Bourret[1], Ruth E. Silversmith
Department of Microbiology & Immunology, University of North Carolina, Chapel Hill, NC, United States
[1]Corresponding author: e-mail address: bourret@med.unc.edu

Contents

Abstract

Two-component regulatory systems (TCSs) are used for signal transduction by organisms from all three phylogenetic domains of the living world. TCSs use transient protein phosphorylation and dephosphorylation reactions to convert stimuli into appropriate responses to changing environmental conditions. Phosphoryl groups flow from ATP to sensor kinases (which detect stimuli) to response regulators (which implement responses) to inorganic phosphate (P_i). The phosphorylation state of response regulators controls their output activity. The rate at which phosphoryl groups are removed

Methods in Enzymology, Volume 607
ISSN 0076-6879
https://doi.org/10.1016/bs.mie.2018.05.006

from response regulators correlates with the timescale of the corresponding biological function. Dephosphorylation reactions are fastest in chemotaxis TCS and slower in other TCS. Response regulators catalyze their own dephosphorylation, but at least five types of phosphatases are known to enhance dephosphorylation of response regulators. In each case, the phosphatases are believed to stimulate the intrinsic autodephosphorylation reaction. We discuss in depth the properties of TCS (particularly the differences between chemotaxis and nonchemotaxis TCS) relevant to designing in vitro assays for TCS phosphatases. We describe detailed assay methods for chemotaxis TCS phosphatases using loss of ^{32}P, change in intrinsic fluorescence as a result of dephosphorylation, or release of P_i. The phosphatase activities of nonchemotaxis TCS phosphatases are less well characterized. We consider how the properties of nonchemotaxis TCS affect assay design and suggest suitable modifications for phosphatases from nonchemotaxis TCS, with an emphasis on the P_i release method.

ABBREVIATIONS

ATP	adenosine triphosphate
CA	catalytic and ATP binding
D+*n*	the position *n* residues C-terminal to the conserved Asp phosphorylation site in response regulators
DHp	dimerization and His phosphorylation
H+*n*	the position *n* residues C-terminal to the conserved His phosphorylation site in sensor kinases
Hpt	histidine phosphotransfer
MESG	2-amino-6-mercapto-7-methylpurine ribonucleoside
MPI	monophosphoimidazole
P_i	inorganic phosphate
SDS-PAGE	SDS polyacrylamide gel electrophoresis
T+*n*	the position *n* residues C-terminal to the conserved Thr/Ser in response regulators
TCS	two-component system

1. OVERVIEW

1.1 Overview of Two-Component Systems

Microorganisms and plants use two-component regulatory systems (TCSs) to accomplish signal transduction via a series of protein phosphorylation and dephosphorylation reactions (reviewed in Zschiedrich, Keidel, & Szurmant, 2016). The simplest TCS consists of a sensor kinase and a response regulator (Fig. 1A). The sensor kinase detects relevant environmental stimuli and encodes that information by autophosphorylation of a conserved His residue using ATP as the phosphodonor. The response regulator catalyzes transfer of

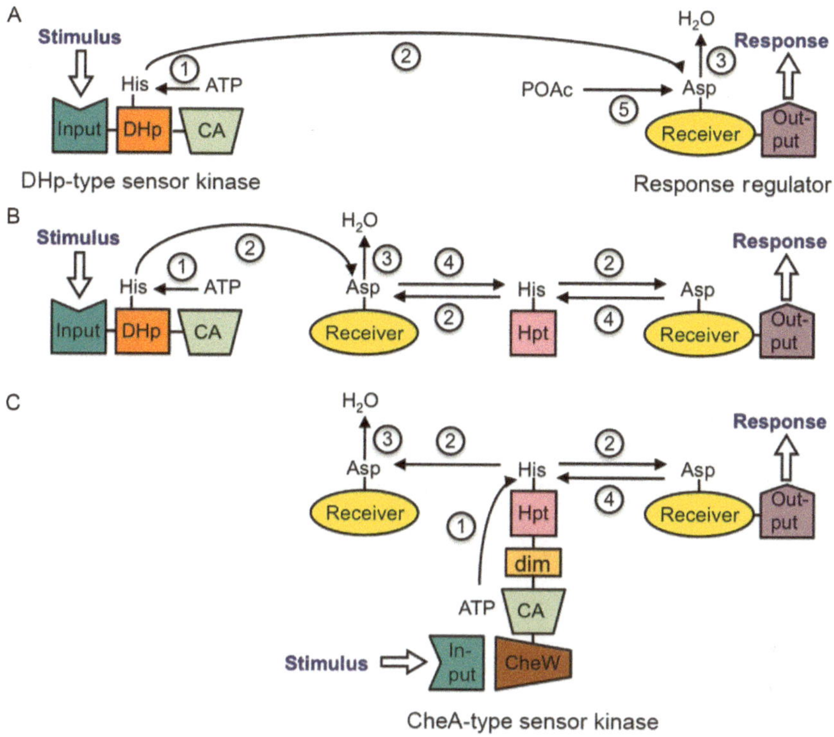

Fig. 1 Domain organization of signaling proteins and phosphoryl group flow in (A) a basic two-component system, (B) a TCS with a multistep phosphorelay, and (C) a chemotaxis TCS with a phosphate sink. Reaction 1 is sensor kinase autophosphorylation with ATP as the phosphodonor. Reaction 2 is His-P to Asp phosphotransfer. Reaction 3 is receiver domain autodephosphorylation with water as the phosphoacceptor. Reaction 4 is Asp-P to His phosphotransfer. Reaction 5 is receiver domain autophosphorylation with a small molecule phosphodonor such as acetyl phosphate (POAc). Actions of TCS phosphatases to stimulate Reaction 3 are not shown for clarity. Conserved domains depicted in the figure and their corresponding Pfam designations are: DHp (HisKA, HisKA_2, HisKA_3, His_Kinase, or HWE_HK), CA (HATPase_c), receiver (Response_reg), Hpt (Hpt), dim (H-kinase_dim), and CheW (CheW).

the phosphoryl group from the sensor kinase to a conserved Asp residue on the response regulator. Phosphorylation controls the output activity of the response regulator. Removal of the phosphoryl group from the response regulator via either autodephosphorylation or the participation of another protein terminates the signal.

Known responses controlled by TCS span a million-fold range of time, from less than a second for directional changes during chemotaxis (Segall, Manson, & Berg, 1982) to days for multicellular development

(O'Connor & Zusman, 1991). The kinetics of the phosphorylation and dephosphorylation reactions are relevant to the timescale of the biological process controlled by a particular TCS. For example, reactions must be sufficiently fast to ensure a response is appropriately synchronized with a fluctuating stimulus, or sufficiently slow to ensure that a response persists once a commitment is made. Therefore, assays of phosphorylation and dephosphorylation reaction kinetics have the potential to provide insight into biological function.

An understanding of key TCS properties is necessary to design appropriate kinetic assays. TCSs use a modular design, which provides substantial functional flexibility. A small number of conserved domains supply the catalytic machinery for the phosphoryl group transactions that are characteristic of TCS, whereas variable domains (Ulrich, Koonin, & Zhulin, 2005) splice numerous stimulus and response capabilities onto the conserved core of phosphotransfer functions. A basic sensor kinase contains one or more variable input domains to detect stimuli, a conserved DHp (dimerization and His phosphorylation) domain containing the phosphorylation site, and a conserved CA (catalytic and ATP binding) domain that catalyzes autophosphorylation in conjunction with the DHp domain (Fig. 1A). A typical response regulator contains a conserved receiver domain that catalyzes phosphorylation and dephosphorylation reactions, and a variable output domain to implement responses.

Although the His to Asp phosphotransfer reaction can run in reverse in vitro (Dutta & Inouye, 1996; Shi, Liu, & Hulett, 1999), physiologically relevant phosphotransfer between DHp and receiver domains is generally believed to be unidirectional. Some TCSs introduce additional complexity by adding a conserved histidine phosphotransfer (Hpt) domain after the DHp and receiver domains (Fig. 1B). Hpt domains participate in bidirectional Asp-P to His and His-P to Asp phosphotransfer reactions with multiple receiver domains to form linear or branched multistep phosphorelays.

Hpt domains can also provide the first phosphorylation site in some TCS. In TCS that control chemotaxis, the structure of the sensor kinase (termed CheA) is altered (Dutta, Qin, & Inouye, 1999) (Fig. 1C). The DHp domain is replaced with two domains—an Hpt domain to provide a phosphorylation site and a separate dimerization domain—but the CA domain is retained. In contrast to DHp-type sensor kinases, CheA-type sensor kinases lack input domains and instead receive information via a CheW scaffold domain that binds to separate chemoreceptor proteins. Some chemotaxis TCSs take advantage of the bidirectional phosphotransfer supported by the Hpt domain to turn off the signal without a phosphatase (Sourjik & Schmitt, 1998). Phosphoryl groups drain off one response regulator through CheA to another response regulator that functions as a phosphate sink.

Receiver domains provide the basic catalytic machinery for four of the five types of phosphoryl group reactions used by TCS (Fig. 1), including His-P to Asp phosphotransfer (Silversmith, Appleby, & Bourret, 1997), response regulator autodephosphorylation with water as the phosphoacceptor (Hess, Bourret, Oosawa, Matsumura, & Simon, 1988), Asp-P to His phosphotransfer (Page, Silversmith, Collins, & Bourret, 2015), and response regulator autophosphorylation with small molecule phosphodonors such as acetyl phosphate (Lukat, McCleary, Stock, & Stock, 1992), which can provide input about metabolic state (Wolfe, 2010). The receiver domain active site consists of five absolutely conserved residues: a pair of acidic residues that coordinate a divalent cation, the Asp site of phosphorylation, a Thr/Ser, and a Lys (reviewed in Bourret, 2010). The cation, Thr/Ser, and Lys each form an ionic or hydrogen bond with a different phosphoryl group oxygen atom in the phosphorylated protein. The same three conserved active-site elements are believed to form similar bonds with the three oxygens of a planar phosphoryl group in the transition states of the response regulator phosphorylation and dephosphorylation reactions.

1.2 Overview of TCS Phosphatases

The reported rate constants for autodephosphorylation reactions of wild-type response regulators span five to six orders of magnitude (Jagadeesan, Mann, Schink, & Higgs, 2009; Porter & Armitage, 2002). Nevertheless, in many cases autodephosphorylation is too slow to support the corresponding biology. For example, the half-life of the phosphoryl group on the *Escherichia coli* response regulator CheY is only 20 s (Mayover, Halkides, & Stewart, 1999), but that is too stable by two orders of magnitude to account for the 0.1 s response time of chemotaxis (Segall et al., 1982). Therefore, many TCSs also contain phosphatases (not shown in Fig. 1) that enhance removal of phosphoryl groups from receiver domains to close the gap between autodephosphorylation and physiological relevance.

There are five known families of phosphatases active against phosphorylated response regulator substrates: CheZ, CheC/CheX/FliY, Rap, Spo0E, and DHp-type sensor kinases (Huynh & Stewart, 2011; Silversmith, 2010). CheZ and CheC/CheX/FliY phosphatases regulate chemotaxis, Rap and Spo0E phosphatases regulate sporulation, and DHp-type sensor kinases regulate a wide variety of processes. The reaction mechanism is well established for CheZ (Zhao, Collins, Bourret, & Silversmith, 2002), CheX (Pazy et al., 2010), and Rap (Parashar, Mirouze, Dubnau, & Neiditch, 2011). In each case, an amide residue protruding from a helix of the phosphatase inserts into the response regulator active site, which contains the phosphoaspartate and the conserved residues that catalyze Asp phosphorylation and dephosphorylation. The phosphatase amide helps properly position a water molecule, which must attack in line with the P-O bond that connects the phosphoryl group to the Asp (Lassila, Zalatan, & Herschlag, 2011). Hydrolysis of the phosphoaspartate releases inorganic phosphate (P_i) as a product. Although not yet definitively determined, DHp-type sensor kinases likely use an analogous mechanism (Huynh, Noriega, & Stewart, 2010) but utilize a broader range of polar amino acids to position the water nucleophile. The catalytically critical residue of Spo0E family phosphatases is an acid rather than an amide (Diaz et al., 2008). Currently available data [including structures of CheC (Park et al., 2004), FliY (Sircar, Greenswag, Bilwes, Gonzalez-Bonet, & Crane, 2013), Spo0E (Grenha et al., 2006), and a DHp-type sensor kinase (Trajtenberg et al., 2016)] are consistent with the hypothesis that all response regulator phosphatases enhance the intrinsic autodephosphorylation activity of receiver domains by positioning the attacking water molecule with a polar residue extended from a helix.

The reaction mechanism for TCS phosphatases has implications for assay design. TCS phosphatases are specific for their partner response regulator substrate. Chemotaxis (Hess, Oosawa, Kaplan, & Simon, 1988) and sporulation (Perego, 2001) phosphatases discriminate between multiple response regulators within the same pathway. Bacteria often possess dozens of different TCS pathways, but the kinase and phosphatase activities of DHp-type sensor kinases are specific for response regulators within the same pathway as the sensor kinase (Skerker, Prasol, Perchuk, Biondi, & Laub, 2005; Willett et al., 2013). This specificity, as well as the absence of a full phosphatase active site in TCS phosphatases, precludes the use of a universal substrate such as *para*-nitrophenyl phosphate, which yields a color change when cleaved by many other types of phosphatases. Instead a different phosphorylated response regulator substrate must be prepared to assay each TCS phosphatase. Furthermore, the reaction mechanism for TCS phosphatases (stimulation of response regulator autodephosphorylation) eliminates the possibility of monitoring a covalent phosphatase-phosphoryl group intermediate during an assay.

The chemical requirements to support response regulator autodephosphorylation necessarily affect assay design for TCS phosphatases. Most importantly, response regulator autodephosphorylation (Lukat, Stock, & Stock, 1990) is absolutely dependent upon the presence of a divalent cation. The cation is usually Mg^{2+}, but Mn^{2+} can often substitute. Therefore, TCS phosphatase activity is also dependent upon the presence of a divalent cation (Lukat et al., 1990; Perego, 2001; Zhu, Qin, Yoshida, & Inouye, 2000). To the best of our knowledge, the metal-binding affinity of a phosphorylated response regulator has not been reported, but is presumably higher than that of a non-phosphorylated response regulator due to direct coordination of the metal ion by a phosphoryl group oxygen atom instead of a water molecule (Bourret, 2010). Care should be taken that the metal ion concentration is not limiting for phosphatase activity.

Metal ions are also essential for response regulator phosphorylation, whether self-catalyzed or via phosphotransfer from other proteins (Creager-Allen, Silversmith, & Bourret, 2013; Lukat et al., 1992). Reported binding affinities for divalent cations vary by one to two orders of magnitude between different unphosphorylated response regulators and different cations (Creager-Allen et al., 2013; Lukat et al., 1990; Sheftic et al., 2012). The diversity of response regulator metal-binding affinities has at least two potential implications for TCS phosphatase assays. First, failure to use saturating concentrations of metal ion could potentially affect yield during preparation of phosphorylated response regulator substrates. Second, in assays where the phosphatase reaction is coupled with a response regulator phosphorylation reaction, subsaturating concentrations of metal ion would decrease the apparent rate of phosphorylation and could affect which reaction is rate limiting.

The rate of autodephosphorylation of the CheY response regulator is largely independent of pH (Silversmith et al., 1997) and ionic strength (Mayover et al., 1999). To the best of our knowledge, the impacts of pH or ionic strength on TCS phosphatase activity have not been evaluated.

An extensive description of assays of response regulator autodephosphorylation kinetics, which may inspire additional assays of TCS phosphatases, can be found in a previous volume of *Methods in Enzymology* (Bourret et al., 2010).

1.3 Chapter Overview

For historical and technical reasons, quantitative assays of chemotaxis phosphatases are more developed than assays of other TCS phosphatases. We first consider the properties of chemotaxis TCS that affect phosphatase assay design and then describe various methods to assay chemotaxis phosphatases in detail. We then turn our attention to the characteristics of non-chemotaxis TCS to motivate suitable assay modifications for corresponding phosphatases.

2. RELEVANT PROPERTIES OF CHEMOTAXIS TCS

Assays of chemotaxis phosphatases are greatly facilitated by two properties. First, chemotaxis phosphatases are cytoplasmic proteins and typically soluble, which aids purification in an active form. Second, chemotaxis phosphatases have a single biochemical activity with regard to phosphoryl groups, namely, stimulation of response regulator dephosphorylation. Conversely, CheA kinases generally do not have phosphatase activity. In contrast, standard TCS proteins (sensor kinases, response regulators, Hpt proteins) can participate in multiple types of reactions involving phosphoryl groups (Fig. 1), which can complicate design of an assay to isolate only the phosphatase reaction.

CheZ and CheX are homodimers, which raises the possibility of cooperative interaction between active sites. Indeed, CheZ phosphatase activity is positively cooperative with respect to CheY-P concentration (Blat, Gillespie, Bren, Dahlquist, & Eisenbach, 1998; Silversmith, Levin, Schilling, & Bourret, 2008). Other chemotaxis phosphatases also have the potential for complex kinetic features. CheC is a monomer but contains two active sites, and CheC phosphatase activity is enhanced by binding to CheD (Park et al., 2004). FliY is a dimer, with two active sites of unequal activity in each monomer (Sircar et al., 2013).

The primary challenge when designing assays of chemotaxis phosphatases is that chemotaxis is a fast process, and autodephosphorylation of the CheY response regulator is correspondingly rapid. The half-lives of the phosphoryl groups on CheY proteins from many different species are typically on the order of tens of seconds (Thomas, Brewster, & Bourret, 2008). Rapid decomposition makes it difficult to purify the CheY-P substrate, or to measure enhanced substrate degradation in the presence of a phosphatase. One solution is to stop autodephosphorylation by chelation of the essential Mg^{2+} ion with EDTA, which permits purification of CheY-P (Hess, Bourret, et al., 1988). Reactions to dephosphorylate CheY-P can be subsequently initiated by addition of excess Mg^{2+}. However, the technique has several disadvantages, including poor yields of CheY-P as originally described, a need for careful buffering of pH when EDTA is released from chelation complexes, and high ionic strength conditions (see *Methods in Enzymology* article by Hess, Bourret, & Simon, 1991). Another strategy is to continuously generate various concentrations of CheY-P during the assay. This is possible because the rate of CheY autophosphorylation is linearly dependent on increasing phosphodonor concentration (Da Re, Deville-Bonne, Tolstykh, Veron, & Stock, 1999).

The similarities between chemotaxis phosphatases imply that it should be feasible to adapt methods used to assay one chemotaxis phosphatase for use with others. Indeed, some assays originally developed for CheZ have been used with CheC, CheX, or FliY (see *Methods in Enzymology* article by Muff & Ordal, 2007).

3. IN VITRO ASSAYS OF CHEMOTAXIS TCS PHOSPHATASES

A given TCS includes multiple possible phosphorylation and dephosphorylation reactions (Fig. 1). Therefore, general principles of assay design include minimizing the number of different reactions that occur during an assay and ensuring that the reaction to be quantified is rate limiting when a system of reactions is present.

In vitro assays require purified proteins. Classical chromatographic techniques for purification of *E. coli* CheA, CheY, and CheZ have been described in detail (Hess et al., 1991). Alternatively, affinity purification using His_6-tags often generates pure, active proteins suitable for phosphatase assays (e.g., *Borrelia burgdorferi* CheA2, CheY3, and CheX; Motaleb et al., 2005). However, the imidazole used to elute proteins from nickel affinity columns must be removed prior to use in assays. Imidazole can directly react with phosphoryl groups on response regulators, resulting in dephosphorylation (Page et al., 2015).

Three general methods to monitor phosphatase activity toward CheY-P have been used extensively. Dephosphorylation of the CheY-P substrate can be observed directly as a decrease in [^{32}P]CheY-P. Dephosphorylation also causes an increase in fluorescence of a Trp residue in the CheY active site. Finally, release of the P_i product can be followed colorimetrically. Each method has different strengths and weaknesses, as described below along with practical tips for implementing each method.

3.1 Assay by Loss of ^{32}P

There are multiple advantages to using radiolabeled phosphoryl groups as the basis for a phosphatase assay, including: (i) ATP serves as a phosphodonor for all TCS, so such assays should be universally applicable; (ii) [γ-^{32}P]ATP is commercially available; (iii) small amounts of radioactivity are easily detected, so assays are very sensitive; (iv) use of a phosphorimager gives the assays a range of several orders of magnitude; and (v) it can be informative to monitor the distribution of radiolabeled phosphoryl groups across different proteins in a signaling pathway. There are two primary disadvantages. First, radioactivity is hazardous. Second, in contrast to optical techniques, radioactivity cannot easily be used to quantify the amount of phosphorylated response regulator in real time. In particular, phosphatase-mediated dephosphorylation of chemotaxis response regulators is typically so fast that it is not possible to manually collect suitable reaction time points. Therefore, assays of chemotaxis TCS phosphatase activity based on radioactivity primarily provide qualitative rather than quantitative information, unless specialized rapid-quench equipment is available for accurate collection of samples at very short time intervals.

3.1.1 Equipment and Supplies

- Equipment for safely handling radioactive materials, including assorted Lucite shielding, spill paper, Geiger counter, and containers for liquid and solid radioactive waste
- Personal protective equipment for working with radioactive materials (lab coat, double gloves, safety glasses, dosimeter)
- Timer
- Pipetmen with aerosol-resistant tips to prevent contamination
- Gels, gel boxes, power supplies for SDS-PAGE, plus gel dryer
- Phosphorimager, storage phosphor screens, and software for quantification (e.g., GE Typhoon and ImageQuant TL)
- Disposable G-100 Sephadex column for purification of [^{32}P]CheA

3.1.2 Chemicals

- [γ-^{32}P]ATP (6000 Ci/mmol, 10 mCi/mL)
- Unlabeled ATP
- KCl, $MgCl_2$, Tris
- Ammonium sulfate
- 2 × SDS sample buffer, SDS-PAGE running buffer

3.1.3 General Procedure for Single Time Point Assay

1. A single time point assay reveals the steady-state distribution of phosphoryl groups in the presence of various combinations of proteins. Samples for a typical experiment would include CheA alone (to confirm sensor kinase autophosphorylation to form CheA-P), CheA + phosphatase (to confirm phosphatase does not act on CheA-P), CheA + CheY (to show generation of CheY-P by phosphotransfer from CheA-P), and CheA + CheY + phosphatase (to assess the effect of phosphatase on CheY-P). Examples of the single time point assay may be found in Appleby and Bourret (1998) and Boesch, Silversmith, and Bourret (2000).
2. Protein concentrations should be in the low μ*M* range. CheA acts catalytically, so using a 10- to 20-fold molar excess of CheY over CheA will maximize the CheY-P signal. The phosphatase should act catalytically as well, and so should be used at substoichiometric amounts compared to CheY. A range of phosphatase concentrations can be tested.
3. There are several factors to consider in choosing reaction conditions. The K_m of *Salmonella* CheA for ATP is ~0.3 m*M* (Wylie, Stock, Wong, & Stock, 1988). [γ-^{32}P]ATP is supplied at 1.7 μ*M* and should be supplemented with sufficient unlabeled ATP to drive the autophosphorylation reaction, but not so much as to unnecessarily diminish specific activity. We typically make a working stock of 1 m*M* [γ-^{32}P]ATP at 10 Ci/mmol by mixing nine volumes of 1.7 μ*M* [γ-^{32}P]ATP at 6000 Ci/mmol with one volume of 10 m*M* unlabelled ATP and then use a final concentration of ~0.4 m*M* [γ-^{32}P]ATP. The CheA His residue must be deprotonated in order to serve as a nucleophile in the autophosphorylation reaction, so reactions should be performed at slightly alkaline pH. We typically use Tris buffer at pH 7.5. CheA autophosphorylation is significantly stimulated by K^+. We typically use 35 m*M* KCl. The cofactor Mg^{2+} is required for all phosphorylation and dephosphorylation reactions. To ensure CheY is saturated with Mg^{2+}, we typically use 10 m*M* $MgCl_2$.
4. The reactions are started by addition of the ATP solution to tubes containing protein, buffer, KCl, and $MgCl_2$. The contents of the tube are mixed by pipetting up and down.
5. At a time sufficient for the system of reactions to reach steady state (typically several minutes with *E. coli* chemotaxis proteins), the reactions are halted by addition of an equal volume of 2 × SDS sample buffer and the contents of the tubes mixed. The SDS disrupts protein structure.

6. His-P and Asp-P bonds are much more labile than Ser-P, Thr-P, or Tyr-P and are sensitive to heat, acid, and/or base. Therefore samples are not boiled prior to SDS-PAGE.
7. The vast majority of the radioactivity will be in the form of ATP or P_i, which are present at approximately mM concentrations. To avoid overwhelming the radioactivity associated with phosphorylated proteins (present at μM concentrations), the dye front and associated small molecules must be physically separated from the remainder of the gel prior to phosphorimaging. The simplest method is to cut off the bottom of the gel and discard in solid radioactive waste. Alternatively, the dye front can be run off the end of the gel, with the result that most of the radioactivity will end up in liquid radioactive waste. Gels are not chemically fixed or stained, but simply dried and exposed to a storage phosphor screen (Fig. 2).

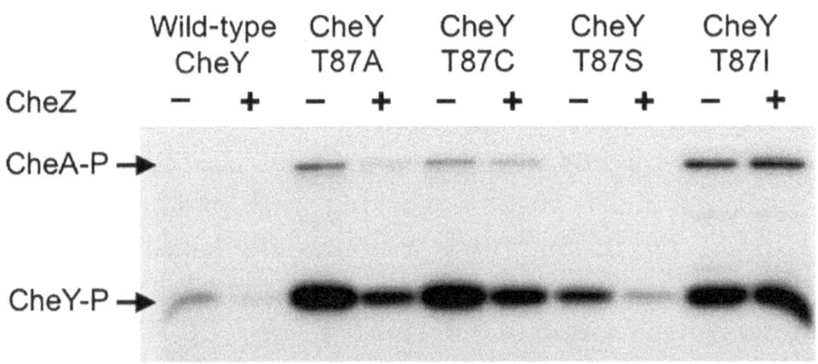

Fig. 2 Sample phosphorimage from a single time point assay. *E. coli* Che A kinase, CheY response regulator, and CheZ phosphatase were incubated in a 1:5:1 stoichiometric ratio with [γ-^{32}P]ATP for 10 min. The reaction was then halted and the products separated by SDS-PAGE. The experiment assessed the properties of variant CheY proteins carrying substitutions for the conserved Thr87. The increased amounts of variant CheY-P in the absence of CheZ compared to wild-type indicates reduced autodephosphorylation of CheY. The resulting reduced flow of phosphoryl groups through the system increases the amount of CheA-P, observable with three variants under these experimental conditions. Comparison of CheY-P in lanes with and without added CheZ reveals that the CheY variants exhibit different degrees of sensitivity to the CheZ phosphatase (e.g., CheY T87I is highly resistant). *Image modified from Appleby, J. L., & Bourret, R. B. (1998). Proposed signal transduction role for conserved CheY residue Thr87, a member of the response regulator active-site quintet.* Journal of Bacteriology, 180*(14), 3563–3569. Copyright © 1998, American Society for Microbiology.*

3.1.4 General Procedure for Time Course Assay

1. The single time point assay involves four simultaneous reactions: CheA autophosphorylation, phosphotransfer from CheA-P to CheY, auto-dephosphorylation of CheY-P, and phosphatase-stimulated dephosphorylation of CheY-P. In contrast, a time course starting with purified CheA-P and followed by synchronous phosphotransfer to CheY allows the kinetics of CheY dephosphorylation to be observed in isolation. Examples of the time course assay may be found in Immormino, Silversmith, and Bourret (2016) and Thomas et al. (2008).
2. [^{32}P]CheA-P is prepared by autophosphorylation with [γ-^{32}P]ATP. Ammonium sulfate precipitation separates CheA from the small molecules that contain the vast majority of the radioactivity in the reaction. Gel filtration then removes residual ATP and P_i. Detailed methods are described in Hess et al. (1991) and Silversmith et al. (1997). [^{32}P]CheA-P can be stored frozen until use.
3. [^{32}P]CheA-P is placed in a microfuge tube with reaction buffer (e.g., 100 m*M* Tris, pH 7.5, 10 m*M* $MgCl_2$) with sufficient volume to remove aliquots at planned time points. Aliquot size (typically 5–10 μL) is determined by the well volume of the gel to be used to separate CheA, CheY, and small molecules. Tubes containing the appropriate volume of 2 × SDS sample buffer are ready for use.
4. A "zero time point" sample is removed into 2 × SDS sample buffer to assess the amount of CheA-P present at the start of the reaction.
5. The reaction is initiated by the addition of a 10- to 20-fold molar excess of CheY, and the contents of the tube mixed by pipetting up and down. This will result in transfer of phosphoryl groups from CheA-P to CheY within 5 s. From this point forward, the only reaction is dephosphorylation of CheY.
6. Time points should be chosen to span four to five half-lives of decrease in CheY-P (the half-life of wild-type *E. coli* CheY-P is ~15 s). It is difficult to remove aliquots less than 10 s apart. It can be helpful to have multiple Pipetmen preloaded with tips, as well as a colleague to record the times at which samples are actually moved into 2 × SDS sample buffer.
7. To assess the effect of phosphatase activity, a second time course is conducted in which the phosphatase is added shortly after CheY. Samples from any time course experiment should be loaded on alternate lanes of the gel, so blank lanes clearly separate the radioactivity associated with each time point.

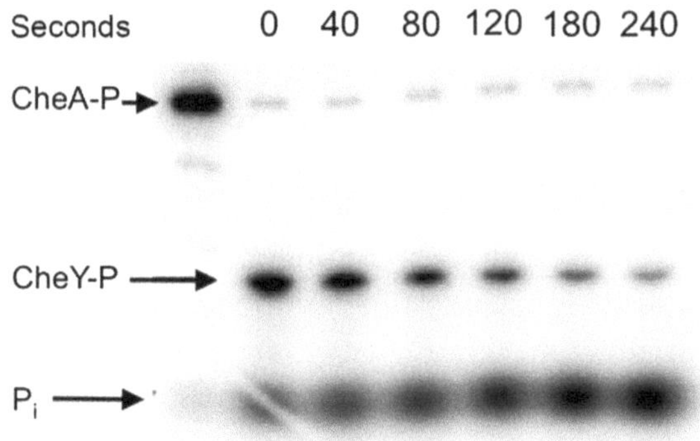

Fig. 3 Sample phosphorimage from a time course assay. *E. coli* [^{32}P]CheA-P was mixed with a molar excess of the CheY N59E E89H variant. Samples were removed at the indicated times, the reaction halted, and the products separated by SDS-PAGE. Almost all the phosphoryl groups transferred from CheA-P to CheY in the few seconds before a sample could be removed ($t = 0$). Phosphoryl groups were subsequently lost from CheY-P via autodephosphorylation with water to generate P_i. *Image courtesy of Stephanie A. Thomas.*

8. All of the radioactivity in the time course assay begins on CheA-P. It can be informative to observe the flow of phosphoryl groups from CheA to CheY to P_i over the time course, so the dye front is retained on the gel. The gel is dried and exposed to a phosphorimager (Fig. 3).

3.2 Assay by Fluorescence

E. coli CheY contains a single Trp residue, fortuitously located adjacent to the phosphorylated Asp. The intensity of Trp fluorescence is reduced when CheY is phosphorylated, allowing determination of the fraction of the CheY population that is phosphorylated at any given time. As an optical method, fluorescence allows continuous monitoring of reaction progress with high temporal resolution. Reactions can be observed under either steady-state or preequilibrium conditions. However, because the method depends on changing the amount of CheY-P, it cannot be used to measure phosphatase reaction rates at fixed substrate (CheY-P) concentrations (e.g., determine a Michaelis–Menton curve). Fluorescence also cannot be used with CheY proteins that lack a fluorophore near their active site.

3.2.1 Equipment and Supplies

- Spectrofluorimeter (e.g., Perkin Elmer LS-50B)
- Cuvette equipped with stir bar
- Pipetmen

3.2.2 Chemicals

- Acetyl phosphate or other small molecule phosphodonor
- Reaction buffer (100 m*M* Tris, pH 7.5, 10 m*M* $MgCl_2$)

3.2.3 General Procedure

1. This general method was first reported by Lukat et al. (1992) and is suitable for qualitative exploration of phosphatase activity.
2. For *E. coli* CheY, the fluorimeter should be set for an excitation wavelength of 295 nm and an emission wavelength of 346 nm. Appropriate wavelengths can be determined empirically for other proteins. The concentration of CheY used should be chosen to produce a strong signal while minimizing the amount of protein used. We typically use ~5 μ*M* CheY. Adjusting slit widths can maximize the strength of the fluorescence signal.
3. With continuous stirring in progress, the reaction is initiated by adding small molecule phosphodonor to CheY in reaction buffer. This creates a system of two reactions—CheY autophosphorylation and CheY autodephosphorylation. The Trp fluorescence will diminish as CheY-P is formed and will stabilize at a value reflecting the equilibrium between the two reactions. Subsequent additions of phosphodonor increase the rate of autophosphorylation without affecting the rate of autodephosphorylation, and thus shifts the equilibrium toward more CheY-P (decreased fluorescence).
4. Adding phosphatase creates a system of three reactions and increases dephosphorylation, which leads to a new steady state with less CheY-P, observable as an increase in fluorescence. An example of a fluorescence trace for such an experiment can be found in fig. 1 of Lukat et al. (1992).

3.3 Assay by P_i Release

P_i released by the phosphatase reaction can be rapidly utilized to generate a colored product using commercially available reagents (EnzChek Phosphate Assay Kit). The assay is based on the reaction of 2-amino-6-mercapto-7-methylpurine ribonucleoside (MESG) and P_i catalyzed by purine nucleoside phosphorylase to form 2-amino-6-mercapto-7-methylpurine and ribose-1-phosphate (Webb, 1992). The MESG substrate has an absorbance maximum at 330 nm, whereas the 2-amino-6-mercapto-7-methylpurine product has an absorbance maximum at 355 nm. Measuring the change in absorbance at 360 nm allows determination of P_i concentration.

Coupling the purine nucleoside phosphorylase reaction to a phosphatase reaction allows simple optical quantification of steady-state phosphatase reaction rates, as long as the phosphorylase reaction is not rate limiting. Interpretation of the data is slightly complicated by the fact that P_i is released from response regulators by autodephosphorylation in addition to phosphatase stimulation. Absorbance of the assay product is linear over a range of 2–150 μM P_i, which constrains the conditions under which the method can be used. To avoid shrinking the usable window of P_i concentration further, phosphodonors that are minimally contaminated with P_i must be used. Calcium monophosphoimidazole (MPI) is preferred, but not commercially available. Typical preparations of acetyl phosphate and phosphoramidate are not suitable (Silversmith, Smith, Guanga, Les, & Bourret, 2001).

3.3.1 Equipment and Supplies

- Spectrophotometer and/or microtiter plate reader
- Cuvettes and/or flat-bottom 96-well microtiter plates suitable for use at 360 nm
- Pipetmen
- Parafilm

3.3.2 Chemicals

- Calcium MPI. Synthesis is described in detail in Silversmith and Bourret (2018). Preparations of calcium MPI have a low concentration of contaminating P_i due to formation of insoluble $CaPO_4$. When making a stock solution of calcium MPI, vigorously vortex to maximally dissolve. Remove $CaPO_4$ prior to use by centrifugation for 10 min at maximum speed in a microfuge.
- Bovine serum albumin
- HEPES, Tris, $MgCl_2$, EDTA
- EnzChek Phosphate Assay Kit (ThermoFisher Scientific)

3.3.3 Optimization of Assay Conditions

1. Construct a standard curve by measuring absorbance at 360 nm of known concentrations of P_i under the experimental conditions to be used in the phosphatase assay. The absorbances of MESG and 2-amino-6-mercapto-7-methylpurine are both pH sensitive, so the extinction coefficient derived from the standard curve to convert A_{360} values to P_i concentration will also be pH sensitive. Reported extinction coefficients include 9100 M^{-1} cm^{-1} at pH 7.0 (Silversmith et al., 2001) and

11,000 M^{-1} cm^{-1} at pH 7.6 (Webb, 1992). The standard curve will also establish the linear range of the assay, which is a consideration in other aspects of assay design such as the appropriate concentration of CheY to use.

2. It is critical to ensure that the reactions releasing P_i are rate limiting in the system of coupled reactions used in this assay. The reactions involved are (i) CheY autophosphorylation with MPI to generate the CheY-P substrate for the phosphatase reaction, (ii) release of P_i from CheY-P by autodephosphorylation and by phosphatase-stimulated dephosphorylation, and (iii) conversion of MESG to 2-amino-6-mercapto-7-methylpurine in a reaction that consumes P_i. To establish combinations of concentrations of MPI, CheY, phosphatase, and EnzChek assay components under which the assay is valid, perform the following exploratory experiments (steps 3–5):
3. In the absence of phosphatase, measure the rate of P_i release from CheY as a function of increasing MPI concentration (Silversmith et al., 2001). The rate of P_i release will initially increase with [MPI], because autophosphorylation is rate limiting. The rate of P_i release will then plateau when autodephosphorylation becomes rate limiting. Subsequent addition of phosphatase should increase the P_i release rate, confirming that the plateau in the absence of phosphatase reflects limitation by autodephosphorylation.
4. Using a concentration of MPI from step 3 that results in limitation by autodephosphorylation, increase the concentration of phosphatase. This will increase the rate of P_i release until autophosphorylation to form the CheY-P substrate becomes rate limiting and the rate of P_i release again reaches a plateau.
5. Confirm that adding more purine nucleoside phosphorylase does not affect the rate of P_i release, i.e., the reaction to generate a colored product is not rate limiting.
6. Different CheY proteins and chemotaxis phosphatases, whether variants or from different species, may have different kinetic parameters. Therefore, the validation procedure in steps 3–5 must be repeated for each set of proteins assayed.

3.3.4 General Procedure for Assay as a Function of [Phosphatase]

1. Measuring the slope of the rate of P_i release as a function of phosphatase concentration can quantify the relative sensitivity of CheY variants to a phosphatase (Silversmith et al., 2003, 2001).

2. The rate of P_i release is measured continuously in a temperature-controlled cuvette. We use 100 m*M* HEPES, pH 7.0, 20 m*M* $MgCl_2$, 0.2 m*M* MESG, and 1.1 U/mL purine nucleoside phosphorylase. The desired amount of MPI (established in step 3 of Section 3.3.3) is added. P_i contaminating the MPI will cause a (hopefully minor) increase in A_{360}. Phosphatase is added, which should not affect A_{360}. Adding CheY starts the assay. After a lag during which the various phosphorylation and dephosphorylation reactions come to equilibrium, A_{360} should increase at a constant rate. Several minutes of observation are sufficient to determine the slope.
3. A_{360} is converted to [P_i] using the standard curve, and the P_i release rate plotted as a function of [phosphatase]. The point at zero [phosphatase] corresponds to the CheY autodephosphorylation rate.

3.3.5 Microtiter Plate Assay

1. In order to consume an order of magnitude less material and achieve higher throughput by simultaneously performing multiple assays, the P_i release assay can be performed in a microtiter plate instead of a cuvette (Roberts, Lee, McCullagh, Silversmith, & Wemmer, 2005; Silversmith et al., 2008). The ability to continuously monitor the reaction is lost, so preliminary experiments to establish the linearity of the reaction are critical.
2. We use 50 μL final reaction volumes containing CheY and phosphatase in 50 m*M* HEPES, pH 7.0, 10 m*M* $MgCl_2$, and 1 mg/mL bovine serine albumin to block adsorption of proteins to the microtiter plate. The chosen pH of 7.0 is optimal for autophosphorylation of CheY with MPI (Silversmith et al., 1997). Control wells lacking protein are included to measure background P_i. Samples of known P_i concentration are included on the same plate to generate a standard curve.
3. Reactions are started by the addition of an appropriate concentration of MPI, established as described in step 3 of Section 3.3.3, to ensure that P_i release is rate limiting. Well contents are mixed by pipetting up and down. Reactions in different wells are initiated with conveniently staggered start time, e.g., at 30 s intervals. The plate is then covered with Parafilm to minimize evaporation during an incubation period chosen to maximize signal without exceeding the linear range of the assay (in our experience, 15–45 min). The incubation period should be sufficiently long that the initial lag (in our experience 10–20 s) does not significantly affect calculation of the P_i release rate from a single time point.

4. After the fixed incubation period, reactions are stopped at staggered intervals by the addition of 150 μL of 25 m*M* EDTA, 250 m*M* Tris, pH 7.5, 0.31 m*M* MESG, and 1.4 U/mL purine nucleoside phosphorylase. The EDTA stops the phosphorylation and dephosphorylation reactions by chelating the required Mg^{2+}. The Tris ensures that the quantification of P_i is performed in the midpoint of the useable 6.5–8.5 pH range. The EnzChek assay components use all P_i to form colored product during a subsequent 15 min incubation.
5. The rate of P_i release observed in each well is calculated by dividing the final P_i concentration (after subtraction of background P_i introduced with the MPI) by the elapsed assay time.

3.3.6 General Procedure for Assay as a Function of [CheY-P]

1. The microtiter plate format can be used to measure phosphatase activity as a function of CheY-P concentration, and thus determine whether the phosphatase has a hyperbolic (Michaelis–Menton) or sigmoidal (cooperative) dependence on substrate (Freeman, Mole, Silversmith, & Bourret, 2011; Silversmith et al., 2008).
2. The useful range of the assay depends on avoiding rate limitation by CheY autophosphorylation. This is particularly problematic when using low concentrations of CheY in the presence of phosphatase. The *E. coli* CheY 113AP variant, which exhibits enhanced rate constants for autophosphorylation yet retains sensitivity to CheZ-stimulated dephosphorylation (Smith et al., 2004), is useful to extend the assay range.
3. The CheY-P substrate is labile and therefore generated in situ by varying CheY concentration in the presence of a fixed concentration of MPI sufficiently high to make dephosphorylation rate limiting, rather than autophosphorylation. The value of [CheY-P] achieved can be calculated as described in steps 4–6 knowing that $[\text{CheY}_{\text{total}}] = [\text{CheY}] + [\text{CheY-P}]$:
4. In the presence of excess phosphatase, CheY-P is dephosphorylated as rapidly as it is formed (i.e., autophosphorylation is rate limiting), so $[\text{CheY}] = [\text{CheY}_{\text{total}}]$. The observed rate of P_i release is given by $k_{\text{phos}}[\text{CheY}] = k_{\text{phos}}[\text{CheY}_{\text{total}}]$, which allows determination of the pseudo-first-order rate constant $k_{\text{phos}} = (P_i \text{ release rate})/[\text{CheY}_{\text{total}}]$ from an experiment using excess phosphatase.
5. Under steady-state conditions where dephosphorylation is rate limiting (no phosphatase or modest concentrations of phosphatase), the rates of CheY phosphorylation and dephosphorylation are equal, i.e., $k_{\text{phos}}[\text{CheY}] = k_{\text{dephos}}[\text{CheY-P}]$. The rate of P_i release is given by

k_{dephos}[CheY-P]. Therefore, [CheY] = (rate of P_i release)/k_{phos} can be calculated from the observed rate of P_i release and k_{phos} as determined in step 4.

6. Because [CheY-P] = [$CheY_{total}$] − [CheY], the concentration of phosphorylated CheY can be calculated from the known concentration of total CheY and the concentration of unphosphorylated CheY is inferred in step 5.
7. The rate of P_i release due to phosphatase is equal to the observed rate of P_i release measured in the presence of phosphatase minus the observed rate of P_i release due to autodephosphorylation (measured in the absence of phosphatase). Both observed P_i release rates are determined after correction for background P_i introduced with MPI. At low [CheY-P], most P_i is released via autodephosphorylation, so the calculated P_i release due to phosphatase stimulation is subject to error as the small difference between two similar numbers.
8. Plotting the phosphatase P_i release rate from step 7 as a function of [CheY-P] from step 6 reveals if phosphatase activity is cooperative. The analysis will yield k_{cat} and K_m if the reaction is noncooperative, and $K_{1/2}$ if the reaction is cooperative.

4. RELEVANT PROPERTIES OF NONCHEMOTAXIS TCS

Characteristics that potentially impact the design of assays for phosphatases from nonchemotaxis TCS are described in this section.

4.1 DHp-Type Sensor Kinases

Because DHp-type sensor kinases are found in most TCS, DHp phosphatases comprise by far the largest group of TCS phosphatases. Five types of DHp domains are known (Pfam designations HisKA, HisKA_2, HisKA_3, His_Kinase, or HWE_HK). Although the phosphatase reaction mechanisms of the different DHp domain types are likely to be similar, this hypothesis has not been confirmed experimentally. In the major HisKA family, the probable catalytic residue is a Thr, Asn, Ser, Gln, or Glu located four positions C-terminal to the His phosphorylation site (H + 4). In the second largest HisKA_3 family, the catalytic residue is likely a Gln or His at H + 5. In the case of His, the phosphatase reaction would presumably be pH dependent, and the nature of the dependence would reveal whether the His serves as a hydrogen bond acceptor or donor to the attacking water molecule. In the three small DHp families, the catalytic residue is likely an Asn at H + 4.

There are two central challenges to designing phosphatase assays for DHp-type sensor kinases. First, about three-quarters of bacterial sensor kinases are transmembrane proteins (Ulrich et al., 2005), which causes practical problems for soluble assay systems. Historically, the most common approach has been to work with a soluble cytoplasmic fragment of the sensor kinase. Such constructs necessarily lack the regulation normally exerted by periplasmic input domains. Furthermore, soluble constructs of sensor kinases often exhibit weak activity in vitro. Trial and error regarding the location of truncations may be necessary to obtain proteins with useful activity. In many published assays, the DHp phosphatase and the response regulator substrate were used at similar stoichiometry, which raises questions about whether or not the phosphatase reaction is catalytic (Kenney, 2010). It is not yet clear whether the low activity is a genuine characteristic of DHp-type phosphatases or an artifact of the protein construct examined. A recent study concluded that the in vivo phosphatase activity of the PhoR sensor kinase is much higher than anticipated from in vitro measurements (Gao & Stock, 2017), suggesting the in vitro measurements are flawed. For more information on quantitative in vivo assays of TCS phosphatases, see the chapter by Rong Gao and Ann Stock in this volume of *Methods in Enzymology* (Gao & Stock, 2018).

A second tactic is to solubilize intact transmembrane sensor kinases in active form. Nanodisc technology provides a soluble disc of lipid bilayer into which transmembrane proteins can be incorporated. At least one sensor kinase has been incorporated into nanodiscs in a functional form that exhibits ligand regulated activity (Wang, Zhao, Novick, & Muir, 2014; Wang et al., 2017).

A final approach to the problem of transmembrane proteins is to instead focus attention on the one-quarter of sensor kinases that lack transmembrane segments and are naturally soluble (i.e., cytoplasmic).

The second major challenge that DHp-type sensor kinases pose to assay design is their ability to participate in three different reactions involving phosphoryl groups: autophosphorylation with ATP, phosphotransfer to a response regulator, and phosphatase activity toward the response regulator. Under some in vitro conditions, reverse phosphotransfer from the response regulator to the sensor kinase can also occur. Many published studies did not take sufficient care to distinguish between reverse phosphotransfer and phosphatase reactions, both of which result in loss of phosphoryl groups from the response regulator, with confusing consequences for interpretation of the resulting data. One solution is to utilize sensor kinase variants lacking the His phosphorylation site, which cannot support autophosphorylation or reverse phosphotransfer. However, such variants preclude investigation of a potential role for the His residue in the phosphatase reaction.

A partial solution to the problem of multiple biochemical activities for sensor kinases is provided by the observation that phosphoryl groups are typically much more stable (half-lives of minutes to hours) on response regulators from nonchemotaxis TCS than on chemotaxis response regulators (Thomas et al., 2008). This potentially makes it feasible to purify phosphorylated response regulator substrate, without the need for continuous generation during the assay itself. The response regulator could be phosphorylated using either the partner sensor kinase and ATP (which is commercially available in radiolabeled form), or autophosphorylated with the commercially available small molecule phosphodonor acetyl phosphate. There are many straightforward schemes to quickly separate the phosphorylated response regulator from other reaction components in either case.

Finally, as the name implies, proteins containing DHp domains form dimers, which leads to the possibility of cooperativity.

4.2 Sporulation Phosphatases

From an assay perspective, the sporulation phosphatases (Rap, Spo0E) combine the best properties of chemotaxis and DHp phosphatases. Like chemotaxis phosphatases, Rap and Spo0E phosphatases are soluble (i.e., lacking transmembrane segments) and do not have complicating kinase activity or phosphorylation sites. Like DHp phosphatases, the response regulator substrates of sporulation phosphatases are stably phosphorylated, with phosphoryl group half-lives of half an hour or more (Perego, 2001; Zapf et al., 1998). This combination of properties should facilitate quantitative characterization of reaction kinetics for sporulation phosphatases.

One potential complexity to keep in mind is oligomeric state. RapH is a dimer (Parashar et al., 2011). Some members of the Spo0E family are also dimers, whereas others are monomers (Grenha et al., 2006).

5. ASSAY MODIFICATIONS FOR NONCHEMOTAXIS TCS PHOSPHATASES

Potential modifications of the three assays described in Section 3 for successful use with phosphatases from nonchemotaxis TCS are considered in this section:

5.1 Assay by Loss of ^{32}P

The most common assays for nonchemotaxis TCS phosphatases employ radiolabeled response regulators. To minimize the number of simultaneously occurring reactions, it is ideal to first purify the [^{32}P]-labeled response regulator and then observe the effect of adding phosphatase. Two general approaches to radiolabeling the response regulator are described below.

5.1.1 Preparation of [^{32}P]Response Regulator by Autophosphorylation

One option is to autophosphorylate the response regulator with [^{32}P]acetyl phosphate or [^{32}P]phosphoramidate. Substrate preparation via autophosphorylation avoids using the partner DHp-type kinase, which typically also has phosphatase activity toward the substrate. Furthermore, one radiolabeled phosphodonor can be used for many different response regulators, avoiding the need to purify multiple sensor kinases in functional form if multiple TCSs are studied. It is simple to rapidly separate the radiolabeled response regulator from radiolabeled small molecules using small centrifugal devices employing either a desalting column or ultrafiltration. To completely remove the large excess of radioactive small molecules, either use two successive desalting columns (e.g., Pierce polyacrylamide spin desalting column) (protein goes through device) or extensive washes during ultrafiltration (protein retained in device).

The major disadvantage of autophosphorylation is that radiolabeled small molecular phosphodonors are not commercially available and must be synthesized by the user. [^{32}P]acetyl phosphate is more commonly used, perhaps because an enzymatic synthesis method is available (Lee, Schramm, Jagadeesan, & Higgs, 2010). Preparation of [^{32}P]phosphoramidate is by chemical synthesis (Buckler & Stock, 2000). An occasional limitation is that some response regulators are unable to use one or the other compound as a phosphodonor.

5.1.2 Preparation of [^{32}P]Response Regulator With a Sensor Kinase

The other option is to take advantage of commercially available [γ-^{32}P]ATP and use a partner sensor kinase to radiolabel the response regulator. There are two central challenges with this strategy. First, the sensor kinase may dephosphorylate the response regulator, reducing yields of the desired product. Using a sensor kinase variant that retains kinase activity but is deficient in phosphatase activity can minimize this problem (Huynh, Noriega, & Stewart, 2013; Willett & Kirby, 2012).

Second, it is necessary to rapidly separate the response regulator from the sensor kinase. A variety of strategies based on affinity tags are available. Tagging the sensor kinase would allow an unmodified response regulator to be used in the assay. After removal of the sensor kinase, the untagged response regulator could be separated from radioactive small molecules as described in Section 5.1.1. If His_6-tags are used, care should be taken that imidazole is not introduced into the assay, because imidazole can dephosphorylate response regulators (Page et al., 2015).

5.2 Assay by Fluorescence

The fluorescence assay method requires a fluorophore that is sensitive to the phosphorylation state of the response regulator. This approach is not broadly applicable, because only ~5% of response regulators have a naturally occurring Trp in their active site (Silversmith & Bourret, 2018). Introduction of a nonnative fluorophore into a response regulator is possible (Madhusudan et al., 1997; Stewart & VanBruggen, 2004), but the possibility of concomitant changes in function (particularly interaction with the phosphatase) complicates interpretation of results.

If a response regulator with a suitable fluorophore is available, it will likely be easy to generate the phosphorylated substrate for the phosphatase assay. Many (but not all) response regulators can autophosphorylate with the physiologically relevant and commercially available phosphodonor acetyl phosphate.

5.3 Assay by P_i Release

To the best of our knowledge, the P_i release assay has not been applied to nonchemotaxis TCS phosphatases, but offers the possibility of obtaining quantitative results, such as determination of K_m and k_{cat}. The ability to measure enzyme parameters could be invaluable in dissecting reaction mechanism, for example, by determining whether specific amino acid substitutions in the phosphatase or response regulator affect substrate binding and/or catalysis. If the phosphorylated response regulator is sufficiently stable, then reaction rates could be measured as a function of substrate concentration simply by using purified substrate of known concentration in the assay. Alternatively, constant concentrations of substrate can be continuously generated as has been done for chemotaxis TCS.

5.3.1 *Assay With Purified Substrate*

It should be possible to synthesize and purify unlabeled phosphorylated response regulators from nonchemotaxis TCS essentially as described in Section 5.1, using either a sensor kinase and ATP, or autophosphorylation with acetyl phosphate. Aliquots could be frozen to prevent dephosphorylation prior to use. The proportion that is phosphorylated can be determined by gel electrophoresis in the presence of the Phos-tag reagent (Barbieri & Stock, 2008), and the total concentration of response regulator (phosphorylated and nonphosphorylated) can be determined spectroscopically. If the proportion of phosphorylated response regulator needs to be increased, then the strategies described in Section 5.3.2 could be utilized.

Using purified substrate allows direct determination of the concentration of phosphorylated response regulator present in the reaction, rather than an indirect calculation that is subject to the errors described in Section 3.3.6. However, the concentration of phosphorylated response regulator will decrease over the course of the assay due to autodephosphorylation and phosphatase-stimulated dephosphorylation. Therefore, it is important to measure initial rates of P_i release before the substrate concentration changes significantly. Thus, P_i release should be measured continuously rather than at a single time point.

5.3.2 *Validating and Optimizing Assay Conditions for Continuous Generation of Substrate*

If continuous generation of phosphorylated response regulator substrate is chosen, the relatively slow autodephosphorylation rate of nonchemotaxis response regulators should substantially reduce background signal when applying the P_i release assay to nonchemotaxis TCS. This feature should make the P_i release assay more reproducible and accurate than when used with chemotaxis TCS.

The P_i release assay for response regulator (RR) dephosphorylation involves three coupled reactions:

Autophosphorylation	RR + monophosphoimidazole → RR-P + imidazole
Autodephosphorylation	RR-P → RR + P_i
Phosphatase-stimulated dephosphorylation	RR-P + phosphatase → RR + P_i + phosphatase

To measure phosphatase activity, the assay must be performed under conditions where the phosphatase reaction is rate limiting for P_i release. The window of rates available to measure phosphatase activity is between the background of P_i release from autodephosphorylation and the point at which phosphatase-stimulated dephosphorylation is so fast that autophosphorylation to generate the RR-P substrate becomes rate limiting.

In contrast to autodephosphorylation, autophosphorylation kinetics have been reported for few response regulators. Therefore, is not known whether autophosphorylation or autodephosphorylation typically will be rate limiting. The rate constants for PhoB (Creager-Allen et al., 2013) suggest it would be possible to make PhoB autophosphorylation faster than PhoB-P autodephosphorylation using reasonable (m*M*) concentrations of phosphodonor. For other response regulators, appropriate conditions must be established empirically. The rate of P_i release should be measured as a function of increasing MPI concentration, until a plateau is reached. The effect of adding phosphatase at saturating MPI concentration should then be determined. An increase in P_i release rate in the presence of phosphatase would confirm that the plateau represents rate limitation by autodephosphorylation. No change would imply either lack of phosphatase activity under the assay conditions or autophosphorylation is rate limiting.

The properties of nonchemotaxis response regulators offer two fairly general strategies to increase the rate of autophosphorylation and thus expand the usable window of the P_i release assay. First, in some response regulators, interactions between the output and receiver domains inhibit autophosphorylation (Barbieri, Mack, Robinson, Miller, & Stock, 2010). In such cases, using the isolated receiver domain in the assay instead of the entire response regulator could increase the rate of autophosphorylation up to two orders of magnitude. Second, about two-thirds of response regulators contain DNA-binding output domains (Galperin, 2010). Binding of specific DNA by the output domain enhances autophosphorylation of the receiver domain (Ames, Frankema, & Kenney, 1999). Therefore, performing the P_i release assay in the presence of appropriate oligonucleotides could increase the useable range of the assay.

A less general possibility is to employ response regulator variants that exhibit enhanced autophosphorylation yet retain sensitivity to the phosphatase, as described in Section 3.3.6 with CheY 113AP. Amino acid substitutions two residues to the C-terminal side of the conserved Asp phosphorylation site (D + 2) and/or the conserved Thr/Ser (T + 2) enhance autophosphorylation of CheY (Thomas, Immormino, Bourret, & Silversmith, 2013). The D + 2 and

T + 2 substitutions likely affect the chemistry of the autophosphorylation reaction and therefore may have similar effects on autophosphorylation of other response regulators. However, D + 2 and T + 2 substitutions in CheY interfere with CheZ function (Silversmith et al., 2003, 2001). Because these positions are proximal to the phosphorylation site, D + 2 and T + 2 substitutions in other response regulators could also interfere with partner phosphatases and therefore may not be useful for a phosphatase assay. Nevertheless, activating substitutions at other positions have been isolated in many TCS (Smith et al., 2004). A substitution that enhances autophosphorylation by altering the conformational equilibrium of a response regulator and is located outside of the active site (even in the output domain) may be available.

5.4 Assay by Separation of Phosphorylated and Nonphosphorylated Response Regulators

Some investigators have avoided the use of radioactivity in assays of TCS phosphatases by using HPLC or Phos-Tag gels to separate phosphorylated and unphosphorylated response regulators (Wang et al., 2013). These methods are useful for qualitative conclusions, e.g., comparing reaction conditions or the activity of variant proteins. However, accurate quantification requires detection of both the phosphorylated and unphosphorylated populations, which may be difficult when the two are very unequal in abundance.

ACKNOWLEDGMENTS

We thank Clay Foster, Emily Kennedy, and Philip Straughn for helpful comments on the manuscript, Jeryl Appleby for Fig. 2, and Stephanie Thomas for Fig. 3.

Investigation of TCS phosphatases in our laboratory is funded by National Institutes of Health Grant GM050860. The content is solely the responsibility of the authors and does not necessarily represent the official views of the National Institute of General Medical Sciences or the National Institutes of Health.

REFERENCES

Ames, S. K., Frankema, N., & Kenney, L. J. (1999). C-terminal DNA binding stimulates N-terminal phosphorylation of the outer membrane protein regulator OmpR from *Escherichia coli*. *Proceedings of the National Academy of Sciences of the United States of America*, *96*(21), 11792–11797.

Appleby, J. L., & Bourret, R. B. (1998). Proposed signal transduction role for conserved CheY residue Thr87, a member of the response regulator active-site quintet. *Journal of Bacteriology*, *180*(14), 3563–3569.

Barbieri, C. M., Mack, T. R., Robinson, V. L., Miller, M. T., & Stock, A. M. (2010). Regulation of response regulator autophosphorylation through interdomain contacts. *Journal of Biological Chemistry*, *285*(42), 32325–32335.

Barbieri, C. M., & Stock, A. M. (2008). Universally applicable methods for monitoring response regulator aspartate phosphorylation both *in vitro* and *in vivo* using Phos-tag-based reagents. *Analytical Biochemistry, 376*(1), 73–82.

Blat, Y., Gillespie, B., Bren, A., Dahlquist, F. W., & Eisenbach, M. (1998). Regulation of phosphatase activity in bacterial chemotaxis. *Journal of Molecular Biology, 284*(4), 1191–1199.

Boesch, K. C., Silversmith, R. E., & Bourret, R. B. (2000). Isolation and characterization of nonchemotactic CheZ mutants of *Escherichia coli*. *Journal of Bacteriology, 182*(12), 3544–3552.

Bourret, R. B. (2010). Receiver domain structure and function in response regulator proteins. *Current Opinion in Microbiology, 13*(2), 142–149.

Bourret, R. B., Thomas, S. A., Page, S. C., Creager-Allen, R. L., Moore, A. M., & Silversmith, R. E. (2010). Measurement of response regulator autodephosphorylation rates spanning six orders of magnitude. *Methods in Enzymology, 471*, 89–114.

Buckler, D. R., & Stock, A. M. (2000). Synthesis of [(32)P]phosphoramidate for use as a low molecular weight phosphodonor reagent. *Analytical Biochemistry, 283*(2), 222–227.

Creager-Allen, R. L., Silversmith, R. E., & Bourret, R. B. (2013). A link between dimerization and autophosphorylation of the response regulator PhoB. *Journal of Biological Chemistry, 288*(30), 21755–21769.

Da Re, S. S., Deville-Bonne, D., Tolstykh, T., Veron, M., & Stock, J. B. (1999). Kinetics of CheY phosphorylation by small molecule phosphodonors. *FEBS Letters, 457*(3), 323–326.

Diaz, A. R., Stephenson, S., Green, J. M., Levdikov, V. M., Wilkinson, A. J., & Perego, M. (2008). Functional role for a conserved aspartate in the Spo0E signature motif involved in the dephosphorylation of the *Bacillus subtilis* sporulation regulator Spo0A. *Journal of Biological Chemistry, 283*(5), 2962–2972.

Dutta, R., & Inouye, M. (1996). Reverse phosphotransfer from OmpR to EnvZ in a kinase^{-}/phosphatase^{+} mutant of EnvZ (EnvZ.N347D), a bifunctional signal transducer of *Escherichia coli*. *Journal of Biological Chemistry, 271*(3), 1424–1429.

Dutta, R., Qin, L., & Inouye, M. (1999). Histidine kinases: Diversity of domain organization. *Molecular Microbiology, 34*(4), 633–640.

Freeman, A. M., Mole, B. M., Silversmith, R. E., & Bourret, R. B. (2011). Action at a distance: Amino acid substitutions that affect binding of the phosphorylated CheY response regulator and catalysis of dephosphorylation can be far from the CheZ phosphatase active site. *Journal of Bacteriology, 193*(18), 4709–4718.

Galperin, M. Y. (2010). Diversity of structure and function of response regulator output domains. *Current Opinion in Microbiology, 13*(2), 150–159.

Gao, R., & Stock, A. M. (2017). Quantitative kinetic analyses of shutting off a two-component system. *MBio, 8*(3), e00412-00417.

Gao, R., & Stock, A. M. (2018). Quantitative analysis of intracellular response regulator phosphatase activity of histidine kinases. *Methods in Enzymology, 607*, 301–320.

Grenha, R., Rzechorzek, N. J., Brannigan, J. A., de Jong, R. N., Ab, E., Diercks, T., et al. (2006). Structural characterization of Spo0E-like protein-aspartic acid phosphatases that regulate sporulation in bacilli. *Journal of Biological Chemistry, 281*(49), 37993–38003.

Hess, J. F., Bourret, R. B., Oosawa, K., Matsumura, P., & Simon, M. I. (1988). Protein phosphorylation and bacterial chemotaxis. *Cold Spring Harbor Symposia on Quantative Biology, 53*(Pt. 1), 41–48.

Hess, J. F., Bourret, R. B., & Simon, M. I. (1991). Phosphorylation assays for proteins of the two-component regulatory system controlling chemotaxis in *Escherichia coli*. *Methods in Enzymology, 200*, 188–204.

Hess, J. F., Oosawa, K., Kaplan, N., & Simon, M. I. (1988). Phosphorylation of three proteins in the signaling pathway of bacterial chemotaxis. *Cell, 53*(1), 79–87.

Huynh, T. N., Noriega, C. E., & Stewart, V. (2010). Conserved mechanism for sensor phosphatase control of two-component signaling revealed in the nitrate sensor NarX. *Proceedings of the National Academy of Sciences of the United States of America*, *107*(49), 21140–21145.

Huynh, T. N., Noriega, C. E., & Stewart, V. (2013). Missense substitutions reflecting regulatory control of transmitter phosphatase activity in two-component signalling. *Molecular Microbiology*, *88*(3), 459–472.

Huynh, T. N., & Stewart, V. (2011). Negative control in two-component signal transduction by transmitter phosphatase activity. *Molecular Microbiology*, *82*(2), 275–286.

Immormino, R. M., Silversmith, R. E., & Bourret, R. B. (2016). A variable active site residue influences the kinetics of response regulator phosphorylation and dephosphorylation. *Biochemistry*, *55*(39), 5595–5609.

Jagadeesan, S., Mann, P., Schink, C. W., & Higgs, P. I. (2009). A novel "four-component" two-component signal transduction mechanism regulates developmental progression in *Myxococcus xanthus*. *Journal of Biological Chemistry*, *284*(32), 21435–21445.

Kenney, L. J. (2010). How important is the phosphatase activity of sensor kinases? *Current Opinion in Microbiology*, *13*(2), 168–176.

Lassila, J. K., Zalatan, J. G., & Herschlag, D. (2011). Biological phosphoryl-transfer reactions: Understanding mechanism and catalysis. *Annual Review of Biochemistry*, *80*, 669–702.

Lee, B., Schramm, A., Jagadeesan, S., & Higgs, P. I. (2010). Two-component systems and regulation of developmental progression in *Myxococcus xanthus*. *Methods in Enzymology*, *471*, 253–278.

Lukat, G. S., McCleary, W. R., Stock, A. M., & Stock, J. B. (1992). Phosphorylation of bacterial response regulator proteins by low molecular weight phospho-donors. *Proceedings of the National Academy of Sciences of the United States of America*, *89*(2), 718–722.

Lukat, G. S., Stock, A. M., & Stock, J. B. (1990). Divalent metal ion binding to the CheY protein and its significance to phosphotransfer in bacterial chemotaxis. *Biochemistry*, *29*(23), 5436–5442.

Madhusudan, M., Zapf, J., Hoch, J. A., Whiteley, J. M., Xuong, N. H., & Varughese, K. I. (1997). A response regulatory protein with the site of phosphorylation blocked by an arginine interaction: Crystal structure of Spo0F from *Bacillus subtilis*. *Biochemistry*, *36*(42), 12739–12745.

Mayover, T. L., Halkides, C. J., & Stewart, R. C. (1999). Kinetic characterization of CheY phosphorylation reactions: Comparison of P-CheA and small-molecule phosphodonors. *Biochemistry*, *38*(8), 2259–2271.

Motaleb, M. A., Miller, M. R., Li, C., Bakker, R. G., Goldstein, S. F., Silversmith, R. E., et al. (2005). CheX is a phosphorylated CheY phosphatase essential for *Borrelia burgdorferi* chemotaxis. *Journal of Bacteriology*, *187*(23), 7963–7969.

Muff, T. J., & Ordal, G. W. (2007). Assays for CheC, FliY, and CheX as representatives of response regulator phosphatases. *Methods in Enzymology*, *423*, 336–348.

O'Connor, K. A., & Zusman, D. R. (1991). Development in *Myxococcus xanthus* involves differentiation into two cell types, peripheral rods and spores. *Journal of Bacteriology*, *173*(11), 3318–3333.

Page, S. C., Silversmith, R. E., Collins, E. J., & Bourret, R. B. (2015). Imidazole as a small molecule analogue in two-component signal transduction. *Biochemistry*, *54*(49), 7248–7260.

Parashar, V., Mirouze, N., Dubnau, D. A., & Neiditch, M. B. (2011). Structural basis of response regulator dephosphorylation by Rap phosphatases. *PLoS Biology*, *9*(2), e1000589.

Park, S. Y., Chao, X., Gonzalez-Bonet, G., Beel, B. D., Bilwes, A. M., & Crane, B. R. (2004). Structure and function of an unusual family of protein phosphatases: The bacterial chemotaxis proteins CheC and CheX. *Molecular Cell*, *16*(4), 563–574.

Pazy, Y., Motaleb, M. A., Guarnieri, M. T., Charon, N. W., Zhao, R., & Silversmith, R. E. (2010). Identical phosphatase mechanisms achieved through distinct modes of binding phosphoprotein substrate. *Proceedings of the National Academy of Sciences of the United States of America*, *107*(5), 1924–1929.

Perego, M. (2001). A new family of aspartyl phosphate phosphatases targeting the sporulation transcription factor Spo0A of *Bacillus subtilis*. *Molecular Microbiology*, *42*(1), 133–143.

Porter, S. L., & Armitage, J. P. (2002). Phosphotransfer in *Rhodobacter sphaeroides* chemotaxis. *Journal of Molecular Biology*, *324*(1), 35–45.

Roberts, A., Lee, S. Y., McCullagh, E., Silversmith, R. E., & Wemmer, D. E. (2005). YbiV from *Escherichia coli* K12 is a HAD phosphatase. *Proteins*, *58*(4), 790–801.

Segall, J. E., Manson, M. D., & Berg, H. C. (1982). Signal processing times in bacterial chemotaxis. *Nature*, *296*(5860), 855–857.

Sheftic, S. R., Garcia, P. P., White, E., Robinson, V. L., Gage, D. J., & Alexandrescu, A. T. (2012). Nuclear magnetic resonance structure and dynamics of the response regulator Sma0114 from *Sinorhizobium meliloti*. *Biochemistry*, *51*(35), 6932–6941.

Shi, L., Liu, W., & Hulett, F. M. (1999). Decay of activated *Bacillus subtilis pho* response regulator, PhoP~P, involves the PhoR~P intermediate. *Biochemistry*, *38*(31), 10119–10125.

Silversmith, R. E. (2010). Auxilliary phosphatases in two-component signal transduction. *Current Opinion in Microbiology*, *13*(2), 177–183.

Silversmith, R. E., Appleby, J. L., & Bourret, R. B. (1997). Catalytic mechanism of phosphorylation and dephosphorylation of CheY: Kinetic characterization of imidazole phosphates as phosphodonors and the role of acid catalysis. *Biochemistry*, *36*(48), 14965–14974.

Silversmith, R. E., & Bourret, R. B. (2018). Fluorescence measurement of kinetics of CheY autophosphorylation with small molecule phosphodonors. *Methods in Molecular Biology*, *1729*, 321–335.

Silversmith, R. E., Guanga, G. P., Betts, L., Chu, C., Zhao, R., & Bourret, R. B. (2003). CheZ-mediated dephosphorylation of the *Escherichia coli* chemotaxis response regulator CheY: Role for CheY glutamate 89. *Journal of Bacteriology*, *185*(5), 1495–1502.

Silversmith, R. E., Levin, M. D., Schilling, E., & Bourret, R. B. (2008). Kinetic characterization of catalysis by the chemotaxis phosphatase CheZ. Modulation of activity by the phosphorylated CheY substrate. *Journal of Biological Chemistry*, *283*(2), 756–765.

Silversmith, R. E., Smith, J. G., Guanga, G. P., Les, J. T., & Bourret, R. B. (2001). Alteration of a nonconserved active site residue in the chemotaxis response regulator CheY affects phosphorylation and interaction with CheZ. *Journal of Biological Chemistry*, *276*(21), 18478–18484.

Sircar, R., Greenswag, A. R., Bilwes, A. M., Gonzalez-Bonet, G., & Crane, B. R. (2013). Structure and activity of the flagellar rotor protein FliY: A member of the CheC phosphatase family. *Journal of Biological Chemistry*, *288*(19), 13493–13502.

Skerker, J. M., Prasol, M. S., Perchuk, B. S., Biondi, E. G., & Laub, M. T. (2005). Two-component signal transduction pathways regulating growth and cell cycle progression in a bacterium: A system-level analysis. *PLoS Biology*, *3*(10), e334.

Smith, J. G., Latiolais, J. A., Guanga, G. P., Pennington, J. D., Silversmith, R. E., & Bourret, R. B. (2004). A search for amino acid substitutions that universally activate response regulators. *Molecular Microbiology*, *51*(3), 887–901.

Sourjik, V., & Schmitt, R. (1998). Phosphotransfer between CheA, CheY1, and CheY2 in the chemotaxis signal transduction chain of *Rhizobium meliloti*. *Biochemistry*, *37*(8), 2327–2335.

Stewart, R. C., & VanBruggen, R. (2004). Phosphorylation and binding interactions of CheY studied by use of Badan-labeled protein. *Biochemistry*, *43*(27), 8766–8777.

Thomas, S. A., Brewster, J. A., & Bourret, R. B. (2008). Two variable active site residues modulate response regulator phosphoryl group stability. *Molecular Microbiology*, *69*(2), 453–465.

Thomas, S. A., Immormino, R. M., Bourret, R. B., & Silversmith, R. E. (2013). Nonconserved active site residues modulate CheY autophosphorylation kinetics and phosphodonor preference. *Biochemistry*, *52*(13), 2262–2273.

Trajtenberg, F., Imelio, J. A., Machado, M. R., Larrieux, N., Marti, M. A., Obal, G., et al. (2016). Regulation of signaling directionality revealed by 3D snapshots of a kinase:regulator complex in action. *eLife*, *5*, e21422.

Ulrich, L. E., Koonin, E. V., & Zhulin, I. B. (2005). One-component systems dominate signal transduction in prokaryotes. *Trends in Microbiology*, *13*(2), 52–56.

Wang, C., Sang, J., Wang, J., Su, M., Downey, J. S., Wu, Q., et al. (2013). Mechanistic insights revealed by the crystal structure of a histidine kinase with signal transducer and sensor domains. *PLoS Biology*, *11*(2), e1001493.

Wang, B., Zhao, A., Novick, R. P., & Muir, T. W. (2014). Activation and inhibition of the receptor histidine kinase AgrC occurs through opposite helical transduction motions. *Molecular Cell*, *53*(6), 929–940.

Wang, B., Zhao, A., Xie, Q., Olinares, P. D., Chait, B. T., Novick, R. P., et al. (2017). Functional plasticity of the AgrC receptor histidine kinase required for staphylococcal virulence. *Cell Chemical Biology*, *24*(1), 76–86.

Webb, M. R. (1992). A continuous spectrophotometric assay for inorganic phosphate and for measuring phosphate release kinetics in biological systems. *Proceedings of the National Academy of Sciences of the United States of America*, *89*(11), 4884–4887.

Willett, J. W., & Kirby, J. R. (2012). Genetic and biochemical dissection of a HisKA domain identifies residues required exclusively for kinase and phosphatase activities. *PLoS Genetics*, *8*(11), e1003084.

Willett, J. W., Tiwari, N., Muller, S., Hummels, K. R., Houtman, J. C., Fuentes, E. J., et al. (2013). Specificity residues determine binding affinity for two-component signal transduction systems. *MBio*, *4*(6), e00420-00413.

Wolfe, A. J. (2010). Physiologically relevant small phosphodonors link metabolism to signal transduction. *Current Opinion in Microbiology*, *13*(2), 204–209.

Wylie, D., Stock, A., Wong, C. Y., & Stock, J. (1988). Sensory transduction in bacterial chemotaxis involves phosphotransfer between Che proteins. *Biochemical and Biophysical Research Communications*, *151*(2), 891–896.

Zapf, J., Madhusudan, M., Grimshaw, C. E., Hoch, J. A., Varughese, K. I., & Whiteley, J. M. (1998). A source of response regulator autophosphatase activity: The critical role of a residue adjacent to the Spo0F autophosphorylation active site. *Biochemistry*, *37*(21), 7725–7732.

Zhao, R., Collins, E. J., Bourret, R. B., & Silversmith, R. E. (2002). Structure and catalytic mechanism of the *E. coli* chemotaxis phosphatase CheZ. *Nature Structural Biology*, *9*(8), 570–575.

Zhu, Y., Qin, L., Yoshida, T., & Inouye, M. (2000). Phosphatase activity of histidine kinase EnvZ without kinase catalytic domain. *Proceedings of the National Academy of Sciences of the United States of America*, *97*(14), 7808–7813.

Zschiedrich, C. P., Keidel, V., & Szurmant, H. (2016). Molecular mechanisms of two-component signal transduction. *Journal of Molecular Biology*, *428*(19), 3752–3775.

CHAPTER TWELVE

Assembly of PGAM5 into Multimeric Complexes Provides a Mechanism for Allosteric Regulation of Phosphatase Activity

Peter Tipton[1], Tong Su, Mark Hannink
University of Missouri, Columbia, MO, United States
[1]Corresponding author: e-mail address: pacole@bwh.harvard.edu

Contents

Abstract

Phosphoglycerate mutase family member 5 (PGAM5) is a serine/threonine phosphatase that has been localized to both inner and outer mitochondrial membranes. PGAM5 has been suggested to regulate multiple aspects of mitochondrial dynamics, including fission/fusion and mitophagy, through phosphatase-dependent and phosphatase-independent mechanisms. Understanding how the phosphatase activity of PGAM5 is

Methods in Enzymology, Volume 607
ISSN 0076-6879
https://doi.org/10.1016/bs.mie.2018.05.017

regulated will provide new insight into signaling mechanisms that link changes in cell physiology with mitochondrial function. In this chapter, we describe methods for obtaining both multimeric and dimeric complexes of PGAM5 and for characterizing their kinetic properties. The ability to purify different PGAM5 complexes and to characterize their kinetic properties will enable detailed biophysical studies of the quaternary structures of the various PGAM5-containing complexes. The phosphatase activity of different PGAM5 complexes varies over three orders of magnitude. We suggest that the ability to generate PGAM5 complexes that have a wide range of phosphatase activities will facilitate screens to identify small molecules that modulate the phosphatase activity of PGAM5.

1. INTRODUCTION

Protein phosphorylation is a widespread posttranslational regulatory mechanism. Altered protein phosphorylation is a contributing factor of diverse diseases, particularly cancer. Considerable effort has been devoted to the development of small-molecule protein kinase inhibitors, with several dozen protein kinase inhibitors in clinical use (Wu, Nielsen, & Clausen, 2015). However, small molecules that target protein phosphatases have received less attention, although there has been recent success in developing such molecules (Bollu, Mazumdar, Savage, & Brown, 2017; De Munter, Kohn, & Bollen, 2013).

The phosphoglycerate mutase family includes well-characterized metabolic enzymes that have mutase activity on small-molecule substrates such as 3-phosphoglycerate (PGAM1 and PGAM2) or fructose-2,6-bisphosphate (PFKPB1–4) (Sadatomi, Tanimura, Ozaki, & Takeda, 2013). However, several members of this family, including PGAM5, UBASH3A, and UBASH3B, are protein phosphatases (Sadatomi et al., 2013). UBASH3A and UBASH3B are specific for phosphotyrosine and are important for the regulation of T-cell receptor signaling (Newman et al., 2014). PGAM5 is specific for phosphoserine and phosphothreonine, and perhaps phosphohistidine (Panda et al., 2016; Takeda et al., 2009; Wilkins, McConnell, Tipton, & Hannink, 2014). Several protein targets for PGAM5 have been identified, including a protein kinase involved in oxidative stress signaling (ASK1), a protein that initiates mitochondrial fission (DRP1), a regulator of cell survival (BCL-XL), and a protein that initiates mitophagy (FUNDC1) (Chen et al., 2014; Takeda et al., 2009; Wang, Jiang, Chen, Du, & Wang, 2012; Wilkins et al., 2014). These proteins localize to the outer mitochondrial membrane, and PGAM5 contains an N-terminal mitochondrial targeting sequence (Lo & Hannink, 2008). PGAM5 has also been localized to the inner mitochondrial membrane, where it has been suggested to regulate the stability of the PINK1 protein kinase in a manner independent of its phosphatase activity (Lu et al., 2014). Understanding how the phosphatase activity of PGAM5 is regulated will provide new insight into signaling mechanisms by which changes in mitochondrial dynamics drive changes in cell physiology.

We have described a novel mechanism by which the phosphatase activity of PGAM5, as measured by k_{cat}/K_m for phosphopeptide substrates, is modulated over three orders of magnitude (Wilkins et al., 2014). This allosteric mechanism operates both within the minimal active unit, which is an α_2 dimer, and between the α_2 dimeric complexes when assembled into a multimeric complex. The elucidation of this mechanism, coupled with recent structural insights into both dimeric and multimeric forms of PGAM5 (Chaikuad et al., 2017), has revealed the importance of quaternary structure as a key determinant of the protein phosphatase activity of PGAM5. In this chapter, we describe methods for the expression and purification of homogenous PGAM5 complexes. We also describe methods for the analysis of the enzymatic activity of PGAM5 that provide insight into the relationship between the quaternary structure of PGAM5 and its phosphatase activity.

2. EXPRESSION AND PURIFICATION OF PGAM5

In humans, the PGAM5 gene is located on chromosome 12 and encodes several alternatively spliced transcripts. The most abundant transcript, expressed in nearly all cells and tissues, contains six exons and encodes a protein of 289 amino acids, termed PGAM5-L (Lo & Hannink, 2006). A second transcript, that lacks the C-terminal 50 amino acids of the PGAM5-L isoform and instead contains 16 unique amino acids derived from an alternatively spliced exon, encodes a protein of 255 amino acids (PGAM5-S) (Lo & Hannink, 2006). Both isoforms contain the same N-terminal mitochondrial targeting sequence (amino acids 1–25). The PGAM5-S protein cannot be heterologously expressed in a soluble form in bacteria but can be refolded after purification under denaturing conditions. However, whereas the phosphatase activity of PGAM5-L is restored upon refolding, the refolded PGAM5-S protein does not have phosphatase activity (unpublished data). Thus, our work has focused on the PGAM5-L isoform. For the sake of simplicity, we will use the acronym PGAM5 to refer to the PGAM5-L isoform (Fig. 1).

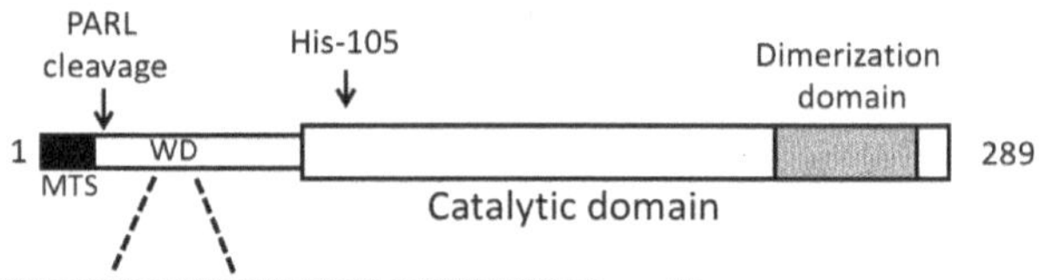

Conservation of the WD motif across species	51	ARPGPGV**WD**PN**WD**RREPLSLINVRKRNVES	Hs
	50	ARAGRGV**WD**TN**WD**RREPLSLINLKKRNVES	Ms
	53	AWPTANG**WD**YN**WD**KREPSSMVNGKRKESTG	Dr
	40	PVDPWAL**WD**TN**WD**CREPRALVRPLRNSQPE	Dm
	49	EHFPRGE**WD**KN**WD**FRDPISLVDKGKWEKAD	Ce
A4 protein:	51	ARPGPGV**AA**PN**AA**RREPLSLINVRKRNVES	Hs
ΔN65 protein:	65	--------------REPLSLINVRKRNVES	Hs

WD peptide: 52–74 RPGPGV**WD**PN**WD**RREPLSLINVR

Fig. 1 Functional elements of PGAM5. The PGAM5-L isoform (289 amino acids) is represented. Important functional elements of PGAM5 include an N-terminal mitochondrial targeting sequence, a PARL proteolytic site (amino acid 28), a WD motif (WDPNWD, amino acids 58–63), a core catalytic domain (amino acids 98–289) that contains residues, including H105, that are required for catalysis, and a C-terminal dimerization domain. Amino acid sequences spanning the WD motif from humans (Hs), mouse (Ms), zebrafish (Dr), flies (Dm), and worms (Ce) illustrate the conserved nature of the WD motif. The amino acid sequence of two variant proteins discussed in this chapter is illustrated, including the PGAM5-A4 protein, which contains four alanine residues substituted for the two WD pairs, and the ΔN65-PGAM5 protein, which lacks the N-terminal 64 amino acids. The sequence of a peptide that contains amino acids 52–74 of PGAM5, including the WD motif, is shown.

PGAM5 contains a single domain of about 190 amino acids that is homologous to the other members of the PGAM family (Fig. 1). Like the other PGAM family members, the active site of PGAM5 contains two histidine residues, H105 and H230, and two arginine residues, R104 and R152 (Chaikuad et al., 2017). A tyrosine, Y108, and a glutamic acid residue, E177, are also present in the active site. Substitution of an alanine residue for H105 completely abolishes catalytic activity.

A PGAM5 protein containing only the core catalytic domain of PGAM5 (amino acids 90–289) can be expressed as a soluble protein in *E. coli* and the X-ray crystal structure has been determined. The structure of the 90–289 region shows that PGAM5 forms a dimer, mediated by residues located in the C-terminal region of the protein (Chaikuad et al., 2017).

The core catalytic domain of PGAM5 has low enzymatic activity, with a k_{cat}/K_m of approximately $1 \times 10^3\ \text{min}^{-1}\ M^{-1}$ when assayed against a phosphopeptide derived from BCL-XL (Wilkins et al., 2014). Analysis of N-terminal truncated proteins revealed that a conserved motif containing two WD pairs (WDPNWD, amino acids 58–63 of the human protein; Fig. 1) is required for robust activity. For example, a protein lacking only the N-terminal mitochondrial targeting sequence (ΔN25) has a k_{cat}/K_m for the BCL-XL phosphopeptide of approximately 1×10^6 min^{-1} M^{-1}. This WD motif is also required for the formation of multimeric complexes of PGAM5. The X-ray crystal structure of an N-terminal truncated PGAM5 protein (amino acids 54–289), which contains the WDPNWD motif, confirms that the two WD pairs within WDPNWD sequence bind into a pocket located in the core catalytic domain (Chaikuad et al., 2017). The full-length PGAM5 protein forms multimeric complexes with enzymatic activity comparable to that of the ΔN25-PGAM5 protein, although the yield of soluble full-length protein following expression in *E. coli* is reduced. In mammalian cells, the mitochondrial targeting sequence can be cleaved by PARL, a protease located in the inner mitochondrial membrane (Sekine et al., 2012). Thus, the bacterial expressed ΔN25-PGAM5 protein is representative of at least a portion of PGAM5 proteins present in mammalian cells.

2.1 Equipment

- Temperature-controlled incubators at 37°C and 18°C for bacterial growth on agar plates and in liquid media.
- Cold box or cold room at 4°C for protein purification.
- Sonicator with microprobe.
- Sorvall centrifuge with appropriate rotors.
- Refrigerated microfuge.

2.2 Buffers and Reagents

- Competent Rosetta™ BL21 (DE3) pLysS *E. coli*.
- Bacterial growth media: LB agar plates with chloramphenicol and ampicillin; LB growth media containing ampicillin and 50 m*M* sodium phosphate, pH 8.
- PGAM5 cDNA cloned into pET15b, with in-frame His-tag sequence.
- 1 *M* Isopropyl β-D-1-thiogalactopyranoside (IPTG).
- Nitrilotriacetic acid (NTA) cell lysis buffer: 20 m*M* Hepes, pH 8.0; 500 m*M* NaCl; 10 m*M* imidazole containing 14 m*M* β-mercaptoethanol added just before use.
- Protease inhibitor (Sigma P8340).
- Benzonase nuclease (Sigma E1014).

- NTA beads (Superflow) equilibrated in lysis buffer.
- NTA wash buffer: 20 m*M* Hepes, pH 8.0; 500 m*M* NaCl; 60 m*M* imidazole containing 14 m*M* β-mercaptoethanol.
- NTA elution buffer: 20 m*M* Hepes, pH 8.0; 500 m*M* NaCl; 500 m*M* imidazole containing 14 m*M* β-mercaptoethanol.
- Dialysis and SEC chromatography buffer: 20 m*M* Hepes, pH 7.5; 300 m*M* NaCl; 1 m*M* imidazole. Add DTT to 1 m*M* before use.
- Spectra/Por™ dialysis tubing, 12–14 kDa molecular weight cutoff. Pretreat dialysis membranes by boiling for 5 m in 10 m*M* sodium bicarbonate followed by boiling for 5 m in 10 m*M* Na_2EDTA. Rinse with distilled water and store pretreated dialysis membranes in 20% EtOH at 4°C.

2.3 Procedure

1. Transform the expression plasmid into competent Rosetta™ BL21 (DE3) pLysS *E. coli*. Spread transformed bacteria onto LB agar plates containing chloramphenicol and ampicillin and incubate plates overnight at 37°C.
2. Scrape colonies into 20 mL LB and measure absorbance at 600 nm. Inoculate 500 mL of buffered LB containing ampicillin with bacteria to a final concentration of 0.01 OD units at 600 nm and incubate in a shaking incubator at 37°C. Grow till OD at 600 nm is 0.6–0.8 (about 3–4 h).
3. When OD of the culture reaches 0.6–0.8, add IPTG to 1 m*M* and incubate at 18°C in a shaking incubator for 18–24 h.
4. Collect bacterial pellet by centrifugation and resuspend in 5 mL of NTA cell lysis buffer containing 100 units benzonase per mL. Aliquot into 1.5-mL microfuge tubes, freeze the samples at −80°C for 30 min followed by incubation in a 37°C water bath to thaw. As soon as the samples are thawed, return to −80°C. Three freeze–thaw cycles are sufficient to achieve complete lysis. Extended incubation times at 37°C during the freeze–thaw cycles will increase proteolytic removal of the N-terminal region of PGAM5.
5. Sonicate on ice for 5–10 s. Repeat as needed to reduce viscosity of the lysate.
6. Centrifuge 15 min at full speed in a microfuge at 4°C and collect the supernatant.

7. Load the cleared lysate onto a gravity-flow NTA column with a 1 mL bed volume, wash with 10 column volumes of NTA cell lysis buffer, and 10 column volumes of NTA wash buffer. Elute with 5 mL of NTA elution buffer, collecting 1-mL fractions.
8. Measure OD280 and determine protein concentration from the molecular mass and extinction coefficient of the purified protein. Dialyze overnight against 4 L dialysis buffer at 4°C.
9. Transfer dialyzed protein into 1.5-mL microfuge tubes and centrifuge 15 min at full speed in a microfuge at 4°C to pellet any protein that has precipitated during dialysis.
10. Measure OD_{280} and determine protein concentration from the molecular mass and extinction coefficient of the purified protein. Protein preparations of PGAM5 can be stored at 4°C for several weeks with minimal loss of activity, although some precipitation of aggregated protein will occur. The addition of glycerol to 25% allows purified PGAM5 proteins to be frozen at −80°C for long-term storage and retain activity upon thawing. Avoid multiple freeze–thaw cycles of the same protein preparation.

2.4 Notes

1. Other bacterial growth media formulations such as 2XYT or TB can also be used, but should be buffered to prevent the pH of the bacterial culture from dropping below pH 7 during the induction period. The use of pH-buffered growth media increases both the overall yield and the amount of multimeric protein purified over the NTA column.
2. The use of benzonase reduces nucleic acid contamination of the purified protein. PGAM5 is a basic protein with a pI of 8.8 and binds nucleic acids in a nonspecific manner.
3. Extended incubation times at 37°C during the freeze–thaw cycles will increase proteolytic removal of the N-terminal region of PGAM5.
4. NTA columns suitable for HPLC can be used instead of gravity flow.
5. We use Hepes as a pH buffer to minimize temperature-dependent changes in pH. PGAM5 can be purified using other buffer systems, e.g., Tris-based or phosphate-based buffers. The enzymatic activity of PGAM5 preparations obtained with phosphate-containing buffers cannot be measured using Malachite Green-based assays.
6. Both multimeric and dimeric forms of PGAM5 form aggregates that slowly precipitate out of solution, particularly in solutions of low ionic

strength. Centrifugation at full speed for 15 min in a microfuge at 4°C will pellet protein aggregates of PGAM5. The protein that remains in solution retains full enzymatic activity. Aggregate formation is minimized, but not eliminated, by including 300 m*M* NaCl to increase the ionic strength of the buffers.

3. ISOLATION OF DIFFERENT PGAM5 COMPLEXES USING SIZE-EXCLUSION CHROMATOGRAPHY

The bacterial expression and purification protocol described in Section 2 yields high levels of PGAM5 protein that is homogenous when analyzed by SDS-polyacrylamide gel electrophoresis (Fig. 2A). However, analysis of purified PGAM5 proteins by size-exclusion chromatography (SEC) revealed that PGAM5 proteins containing an intact WD motif displayed considerable heterogeneity in their quaternary structure (Fig. 2C and D).

As seen in Fig. 2B, the ΔN65-PGAM5 protein, which starts at amino acid 65 immediately downstream of the WD motif, elutes after 15 mL. In contrast, the ΔN25-PGAM5 protein, which contains an intact WD motif, elutes over a wide volume range, from 7 to 15 mL (Fig. 2C). As the ΔN25-PGAM5 is a single molecular species when analyzed by SDS-polyacrylamide gel electrophoresis, this result suggests that PGAM5 can assemble into multiple oligomeric forms.

To obtain homogenous preparations of PGAM5 proteins containing the WD motif, we have tested a number of different protocols for expression and purification of PGAM5, including ion-exchange chromatography and different SEC columns. We find that overnight dialysis at pH 7.5 followed by SEC will reproducibly generate high molecular weight forms of either full-length or ΔN25-PGAM5 proteins that elute in the void volume of either Superdex 200 columns, with a void volume of 500 kDa, or Superose 6 columns, which have a void volume of 5000 kDa (Fig. 2D). The high molecular mass of multimeric ΔN25-PGAM5 complexes has been confirmed through chemical cross-linking, dynamic light scattering (unpublished data), and analytical ultracentrifugation (Dr. Henzl, personal communication).

The WD motif is required for assembly of multimeric complexes, as variant PGAM5 proteins that lack an intact WD motif can only form dimeric complexes. Furthermore, multimeric PGAM5 complexes can be dissociated into dimeric PGAM5 complexes by adding, in *trans*, a peptide containing

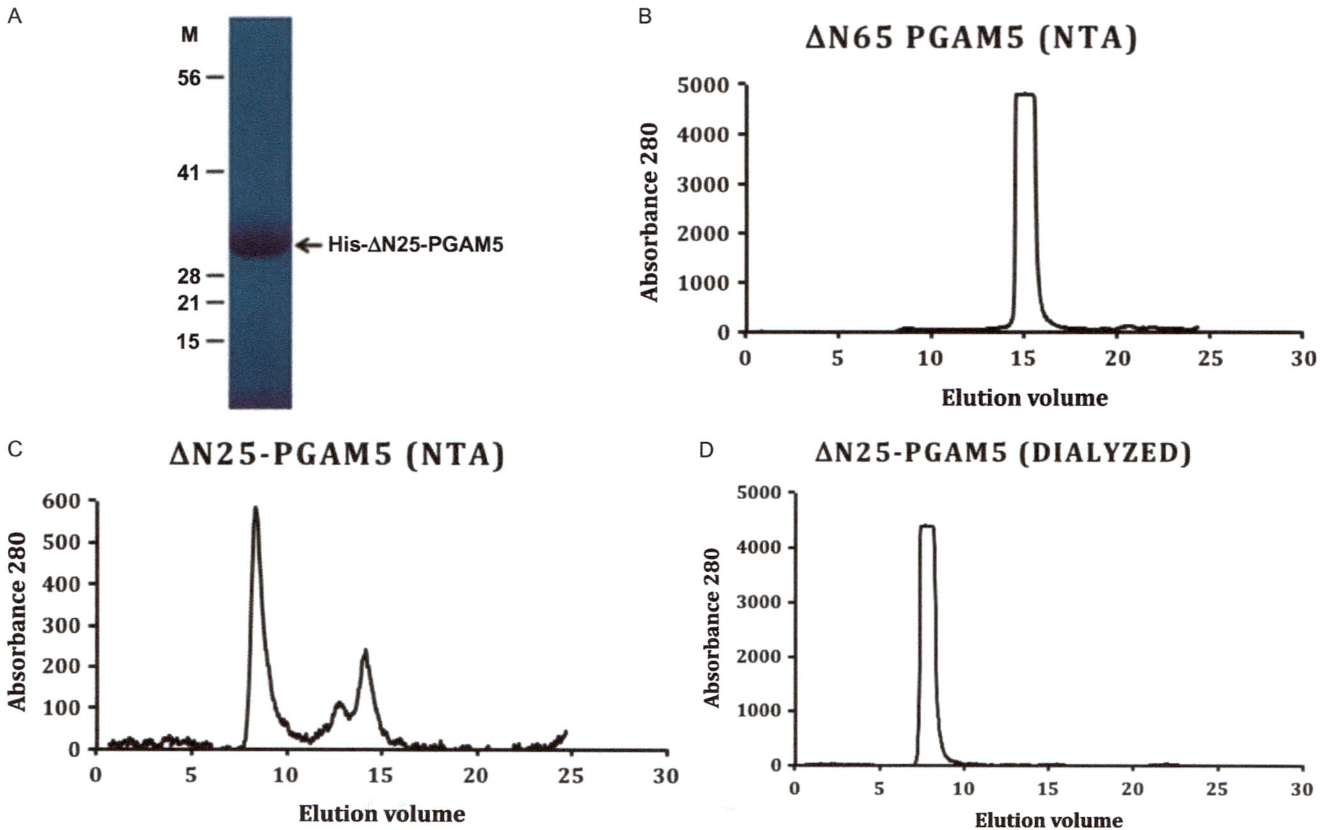

Fig. 2 See legend on opposite page.

the intact WD motif (Wilkins et al., 2014). The sequence of a 23-mer peptide capable of dissociating multimeric PGAM5 complexes is shown in Fig. 1. Thus, it is likely that these high molecular mass complexes of PGAM5 are multimers of dimers. The dynamics that underlie assembly of PGAM5 dimers into multimers and, conversely, the disassembly of multimeric PGAM5 complexes into dimers are poorly understood.

3.1 Equipment

- HPLC system with fraction collector and UV detector at 4°C
- Superdex 200 Increase 10/300 size-exclusion column (or similar)

3.2 Buffers and Reagents

- SEC buffer: 20 m*M* Hepes, pH 7.5; 300 m*M* NaCl; 1 m*M* imidazole

3.3 Procedure

1. Equilibrate the size-exclusion column with SEC buffer. Typically, the used dialysis buffer is passed through a 0.22-μm filter and used as the SEC buffer.
2. Use an appropriate sized loading loop to load the dialyzed protein onto the column. Do not overload the capacity of the column, as that will result in decreased ability to separate different oligomeric complexes of PGAM5.

Fig. 2 Purification and characterization of PGAM5 proteins. (A) The ΔN25-PGAM5 protein containing an N-terminal His-tag was purified from *E. coli* by NTA-affinity chromatography and analyzed by SDS-polyacrylamide gel electrophoresis. The protein was visualized by Coomassie Blue staining and is indicated by the *arrow*. The location of molecular size markers (M) are indicated on the *left side*. (B) The ΔN65-PGAM5 containing an N-terminal His-tag was purified from *E. coli* by NTA-affinity chromatography. Five nanomoles of protein was analyzed by size-exclusion chromatography over a Superdex 200 10/300 column. Absorbance at 280 (in milliabsorbance units) is indicated on the vertical axis and elution volume is indicated on the horizontal axis. (C) The ΔN25-PGAM5 containing an N-terminal His-tag was purified from *E. coli* by NTA-affinity chromatography. Five nanomoles of protein was analyzed by size-exclusion chromatography over a Superdex 200 10/300 column. Absorbance at 280 (in milliabsorbance units) is indicated on the vertical axis and elution volume is indicated on the horizontal axis. (D) The ΔN25-PGAM5 containing an N-terminal His-tag was purified from *E. coli* by NTA-affinity chromatography and dialyzed overnight. Five nanomoles of protein was analyzed by size-exclusion chromatography over a Superdex 200 10/300 column. Absorbance at 280 (in milliabsorbance units) is indicated on the vertical axis and elution volume is indicated on the horizontal axis.

3. Collect 0.5- or 1.0-mL fractions, measure OD_{280}, and determine protein concentration from the molecular mass and extinction coefficient of the purified protein. Store at 4°C.

3.4 Notes

1. Dimeric PGAM5 complexes are stable to repeated SEC analysis. Thus, when fractions containing the dimeric PGAM5 protein are reanalyzed by SEC, the elution profile of the protein is not altered (Wilkins et al., 2014).
2. When multimeric ΔN25-PGAM5 protein complexes obtained from early elution fractions (e. g., the void volume peak at 7 mL elution volume) are reanalyzed by SEC, the elution profile broadens and is shifted toward later elution volumes. For example, if 1 mL volumes are collected and fraction 7, containing the void volume, is reanalyzed over the same SEC column, the protein will elute in a broad peak over fractions 7–10 (unpublished data).
3. A truncated PGAM5 protein that lacks the N-terminal 54 amino acids of PGAM5 (ΔN55-PGAM5) is capable of forming a multimeric protein complex that is stable to repeated SEC analysis. Typically about 20%–30% of the total ΔN55-PGAM5 protein obtained either immediately after NTA purification or after dialysis of NTA-purified protein is present in this stable complex. The molecular mass of the stable ΔN55-PGAM5 complex, as determined by light scattering, is consistent with the formation of a dodecameric complex (Wilkins et al., 2014). The relationship of this stable 12-mer complex to the high molecular mass multimeric complexes formed with either the full-length protein or the ΔN25-PGAM5 protein is not fully understood.
4. Consistent with the conservation of the WD motif across species (Fig. 1), PGAM5 proteins from other species also form high molecular mass complexes when analyzed by SEC (unpublished data).

4. PHOSPHATASE ASSAY USING SYNTHETIC PHOSPHOPEPTIDES

The phosphatase activity of PGAM5 can be measured using either phosphorylated proteins or phosphorylated peptides as substrates. For quantitative determination of enzymatic activity, we prefer the use of synthetic phosphopeptides. Synthetic phosphopeptides that contain a single phosphorylated residue can be obtained in high purity. In contrast, the known or suspected substrates of PGAM5 are not readily available as phosphoproteins. Furthermore, site-specific phosphorylation of proteins, whether purified from cells or following phosphorylation in vitro by a kinase, is difficult to control in a way that provides a known concentration of protein with a single phosphorylation site. Using a range of synthetic phosphopeptides, we have demonstrated that phosphopeptides that have one or more acidic residues flanking the phosphorylated serine have lower K_m values and higher V_{max} values compared to phosphorylated sites that are flanked by neutral or basic amino acids.

A caveat with the use of synthetic phosphopeptides is that the K_m and V_{max} values determined with a synthetic phosphopeptide from a candidate substrate protein may not reflect the propensity of PGAM5 to dephosphorylate that protein in vivo. For example, the DRP1 protein is an important regulator of mitochondrial fission and is regulated, in part, by the phosphorylation status of Ser637. Phosphorylation of Ser637 has been suggested to inhibit mitochondrial fission (Chang & Blackstone, 2007; Cribbs & Strack, 2007). Although PGAM5 can dephosphorylate a peptide corresponding to the sequence of Ser637 of DRP1, the phosphopeptide has a K_m (~400 μ*M*) that is more than 10 times higher compared to phosphopeptides derived from BCL-XL or ASK1 (both ~25 μ*M*). However, PGAM5 is able to dephosphorylate a phosphopeptide corresponding to the sequence of the Ser548 site in DRP1, with a respectable K_m of 114 μ*M*. Thus, although PGAM5 has been suggested to be a positive regulator of mitochondrial fission through its ability to dephosphorylate DRP1 (Wang et al., 2012), it is not clear which site(s) in DRP1 are the in vivo substrate(s) of PGAM5. Although the use of synthetic phosphopeptides can provide valuable information, determination of the kinetic parameters of PGAM5 against native protein substrate(s) will be necessary to fully understand the biological relevance of PGAM5's phosphatase activity.

4.1 Equipment

- ½ size 96-well plates
- Plate reader capable of measuring absorbance at 620 nm

4.2 Buffers and Reagents

- Phosphatase assay buffer: 50 m*M* Hepes, pH 7.5, 150 m*M* NaCl, 2.5 m*M* EDTA, 0.5 m*M* DTT.
- Phosphate analysis buffer: Biomol Green (BML-AK111), Enzo Life Sciences.

- Substrate peptides: The sequence of phosphopeptide substrates that we typically use are: NFEDH(S)APPSP, containing phosphoserine at Ser1029 of ASK1; NFEDH(S)APPSP, containing phosphoserine corresponding to Ser62 in BCL-XL; and HSAPP(S)PEEKD, containing phosphoserine corresponding to Ser1033 of ASK1. Peptides were obtained from GenScript or Lifetein at 95% or greater purity. Peptides were resuspended in water, aliquoted, and frozen at −80°C and thawed as needed.
- WD-containing peptide: A peptide containing amino acids 52–74 of human PGAM5 (RPGPGVWDPNWDRREPLSLINVR) was obtained from Lifetein at 95% purity and passed over a PD-10 desalting column (GE Healthcare) prior to use.
- PGAM5 protein preparations obtained following NTA purification, dialysis, and subsequent SEC purification.

4.3 Procedure

1. Make serial dilutions of phosphopeptide substrate in phosphatase reaction buffer into individual microfuge tubes. Typical reactions are carried out with 45 μL of diluted phosphopeptide substrate added to 5 μL of enzyme. The concentration range of substrate is chosen to span the estimated K_m of the substrate, based on preliminary experiments. The K_m of multimeric PGAM5 for a good substrate is around 30 μ*M*, while the K_m of dimeric PGAM5 is approximately 500 μ*M*.
2. Dilute the enzyme into phosphatase reaction buffer to 100–200 n*M* for multimeric PGAM5 and 5–10 μ*M* for the dimeric PGAM5, for a final concentration of PGAM5 of 10–20 n*M* (multimeric) or 0.5–1 μ*M* (dimeric). Aliquot 5 μL into individual wells of a ½ size 96-well plate. WD-containing peptides from PGAM5 are preincubated with the enzyme for 10 min at room temperature prior to initiating the reaction by addition of appropriate dilutions of phosphopeptide substrate. The WD-containing peptides should be evaluated over an appropriate concentration range spanning their estimated affinity for PGAM5.
3. To start the reaction, aliquot 45 μL of diluted phosphopeptide substrate in phosphatase reaction buffer into enzyme individual wells of a ½ size 96-well plate.
4. For the multimeric PGAM5 protein, the reaction is typically linear over a 10-min time period, while reactions with the dimeric PGAM5 typically take 60 min to generate sufficient free phosphate. Multiple time

points per sample should be taken to confirm that each reaction is linear with respect to product formation.

5. Quench the reaction with 100 μL of Biomol Green, allow 20 min for the phosphate-binding reaction to go to completion, and read the absorbance at 620 nm with a plate reader. For each sample, calculate moles of free phosphate released during the reaction, using a standard curve prepared with known concentrations of phosphate.

4.4 Notes

1. Following purification, PGAM5 proteins are stored in 20 m*M* Hepes, pH 7.5, 300 m*M* NaCl, 1 m*M* imidazole, and 1 m*M* DTT at 4°C, at concentrations ranging from 10 to 100 μ*M*. Any precipitated protein is removed by centrifugation in a 4°C microfuge at full speed for 10 min prior to use. Proteins can be stored at 4°C with minimal loss of activity of soluble protein for several weeks.

5. ANALYSIS OF KINETIC DATA

Kinetic studies provide several insights into the functional behavior of PGAM5 and can guide experimental investigations into its biological roles. Full-length PGAM5 exhibits classical Michaelis–Menten behavior when it is used to dephosphorylate model phosphopeptides. Initial velocity data are fitted to Eq. (1) to determine values of k_{cat}/K_m. The k_{cat}/K_m parameter is the second-order rate constant for productive encounter between PGAM5 and its substrate and is interpreted as an indicator of substrate specificity (Purich, 2010). When comparing different model phosphopeptides, better substrates are described by higher values of k_{cat}/K_m. The measured k_{cat}/K_m values for different substrates span three orders of magnitude, with the highest values for phosphopeptides containing negatively charged residues located near the phosphorylated residue. The top-ranked phosphopeptide substrates were derived from ASK1 (Ser1029) and BCL-XL (Ser62) (Wilkins et al., 2014).

$$v/[\mathrm{E}] = \frac{k_{cat} \cdot [\mathrm{A}]}{K_m + [\mathrm{A}]} \qquad (1)$$

Truncated and site-specific variants of PGAM5 have distinct physical properties, as described above, and also exhibit unique kinetic behavior. In particular, a WD motif located in the N-terminal region of PGAM5, outside of the core catalytic domain, was found to have a profound effect on the catalytic activity of PGAM5. Deletion or mutation of this WD motif reduces the k_{cat}/K_m of PGAM5 for the BCL-XL phosphopeptide by three orders of magnitude, driven largely by a decrease in k_{cat} of more than two orders of magnitude.

We suggest that the N-terminal portion of PGAM5 mediates protein–protein interactions that are required for multimer assembly, as well as allosteric effects that are independent of multimer formation (Fig. 3). It is likely that other proteins, involved in other signal transduction pathways, exhibit a similar behavior in which protein–protein interactions that mediate assembly of oligomeric complexes also mediate allosteric effects that directly alter catalytic activity. The kinetic analysis that we describe below may be broadly applicable.

The experiments that helped us characterize kinetic behavior of multimeric and dimeric PGAM5 proteins used either the multimeric ΔN25-PGAM5 protein or the dimeric ΔN25-A4-PGAM5 protein, along with

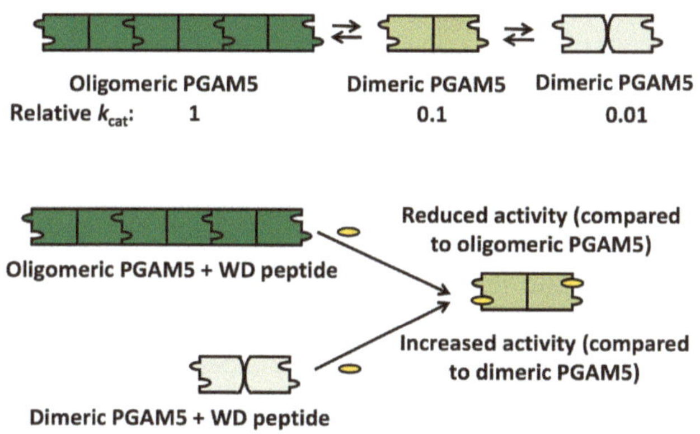

Fig. 3 The interplay between quaternary structure and allosteric regulation of PGAM5 is illustrated in this schematic model. (A) Three conformations of PGAM5, including a dimer with low activity (*right*), a dimer with moderate activity (*middle*), and a multimer with high activity (*left*), are indicated. The relative activity (k_{cat}) is indicated below each conformation. A given population of PGAM5 proteins will be distributed between these conformations at equilibrium. The rate constants that govern the equilibrium distribution of PGAM5 between different complexes are not known. (B) Both quaternary structure and allosteric regulation are modulated by a *trans*-acting peptide that contains the WD motif. The WD peptide (*yellow oval*) can disrupt the multimeric complex when it binds into the allosteric regulatory sites. The WD peptide can also bind to the empty allosteric sites present in a dimeric PGAM5 protein that is unable to assemble into a multimeric complex. Both reactions lead to the formation of a dimeric PGAM5 complex with two WD peptides bound. The activity of the resultant dimeric complex, in terms of k_{cat}, is approximately 25% of the maximal activity exhibited by the multimeric complex.

the WD peptide and a phosphopeptide substrate. Kinetic data were collected over a range of concentrations for both the WD peptide and the substrate phosphopeptide. The different combinations of protein and peptide that were used are shown in Fig. 3.

In one set of kinetic experiments, multimeric complexes containing the ΔN25-PGAM5 protein were incubated with the WD peptide in *trans*, followed by determination of the kinetic properties of the ΔN25-PGAM5 protein. Quantitative analysis of the kinetic data was used to provide mechanistic insight into the observed behavior.

Multimeric forms of either full-length or ΔN25-PGAM5 proteins exhibit maximal activity. When the WD peptide is provided in *trans* to the multimeric ΔN25-PGAM5 protein, the catalytic activity of the protein is reduced, indicating that the WD peptide acts as an inhibitor. The WD peptide is not competitive with respect to the substrate peptide. Importantly, the catalytic activity of the ΔN25-PGAM5 protein is not fully inhibited by the WD peptide, such that the k_{cat} of the maximally inhibited protein is about 25% that of the ΔN25-PGAM5 protein alone. This behavior is indicative of partial, hyperbolic noncompetitive inhibition. The "hyperbolic" part of the name comes from the fact that a slope replot vs inhibitor concentration is hyperbolic, rather than linear. (In a slope replot, the slope of each line in the double reciprocal plot of the data gathered at different inhibitor concentrations is plotted as a function of the inhibitor concentration.) Partial hyperbolic noncompetitive inhibition is described by Eq. (2) (Segel, 1975). The β term in Eq. (2) is the factor by which the rate constant for product formation is diminished in the enzyme–inhibitor–substrate complex, relative to the enzyme–substrate complex. For ΔN25-PGAM5 protein in the presence of the WD peptide, β is 0.25 ± 0.03.

$$v/[\mathrm{E}] = \frac{k_{cat} \cdot [\mathrm{A}]}{\alpha K_a \left(\frac{[\mathrm{I}] + K_i}{\beta[\mathrm{I}] + \alpha K_i} \right) + [\mathrm{A}] \left(\frac{[\mathrm{I}] + \alpha K_i}{\beta[\mathrm{I}] + \alpha K_i} \right)} \tag{2}$$

In a second set of kinetic experiments, dimeric complexes containing the ΔN25-A4-PGAM5 protein were incubated with the WD peptide in *trans*, followed by determination of the kinetic properties of the ΔN25-A4-PGAM5 protein. The ΔN25-A4-PGAM5 is a mutated form of the ΔN25-PGAM5 protein, in which both WD pairs in the WD motif have been substituted with alanine residues. The ΔN25-A4-PGAM5 can only form a dimeric complex. The ΔN25-A4-PGAM5 has low catalytic activity, with a k_{cat} approximately 1% compared to the wild-type ΔN25-PGAM5 protein. However, the WD peptide, in *trans*, markedly stimulated the catalytic activity of the ΔN25-A4-PGAM5 protein, increasing k_{cat} more than 50-fold. Maximal activation of the dimeric N25-A4-PGAM5 by the WD peptide yielded a k_{cat} that is approximately 25% that of the multimeric N25-PGAM5 protein.

There is no evidence that the exogenously added peptide stimulates oligomerization. Rather, the gain in activity can be attributed to the WD peptide acting as an allosteric activator (Fig. 3). Kinetic data obtained by varying the concentration of substrate peptide at different levels of the WT peptide adhere to the rapid equilibrium-ordered mechanism, which is described by Eq. (3) (Cook & Cleland, 2007). In this equation [A] refers to the WD peptide and [B] is the substrate phosphopeptide from BCL-XL.

$$v/[\mathrm{E}] = \frac{k_{cat} \cdot [\mathrm{A}][\mathrm{B}]}{K_{ia}K_b + K_b[\mathrm{A}] + [\mathrm{A}][\mathrm{B}]} \quad (3)$$

The data suggest that the WD peptide must bind before the substrate peptide is able to bind. Taken together, our analysis of the kinetic behavior of the multimeric and dimeric PGAM5 proteins in the presence of the WD peptide is consistent with the idea that the WD peptide is an allosteric effector that binds into an allosteric regulatory site within the catalytic domain of PGAM5.

6. SUMMARY AND CONCLUSION

The phosphatase activity of PGAM5, as measured by k_{cat}/K_m for phosphopeptide substrates, can be modulated over three orders of magnitude. Changes in the phosphatase activity of PGAM5 are closely linked to changes in its quaternary structure. In addition, both our kinetic data and a recently published structural study have provided evidence that the phosphatase activity of PGAM5 is also regulated by an allosteric mechanism (Chaikuad et al., 2017; Wilkins et al., 2014). The quaternary structure of PGAM5 is closely linked to allosteric regulation, as the key interactions that mediate assembly of multimeric complexes with robust phosphatase activity also mediate allosteric interactions that modulate the modest activity of dimeric complexes.

PGAM5 provides a striking example of how quaternary structure can influence enzymatic activity. Thus, when initiating a series of experiments directed toward developing small molecules that modulate PGAM5 activity, it is of critical importance to have well-developed methods for the isolation and characterization of homogenous preparations of the protein. In the case of PGAM5, the protein that is obtained from a standard NTA-affinity column is remarkably heterogeneous when analyzed by SEC. SEC provides both an analytical tool to characterize quaternary structure and a preparative tool to isolate specific complexes. Subsequent analysis of the kinetic properties of the different PGAM5 complexes revealed large differences in their activity and the presence of an allosteric regulatory mechanism.

Fig. 3 provides a model that summarizes our perspective on the interplay between quaternary structure and allosteric regulation of PGAM5. There are three aspects to this model.

First, as illustrated in Fig. 3A, we propose that the dimeric PGAM5 complexes can have two different conformations, one with very low activity and one with moderate activity. The presence of dimeric PGAM5 complexes with different conformations is supported by the recent crystallographic work of Stefan Knapp and colleagues, in which changes in the active site upon binding a phosphate molecule were linked to changes in the $\beta3$-$\alpha3$ loop located near the substrate-binding site (Chaikuad et al., 2017).

Second, also illustrated in Fig. 3A, we propose that dimeric PGAM5 complexes can assemble into multimeric complexes that have maximal phosphatase activity. While it is clear that PGAM5 can assemble into multimeric complexes, both in vitro and in vivo (Wilkins et al., 2014), the mechanisms that govern the assembly and disassembly of PGAM5 complexes are poorly understood. As the multimeric and dimeric PGAM5 complexes may have different biological roles, understanding the mechanisms that govern the assembly and disassembly of PGAM5 complexes is of critical importance.

Finally, as illustrated in Fig. 3B, we propose that each dimeric unit of PGAM5 contains two allosteric regulatory sites. In the multimeric protein, these sites are occupied by the WD motif present in adjacent dimer subunits. Thus, adding an excess of a WD-containing peptide in *trans* competes for binding to these sites, resulting in dissociation of the multimer into dimeric subunits, with each allosteric site occupied by a WD peptide. As the allosteric sites are occupied, the resultant dimeric complexes have considerable phosphatase activity. Likewise, the WD peptide can bind to a dimeric complex that has unoccupied allosteric sites, resulting in marked stimulation of activity.

This schematic highlights several opportunities for developing drugs that regulate the phosphatase activity of PGAM5. One route would be to directly target the active site, using a highly active multimeric PGAM5 complex in an activity-based screen. A second route would be to directly target the allosteric regulatory site, using the multimeric complex in a protein interaction-based screen. Alternatively, the dimeric complex could be used in an activity-based screen for molecules that increased the phosphatase activity of the dimeric complex. Given the emerging biological role of PGAM5 as a key regulator of mitochondrial dynamics, the identification of small molecules that selectively modulate PGAM5 quaternary structure or phosphatase activity is likely to yield clinically useful drugs.

REFERENCES

Bollu, L. R., Mazumdar, A., Savage, M. I., & Brown, P. H. (2017). Molecular pathways: Targeting protein tyrosine phosphatases in cancer. *Clinical Cancer Research*, *23*(9), 2136–2142. https://doi.org/10.1158/1078-0432.CCR-16-0934.

Chaikuad, A., Filippakopoulos, P., Marcsisin, S. R., Picaud, S., Schroder, M., Sekine, S., et al. (2017). Structures of PGAM5 provide insight into active site plasticity and multimeric assembly. *Structure*, *25*(7), 1089–1099, e1083. https://doi.org/10.1016/j.str.2017.05.020.

Chang, C. R., & Blackstone, C. (2007). Cyclic AMP-dependent protein kinase phosphorylation of Drp1 regulates its GTPase activity and mitochondrial morphology. *The Journal of Biological Chemistry*, *282*(30), 21583–21587. https://doi.org/10.1074/jbc.C700083200.

Chen, G., Han, Z., Feng, D., Chen, Y., Chen, L., Wu, H., et al. (2014). A regulatory signaling loop comprising the PGAM5 phosphatase and CK2 controls receptor-mediated mitophagy. *Molecular Cell*, *54*(3), 362–377. https://doi.org/10.1016/j.molcel.2014.02.034.

Cook, P. F., & Cleland, W. W. (2007). *Enzyme kinetics and mechanism*. London; New York: Garland Science.

Cribbs, J. T., & Strack, S. (2007). Reversible phosphorylation of Drp1 by cyclic AMP-dependent protein kinase and calcineurin regulates mitochondrial fission and cell death. *EMBO Reports*, *8*(10), 939–944. https://doi.org/10.1038/sj.embor.7401062.

De Munter, S., Kohn, M., & Bollen, M. (2013). Challenges and opportunities in the development of protein phosphatase-directed therapeutics. *ACS Chemical Biology*, *8*(1), 36–45. https://doi.org/10.1021/cb300597g.

Lo, S. C., & Hannink, M. (2006). PGAM5, a Bcl-XL-interacting protein, is a novel substrate for the redox-regulated Keap1-dependent ubiquitin ligase complex. *The Journal of Biological Chemistry*, *281*(49), 37893–37903. https://doi.org/10.1074/jbc.M606539200.

Lo, S. C., & Hannink, M. (2008). PGAM5 tethers a ternary complex containing Keap1 and Nrf2 to mitochondria. *Experimental Cell Research*, *314*(8), 1789–1803. https://doi.org/10.1016/j.yexcr.2008.02.014.

Lu, W., Karuppagounder, S. S., Springer, D. A., Allen, M. D., Zheng, L., Chao, B., et al. (2014). Genetic deficiency of the mitochondrial protein PGAM5 causes a Parkinson's-like movement disorder. *Nature Communications*, *5*, 4930. https://doi.org/10.1038/ncomms5930.

Newman, T. N., Liverani, E., Ivanova, E., Russo, G. L., Carpino, N., Ganea, D., et al. (2014). Members of the novel UBASH3/STS/TULA family of cellular regulators suppress T-cell-driven inflammatory responses in vivo. *Immunology and Cell Biology*, *92*(10), 837–850. https://doi.org/10.1038/icb.2014.60.

Panda, S., Srivastava, S., Li, Z., Vaeth, M., Fuhs, S. R., Hunter, T., et al. (2016). Identification of PGAM5 as a mammalian protein histidine phosphatase that plays a central role to negatively regulate CD4(+) T cells. *Molecular Cell*, *63*(3), 457–469. https://doi.org/10.1016/j.molcel.2016.06.021.

Purich, D. L. (2010). *Enzyme kinetics: Catalysis & control*. Elsevier, (pp. 1–892).

Sadatomi, D., Tanimura, S., Ozaki, K., & Takeda, K. (2013). Atypical protein phosphatases: Emerging players in cellular signaling. *International Journal of Molecular Sciences*, *14*(3), 4596–4612. https://doi.org/10.3390/ijms14034596.

Segel, I. H. (1975). *Enzyme kinetics: Behavior and analysis of rapid equilibrium and steady state enzyme systems*. New York: Wiley.

Sekine, S., Kanamaru, Y., Koike, M., Nishihara, A., Okada, M., Kinoshita, H., et al. (2012). Rhomboid protease PARL mediates the mitochondrial membrane potential loss-induced cleavage of PGAM5. *The Journal of Biological Chemistry*, *287*(41), 34635–34645. https://doi.org/10.1074/jbc.M112.357509.

Takeda, K., Komuro, Y., Hayakawa, T., Oguchi, H., Ishida, Y., Murakami, S., et al. (2009). Mitochondrial phosphoglycerate mutase 5 uses alternate catalytic activity as a protein serine/threonine phosphatase to activate ASK1. *Proceedings of the National Academy of Sciences of the United States of America*, *106*(30), 12301–12305. https://doi.org/10.1073/pnas.0901823106.

Wang, Z., Jiang, H., Chen, S., Du, F., & Wang, X. (2012). The mitochondrial phosphatase PGAM5 functions at the convergence point of multiple necrotic death pathways. *Cell*, *148*(1–2), 228–243. https://doi.org/10.1016/j.cell.2011.11.030.

Wilkins, J. M., McConnell, C., Tipton, P. A., & Hannink, M. (2014). A conserved motif mediates both multimer formation and allosteric activation of phosphoglycerate mutase 5. *The Journal of Biological Chemistry*, *289*(36), 25137–25148. https://doi.org/10.1074/jbc.M114.565549.

Wu, P., Nielsen, T. E., & Clausen, M. H. (2015). FDA-approved small-molecule kinase inhibitors. *Trends in Pharmacological Sciences*, *36*(7), 422–439. https://doi.org/10.1016/j.tips.2015.04.005.

CHAPTER THIRTEEN

Purification of Lipin and Measurement of Phosphatidic Acid Phosphatase Activity from Liposomes

Mitchell E. Granade, Thurl E. Harris[1]
Department of Pharmacology, University of Virginia, Charlottesville, VA, United States
[1]Corresponding author: e-mail address: teh3c@eservices.virginia.edu

Contents

Abstract

The lipin family of enzymes are phosphatidic acid (PA) phosphatases responsible for converting PA to diacylglycerol (DAG). Lipins therefore occupy a central node in the synthesis of triacylglycerol (TAG) and phospholipids, and may play a role in regulating the levels of PA and DAG as signaling molecules. Some enzymatic assays used to measure PA phosphatase activities use detergents above their critical micelle concentration to present substrate; however, these methods do not represent the physiological membrane bilayers found in cells and these conditions can drastically alter phosphatase activities. Other assays use poorly defined mixtures of phosphatidylcholine (PC), PA, and high concentrations of BSA to present substrate. In this chapter, we describe methods for affinity purification of FLAG-tagged lipin proteins, and an alternative enzymatic assay using small unilamellar vesicles, also known as liposomes, to investigate specific activities of PA phosphatases. These activities are measured using an acidified Bligh–Dyer extraction to separate the water-soluble, radiolabeled, inorganic phosphate released during the assay from the chloroform-soluble PA.

Methods in Enzymology, Volume 607
ISSN 0076-6879
https://doi.org/10.1016/bs.mie.2018.04.028

ABBREVIATIONS

DAG diacylglycerol
PA phosphatidic acid
PAP phosphatidic acid phosphatase
PC phosphatidylcholine
PE phosphatidylethanolamine
TAG triacylglycerol
TLC thin-layer chromatography

1. INTRODUCTION

Lipins are a class of mammalian Mg^{2+}-dependent phosphatidic acid phosphatases (PAPs) that are centrally involved in the synthesis of triacylglycerol and phospholipids (Carman & Han, 2009; Han, Wu, & Carman, 2006; Péterfy, Phan, Xu, & Reue, 2001). The lipin family consists of three members (lipins 1–3) which convert phosphatidic acid (PA) to diacylglycerol (DAG) (Carman & Han, 2006, 2009; Reue & Zhang, 2008). Additionally, lipin 1 plays a role as a transcriptional coactivator in the activation of fatty acid oxidation genes (Finck et al., 2006). The lipin family members are amphiphilic, and their translocation from the cytosol to various membranes is required for their enzymatic activity (Brindley, 1985; Grimsey et al., 2008). The subcellular localization of the lipins is regulated by both nutrient and hormonal signaling and is thought to occur via posttranslation regulation (Elabbadi, Day, Virden, & Yeaman, 2002; Harris et al., 2007; Péterfy, Harris, Fujita, & Reue, 2010; Peterson et al., 2011). Phosphorylation has been shown to both regulate lipin subcellular localization (Harris et al., 2007; Péterfy et al., 2010; Peterson et al., 2011) and to negatively regulate the enzymatic activity of lipin 1, however, the mechanisms behind this regulation remain unclear (Boroda, Takkellapati, et al., 2017; Eaton, Mullins, Brindley, & Harris, 2013).

Amphiphilic lipid-modifying enzymes exhibit surface dilution kinetics due to the 2D surface nature of cellular membranes. Amphiphilic lipid-modifying enzymes depend not only on the bulk concentration of substrate, but also the surface concentration of substrate within the two-dimensional membrane surface (Carman, Deems, & Dennis, 1995). Properly assaying such enzymes requires the determination of kinetics for both the bulk step, in which the enzyme binds to the lipid surface, as well as the surface step, in which the enzyme binds to its lipid substrate molecule. The use of nonionic detergent micelles, particularly Triton X-100, has been suggested as an ideal method for such assays (Carman et al., 1995). Triton X-100 can form homogenous mixed micelles with a variety of lipids, and surface dilution experiments may be performed with minimal effects to the surface characteristics such as size, charge, or area per molecule. However, while micelles present uniform characteristics ideal for such kinetic analyses, they do not represent the native membrane bilayers that are the targets of lipid-modifying enzymes. It is therefore possible that various aspects of enzymatic activity may be constrained or overemphasized with this method of substrate presentation.

Because the lipin family members are amphiphilic, the method used to present the PA substrate can have drastic effects on specific activity. Indeed, the lipin family displays significantly higher specific activity when assayed in Triton micelles as compared to liposomes, and the molecular basis for this observation remains elusive (Boroda, Takkellapati, et al., 2017; Eaton et al., 2013). In addition, both the specific activity and the posttranslational regulation of lipin are clearly dependent on the context of substrate presentation (Boroda, Takkellapati, et al., 2017; Eaton et al., 2013, 2014). At physiological pH, PA exists as a monoanionic lipid species, yet lipin 1 has been shown to have preferential binding to dianionic PA (Eaton et al., 2013; Kooijman et al., 2005, 2007). The hydrogen bond switch mechanism has been proposed as the method by which positively charged lipids such as phosphatidylethanolamine (PE) stabilize dianionic PA at physiological pH to enhance lipin binding (Kooijman et al., 2007). At concentrations of 30 mol% PE, representative of the PE concentration in the endoplasmic reticulum leads to enhancement of lipin activity. Furthermore, phosphorylation has been shown to negatively regulate lipin 1 enzymatic activity only in the presence of PE (Eaton et al., 2013; Gomez-Muñoz et al., 1992). In this chapter, we describe a method for preparing unilamellar liposomes and their use for assaying PAP specific activity. Similar to previously described micelle methods, these activities are measured by extracting the water-soluble radiolabeled inorganic phosphates released during the reaction.

2. PURIFICATION OF LIPINS

Equipment

1. Disposable filter column (Fisher Scientific, 11-387-50)

Buffers and Reagents

1. Anti-FLAG M2 Affinity Gel beads (Sigma, A2220)
2. FLAG peptide (Sigma, F3290)

3. λ-Phosphatase (NEB, P0753L)
4. Phosphate buffered saline (PBS)
5. Lysis Buffer A (150 m*M* NaCl, 20 m*M* HEPES, 0.1% Brij 35, 10 μg/mL Leupeptin, 10 μg/mL Pepstatin A, 1 m*M* PMSF, 0.5 m*M* dithiothreitol, pH 7.2)
6. Phosphatase Buffer B (100 m*M* NaCl, 50 m*M* HEPES, 2 m*M* dithiothreitol, 1 m*M* $MnCl_2$, pH 7.0)
7. Wash Buffer C (150 m*M* NaCl, 20 m*M* HEPES, 10 μg/mL Leupeptin, 10 μg/mL Pepstatin A, 1 m*M* PMSF, 0.5 m*M* dithiothreitol, pH 7.2)
8. Elution Buffer D (150 m*M* NaCl, 20 m*M* HEPES, FLAG peptide 1.0 mg/mL, pH 7.2)

The cloning of the mammalian PAP enzymes allows assays to be run with purified, recombinant enzyme to enable accurate calculations of specific activity. This method may also be used with cleared cell lysates for qualitative comparisons of specific activity. The addition of the FLAG tag (Sigma, DYKDDDDK) allows for single-step affinity purification to a high degree of purity, as has been previously described (Gerace & Moazed, 2015; Tomomori-Sato, Sato, Conaway, & Conaway, 2013).

The lipin family are highly phosphorylated, and the phosphorylation status of lipin 1 correlates with its subcellular localization (Harris et al., 2007). As phosphorylation of the lipin family may impact enzymatic activity, it may be of interest to isolate lipin proteins in a nonphosphorylated form. This can be achieved postisolation by phosphatase treatment of the protein during purification.

1. For a typical purification of 100–500 μg of lipin divided into phosphatase treated and untreated, infect 40 × 150-mm plates of HeLa cells with an adenovirus containing an N-terminal FLAG tagged (DYKDDDDK) lipin construct. For best expression, we seek >90% infection efficiency. Transfection of HEK-293T cells may also be used (see note 1).
2. Allow the cells to grow for 48–72 h to achieve maximal expression. Add fresh media 24 h prior to harvesting.
3. Prior to harvesting cells, cool 500 mL of PBS on ice and place two sterile 50-mL conical tubes on ice. Additionally, place culture dishes of the infected HeLa cells on ice immediately prior to harvesting (see note 2).
4. Aspirate off all media and rinse the cells with 5 mL of cold PBS.
5. Aspirate off all PBS, then add another 2 mL of cold PBS. Use a scraper to dislodge all cells from the surface of the culture dish and pipette them into the sterile 50-mL conical tubes on ice. Centrifuge the harvested

cells at 200 × *g* for 5 min to pellet, then aspirate off all supernatant (see note 3).

6. Lysis Buffer A (150 m*M* NaCl, 20 m*M* HEPES, 0.1% Brij 35, pH 7.2) may be prepared in advance and stored at 4°C. Add protease inhibitors (10 μg/mL Leupeptin, 10 μg/mL Pepstatin A, 1 m*M* PMSF) and 0.5 m*M* dithiothreitol to Buffer A immediately prior to use. To each pellet, add 500 μL of lysis Buffer A per 150-mm plate harvested in the pellet.
7. Pass the cells through a 22-gauge needle 6–7 times using a 5-mL syringe to homogenize the cells and ensure complete shearing of cellular membranes.
8. Centrifuge homogenates at 16,000 × *g* at 4°C for 10 min.
9. During the 10 min centrifugation, prepare Anti-FLAG M2 Affinity Gel beads (Sigma, A2220). Use 15 μL compacted gel volume of beads per 150-mm plate harvested and centrifuge at 1000 × *g* for 1 min to pellet. Aspirate off the storage buffer from the beads and suspend the beads in 500 μL lysis Buffer A.
10. Combine homogenate supernatants into a 50-mL conical tube and add the Anti-FLAG M2 Affinity Gel beads to the homogenate. Use the homogenate to rinse the beads out of the microcentrifuge tube to collect all the beads.
11. Place the combined homogenate with affinity gel beads on a rotating wheel/with agitation at 4°C for 2–4 h.
12. During the incubation step, prepare phosphatase Buffer B (100 m*M* NaCl, 50 m*M* HEPES, pH 7.0). Add 2 m*M* dithiothreitol and 1 m*M* $MnCl_2$ immediately prior to use. Thaw FLAG peptide (Sigma, F3290) on ice during this period as well (see note 4).
13. Remove the homogenate from the rotator and centrifuge for 1 min at 200 × *g* to pellet the beads.
14. Aspirate off the supernatant, taking care not to remove any beads. Suspend the beads in 2 mL of lysis Buffer A and split the beads evenly between two 1.5-mL microcentrifuge tubes. Designate one tube for λ-phosphatase treatment and one for control. Label the tubes accordingly and do not mix the two groups for the remainder of the protocol.
15. Centrifuge the tubes for 1 min at 200 × *g* and check the beads to ensure they are evenly split. Aspirate off the lysis Buffer A and add 500 μL of phosphatase Buffer B to each tube to suspend the beads.

16. Centrifuge for 1 min at 200 × *g*, aspirate off the supernatant, and add 200 μL of phosphatase Buffer B to each tube.
17. Add 5 μL (2000 units) of λ-phosphatase (NEB, P0753L) to the tube designated for λ-phosphatase treatment and none to the control. Incubate both tubes with agitation at 30°C for 25 min (see note 5).
18. During the incubation step, prepare the following:
    a. Wash Buffer C (150 m*M* NaCl, 20 m*M* HEPES, pH 7.2). Immediately before use add 10 μg/mL Leupeptin, 10 μg/mL Pepstatin A, 1 m*M* PMSF, and 0.5 m*M* dithiothreitol
    b. Elution Buffer D (150 m*M* NaCl, 20 m*M* HEPES, pH 7.2), immediately before use add FLAG peptide to a final concentration of 1.0 mg/mL
19. Transfer the beads to two separate 15-mL conical tubes and add 7 mL of lysis Buffer A to each. Gently mix the beads by inverting the tubes several times. Centrifuge for 1 min at 200 × *g* to pellet the beads and aspirate off the supernatants. Repeat the wash with another 7 mL of lysis Buffer A to each and centrifuge to pellet.
20. For the final wash step, aspirate off the supernatants and add 7 mL of wash Buffer C to each tube. Gently mix the beads by inverting the tubes several times. Centrifuge for 1 min at 200 × *g* to pellet the beads.
21. Aspirate the supernatants down to a volume of around 1 mL, then transfer the beads to two separate 1.5-mL microcentrifuge tubes. Centrifuge for 1 min at 200 × *g* to pellet the beads and aspirate off all remaining wash Buffer C from each.
22. Add 500 μL of elution Buffer D to each tube. Incubate tubes with agitation at 4°C for 15 min to elute. While incubating, equilibrate two screening columns with lysis Buffer A and label appropriately. Place each equilibrated screening column over an appropriately labeled 1.5-mL microcentrifuge tube.
23. Remove the microcentrifuge tubes from agitation and centrifuge for 1 min at 200 × *g*. Carefully transfer the supernatant of each tube to the appropriate screening column, allow to flow through, and catch the eluate in the microcentrifuge tubes. Place the eluates on ice.
24. Add a second fraction of 500 μL of elution Buffer D to each of the bead pellets and repeat the incubation and elution in step 22. This time transfer the beads and eluate to the appropriate screening columns. The second elution can be combined with the first or kept separate. If combining, briefly vortex the combined fractions to mix. If desired the eluted protein can be dialyzed against Buffer C.

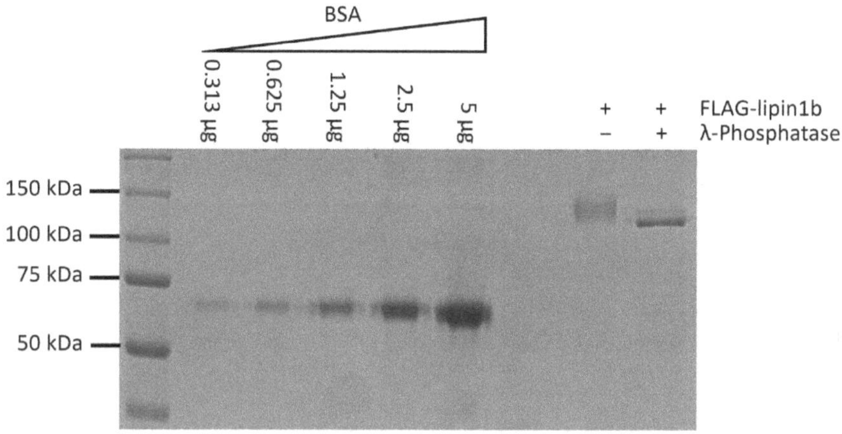

	Concentration(µg/mL)	Purity
-λ	79.7	92.8%
+λ	47.2	92.3%

Fig. 1 SDS-PAGE gel stained with Coomassie Brilliant Blue G-250 showing a BSA standard curve and a typical yield from 20 × 150-mm plates of HeLa cells infected with lipin 1 adenovirus and split between untreated sample and λ-phosphatase treated sample. Concentration and purity values are listed for a 1 mL yield of purified protein.

25. Divide the purified protein into aliquots (20–50 µL). Store the aliquots at −80°C. Freeze–thaw cycles will reduce specific activity, and the aliquots should be used or discarded after thawing (see note 6).
26. Proteins may be quantified by running the purified proteins on a polyacrylamide gel with a standard curve of serially diluted BSA standards, and staining with Coomassie Brilliant Blue G-250, or other appropriate protein stain (Fig. 1).

Notes

1. For large-scale purification, we have found that overexpression of lipin family members in a fast-growing cell line like HeLa or HEK-293T cells is optimal. Lipin family proteins can be efficiently and economically expressed by adenovirus expression or via transient transfection using previously described protocols (Boussif et al., 1995; Ehrhardt et al., 2006; He et al., 1998; Luo et al., 2007). Transfections are allowed 48–72 h of expression before harvesting.
2. Lipins are heat sensitive and are quickly inhibited by incubation at higher temperatures (Han & Carman, 2010). Lysates and proteins should be kept at 4°C throughout the purification process.

3. Harvested cell pellets may be stored at −80°C prior to immunoprecipitation.
4. Flag peptide (Sigma, F3290) is suspended at a stock concentration of 5 mg/mL in elution Buffer D and stored at −20°C.
5. We have found that degradation of the lipin protein occurs when treated with λ-phosphatase at 30°C for a prolonged time.
6. We have observed that prolonged storage of purified lipin protein, even when stored at −80°C, also reduces specific activity. For best results, use purified enzyme within 3 months.

3. PAP ASSAY

Equipment

1. Thin-layer chromatography (TLC) plates, silica gel 60, 20 cm × 20 cm
2. Mini-Extruder set (Avanti, 610000)
3. Polycarbonate membrane, 0.1 μm pore size (Sigma, WHA800309)
4. Filter supports (Sigma, WHA230300)

Buffers and Reagents

1.5 × DGK buffer (250 m*M* imidazole-HCl, 250 m*M* octyl-β-D-glucopyranoside, 250 m*M* NaCl, 62.5 m*M* $MgCl_2$, 5 m*M* EGTA, pH 6.6)

1. *Escherichia coli* DAG kinase, 1 mg/mL, 2 U/mg (Sigma, D3065)
2. Cardiolipin
3. Dioleoyl-diacylglycerol
4. Dioleoyl-phosphatidic acid (DOPA)
5. Dioleoyl-phosphatidylcholine (DOPC)
6. Dioleoyl-phosphatidylethanolamine (DOPE)
7. β-Mercaptoethanol
8. 1 *M* $MgCl_2$
9. 100 m*M* ATP
10. [γ-^{32}P]ATP (6000 Ci/mmol, 150 mCi/mL)
11. 500 m*M* Tris–HCl, pH 7.5
12. PAP Buffer E (50 m*M* Tris–HCl, 0.5 m*M* $MgCl_2$, 10 m*M* β-mercaptoethanol, pH 7.5)
13. Chloroform/methanol/1 N HCl (1:2:0.8, v/v)
14. Chloroform/methanol/water (65:25:4, v/v)
15. 0.1 N HCl
16. 0.1 N HCl in methanol

3.1 Preparation of PA Substrate

To prepare the ^{32}P radiolabeled PA substrate, *E. coli* DAG kinase is used to enzymatically phosphorylate DAG using [γ-^{32}P]ATP as previously described (Carman & Lin, 1991; Han & Carman, 2004). PA is then isolated by TLC.

Caution: The generation of [^{32}P]PA substrate and its subsequent incorporation into liposomes involves multiple steps using ^{32}P with a relatively high-specific activity. Be sure to minimize exposure by keeping manipulations behind a ½″ or greater acrylic shield. Use of a radioactive material handling glove box is recommended.

1. Mix 1 μmol of DAG and 9 μg of cardiolipin in a 1.5-mL microcentrifuge tube and dry down under N_2 gas (see note 1).
2. To the dried lipids, add 20 μL of 5 × DGK buffer (250 m*M* imidazole-HCl, 250 m*M* octyl-β-D-glucopyranoside, 250 m*M* NaCl, 62.5 m*M* $MgCl_2$, 5 m*M* EGTA, pH 6.6), 10 μL of 100 m*M* β-mercaptoethanol, 5 μL of 100 m*M* ATP, 5 μL [γ-^{32}P]ATP (6000 Ci/mmol, 150 mCi/mL), and 40 μL of water (see note 2).
3. Add 20 μL of *E. coli* DAG kinase to start the reaction. Mix and incubate with agitation for 45 min at 30°C.
4. Stop the reaction with the addition of 0.5 mL of 0.1 N HCl in methanol and transfer to a 15-mL polypropylene tube.
5. Add 1.0 mL of chloroform, followed by 1.5 mL of 1 *M* $MgCl_2$. Vortex and centrifuge for 1 min at 100 × *g*.
6. Transfer the organic (bottom) phase to a new polypropylene tube, add 0.5 mL of 0.1 N HCl in methanol followed by 1.5 mL of 1 *M* $MgCl_2$. Vortex and centrifuge for 1 min at 100 × *g*.
7. Transfer the organic (bottom) phase to a glass tube and dry down under N_2 gas to around 100 μL total volume.
8. Spot the sample onto a TLC plate. Rinse the tube with 50 μL of chloroform and spot the sample again.
9. Develop the TLC plate in a TLC chamber with the chloroform/methanol/water (65:25:4, v/v) running buffer. Allow the buffer to run until 2–3 cm from the top of the plate, approximately 1–2 h.
10. Remove the TLC plate from the chamber and allow to dry in a fume hood (5–10 min).
11. Wrap the TLC plate in plastic wrap and expose to an autoradiography film for approximately 1 min.
12. Develop the film.

13. Remove the plastic wrap from the TLC plate and align the developed film. Mark the region of [^{32}P]PA through the film using a razor blade (see note 3).
14. Moisten the radioactive region with water, place a sheet of weighing paper underneath the TLC plate, and use the razor blade to scrape the silica within the region off the plate and onto the weighing paper.
15. Transfer the silica into a 15-mL polypropylene tube and add 1.0 mL of chloroform/methanol/1 N HCl (1:2:0.8, v/v) (see note 4).
16. Vortex vigorously and centrifuge for 1 min at $100 \times g$. Transfer the supernatant into a new 15-mL polypropylene tube.
17. Repeat the extraction in steps 15 and 16 with the silica scrapings and combine the extractions.
18. Add 0.5 mL of chloroform followed by 0.6 mL of 0.1 N HCl (in water) to the combined extractions. Vortex and centrifuge for 1 min at $100 \times g$.
19. Transfer the organic (bottom) phase to a new 15-mL polypropylene tube. Add 1 mL of methanol and 1 mL of 0.1 N HCl (in water) to combined extractions, vortex, and centrifuge for 1 min at $200 \times g$.
20. Transfer the organic (bottom) phase to a 2-mL glass tube with cap. Store at −20°C.
21. Add 5–10 μL of the [^{32}P]PA to scintillation vials in duplicate and count with a scintillation counter to determine the radioactive concentration (cpm/μL). Radioactive concentration will need to be determined each day that liposomes are prepared (see Section 3.2) to calculate the volume of [^{32}P]PA to incorporate into liposomes.

Notes

1. Cardiolipin is added along with other components of the DG kinase buffer to provide optimal conditions for *E. coli* DG kinase (Walsh & Bell, 1986).
2. 5 × DGK buffer may be prepared in advance and stored at −20°C.
3. During a typical separation by TLC, PA migrates to around 4–5 cm from the solvent front. A good quality separation of PA should be free of smearing.
4. The chloroform/methanol/1 N HCl (1:2:0.8, v/v) mixture should not phase separate.

3.2 Preparation of Liposomes

PA phosphatase activities are measured using 100 nm liposomes containing 10 mol% PA at a bulk concentration of 1 m*M*. Lipin activity is induced by dianionic PA, and the phosphoregulation of lipin 1 activity only becomes apparent in the presence of dianionic PA. The use of liposomes containing PE stabilizes dianionic PA at a physiological pH (Kooijman et al., 2007). In this protocol, 100 nm liposomes are generated using the mini-extruder from Avanti Lipids Polar, Inc. (Avanti, 610000) with 100 nm pore size polycarbonate membranes according to manufacturer recommendations. The preparation of liposomes using extrusion techniques has been extensively covered in previous protocols (Mui, Chow, & Hope, 2003; Zhang, 2017). Liposomes should be used for phosphatase assays immediately after they are generated.

1. At room temperature, prepare PA phosphatase Buffer E (50 m*M* Tris–HCl, 0.5 m*M* $MgCl_2$, 10 m*M* β-mercaptoethanol, pH 7.5) (see note 1).
2. In a glass tube, mix 9 μmol of DOPC with 1 μmol of DOPA dissolved in chloroform. This tube will be used to generate PC:PA (9:1) liposomes. In a second glass tube, mix 6 μmol of DOPC, 3 μmol of DOPE, and 1 μmol of DOPA dissolved in chloroform. This tube will be used to generate PC:PE:PA (6:3:1) liposomes.
3. To each tube, add $[^{32}P]$PA to 5000 cpm/nmol DOPA. Mix and use N_2 gas to dry down each tube to a thin film (see note 2).
4. Suspend the lipid film in 1 mL of PA phosphatase Buffer E. Pipette up and down, then vortex vigorously to ensure complete suspension. Transfer the lipid suspensions into a 1.5-mL microcentrifuge tube (see note 3).
5. To disrupt large multilamellar vesicles, snap freeze the lipids in liquid N_2, then place in a room temperature water bath and allow to thaw. Vortex vigorously when completely thawed. Repeat to perform five freeze/thaw cycles.
6. Use a 1 mL mini-extruder to pass the lipid suspension through a polycarbonate filter membrane with a 0.1 μm pore size. Pass through the membrane 11 times.
7. Carefully transfer the prepared liposomes into a 1.5-mL microcentrifuge tube and place on ice while preparing phosphatase assays.

Notes

1. $MgCl_2$ and β-mercaptoethanol should be added immediately prior to use.
2. Generating a thin, even film of lipids will reduce aggregation during subsequent lipid suspensions and liposome preparations. While drying under N_2 gas, angle the glass tubes to the side and rotate slowly to dry the lipids in a thin film around the bottom of the tube.

3. Suspending lipids using vortexing alone may cause the lipids to aggregate, particularly in mixtures containing PE. Mixing up and down with a pipette will help to disperse any lipid aggregates.

3.3 PA Phosphatase Assay

In this assay, specific activity is determined by measuring the radiolabeled, inorganic phosphates released by the enzyme during the reaction. An acidified Bligh–Dyer extraction is used to separate the water-soluble inorganic phosphates from the chloroform-soluble PA. For example, Michaelis–Menten steady-state kinetic constants may be determined by increasing bulk concentration of PA, while maintaining a constant PA surface concentration. All assay conditions are performed in triplicate.

1. In 1.5-mL microcentrifuge tubes, combine PA phosphatase Buffer E and appropriate liposomes to a total volume of 90 μL. For example, create a range of bulk concentrations of PA by varying the amount of liposomes added to generate a binding curve of velocity vs substrate concentration (Fig. 2) (see note 1).
2. Start each phosphatase reaction with the addition of 10 μL of purified enzyme (see note 2).
3. Incubate each reaction with agitation for 20 min at 30°C.
4. Stop each reaction with the addition of 0.5 mL of 0.1 N HCl in methanol. Transfer to a polypropylene centrifuge tube (see note 3).
5. Add 1 mL of chloroform followed by 1 mL of 1 *M* $MgCl_2$. Mix by gentle vortexing and centrifuge for 1 min at 100 × *g*.

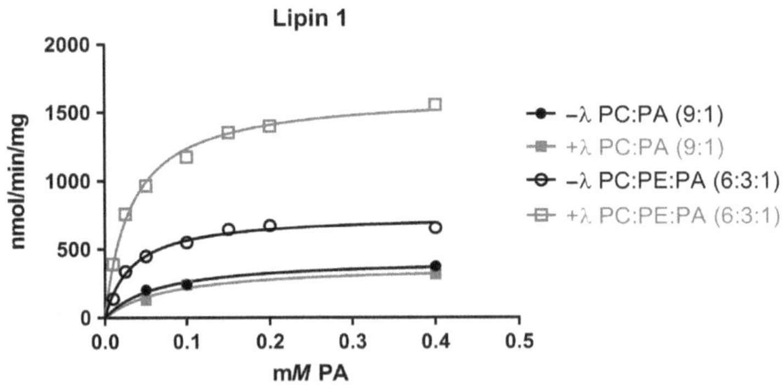

Fig. 2 Velocity vs substrate concentration for lipin 1 PAP activity in PC:PA (9:1) and PC:PE:PA (6:3:1) liposomes. Binding curves are shown for control treated (−λ) and λ-phosphatase treated (+λ) lipin 1 (Eaton et al., 2013).

6. Transfer 0.5 mL of the aqueous (top) phase to a scintillation vial. Samples may then be measured directly or with the addition of 4 mL of scintillation fluid for aqueous samples.
7. Prepare scintillation vials with direct aliquots of liposomes and measure radioactivity (specific label) for data analysis.

Notes

1. A typical range of bulk concentrations of PA may start at 0.025 m*M* PA and increase as high as 0.8 m*M* PA.
2. As always, it is crucial to ensure that the reaction is linear with respect to time and concentration. Thus, the amount of purified enzyme added may need to be adjusted. For the lipin family, we have found that 30–50 ng usually falls within the linear range.
3. To ensure consistent reaction times, stop each reaction in the same order in which they were begun.

3.4 Analysis of Data

Specific activity for an enzyme is expressed as U/mg of protein, or nmol of product/min/mg. Calculate the specific activity using the following:

$$\text{Specific activity}(\text{nmol}/\min/\text{mg}) = \frac{3.2\ (\text{correction factor}) * \text{cpm}\ (\text{corrected for background})}{\text{specific label}\left(\frac{\text{cpm}}{\text{nmol}}\right) * \text{time}\ (\text{mins}) * \text{purified enzyme}\ (\text{mg})}$$

Notes

1. A correction factor of 3.2 is applied for measuring 0.5 mL of aqueous phase from a total 1.6 mL volume of aqueous phase in step 6 under the PA Phosphatase assay.
2. Specific label, measured as described by step 7 under the PA Phosphatase assay, refers to the counts per min per nmol of PA within the liposomes.

4. SUMMARY AND CONCLUSION

Amphiphilic lipid enzymes must translocate to membrane surfaces in order to access substrate and catalyze their reactions. Translocation between cytosolic and membrane fractions is regulated for many such amphiphilic enzymes and seems to target them to specific cellular membranes (Cornell & Northwood, 2000; Harris et al., 2007; Merino, Sanjuán, Moraga, Ciprés, & Mérida, 2007). The method of substrate presentation can play a critical role in the regulation and specificity of translocation, as is the case for lipin 1 in which the presence of PE is necessary for appropriate phosphoregulation of PAP activity at physiological pH (Fig. 2) (Eaton et al., 2013). Current methods using detergent micelles or BSA to present lipid substrates display drastically altered in vitro enzyme kinetics when compared with lipid bilayers, and do not represent the native presentation of lipid substrates in vivo (Boroda, Takkellapati, et al., 2017; Eaton et al., 2013). Liposomes provide an ideal method for a physiological presentation of substrate by enabling the generation of well-defined lipid bilayers. We have described a method for using liposomes to assay lipin PAP activity, however, this method was also used to measure the specific activity of DAG kinases by making minor alterations to the reaction buffer and lipid composition, and measuring the incorporation of ^{32}P into PA in the organic phase (Boroda, Niccum, Raje, Purow, & Harris, 2017). This method may therefore be more broadly applied to other amphiphilic lipid enzymes with minor modifications.

REFERENCES

Boroda, S., Niccum, M., Raje, V., Purow, B. W., & Harris, T. E. (2017). Dual activities of ritanserin and R59022 as DGKα inhibitors and serotonin receptor antagonists. *Biochemical Pharmacology*, *123*, 29–39. https://doi.org/10.1016/j.bcp.2016.10.011.

Boroda, S., Takkellapati, S., Lawrence, R. T., Entwisle, S. W., Pearson, J. M., Granade, M. E., et al. (2017). The phosphatidic acid-binding, polybasic domain is responsible for the differences in the phosphoregulation of lipins 1 and 3. *Journal of Biological Chemistry*, *292*(50), 20481–20493. https://doi.org/10.1074/jbc.M117.786574.

Boussif, O., Lezoualc'h, F., Zanta, M. A., Mergny, M. D., Scherman, D., Demeneix, B., et al. (1995). A versatile vector for gene and oligonucleotide transfer into cells in culture and in vivo: Polyethylenimine. *Proceedings of the National Academy of Sciences*, *92*(16), 7297–7301. https://doi.org/10.1073/pnas.92.16.7297.

Brindley, D. N. (1985). Intracellular translocation of phosphatidate phosphohydrolase and its possible role in the control of glycerolipid synthesis. *Progress in Lipid Research*, *23*, 115–133.

Carman, G. M., & Han, G. S. (2006). Roles of phosphatidate phosphatase enzymes in lipid metabolism. *Trends in Biochemical Sciences*, *31*(12), 694–699. https://doi.org/10.1016/j.tibs.2006.10.003.

Carman, G. M., & Han, G. S. (2009). Phosphatidic acid phosphatase, a key enzyme in the regulation of lipid synthesis. *Journal of Biological Chemistry*, *284*(5), 2593–2597. https://doi.org/10.1074/jbc.R800059200.

Carman, G. M., & Lin, Y.-P. (1991). Phosphatidate phosphatase from yeast. *Methods in Enzymology*, *197*(1991), 548–553. https://doi.org/10.1128/EC.2.6.1304.

Carman, G. M., Deems, A. D., & Dennis, E. A. (1995). Lipid signaling enzymes and surface dilution kinetics. *Journal of Biological Chemistry*, *270*(32), 18711–18714.

Cornell, R. B., & Northwood, I. C. (2000). Regulation of CTP: Phosphocholine cytidylyltransferase by amphitropism and relocalization. *Trends in Biochemical Sciences*, *25*, 441–447. https://doi.org/10.1016/s0968-0004(00)01625-x.

Eaton, J. M., Mullins, G. R., Brindley, D. N., & Harris, T. E. (2013). Phosphorylation of lipin 1 and charge on the phosphatidic acid head group control its phosphatidic acid phosphatase activity and membrane association. *Journal of Biological Chemistry*, *288*(14), 9933–9945. https://doi.org/10.1074/jbc.M112.441493.

Eaton, J. M., Takkellapati, S., Lawrence, R. T., McQueeney, K. E., Boroda, S., Mullins, G. R., et al. (2014). Lipin 2 binds phosphatidic acid by the electrostatic hydrogen bond switch mechanism independent of phosphorylation. *Journal of Biological Chemistry*, *289*(26), 18055–18066. https://doi.org/10.1074/jbc.M114.547604.

Ehrhardt, C., Schmolke, M., Matzke, A., Knoblauch, A., Will, C., Wixler, V., et al. (2006). Polyethylenimine, a cost-effective transfection reagent. *Signal Transduction*, *6*(3), 179–184. https://doi.org/10.1002/sita.200500073.

Elabbadi, N., Day, C. P., Virden, R., & Yeaman, S. J. (2002). Regulation of phosphatidic acid phosphohydrolase 1 by fatty acids. *Lipids*, *37*(1), 69–73. https://doi.org/10.1007/s11745-002-0865-7.

Finck, B. N., Gropler, M. C., Chen, Z., Leone, T. C., Croce, M. A., Harris, T. E., et al. (2006). Lipin 1 is an inducible amplifier of the hepatic PGC-1α/PPARα regulatory pathway. *Cell Metabolism*, *4*(3), 199–210. https://doi.org/10.1016/j.cmet.2006.08.005.

Gerace, E., & Moazed, D. (2015). Affinity pull-down of proteins using anti-FLAG M2 agarose beads. *Methods in Enzymology*, *559*, 99–110. https://doi.org/10.1016/bs.mie.2014.11.010.

Gomez-Muñoz, A., Hatch, G. M., Martin, A., Jamal, Z., Vance, D. E., & Brindley, D. N. (1992). Effects of okadaic acid on the activities of two distinct phosphatidate phosphohydrolases in rat hepatocytes. *FEBS Letters*, *301*(1), 103–106. https://doi.org/10.1016/0014-5793(92)80219-7.

Grimsey, N., Han, G. S., O'Hara, L., Rochford, J. J., Carman, G. M., & Siniossoglou, S. (2008). Temporal and spatial regulation of the phosphatidate phosphatases lipin 1 and 2. *Journal of Biological Chemistry*, *283*(43), 29166–29174. https://doi.org/10.1074/jbc.M804278200.

Han, G.-S., & Carman, G. M. (2004). Assaying lipid phosphate phosphatase activities. In R. C. Dickson & M. D. Mendenhall (Eds.), *Methods in molecular biology: Vol. 284. Signal transduction protocols:* Humana Press. https://doi.org/10.1385/1-59259-816-1:209.

Han, G. S., & Carman, G. M. (2010). Characterization of the human LPIN1-encoded phosphatidate phosphatase isoforms. *Journal of Biological Chemistry*, *285*(19), 14628–14638. https://doi.org/10.1074/jbc.M110.117747.

Han, G. S., Wu, W. I., & Carman, G. M. (2006). The Saccharomyces cerevisiae lipin homolog is a Mg2+-dependent phosphatidate phosphatase enzyme. *Journal of Biological Chemistry*, *281*(14), 9210–9218. https://doi.org/10.1074/jbc.M600425200.

Harris, T. E., Huffman, T. A., Chi, A., Shabanowitz, J., Hunt, D. F., Kumar, A., et al. (2007). Insulin controls subcellular localization and multisite phosphorylation of the phosphatidic acid phosphatase, lipin 1. *Journal of Biological Chemistry*, *282*(1), 277–286. https://doi.org/10.1074/jbc.M609537200.

He, T.-C., Zhou, S., da Costa, L. T., Yu, J., Kinzler, K. W., & Vogelstein, B. (1998). A simplified system for generating recombinant adenoviruses. *Proceedings of the National Academy of Sciences of the United States of America*, *95*(5), 2509–2514. https://doi.org/10.1073/pnas.95.5.2509.

Kooijman, E. E., Carter, K. M., Van Laar, E. G., Chupin, V., Burger, K. N. J., & De Kruijff, B. (2005). What makes the bioactive lipids phosphatidic acid and lysophosphatidic acid so special? *Biochemistry*, *44*(51), 17007–17015. https://doi.org/10.1021/bi0518794.

Kooijman, E. E., Tieleman, D. P., Testerink, C., Munnik, T., Rijkers, D. T. S., Burger, K. N. J., et al. (2007). An electrostatic/hydrogen bond switch as the basis for the specific interaction of phosphatidic acid with proteins. *Journal of Biological Chemistry*, *282*(15), 11356–11364. https://doi.org/10.1074/jbc.M609737200.

Luo, J., Deng, Z. L., Luo, X., Tang, N., Song, W. X., Chen, J., et al. (2007). A protocol for rapid generation of recombinant adenoviruses using the AdEasy system. *Nature Protocols*, *2*(5), 1236–1247. https://doi.org/10.1038/nprot.2007.135.

Merino, E., Sanjuán, M. A., Moraga, I., Ciprés, A., & Mérida, I. (2007). Role of the diacylglycerol kinase α-conserved domains in membrane targeting in intact T cells. *Journal of Biological Chemistry*, *282*(48), 35396–35404. https://doi.org/10.1074/jbc.M702085200.

Mui, B., Chow, L., & Hope, M. J. (2003). Extrusion technique to generate liposomes of defined size. *Methods in Enzymology*, *367*(1980), 3–14. https://doi.org/10.1016/S0076-6879(03)67001-1.

Péterfy, M., Harris, T. E., Fujita, N., & Reue, K. (2010). Insulin-stimulated interaction with 14-3-3 promotes cytoplasmic localization of lipin-1 in adipocytes. *Journal of Biological Chemistry*, *285*(6), 3857–3864. https://doi.org/10.1074/jbc.M109.072488.

Péterfy, M., Phan, J., Xu, P., & Reue, K. (2001). Lipodystrophy in the fld mouse results from mutation of a new gene encoding a nuclear protein, lipin. *Nature Genetics*, *27*(1), 121–124. https://doi.org/10.1038/83685.

Peterson, T. R., Sengupta, S. S., Harris, T. E., Carmack, A. E., Kang, S. A., Balderas, E., et al. (2011). MTOR complex 1 regulates lipin 1 localization to control the SREBP pathway. *Cell*, *146*(3), 408–420. https://doi.org/10.1016/j.cell.2011.06.034.

Reue, K., & Zhang, P. (2008). The lipin protein family: Dual roles in lipid biosynthesis and gene expression. *FEBS Letters*, *582*(1), 90–96. https://doi.org/10.3174/ajnr.A1256. Functional.

Tomomori-Sato, C., Sato, S., Conaway, R. C., & Conaway, J. W. (2013). Immunoaffinity purification of protein complexes from mammalian cells. *Methods in Molecular Biology*, *977*, 273–287. https://doi.org/10.1007/978-1-62703-284-1.

Walsh, J. P., & Bell, R. M. (1986). 1,2-Diacylglycerol kinase of Escherichia coli. *Journal of Biological Chemistry*, *261*(14), 6239–6247.

Zhang, H. (2017). Thin-film hydration followed by extrusion method for liposome preparation. In G. D'Souza (Ed.), *Methods in molecular biology: Vol. 1522* (2nd ed., pp. 17–22). New York, NY: Humana Press. https://doi.org/10.1007/978-1-4939-2917-7.

CHAPTER FOURTEEN

Combining SAMDI Mass Spectrometry and Peptide Arrays to Profile Phosphatase Activities

Lindsey C. Szymczak*, Che-Fan Huang*, Eric J. Berns†, Milan Mrksich*,†,1

*Department of Chemistry, Northwestern University, Evanston, IL, United States
†Department of Biomedical Engineering, Northwestern University, Evanston, IL, United States
[1]Corresponding author: e-mail address: milan.mrksich@northwestern.edu

Contents

Abstract

Phosphatases, the enzymes responsible for dephosphorylating proteins, play critical roles in many cellular processes. While their importance is widely recognized, phosphatase activity and regulation remain poorly understood. Currently, there are few assays available that are capable of directly measuring phosphatase activity and specificity. We have previously introduced SAMDI (self-assembled monolayers on gold for matrix-assisted laser desorption/ionization) mass spectrometry as a technique to profile the substrate specificities of enzymes. SAMDI mass spectrometry assays are well suited to examine phosphatase activities and offer many advantages over current methods. This technique uses monolayers that terminate with a peptide or molecular enzyme substrate and allows for enzyme reactions to be performed on a surface that can easily

Methods in Enzymology, Volume 607
ISSN 0076-6879
https://doi.org/10.1016/bs.mie.2018.04.021

be rinsed and analyzed by mass spectrometry without the need for analyte labeling. In this chapter, we describe the process of combining SAMDI mass spectrometry with peptide arrays to study the substrate specificities of two protein tyrosine phosphatases.

ABBREVIATIONS

AUC area under the curve
DMF dimethylformamide
MALDI matrix-assisted laser desorption/ionization
MS mass spectrometry
PTP protein tyrosine phosphatase
SAM self-assembled monolayer
SAMDI self-assembled monolayers of alkanethiolates on gold for matrix-assisted laser desorption/ionization
TFA trifluoroacetic acid
TOF time-of-flight

1. INTRODUCTION

Phosphorylation is an important posttranslational modification that plays a critical role in most signaling pathways. Kinases and phosphatases act in dynamic balance to maintain the phosphoproteome and regulate cellular processes. Just as with kinases, mutation and disruption of phosphatase activity have been implicated in many human diseases (Hendriks et al., 2013; Julien, Dubé, Hardy, & Tremblay, 2011; Peng & Maller, 2010). However, phosphatases remain poorly understood relative to kinases. A significant factor in this discrepancy is the lack of convenient and general assays that can measure phosphatase activity. Unlike kinases, for which many effective assay formats are available, it remains difficult to measure phosphatase activity, particularly in complex samples such as cell lysates (Geladopoulos, Sotiroudis, & Evangelopoulos, 1991; Takai & Mieskes, 1991; Welte et al., 2005). Current assays for studying phosphatase activity rely on measuring decreases in the amount of phosphosubstrate—typically with ELISA or a ^{32}P-labeled substrate—or by measuring the phosphate ion by-product (Bose & Janes, 2013; Killilea, Cheng, & Wang, 1998; McAvoy & Nairn, 2010).

Our laboratory has developed SAMDI-MS (self-assembled monolayers of alkanethiolates on gold for matrix-assisted laser desorption/ionization mass spectrometry), a label-free, high-throughput analytical method for measuring enzyme activities, including phosphatase activity

(Berns, Cabezas, & Mrksich, 2016; Gurard-Levin, Scholle, Eisenberg, & Mrksich, 2011; Mrksich, 2008; Su & Mrksich, 2002). SAMDI-MS uses self-assembled monolayers (SAMs) of alkanethiolates on gold that present a peptide (or other molecule) substrate for the enzyme of interest. The monolayers are prepared by immersing a gold-coated plate in a solution containing terminally substituted dialkyl disulfides. Most commonly, the monolayers present maleimide groups at a density of 10% against a tri(ethylene glycol) background (Fig. 1). Cysteine-terminated peptides can then be immobilized through conjugate addition to the maleimide groups. The tri(ethylene glycol) groups serve the important role of preventing nonspecific protein adsorption to the monolayer.

To perform the SAMDI assay, an enzyme solution is directly applied to the monolayer. After the enzyme reaction is complete, the SAMDI plate is simply rinsed with water and then ethanol. The MALDI matrix is applied, and the plate is analyzed in an AB Sciex 5800 MALDI TOF/TOF. In the mass spectrometer, irradiation of the monolayer with the laser leads to cleavage of the gold–thiolate bond and release of the intact peptide–alkanethiolate (or corresponding disulfide) conjugates, whose molecular masses are then determined by time-of-flight (TOF). The SAMDI spectrum includes m/z peaks that correspond to the substrate and product(s). For phosphatase activity, the product appears at an m/z that is 80 Da lower than the substrate (Fig. 2).

Peptide arrays have emerged as an approach for studying the substrate specificities of a variety of enzymes (Arsenault, Griebel, & Napper, 2011;

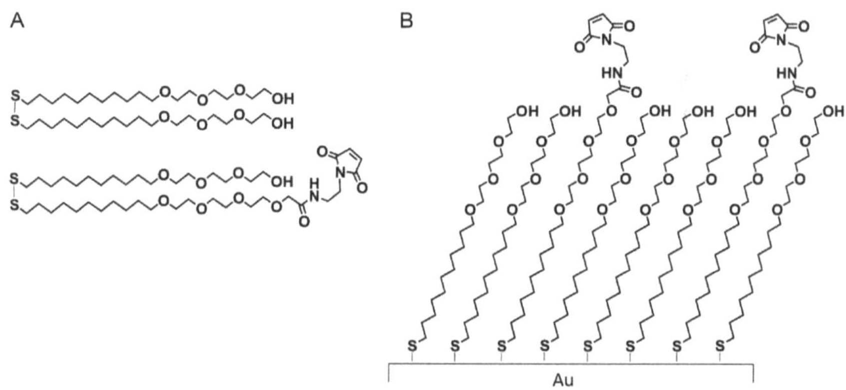

Fig. 1 Formation of self-assembled monolayers on gold. (A) Chemical structures of tri(ethylene glycol) disulfide and tri(ethylene glycol)-maleimide disulfide. (B) Alkanethiolate monolayer self-assembled on gold. The SAMs present functional maleimide groups with a background of tri(ethylene glycol) groups.

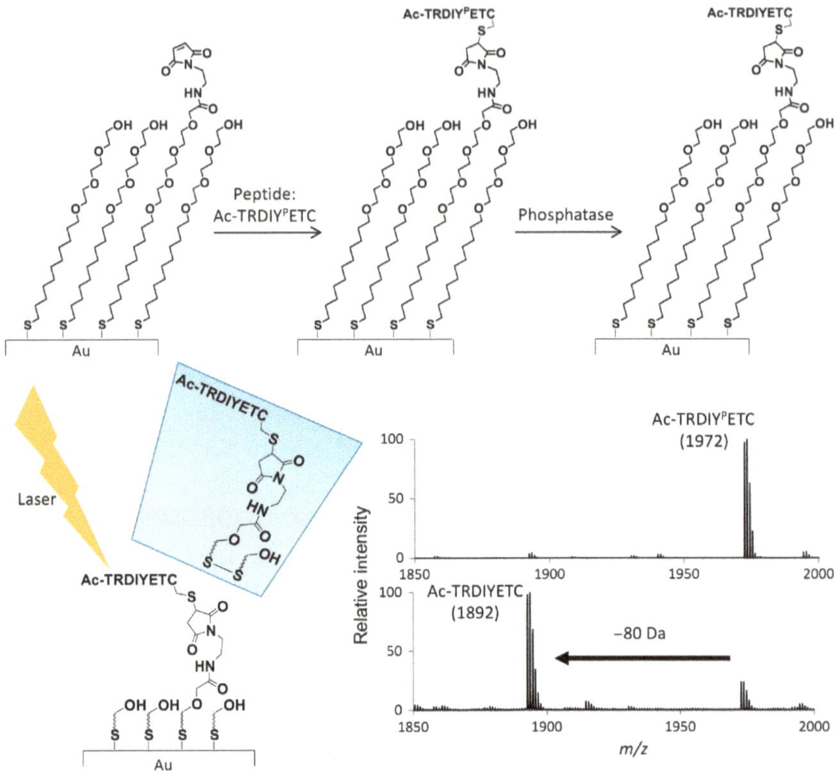

Fig. 2 SAMDI assay workflow. Peptides are immobilized on self-assembled monolayers. After immobilization, peptides are treated with phosphatase solutions. After washing and matrix application, SAMDI plates are analyzed via MALDI mass spectrometry.

Foong, Fu, Yao, & Uttamchandani, 2012; Katz et al., 2011; Szymczak, Kuo, & Mrksich, 2017; Thiele, Stangl, & Schutkowski, 2011). Peptide arrays allow enzyme activities to be measured on a broad range of substrates, giving a more complete understanding of specificity. A SAMDI assay is typically performed on an array plate of 384 (or 1536) gold spots (Fig. 3), where a different peptide is immobilized onto each spot. A solution containing the enzyme of interest is applied to each gold spot, and after reaction and analysis by SAMDI-MS, the extent of enzyme activity on each peptide in the array is quantified in a single experiment (Fig. 4).

We have combined SAMDI-MS and peptide arrays to profile enzyme substrate specificities using large libraries of peptides or other substrates. These examples include profiling lysine deacetylase (KDAC) specificities (Gurard-Levin, Kilian, Kim, Bahr, & Mrksich, 2010; Gurard-Levin,

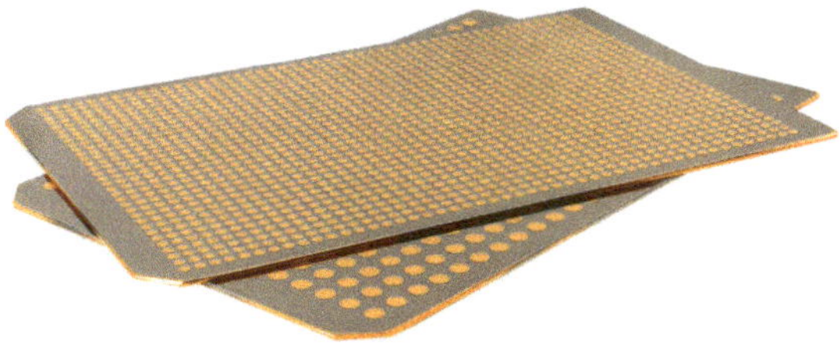

Fig. 3 Gold-evaporated SAMDI array plates: 384 (*bottom*)- and 1536 (*top*) formats.

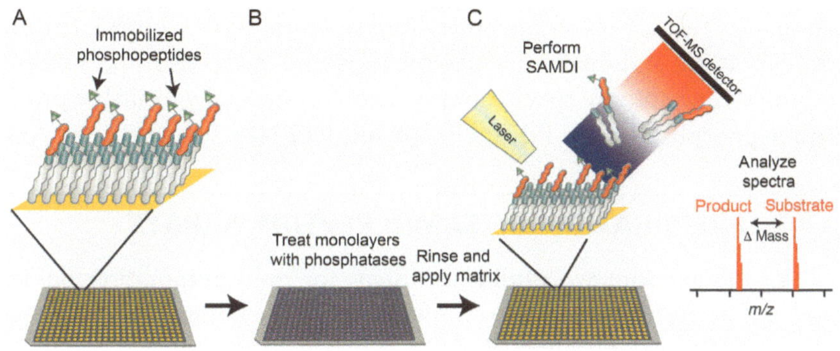

Fig. 4 SAMDI array workflow. (A) Peptides in a peptide library are immobilized on self-assembled monolayers. (B) Peptides on the surface are treated with phosphatase solutions. (C) After rinsing and matrix application, plates are analyzed via MALDI mass spectrometry.

Kim, & Mrksich, 2009), as well as the substrate specificity of the protease OmpT (Wood et al., 2017). Additionally, we have used SAMDI and peptide arrays to optimize substrates for glycosyl transferases, and we have used carbohydrate arrays to discover new glycosyl transferase activities (Ban et al., 2012; Kightlinger et al., 2018). SAMDI peptide arrays are also compatible with measuring enzyme activities in cell lysates, as shown in a study observing distinct profiles of deacetylase activity associated with differentiation (Kuo, DeLuca, Miller, & Mrksich, 2013).

There are several advantages to the SAMDI method that allow for straightforward high-throughput experiments to study enzyme activities. The use of the protein-resistant monolayer is important for experiments using complex samples because it ensures that the immobilized peptides are accessible to the enzyme and removes the need for blocking steps or analyte purification after reaction completion, allowing for a direct measurement of enzyme activity. After running a SAMDI experiment, analytes are not processed or transferred to another platform for analysis; the monolayer array plate is used both to perform the reactions and to measure the reaction product. Additionally, the use of mass spectrometry for analysis yields quantitative measurements of enzyme activity without the need for substrate or product labeling. These benefits make the SAMDI assay amenable to measuring enzyme activity in isolation or in cell lysates on many different peptide substrates.

In this chapter, we describe how to prepare and perform a SAMDI experiment with a phosphotyrosine peptide library to profile phosphatase activity. We will detail the steps of evaporation of gold onto steel plates, formation of the monolayer, peptide library design and synthesis, peptide immobilization, and finally the enzyme assay and analysis. In this example, we profile the substrate specificities of two commercially available protein tyrosine phosphatases (PTPs), PTPN14 and PTP1B.

2. PREPARATION OF SAMDI PEPTIDE ARRAYS

In this section, we outline the steps for gold evaporation on steel plates, self-assembled monolayer formation, peptide library design and synthesis, and peptide immobilization.

2.1 Gold Evaporation on Steel Plates

Equipment

- 12.25 cm × 8 cm Steel plates (the same size as standard MALDI loading plates)
- Thermionics e-beam evaporator with a water chiller and turbo pump
- Patterned masks with 384 spots for steel plates

Buffers and reagents

- Hexanes
- Ethanol
- Milli-Q H_2O
- Titanium
- Gold

Procedure

Washing Steel Plates

1. Remove protective coating from six steel plates.
2. Wash steel plates in hexanes and then ethanol.
3. Let plates soak in water and then ethanol for 5 min each.
4. Dry plates under N_2.

Evaporation

1. Place six steel plates into a plate holder with masks that pattern the plates to have 384 spots.
2. Load the plate holder into a Thermionics e-beam evaporator.
3. Seal the evaporator, vacuum-pressurize to 1×10^{-6} Torr, and chill to 10°C.
4. Deposit titanium onto the plates at a rate of 0.2 Å/s until 50 Å have been deposited.
5. Deposit gold onto the titanium-coated plates at a rate of 0.5 Å/s until 300 Å have been deposited.
6. Depressurize the evaporator, warm back to room temperature, and remove the plates.
7. Store the plates in vacuum-sealed packs for up to 6 months.

2.2 Self-Assembled Monolayer Formation

Buffers and reagents

- 1 m*M* total disulfide solution of 0.8 m*M* tri(ethylene glycol)-terminated C_{11}-alkane disulfide and 0.2 m*M* C_{11}-alkane disulfide with one terminal tri(ethylene glycol) and one terminal maleimide in ethanol (Fig. 1A)
- Ethanol
- Milli-Q H_2O

Procedure

1. Prepare a 1 m*M* total disulfide solution of 0.8 m*M* tri(ethylene glycol) disulfide and 0.2 m*M* tri(ethylene glycol)-maleimide disulfide in ethanol, which will produce a monolayer surface with 10% maleimide density (Fig. 1).
2. Soak plates in monolayer solution for 48 h at 4°C to allow monolayers to self-assemble onto the gold surfaces on the steel plates by forming gold–thiolate bonds.
3. Remove plates from monolayer solution and wash with ethanol, then water, and ethanol again. Dry under N_2. Store in vacuum-sealed bag at −20°C for up to 1 month or use immediately.

Notes

- The plates now have self-assembled monolayers on 384 gold spots presenting a functional maleimide group at 10% density. The maleimide group undergoes 1,4-Michael addition with thiol-containing molecules, allowing for immobilization of 384 unique thiol-containing molecules. The tri(ethylene glycol)background at 90% density functions to prevent nonspecific protein adsorption to the surface.

2.3 Phosphotyrosine Peptide Library Design

A peptide from the IR5 insulin receptor (TRDIY^{P}ETDYYRK) with sequence surrounding pY1146 has been used widely for kinase and phosphatase research (Keane et al., 1994; Madden, Bird, Man, Raven, & Myles, 1991; Stadtmauer & Rosen, 1986; Yu et al., 2013). We designed the pY library to study the importance of the specific amino acids surrounding pY1146. Our library consists of 361 peptides with the sequence Ac-TRD*X*Y^{P}*Z*TC-NH_2, where the *X* and *Z* positions represent all 19 amino acids, excluding cysteine. A cysteine is added to the original sequence at the C-terminus for peptide–maleimide immobilization to the SAM surface, as described earlier. We choose to make the peptides eight amino acids in length. Variable amino acid positions are placed at the +1 and −1 positions surrounding the phosphorylated tyrosine to study the effects and roles of those positions in phosphatase-substrate recognition. SAMDI peptide libraries can be designed to study any of the amino acid positions of interest (Gurard-Levin, Kilian, Kim, Bahr, & Mrksich, 2010; Kuo et al., 2013).

2.4 Peptide Synthesis

Cysteine-containing peptides are synthesized using standard solid-phase peptide synthesis methods (Høeg-Jensen, Havsteen Jakobsen, Olsen, & Holm, 1991). Briefly, Fmoc-Rink Amide MBHA resin is purchased from Anaspec, and Fmoc-protected amino acids are available from multiple commercial sources. The Fmoc-Rink Amide MBHA resin is swelled in dimethylformamide (DMF) for 30 min. The resin is then Fmoc-deprotected with 20% piperidine in DMF for 20 min. The resin is rinsed with DMF by vacuum filtration. The first amino acid is coupled to the resin with PyBop and *N*-methylmorpholine in DMF at a 4:4:8 molar excess, respectively, for 20 min. The resin is washed with DMF. The deprotection, wash, coupling, and wash steps are then repeated until the last amino acid is coupled. The Fmoc group on the last amino acid is deprotected as described previously (Høeg-Jensen et al., 1991), and the N-terminus is acetylated with 10% acetic anhydride in DMF for 60 min. The resin is washed with DMF and dichloromethane and dried in a vacuum desiccator. The peptides are cleaved from the resin with a solution of 95% trifluoroacetic acid (TFA), 2.5% triethylsilane, and 2.5% water for 16 h—many peptide cleavage protocols will suggest shorter times, but we have found that longer cleavage times result in more complete deprotection of the pY residue. The cleavage solution is evaporated using N_2. The peptides are resuspended in 0.1% TFA in water, lyophilized, and resuspended again in 0.1% TFA in water to a peptide concentration of 200 μM.

2.5 Peptide Immobilization

Equipment

- Tecan Robotic Liquid Handler (or pipette by hand)
- Humidifying chamber (to prevent evaporation—simple option: use an empty pipette tip box with water in the bottom)
- Plate centrifuge

Buffers and reagents

- 200 μ*M* peptide in 0.1% TFA in H_2O
- TCEP beads
- 50 m*M* Tris buffer pH 7.5

Procedure

1. Dilute 5 μL of each peptide in 0.1% TFA in water into 40 μL 50 m*M* Tris buffer pH 7.5 and 5 μL of a TCEP bead slurry to prevent the formation of disulfides. The peptides will now be at a concentration of 20 μ*M*.
2. Spin the plate at 3000 rpm for 1 min to sediment the TCEP beads.
3. Pipette or use a robotic liquid handler to transfer each peptide to an individual spot on the SAMDI array plate.
4. Put the SAMDI array plate in a humidifying chamber and incubate for 1 h at 37°C to immobilize peptides.
5. Wash the SAMDI array plate with water followed by ethanol. Dry under N_2. Store in vacuum-sealed bag at −20°C.

Notes

- The cysteine-containing peptides are immobilized through conjugate addition of the thiol side chain of cysteine and the maleimide groups of the monolayer.

3. PROFILING PHOSPHATASE ACTIVITY USING SAMDI-MS

In this section, we describe the steps of running and analyzing a phosphatase assay using SAMDI-MS.

3.1 SAMDI Phosphatase Assay

Equipment

- Multidrop combi and small tube plastic tip dispensing cassette (Thermo Scientific)

Buffers and reagents

- PTPN14, active human recombinant protein 5 mg/mL (Sigma-Aldrich)
- PTP1B full-length, active human 0.5 mg/mL (Sigma-Aldrich)

- Phosphatase buffer: 100 m*M* Tris–HCl (Tris(hydroxymethyl)-aminomethane hydrochloride) pH 7.5, 50 m*M* NaCl, 100 μ*M* TCEP (Tris(2-carboxyethyl)phosphine) made fresh

Procedure

1. Thaw the PTPN14 and PTP1B stock (aliquots stored at −80°C) on ice.
2. Dilute 4 μL PTPN14 into 2 mL phosphatase buffer (final concentration: 1 μg/mL). Gently mix with pipette.
3. Rinse the combi cassette with 10 mL deionized water and 10 mL phosphatase buffer.
4. Use the combi to apply 2 μL PTPN14 solution to each gold spot on the SAMDI plate.
5. Place the plate in a humidifying chamber to avoid evaporation and incubate at 37°C for 6 h.
6. Quench the reaction by rinsing the plate with deionized water (>50 mL).
7. Rinse the plate with ethanol and dry under N_2. Analyze by MALDI-MS immediately or place the plate in a vacuum-sealed bag and store at −20°C for up to 2 weeks.
8. For PTP1B: Dilute 1 μL PTP1B into 2.5 mL phosphatase buffer (final concentration: 0.2 μg/mL). Mix and repeat steps 4–7 with PTP1B solution. The reaction incubation time for PTP1B is 1 h.

Notes

- The SAM surface is sensitive to reducing agents. The TCEP concentration should not exceed 500 μ*M*.
- The concentration of enzyme and time of reaction should be determined in advance of the array experiment. We use a standard peptide, Ac-IY^{P}ERC-NH_2, immobilized on the surface, and examine the extent of phosphatase activity at various enzyme concentrations to determine the optimal conditions for the peptide-array experiment.

3.2 SAMDI-MS

Equipment

- MALDI (AB Sciex TOF/TOF 5800 System)
- Multidrop combi and small tube plastic tip dispensing cassette (Thermo Scientific)

Buffers and reagents

- Phosphopeptide matrix solution: 10 mg/mL THAP (2,4,6-trihydroxyacetophenone) and 5 mg/mL ammonium citrate dibasic in 50% acetonitrile, 50% deionized water, and 0.1% phosphoric acid.

Procedure

1. If SAMDI plate was stored at −20°C, thaw at room temperature for 15 min, rinse the plate with ethanol, and dry under N_2.
2. Rinse the combi cassette with 50% acetonitrile and 50% deionized water.
3. Use the combi to apply 1 μL phosphopeptide matrix solution to each gold spot on the plate.
4. Allow the matrix to dry and crystallize for 20 min.
5. Read the plate by MALDI-MS using MS positive reflector mode. Spectra are collected in the mass range of 1300–2500 Da with the focus on 1900 Da. A 355 nm Nd:YAG laser is used as a desorption/ionization source with intensity set at 5200; 900 laser shots are collected per spectrum on each gold spot. Detector voltage multiplier is set at 0.72.

Notes

- The parameters for MALDI settings vary from day to day. It is best to calibrate the parameters using a standard peptide for optimal SAMDI signal.
- Addition of ammonium citrate to the MALDI matrix has been shown to enhance the ionization efficiency of phosphopeptides (Asara & Allison, 1999).

3.3 Analyzing the SAMDI Array

Procedure

1. Identify the dialkyl disulfide substrate and product peaks (−80 m/z) in each of the 361 spectra. The dialkyl disulfide peak is the original peptide mass plus 850.5 m/z for the H^+ adduct. Other adduct peaks, including Na^+ and K^+, are often observed and can be included in the analysis.
2. Integrate the area under the curve (AUC) for each monoisotopic peak using the software provided by the mass spectrometer vendor (Fig. 5).
3. Calculate activities for each peptide according to the definition Activity $(\%) = AUC_{product}/(AUC_{substrate} + AUC_{product}) \times 100\%$.
4. Using Microsoft Excel, arrange the activities into a 19 × 19 matrix where each column represents an amino acid at *X* position and each row represents an amino acid at *Z* position in the pY library so that every box is the PTP activity on a unique combination of *X* and *Z*.
5. Create a heatmap by transforming the numeric chart into gray scale (conditional formatting color scales). Define 100% activity as black and 0% activity as white.

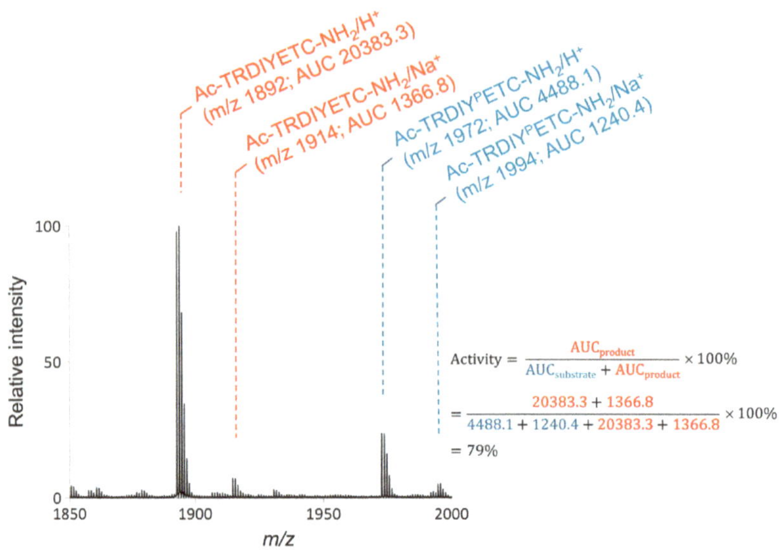

Fig. 5 Analysis of the mass spectrum to measure phosphatase activity on a peptide. The mass spectrum quantifies the amounts of unreacted phosphorylated substrate and reacted dephosphorylated product. We calculate the area under the curve (AUC) to determine the extent of dephosphorylation of the peptide. The spectra above shows both the substrate and product peaks as well as their integrated areas. To calculate percent activity, we divide the AUC of the product peak(s) by the AUC of both the substrate and product peaks.

Notes

- Methionine (M) and Tryptophan (W) residues are often found oxidized in MALDI measurements. Therefore, oxidation peaks (+17 m/z for M and +33 m/z for W) should be included for peptides containing these residues.
- We use custom software to automate the process described earlier and are able to analyze the data from a library in less than 5 min.

3.4 Data Interpretation

The heatmaps of PTPN14 and PTP1B on the pY library are shown in Fig. 6. The heatmaps show that PTPN14 prefers substrates with a glycine (G) in the + 1 (Z) position and a tryptophan (W) in the −1 (X) position. PTP1B shows different specificity disfavoring substrates with a proline (P) in the +1 (Z) position. These different heatmap profiles help us determine the substrate specificity of phosphatases and allow us to begin to understand recognition and the active-site structure of the enzyme.

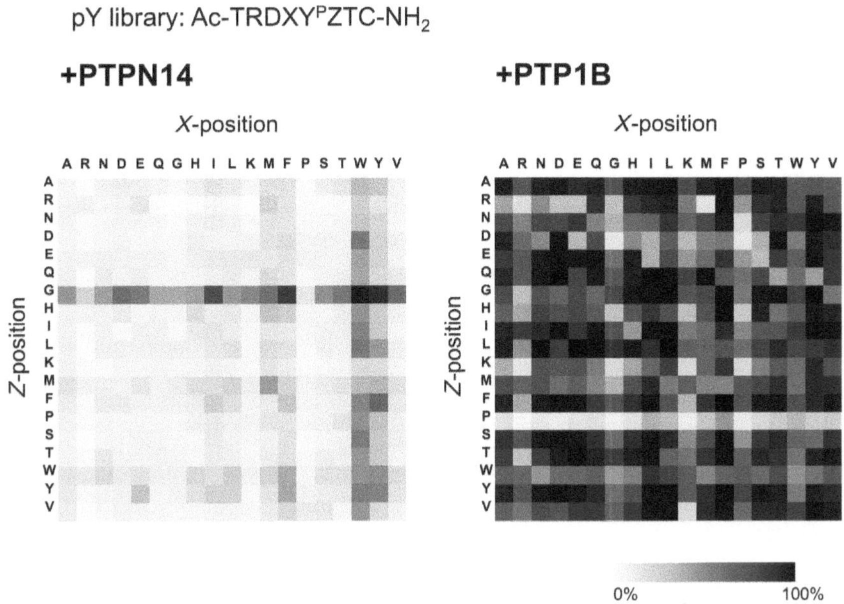

Fig. 6 Heatmaps of PTPN14 (*left*) and PTP1B (*right*) specificity on pY library peptide array. PTPN14 exhibits a clear preference for glycine (G) in the *Z* (+1) position and tryptophan (W) in the *X* (−1) position. PTP1B exhibits different sequence selectivity, where proline (P) is disfavored at *Z* (+1) position.

4. SUMMARY AND CONCLUSION

Insight into the specificities of phosphatase enzymes is important for a better understanding of the regulation of the phosphoproteome and its role in various cellular processes. This work describes the use of peptide arrays and SAMDI-MS to profile phosphatase activities. SAMDI-MS is a powerful tool for understanding the specificities and regulation of various enzymes, but is particularly significant for phosphatases as there are few other existing assays capable of directly measuring dephosphorylation. The advantages of the SAMDI method—including the use of an inert surface background preventing nonspecific protein adsorption, the lack of a need for analyte labeling, the compatibility with high-throughput experiments, and the use of mass spectrometry for direct and quantitative data—make it easily extendable to measuring phosphatase activities in complex samples such as cell lysates.

ACKNOWLEDGMENTS

Our recent work with phosphatases has been supported by the National Cancer Institute of the National Institute of Health under Award Number U54CA199091; the Department of Defense, Defense Threat Reduction Agency under Award Number HDTRA1-15-1-0052; the Air Force Research Laboratory under Award Number FA8650-15-2-5518; and the NU-NTU Institute for Nano Medicine located at the International Institute for Nanotechnology, Northwestern University, USA; and the Nanyang Technological University, Singapore, under Award Number Agmt10/20/14.

REFERENCES

Arsenault, R., Griebel, P., & Napper, S. (2011). Peptide arrays for kinome analysis: New opportunities and remaining challenges. *Proteomics*, *11*(24), 4595–4609.

Asara, J. M., & Allison, J. (1999). Enhanced detection of phosphopeptides in matrix-assisted laser desorption/ionization mass spectrometry using ammonium salts. *Journal of the American Society for Mass Spectrometry*, *10*(1), 35–44.

Ban, L., Pettit, N., Li, L., Stuparu, A. D., Cai, L., Chen, W., et al. (2012). Discovery of glycosyltransferases using carbohydrate arrays and mass spectrometry. *Nature Chemical Biology*, *8*(9), 769–773.

Berns, E. J., Cabezas, M. D., & Mrksich, M. (2016). Cellular assays with a molecular endpoint measured by SAMDI mass spectrometry. *Small*, *12*(28), 3811–3818.

Bose, A. K., & Janes, K. A. (2013). A high-throughput assay for phosphoprotein-specific phosphatase activity in cellular extracts. *Molecular & Cellular Proteomics*, *12*(3), 797–806.

Foong, Y. M., Fu, J., Yao, S. Q., & Uttamchandani, M. (2012). Current advances in peptide and small molecule microarray technologies. *Current Opinion in Chemical Biology*, *16*(1–2), 234–242.

Geladopoulos, T. P., Sotiroudis, T. G., & Evangelopoulos, A. E. (1991). A malachite green colorimetric assay for protein phosphatase activity. *Analytical Biochemistry*, *192*(1), 112–116.

Gurard-Levin, Z. A., Kilian, K. A., Kim, J., Bahr, K., & Mrksich, M. (2010). Peptide arrays identify isoform-selective substrates for profiling endogenous lysine deacetylase activity. *ACS Chemical Biology*, *5*(9), 863–873.

Gurard-Levin, Z. A., Kim, J., & Mrksich, M. (2009). Combining mass spectrometry and peptide arrays to profile the specificities of the histone deacetylases. *Chembiochem: A European Journal of Chemical Biology*, *10*(13), 2159–2161.

Gurard-Levin, Z. A., Scholle, M. D., Eisenberg, A. H., & Mrksich, M. (2011). High-throughput screening of small molecule libraries using SAMDI mass spectrometry. *ACS Combinatorial Science*, *13*(4), 347–350.

Hendriks, W. J. A. J., Elson, A., Harroch, S., Pulido, R., Stoker, A., & den Hertog, J. (2013). Protein tyrosine phosphatases in health and disease. *FEBS Journal*, *280*(2), 708–730.

Høeg-Jensen, T., Havsteen Jakobsen, M., Olsen, C. E., & Holm, A. (1991). Formation of peptide thioamides by use of Fmoc amino monothioacids and PyBOP. *Tetrahedron Letters*, *32*(51), 7617–7620.

Julien, S. G., Dubé, N., Hardy, S., & Tremblay, M. L. (2011). Inside the human cancer tyrosine phosphatome. *Nature Reviews Cancer*, *11*, 35–49.

Katz, C., Levy-Beladev, L., Rotem-Bamberger, S., Rito, T., Rudiger, S. G. D., & Friedler, A. (2011). Studying protein-protein interactions using peptide arrays. *Chemical Society Reviews*, *40*(5), 2131–2145.

Keane, N. E., Chavanieu, A., Quirk, P. G., Evans, J. S., Levine, B. A., Calas, B., et al. (1994). Structural determinants of substrate selection by the human insulin-receptor protein-tyrosine kinase. *European Journal of Biochemistry*, *226*(2), 525–536.

Kightlinger, W., Lin, L., Rosztoczy, M., Li, W., DeLisa, M. P., Mrksich, M., et al. (2018). Design of glycosylation sites by rapid synthesis and analysis of glycosyltransferases. *Nature Chemical Biology*. https://doi.org/10.1038/s41589-018-0051-2.

Killilea, S. D., Cheng, Q., & Wang, Z.-X. (1998). Protein phosphatase type 1 and type 2A assays. In J. W. Ludlow (Ed.), *Protein phosphatase protocols* (pp. 23–33). Totowa, NJ: Humana Press.

Kuo, H.-Y., DeLuca, T. A., Miller, W. M., & Mrksich, M. (2013). Profiling deacetylase activities in cell lysates with peptide arrays and SAMDI mass spectrometry. *Analytical Chemistry*, *85*(22), 10635–10642.

Madden, J. A., Bird, M. I., Man, Y., Raven, T., & Myles, D. D. (1991). Two nonradioactive assays for phosphotyrosine phosphatases with activity toward the insulin receptor. *Analytical Biochemistry*, *199*(2), 210–215.

McAvoy, T., & Nairn, A. C. (2010). Serine/threonine protein phosphatase assays. *Current Protocols in Molecular Biology*, *92*. 18.18.11-18.18.11.

Mrksich, M. (2008). Mass spectrometry of self-assembled monolayers: A new tool for molecular surface science. *ACS Nano*, *2*(1), 7–18.

Peng, A., & Maller, J. L. (2010). Serine/threonine phosphatases in the DNA damage response and cancer. *Oncogene*, *29*(45), 5977–5988.

Stadtmauer, L., & Rosen, O. M. (1986). Increasing the cAMP content of IM-9 cells alters the phosphorylation state and protein kinase activity of the insulin receptor. *Journal of Biological Chemistry*, *261*(7), 3402–3407.

Su, J., & Mrksich, M. (2002). Using mass spectrometry to characterize self-assembled monolayers presenting peptides, proteins, and carbohydrates. *Angewandte Chemie International Edition*, *41*(24), 4715–4718.

Szymczak, L. C., Kuo, H.-Y., & Mrksich, M. (2018). Peptide arrays: Development and application. *Analytical Chemistry*, *90*, 266–282.

Takai, A., & Mieskes, G. (1991). Inhibitory effect of okadaic acid on the p-nitrophenyl phosphate phosphatase activity of protein phosphatases. *The Biochemical Journal*, *275*(Pt. 1), 233–239.

Thiele, A., Stangl, G. I., & Schutkowski, M. (2011). Deciphering enzyme function using peptide arrays. *Molecular Biotechnology*, *49*(3), 283–305.

Welte, S., Baringhaus, K. H., Schmider, W., Muller, G., Petry, S., & Tennagels, N. (2005). 6,8-Difluoro-4-methylumbiliferyl phosphate: A fluorogenic substrate for protein tyrosine phosphatases. *Analytical Biochemistry*, *338*(1), 32–38.

Wood, S. E., Sinsinbar, G., Gudlur, S., Nallani, M., Huang, C.-F., Liedberg, B., et al. (2017). A bottom-up proteomic approach to identify substrate specificity of outer-membrane protease OmpT. *Angewandte Chemie International Edition*, *56*(52), 16531–16535.

Yu, B., Liu, W., Yu, W.-M., Loh, M. L., Alter, S., Guvench, O., et al. (2013). Targeting protein tyrosine phosphatase SHP2 for the treatment of PTPN11-associated malignancies. *Molecular Cancer Therapeutics*, *12*(9), 1738–1748.

CHAPTER FIFTEEN

Protein Chemical Approaches to Understanding PTEN Lipid Phosphatase Regulation

Daniel R. Dempsey[*,†,‡], **Philip A. Cole**[*,†,‡,1]
[*]Division of Genetics, Brigham and Women's Hospital, Boston, MA, United States
[†]Department of Medicine, Harvard Medical School, Boston, MA, United States
[‡]Department of Biological Chemistry and Molecular Pharmacology, Harvard Medical School, Boston, MA, United States
[1]Corresponding author: e-mail address: pacole@bwh.harvard.edu

Contents

Abstract

Since the discovery of C-tail phosphorylation of PTEN almost 20 years ago, much progress has been made in understanding its regulatory influences on the cellular function of PTEN. Phosphorylation of Ser380, Thr382, Thr383, and Ser385 drives a PTEN conformational change from an open to closed state where catalytic function is impaired, plasma membrane binding is reduced, and cellular stability is enhanced. Despite these advances, a detailed structural and mechanistic model of how these phosphorylations impact PTEN function is lacking. We discuss here several recent approaches to analyzing PTEN phosphorylation and highlight several insights that have come from this work. We also discuss remaining challenges for the PTEN regulation field and potential directions for future research.

Methods in Enzymology, Volume 607
ISSN 0076-6879
https://doi.org/10.1016/bs.mie.2018.05.007

ABBREVIATIONS

AP alkaline phosphatase
Bpa benzoyl-L-phenylalanine
E. coli *Escherichia coli*
EPL expressed protein ligation
MESNA sodium 2-mercaptoethanesulfonate
NCL native chemical ligation
PIP2 phosphatidylinositol (4,5)-trisphosphate
PIP3 hosphatidylinositol (3,4,5)-trisphosphate
PTEN phosphatase and tensin homolog
PTMs posttranslational modifications
Ub ubiquitin

1. INTRODUCTION

Phosphatase and tensin homolog (PTEN) functions as a tumor suppressor by negatively regulating the PI3-kinase/AKT kinase signaling pathway (Fig. 1). It is frequently mutated in many cancers resulting in a loss of function leading to unregulated cell growth, proliferation, and protein synthesis. Originally, PTEN was thought to be a protein tyrosine phosphatase

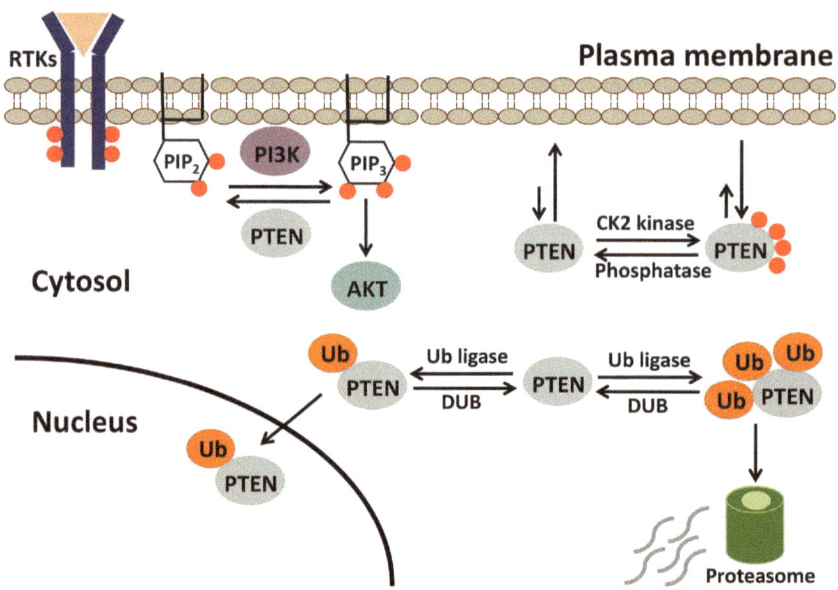

Fig. 1 Posttranslational regulation of PTEN and its cellular influence.

based on its conserved catalytic motif of HCxxGxxR (Li & Sun, 1997; Li et al., 1997; Myers et al., 1997); however, in 1998 Maehama and Dixon elegantly demonstrated that PTEN is a lipid phosphatase that dephosphorylates the 3′-phosphate of phosphatidylinositol (3,4,5)-trisphosphate, providing the molecular basis for antagonism to PI3-kinase (Maehama & Dixon, 1998; Myers et al., 1998; Sun et al., 1999; Worby & Dixon, 2014).

PTEN is a 403-amino acid protein that is localized to the cytosol, plasma membrane, and nucleus. The architecture of PTEN consists of an N-terminal phosphatase domain followed by a C2 domain and a flexible C-terminal tail (Fig. 2) (Lee et al., 1999). These domains enable and regulate the catalytic function of PTEN to dephosphorylate PIP3 at the plasma membrane. The N-terminal phosphatase domain spans residues 1–185 and is responsible for plasma membrane binding and catalysis. It contains an N-terminal PIP2-binding motif believed to be important for PTEN activation and plasma membrane binding that works together with the phosphatase domain to achieve normal enzymatic function (Campbell, Liu, & Ross, 2003; McConnachie, Pass, Walker, & Downes, 2003; Redfern et al., 2008). The C2 domain is important for binding to the phospholipid bilayer in a Ca^{2+}-independent manner. The C2 domain also makes extensive contacts with the phosphatase domain that stabilize the overall protein

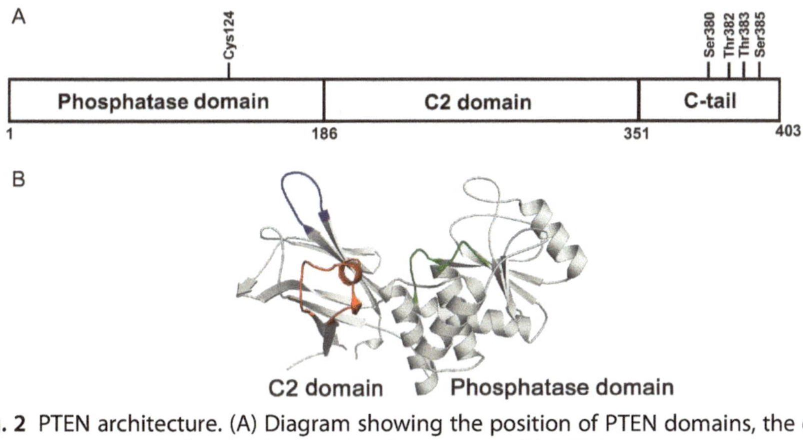

Fig. 2 PTEN architecture. (A) Diagram showing the position of PTEN domains, the catalytic cysteine, and the C-tail phosphorylation sites. (B) PTEN X-ray crystal structure (1D5R) highlighting the location of the active site (*green*), CBRIII loop (*blue*), and Cα2 loop (*red*). The crystal structure displays the phosphatase domain (aa 14–186) and C2 domain (aa 186–351) but structural information for the N-terminus (Δaa 1–13), D loop (Δaa 282–312), and C-terminal tail (Δaa 352–403) is missing. The CBRIII and Cα2 loops are highlighted because of their importance to membrane binding and the closed conformation of PTEN as discussed in Section 4.

fold (Lee et al., 1999). The C-terminal tail is considered to be intrinsically disordered and is suggested to influence the stability, localization, protein–protein interactions, and enzymatic activity of PTEN. Scattered through these domains are numerous posttranslational modifications (PTMs) including phosphorylation (Al-Khouri, Ma, Togo, Williams, & Mustelin, 2005; Cordier et al., 2012; Miller, Lou, Seldin, Lane, & Neel, 2002; Torres & Pulido, 2001; Vazquez, Ramaswamy, Nakamura, & Sellers, 2000), acetylation (Ikenoue, Inoki, Zhao, & Guan, 2008; Meng, Jia, & Gan, 2016; Okumura et al., 2006), sumoylation (Huang et al., 2012), ubiquitination (Wang et al., 2007), and oxidation (Lee et al., 2002) which are suggested to modulate PTEN in various ways. A cluster of phosphorylation sites on the C-terminal tail at positions Ser380, Thr382, Thr383, and Ser385 have been the focus of numerous studies. Casein kinase 2 or glycogen synthase kinase 3β has been reported to catalyze these PTEN phosphorylations (Al-Khouri et al., 2005; Torres & Pulido, 2001). Phosphorylation of these residues seems to convert PTEN to a more closed conformational state characterized by lower enzymatic activity, reduced plasma membrane binding, and increased stability (Bolduc et al., 2013; Das, Dixon, & Cho, 2003; Masson, Perisic, Burke, & Williams, 2016; Nguyen, Afkari, et al., 2014; Odriozola, Singh, Hoang, & Chan, 2007; Rahdar et al., 2009; Vazquez et al., 2000). Key to understanding the regulation of PTEN by these phosphorylations requires access to PTEN protein possessing stoichiometric and site-specific modifications for mechanistic studies. Recent efforts have led to a pair of semisynthetic strategies to install phosphorylations at each site, classical expressed protein ligation (EPL) (Bolduc et al., 2013; Muir, Sondhi, & Cole, 1998), and a related technique enzyme-catalyzed EPL (Henager et al., 2016). In Sections 2–5 we discuss these methods and the analysis of modified semisynthetic PTENs using biochemical and biophysical approaches.

2. PRODUCTION OF SEMISYNTHETIC PTENs

Semisynthetic PTEN production has involved using insect cell expression of the N-terminal recombinant fragment containing a C-terminal thioester for ligation to C-terminal synthetic peptides. PTEN constructs (aa 1–378 or aa 1–377) were subcloned into a pFastBac1 vector which also harbors a C-terminal mxe gyrA intein fused to a chitin-binding domain (CBD). The plasmids were then transformed into competent DH10Bac *Escherichia coli* cells for bacmid production. The bacmid was then transfected into SF21 insect cells using 1–2 ng/well in one six-well plate containing 800,000 cells and incubated at 27°C for 72–96 h until viral infection was observed. The resulting P1 baculovirus was then used to infect additional SF21 insect cells in a T75 flask at a density of 1 million/mL at 27°C for 48–72 h to generate a P2 virus with a higher viral titer. The resulting P2 virus was then used to infect High Five insect cells in liquid suspension at a MOI = 1 and incubated at 27°C for 48 h. Typical expression yields of PTEN fusion proteins are 5–10 mg/L culture (Bolduc et al., 2013; Chen, Dempsey, et al., 2016).

2.1 Classical Expressed Protein Ligation

Classical EPL takes advantage of a heterozygously expressed protein with a truncated C-terminus fused to an intein and CBD (Muir et al., 1998; Schwarzer & Cole, 2005; Vila-Perello & Muir, 2010). The CBD is used to immobilize the protein–intein fusion to chitin resin for purification purposes. The intein facilitates a reversible N → S acyl shift that is released after treatment with a small molecule thiol (typically MESNA) resulting in the free protein thioester fragment (Fig. 3). This protein thioester is then treated with peptide containing an N-terminal cysteine, which undergoes a transthioesterification followed by intramolecular rearrangement to generate a native amide bond with a cysteine at the ligation point. This method is particularly useful for proteins that harbor a cysteine near the C-terminus; however, PTEN has no such cysteine. Therefore, Tyr379 was replaced with a Cys for the ligation step (Bolduc et al., 2013). This Y379C PTEN mutation appeared to be tolerated as it had similar catalytic activity as wild-type (wt) PTEN in hydrolyzing diC6-PIP3, a six-carbon fatty-acid containing PIP3 analog diC6-PIP3 commonly used in biochemical studies as a model substrate because it is soluble and easy to work with unlike natural PIP3 which contains longer fatty acids and forms vesicles. This PTEN hydrolysis of diC6-PIP3 was assayed by measuring the release of inorganic phosphate using the Malachite green assay (Van Veldhoven & Mannaerts, 1987). The tail peptides used for ligation were synthesized by Fmoc solid-phase synthesis strategies. The synthesis of the tetraphosphorylated 25mer peptide (aa 379–403) proved challenging, presumably because the four phosphorylated residues clustered in a hexa-amino acid segment were monoprotected but still negatively charged and bulky. As peptides are typically synthesized from the C-terminus to the N-terminus, the amino acid couplings starting at pSer385 were performed manually with six equivalents

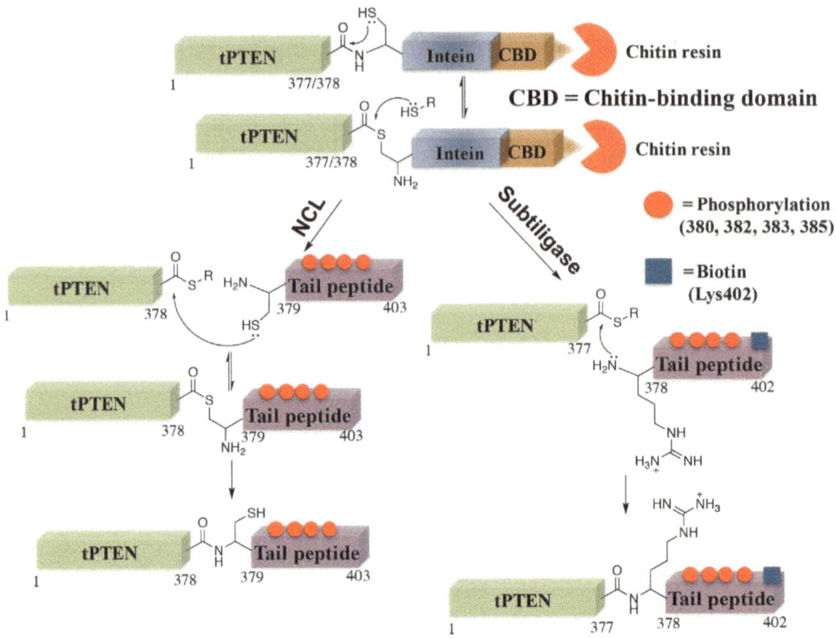

Fig. 3 Classical and enzyme-catalyzed expressed protein ligation (EPL). Truncated PTEN (aa 1–377) is expressed with the GyrA intein and chitin-binding domain (CBD) fused to its C-terminus. PTEN is then immobilized on to chitin resin for purification. The intein facilitates the reversible formation of a thioester which is treated with a small molecule thiol (MESNA) that undergoes a transthioesterification reaction to generate a PTEN-MESNA thioester. For classical EPL, the truncated PTEN-MESNA thioester is then incubated with an N-Cys-containing synthetic peptide, which undergoes another transthioesterification followed by intramolecular rearrangement to generate the full-length protein with a Cys at the ligation point and PTMs at their relevant positions. For enzyme-catalyzed EPL, a synthetic peptide is ligated to the isolated PTEN-thioester by subtiligase to generate the full-length wild-type protein with relevant PTMs.

of Fmoc residues for extended times using a 1:1 ratio of dichloromethane: dimethylformamide to maximize yields by semiquantitative ninhydrin tests. Furthermore, characterizing the phosphopeptide products using matrix-assisted laser desorption/ionization time-of-flight mass spectrometry was also challenging because of the suppressing effects of clustered phosphates; in some cases, negative ion mode with α-cyano-4-hydroxycinnamic acid as the matrix in 50% acetonitrile and 5 m*M* ammonium citrate enhanced detection.

As described in Section 2, the recombinant PTEN (aa 1–378) fusion protein was produced in High Five insect cells. Insect cells express the enzyme chitinase that can interfere with the standard chitin-resin affinity purification. To remove chitinase, the cell lysates are preincubated with fibrous cellulose prior to immobilizing the extracts on chitin resin. Ligations were carried out using a "one-pot" strategy where 400 m*M* MESNA and 2 m*M* N-Cys-containing peptide in aqueous buffer (pH 7.2) are added to the immobilized PTEN-intein-CBD and incubated at room temperature for 48–72 h until >80% conversion monitored by Coomassie-stained SDS-PAGE (Bolduc et al., 2013). The semisynthetic PTEN was further purified using anion exchange chromatography (MonoQ) with a 240-mL linear gradient from 50 m*M* HEPES pH 8.0, 5 m*M* NaCl, 10 m*M* DTT to 50 m*M* HEPES pH 8.0, 500 m*M* NaCl, 10 m*M* DTT. Initial studies afforded semisynthetic tetraphosphorylated (Ser380, Thr382, Thr383, and Ser385) and unphosphorylated PTEN. Remarkably, the elution pattern of these proteins on an anion exchange resin was reversed compared to the predicted outcome. The unphosphorylated PTEN eluted at higher ionic strength relative to tetraphosphorylated PTEN, despite the fact that the four phosphates were expected to add eight negative charges. This paradox was ultimately rationalized as resulting from a conformational closure of tetraphosphorylated PTEN driven by an intramolecular interaction between the tail and core of PTEN that effectively masks the anionic phosphates from contributing to interaction with the positively charged quaternary ammonium resin. It should be noted that even the unphosphorylated PTEN C-tail is anionic and is presumably more available for binding the positively charged resin. Following anion exchange chromatography, the semisynthetic PTEN proteins were further purified by size-exclusion chromatography with either a Superdex 200 or Superdex 75 column with a mobile phase of 50 m*M* Tris pH 8.0, 150 m*M* NaCl, 10 m*M* DTT. Purified fractions were combined and concentrated to 1–5 mg/mL and purity assed to be >95% based on Coomassie-stained SDS-PAGE.

2.2 Enzyme-Catalyzed Expressed Protein Ligation

It was subsequently learned that Y379C was not a silent mutation in tetraphosphorylated PTEN (see Section 4) (Nguyen, Afkari, et al., 2014). We thus adopted a newly developed semisynthetic method termed enzyme-catalyzed EPL to produce PTEN containing a natural Tyr379. This enzyme-catalyzed method avoids the Cys requirement by use of an engineered protease called subtiligase in which the catalytic S221C along with a P225A is mutated to promote peptide ligations over proteolysis (Abrahmsen et al., 1991; Braisted, Judice, & Wells, 1997; Chang, Jackson, Burnier, &

Wells, 1994; Henager et al., 2016). It was shown that enzyme-catalyzed EPL can accommodate a diverse array of amino acids near the ligation junction, obviating the specific need for Cys, which is poorly represented in proteomes. For this alternative ligation strategy, the expression and generation of the protein thioester fragment are the same as the classical EPL; however, instead of a one-pot reaction to ligate the peptide, the protein thioester is isolated after treating the resin overnight with 300 m*M* MESNA in HEPES pH 7.5 containing buffer at room temperature (Fig. 3) (Henager et al., 2016). Subtiligase is expressed and secreted from *Bacillus subtilis* and purified by successive ammonium sulfate precipitation, ethanol precipitation, cation-exchange (Mono S) chromatography, and finally stored in 100 m*M* BICINE pH 8.0, 5 m*M* DTT (Abrahmsen et al., 1991; Braisted et al., 1997; Chang et al., 1994; Jackson et al., 1994). Ligation reactions are typically carried out in 100 m*M* BICINE, 5 m*M* $CaCl_2$, pH 8.0, 0.55 μ*M* subtiligase, 40 μ*M* protein thioester, with 3 m*M* peptide at 25°C. Reaction conversions for enzyme-catalyzed EPL usually range from 50% to 70% depending on the P1, P4, P1′, and P2′ residues surrounding the ligation junction; however, a reduction in yield was observed for ligations that contain an acidic residue at either P4 or P1′. Furthermore, high concentrations of MESNA were noted to inhibit subtiligase-mediated ligations so that a buffer exchange is required to remove MESNA from the protein sample prior to ligations.

For PTEN, a ligation between Tyr377 and Arg378 was selected because the peptide was still manageable in size and is compatible with subtiligase preferences. Ligation of the C-tail peptide was carried out in 100 m*M* BICINE, 110 m*M* $CaCl_2$, pH 8.0, 25 μ*M* subtiligase, 20 μ*M* $PTEN_{aa\ 1–377}$-thioester, 3 m*M* C-tail $peptide_{aa\ 378–402\text{-biotin}}$ at 25°C for 4 h (Henager et al., 2016). The especially high Ca concentration was used in case the phosphopeptide was chelating the metal. Ligation yields for the tetraphosphorylated and unphosphorylated peptides were 30% and 50%, respectively. A biotinylated lysine at amino acid 402 was appended to expedite purification by avidin affinity chromatography of the ligated semi-synthetic PTEN product. The lower ligation yield for the phosphorylated peptide may relate to the phosphorylated serine at P3′ although P3′ is normally a flexible position.

3. C-TERMINAL PHOSPHORYLATION IMPACT ON ENZYMATIC FUNCTION OF PTEN

In initial experiments, semisynthetic PTEN prepared by classical EPL containing a Y379C mutation was analyzed in enzyme assays. Using the physiologically relevant vesicle bound ^{32}P-PIP3 containing palmitoyl chains as substrate, PTEN forms were added and interfacial enzyme kinetic analysis was performed which considers both the bulk and surface concentrations of substrate when measuring kinetic constants (Bolduc et al., 2013; Deems, Eaton, & Dennis, 1975; Hendrickson & Dennis, 1984; McConnachie et al., 2003). These interfacial enzymatic assays, which quantify ^{32}P-inorganic phosphate release, do not follow classical Michaelis–Menten kinetics as two compartments are involved and the rate of reaction depends on both bulk and surface substrate concentrations in contrast to standard solution phase monodisperse substrate assays. To make interfacial enzyme kinetic measurements, the assays are performed by separately varying the bulk and surface substrate concentrations. The bulk dilution experiment varies both the carrier lipid and substrate bulk concentration proportionally, whereas the substrate surface concentration is held constant. The surface dilution experiment holds the bulk concentration of substrate constant while varying the surface substrate by varying the carrier lipid concentration. In these assays, the ratio of vesicles to PTEN molecules was maintained at 4:1 to increase the chances that only one molecule of PTEN is bound to an individual vesicle at a time. These experiments revealed that the iK_m (interfacial K_m, a measure of the K_m in the context of two-dimensional diffusion on the vesicle surface) and k_{cat} were similar for nonphosphorylated (n-PTEN) and tetraphosphorylated PTEN (4p-PTEN) (Bolduc et al., 2013). In contrast, the K_s (effectively the K_d of the PTEN binding to the vesicles) was threefold higher for 4p-PTEN relative to n-PTEN. These results suggest that 4p-PTEN relative to n-PTEN has reduced affinity for lipid vesicles. The effect was most pronounced at low PIP3 concentrations (mol% $<$ 1%) which is thought to be more biologically relevant as PIP3 concentrations in the cell are less than 0.001 mol%.

In addition, a vesicle sedimentation assay was adapted to determine the affinity of semisynthetic PTENs for PIP2 or phosphatidylserine-containing vesicles. These data indicated that the anionic lipids stimulate binding of 4p-PTEN and n-PTEN; however, 4p-PTEN required a greater concentration of the anionic lipid to mimic the binding of n-PTEN, confirming that C-tail phosphorylation reduces membrane binding (Bolduc et al., 2013).

The enzymatic activities with semisynthetic PTENs were also investigated using the soluble diC6-PIP3 substrate. Although the physiologic significance of this enzymatic assay is unclear, it has proven to be a reliable predictor of conformational state (Section 4). The diC6-PIP3 hydrolysis assay is much easier to perform compared to the interfacial enzymatic assay as it avoids the technically challenging synthesis of radiolabeled PIP3 substrate, impregnating vesicles with this substrate or isolation of released radiolabeled inorganic phosphate following the enzymatic assay for analysis. Furthermore, the half-life of ^{32}P is short so all interfacial enzyme assays must be complete within about 2 weeks of receiving the radiolabeled ATP used to make ^{32}P-PIP3 (Goodier & Pritchard, 1966). Enzymatic analysis using the soluble PIP3 assay showed that the 4p-PTEN decreases the catalytic efficiency (k_{cat}/K_m) approximately 10-fold relative to n-PTEN with a significant increase in the diC6-PIP K_m for 4p-PTEN (Bolduc et al., 2013; Chen, Dempsey, et al., 2016). Notably, it was shown that several of the enzymatic properties identified with 4p-PTEN were recapitulated with intermolecular experiments. In particular, C-tail truncated recombinant PTEN diC6-PIP3 phosphatase activity was potently inhibited by the synthetic tetraphosphorylated tail peptide with an apparent K_d of 1 μM (Bolduc et al., 2013). The tetraphosphorylated peptide also inhibited binding of C-tail truncated recombinant PTEN to lipid vesicles. These data indicated that the mechanism of impact of the phospho-tail on PTEN involves an interaction between the phospho-C-tail and the core of PTEN.

4. STRUCTURAL ANALYSIS OF PHOSPHO-PTEN

It was shown by coimmunoprecipitation from cell extracts that the core of PTEN could interact with a recombinant C-tail; however, these experiments had uncertain phosphorylation stoichiometry and relied on impure mixtures which could complicate interpretation (Odriozola et al., 2007; Rahdar et al., 2009). Consistent with an intramolecular interaction, however, both trypsin resistance and small angle X-ray scattering support a more compact conformation of semisynthetic 4p-PTEN (Bolduc et al., 2013). Semisynthetic phospho-PTENs were subsequently prepared using classical EPL to study the specific contributions of individual phospho-sites to intramolecular interactions. A commercially available phospho-C-tail-specific antibody of PTEN (Cell Signaling Technology, Cat # 9549) was shown to be selective for phospho-Ser380 (Chen, Dempsey, et al., 2016). An assay with alkaline phosphatase was developed to examine the exposure of the phospho-C-tail with different PTEN forms (Fig. 4). This alkaline phosphatase sensitivity assay probes the conformational state of PTEN by catalyzing the dephosphorylation of the C-terminal tail which is monitored by Western blot. If PTEN is in a conformationally open state, then the rate of dephosphorylation should be greater because the phosphate(s) are more accessible to alkaline phosphatase. If PTEN is in a

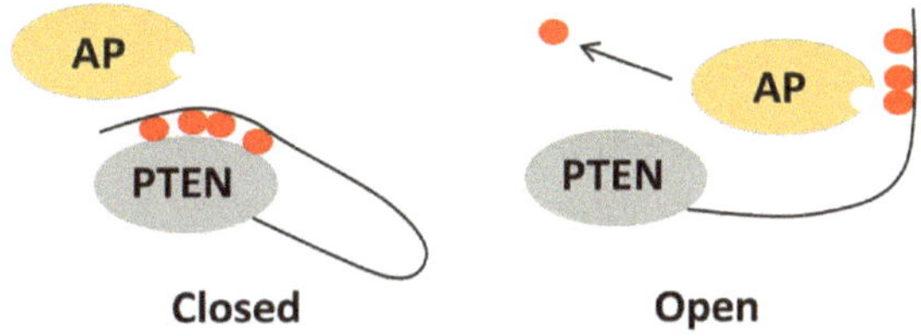

Fig. 4 Phosphate protection assay to sample conformational closure. This assay probes the conformational state of PTEN by evaluating the accessibility of the C-tail phosphatase to alkaline phosphatase (AP). If PTEN is in the closed conformation, then the tail phosphates are masked from AP by interacting with the core of the protein which results in their slower dephosphorylation. If PTEN is in the open conformation, then the tail phosphates are more accessible to AP and their dephosphorylation is much faster. *Red circles* represent phosphorylation sites.

conformationally closed state, then the rate of dephosphorylation should be slower as tail:core interactions will shield the C-tail phosphates from alkaline phosphatase. These experiments revealed that each phosphorylations partially contributes to conformational closure and no individual phosphorylation site of the four stood out as being dominant.

Currently, there is no high-resolution structure that shows the position of the phospho-tail in the conformationally closed state. Mutagenesis of various regions in the core of PTEN has been carried out to investigate potential contacts for phospho-tail interaction (Nguyen, Afkari, et al., 2014; Nguyen, Yang, et al., 2014; Odriozola et al., 2007; Rahdar et al., 2009). A significant rescuing of 4p-PTEN catalytic function and greater sensitivity to tail dephosphorylation by alkaline phosphatase were observed when the basic CBRIII loop residues (Lys260/Lys263/Lys266/Lys267/Lys269) were all mutated to Asp, suggesting a role for the CBRIII loop in stabilizing the closed conformation (Bolduc et al., 2013). Notably, when these CBRIII Lys residues were replaced with Ala, the conformation of 4p-PTEN led to wt-like behavior. These data suggest that the role of the CBRIII loop may be indirect in driving conformational closure.

H/D exchange experiments on phosphorylated PTEN indicated that the Cα2 loop (aa 319–342) might be important in stabilizing the closed conformation (Masson et al., 2016). Indeed, mutagenesis of the basic residues of the Cα2 loop (Lys327, Lys330, Lys332, and Arg335) revealed that both the aspartate and alanine replacements led to opening of 4p-PTEN assessed by alkaline phosphatase sensitivity and diC6-PIP3 processing efficiency (Chen, Dempsey, et al., 2016). These results are consistent with a model where the basic residues of the Cα2 loop directly mediate interactions with one or more C-tail phosphates in PTEN. However, these results cannot exclude an indirect role.

One technique that can establish spatial proximity between two protein surfaces involves photocrosslinking. To apply this method to phospho-PTEN, a photoactivatable crosslinking group was installed near the phospho-cluster in the C-tail using EPL. The crosslinking residue selected was benzoyl L-phenylalanine (Bpa) in place of Phe392 in the synthetic tail peptide. Bpa is structurally somewhat similar to Phe and it can form stable crosslinks to a range of amino acids. Irradiation of the Bpa phospho-PTEN sample (also tagged in the C-terminus with a biotin) at 365 nm for 3 h at 4°C was performed (Fig. 5). It proved necessary to dephosphorylate the sample with alkaline phosphatase prior to trypsin digestion and avidin chromatography to enhance the LC/MS/MS analysis. Using this method, a crosslinked peptide was identified which corresponded to the C-terminus connected to the N-terminal region of PTEN within amino acids 42–50 (Chen, Dempsey, et al., 2016; Forne, Ludwigsen, Imhof, Becker, & Mueller-Planitz, 2012; Mueller-Planitz, 2015). Site-directed mutagenesis was performed to validate the relevance of this crosslink and the phosphorylated R41D/R47D/R74D PTEN variant was found to be modestly more open compared to native 4p-PTEN based on alkaline phosphatase sensitivity.

The development of enzyme-catalyzed EPL allowed for the generation of 4p-PTEN possessing a natural Tyr379 instead of the Cys379 mutation introduced for classical EPL. It was found that the Tyr379 4p-PTEN equilibrium between open and closed conformations favored the closed state

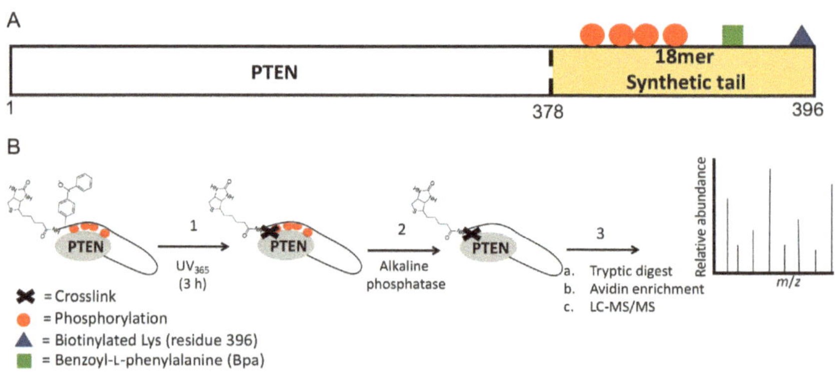

Fig. 5 Photocrosslinking studies to map location of the C-tail in conformationally closed state. (A) Semisynthetic PTEN generated for Bpa UV crosslinking. (B) Diagram of experimental steps from UV irradiation to analysis of Bpa crosslinks.

by about 10-fold relative to the corresponding equilibrium for Cys379 4p-PTEN based both on alkaline phosphatase sensitivity and diC6-PIP3 PTEN processing (Henager et al., 2016). Notably, the natural Tyr379 4p-PTEN confers reduced conformational opening in the presence of high salt buffer. These experiments helped clarify the significance of the Tyr379 mutational sensitivity in cellular studies. Moreover, accessing natural Tyr379 4p-PTEN facilitated more accurate quantification of cellular 4p-PTEN stoichiometry on endogenous protein measured by Western blots because the Tyr379 also proved critical to phospho-PTEN antibody recognition (Henager et al., 2016). These experiments underscore that avoiding Cys mutation is a valuable attribute of enzyme-catalyzed EPL.

5. PHOSPHO-PTEN AND PROTEIN UBIQUITINATION

The protein stability of PTEN in the cell is a factor in its signaling properties and PIP3 control. When PTEN C-terminal phosphorylation was discovered, it was also found that there was a correlation between its phosphorylation state and protein stability (Vazquez et al., 2000). This behavior is somewhat paradoxical as the enzymatic activity of phospho-PTEN is impaired which means that the more stable PTEN form is also less functional in PIP3 hydrolysis. Furthermore, the molecular basis of PTEN protein destruction was not yet clear in this early work. Cellular protein clearance often involves Lys ubiquitination which then targets the protein for proteasome-mediated degradation. Ubiquitination is a three-step process involving E1, E2, and E3 ubiquitin transferring enzymes. Several ubiquitin E3 ligases for PTEN have been reported, and the two most extensively studied have been NEDD4-1 (Wang et al., 2007) and WWP2 (Maddika et al., 2011). NEDD4-1 and WWP2 are both HECT domain E3 ligases and are composed of an N-terminal C2 domain followed by four WW domains and culminating in catalytic HECT domains.

In principle, phosphorylation of PTEN could inhibit its ubiquitination by the requisite E3 ligase as a mechanism to enhance PTEN protein stability. This is difficult to study in live cells. To address whether NEDD4-1 or WWP2 might be affected by PTEN phosphorylation, in vitro ubiquitination assays involving E1 and E2 enzymes along with NEDD4-1 and WWP2 were performed (Chen, Thomas, et al., 2016). It was shown that WWP2 was much slower to ubiquitinate 4p-PTEN relative to n-PTEN. In contrast, NEDD4-1 did not distinguish between the modified and unmodified PTENs and ubiquitinated them with similar rates. These data suggest that WWP2 might be the more important PTEN ubiquitin ligase, at least in some cell types, as the in vitro selectivity of the ligase is consistent with the cellular behavior of PTEN stability.

These findings also suggest that factors that can block WWP2 catalytic activation might allow for enhanced PTEN stability. A recent study revealed that WWP2 is autoinhibited by a centrally located 30 aa peptide linker and that tyrosine phosphorylation of this peptide linker on either of two Tyr residues, aa 369 and aa 392, can relieve this ubiquitin ligase inhibition (Chen et al., 2017). It was subsequently shown that mutation of these Tyr sites to Glu residues could activate WWP2 activity in cells and this led to reduced PTEN levels. These studies further confirm the WWP2-PTEN physiologic connection and suggest that protein tyrosine kinase inhibitors that block WWP2 phosphorylation could drive PTEN cellular stability even in the unphosphorylated state of PTEN.

6. SUMMARY AND FUTURE DIRECTIONS

This chapter has described how Ser/Thr phosphorylation of PTEN at residues 380, 382, 383, and 385 can influence PTEN structure and function. Phosphorylation of these residues converts PTEN from an open to closed conformation resulting in a decrease in enzymatic function and plasma membrane binding, but also protects the protein from cellular destruction. This chapter also highlights the importance of modern chemical tools in investigating PTEN regulation by phosphorylation. Despite the progress, many gaps in our knowledge remain about PTEN regulation. A high-resolution picture revealing the position of the phosphorylated tail in the closed conformational is still lacking. It is unclear how C-tail phosphorylation renders PTEN more resistant to ubiquitination by WWP2. It will be important to define the molecular basis of PTEN recognition by WWP2 and whether there are surfaces that might be druggable. Moreover, it would be a great leap forward if allosteric modulators of phospho-PTEN itself could be identified that would restore its catalytic function (Qiao, Molina, Pandey, Zhang, & Cole, 2006). Such PTEN reactivation might also be achieved in a more traditional fashion by blocking CK2 protein kinase and/or the not yet identified tyrosine kinase that activates WWP2. In the longer term, a full inventory and analysis of the impact on the wide array of PTEN PTMs are needed. We believe that the next two decades of PTEN research promise to be at least as exciting as the first two.

ACKNOWLEDGMENTS

We thank NIH CA74305 and NIH GM120855. We appreciate the helpful advice and collaboration of David Bolduc, Zan Chen, Sam Henager, Nam Chu, Stefani Thomas, Dawn Hayward, Sandra Gabelli, Peter Devreotes, Mario Amzel, and Miho Iijima.

REFERENCES

Abrahmsen, L., Tom, J., Burnier, J., Butcher, K. A., Kossiakoff, A., & Wells, J. A. (1991). Engineering subtilisin and its substrates for efficient ligation of peptide bonds in aqueous solution. *Biochemistry*, *30*(17), 4151–4159.

Al-Khouri, A. M., Ma, Y., Togo, S. H., Williams, S., & Mustelin, T. (2005). Cooperative phosphorylation of the tumor suppressor phosphatase and tensin homologue (PTEN) by casein kinases and glycogen synthase kinase 3beta. *The Journal of Biological Chemistry*, *280*(42), 35195–35202. https://doi.org/10.1074/jbc.M503045200.

Bolduc, D., Rahdar, M., Tu-Sekine, B., Sivakumaren, S. C., Raben, D., Amzel, L. M., et al. (2013). Phosphorylation-mediated PTEN conformational closure and deactivation revealed with protein semisynthesis. *Elife*, *2*, e00691. https://doi.org/10.7554/eLife.00691.

Braisted, A. C., Judice, J. K., & Wells, J. A. (1997). Synthesis of proteins by subtiligase. *Methods in Enzymology*, *289*, 298–313.

Campbell, R. B., Liu, F., & Ross, A. H. (2003). Allosteric activation of PTEN phosphatase by phosphatidylinositol 4,5-bisphosphate. *The Journal of Biological Chemistry*, *278*(36), 33617–33620. https://doi.org/10.1074/jbc.C300296200.

Chang, T. K., Jackson, D. Y., Burnier, J. P., & Wells, J. A. (1994). Subtiligase: A tool for semisynthesis of proteins. *Proceedings of the National Academy of Sciences of the United States of America*, *91*(26), 12544–12548.

Chen, Z., Dempsey, D. R., Thomas, S. N., Hayward, D., Bolduc, D. M., & Cole, P. A. (2016). Molecular features of phosphatase and tensin homolog (PTEN) regulation by C-terminal phosphorylation. *The Journal of Biological Chemistry*, *291*(27), 14160–14169. https://doi.org/10.1074/jbc.M116.728980.

Chen, Z., Jiang, H., Xu, W., Li, X., Dempsey, D. R., Zhang, X., et al. (2017). A tunable brake for HECT ubiquitin ligases. *Molecular Cell*, *66*(3), 345–357, e346. https://doi.org/10.1016/j.molcel.2017.03.020.

Chen, Z., Thomas, S. N., Bolduc, D. M., Jiang, X., Zhang, X., Wolberger, C., et al. (2016). Enzymatic analysis of PTEN ubiquitylation by WWP2 and NEDD4-1 E3 ligases. *Biochemistry*, *55*(26), 3658–3666. https://doi.org/10.1021/acs.biochem.6b00448.

Cordier, F., Chaffotte, A., Terrien, E., Prehaud, C., Theillet, F. X., Delepierre, M., et al. (2012). Ordered phosphorylation events in two independent cascades of the PTEN C-tail revealed by NMR. *Journal of the American Chemical Society*, *134*(50), 20533–20543. https://doi.org/10.1021/ja310214g.

Das, S., Dixon, J. E., & Cho, W. (2003). Membrane-binding and activation mechanism of PTEN. *Proceedings of the National Academy of Sciences of the United States of America*, *100*(13), 7491–7496. https://doi.org/10.1073/pnas.0932835100.

Deems, R. A., Eaton, B. R., & Dennis, E. A. (1975). Kinetic analysis of phospholipase A2 activity toward mixed micelles and its implications for the study of lipolytic enzymes. *The Journal of Biological Chemistry*, *250*(23), 9013–9020.

Forne, I., Ludwigsen, J., Imhof, A., Becker, P. B., & Mueller-Planitz, F. (2012). Probing the conformation of the ISWI ATPase domain with genetically encoded photoreactive crosslinkers and mass spectrometry. *Molecular & Cellular Proteomics*, *11*(4), M111 012088. https://doi.org/10.1074/mcp.M111.012088.

Goodier, I. W., & Pritchard, D. H. (1966). The half-life of P32. *The International Journal of Applied Radiation and Isotopes*, *17*(2), 121–123.

Henager, S. H., Chu, N., Chen, Z., Bolduc, D., Dempsey, D. R., Hwang, Y., et al. (2016). Enzyme-catalyzed expressed protein ligation. *Nature Methods*, *13*(11), 925–927. https://doi.org/10.1038/nmeth.4004.

Hendrickson, H. S., & Dennis, E. A. (1984). Analysis of the kinetics of phospholipid activation of cobra venom phospholipase A2. *The Journal of Biological Chemistry*, *259*(9), 5740–5744.

Huang, J., Yan, J., Zhang, J., Zhu, S., Wang, Y., Shi, T., et al. (2012). SUMO1 modification of PTEN regulates tumorigenesis by controlling its association with the plasma membrane. *Nature Communications*, *3*, 911. https://doi.org/10.1038/ncomms1919.

Ikenoue, T., Inoki, K., Zhao, B., & Guan, K. L. (2008). PTEN acetylation modulates its interaction with PDZ domain. *Cancer Research*, *68*(17), 6908–6912. https://doi.org/10.1158/0008-5472.CAN-08-1107.

Jackson, D. Y., Burnier, J., Quan, C., Stanley, M., Tom, J., & Wells, J. A. (1994). A designed peptide ligase for total synthesis of ribonuclease A with unnatural catalytic residues. *Science*, *266*(5183), 243–247.

Lee, J. O., Yang, H., Georgescu, M. M., Di Cristofano, A., Maehama, T., Shi, Y., et al. (1999). Crystal structure of the PTEN tumor suppressor: Implications for its phosphoinositide phosphatase activity and membrane association. *Cell*, *99*(3), 323–334.

Lee, S. R., Yang, K. S., Kwon, J., Lee, C., Jeong, W., & Rhee, S. G. (2002). Reversible inactivation of the tumor suppressor PTEN by H_2O_2. *The Journal of Biological Chemistry*, *277*(23), 20336–20342. https://doi.org/10.1074/jbc.M111899200.

Li, D. M., & Sun, H. (1997). TEP1, encoded by a candidate tumor suppressor locus, is a novel protein tyrosine phosphatase regulated by transforming growth factor beta. *Cancer Research*, *57*(11), 2124–2129.

Li, J., Yen, C., Liaw, D., Podsypanina, K., Bose, S., Wang, S. I., et al. (1997). PTEN, a putative protein tyrosine phosphatase gene mutated in human brain, breast, and prostate cancer. *Science*, *275*(5308), 1943–1947.

Maddika, S., Kavela, S., Rani, N., Palicharla, V. R., Pokorny, J. L., Sarkaria, J. N., et al. (2011). WWP2 is an E3 ubiquitin ligase for PTEN. *Nature Cell Biology*, *13*(6), 728–733. https://doi.org/10.1038/ncb2240.

Maehama, T., & Dixon, J. E. (1998). The tumor suppressor, PTEN/MMAC1, dephosphorylates the lipid second messenger, phosphatidylinositol 3,4,5-trisphosphate. *The Journal of Biological Chemistry*, *273*(22), 13375–13378.

Masson, G. R., Perisic, O., Burke, J. E., & Williams, R. L. (2016). The intrinsically disordered tails of PTEN and PTEN-L have distinct roles in regulating substrate specificity and membrane activity. *The Biochemical Journal*, *473*(2), 135–144. https://doi.org/10.1042/BJ20150931.

McConnachie, G., Pass, I., Walker, S. M., & Downes, C. P. (2003). Interfacial kinetic analysis of the tumour suppressor phosphatase, PTEN: Evidence for activation by anionic phospholipids. *The Biochemical Journal*, *371*(Pt. 3), 947–955. https://doi.org/10.1042/BJ20021848.

Meng, Z., Jia, L. F., & Gan, Y. H. (2016). PTEN activation through K163 acetylation by inhibiting HDAC6 contributes to tumour inhibition. *Oncogene*, *35*(18), 2333–2344. https://doi.org/10.1038/onc.2015.293.

Miller, S. J., Lou, D. Y., Seldin, D. C., Lane, W. S., & Neel, B. G. (2002). Direct identification of PTEN phosphorylation sites. *FEBS Letters*, *528*(1–3), 145–153.

Mueller-Planitz, F. (2015). Crossfinder-assisted mapping of protein crosslinks formed by site-specifically incorporated crosslinkers. *Bioinformatics*, *31*(12), 2043–2045. https://doi.org/10.1093/bioinformatics/btv083.

Muir, T. W., Sondhi, D., & Cole, P. A. (1998). Expressed protein ligation: A general method for protein engineering. *Proceedings of the National Academy of Sciences of the United States of America*, *95*(12), 6705–6710.

Myers, M. P., Pass, I., Batty, I. H., Van der Kaay, J., Stolarov, J. P., Hemmings, B. A., et al. (1998). The lipid phosphatase activity of PTEN is critical for its tumor suppressor function. *Proceedings of the National Academy of Sciences of the United States of America*, *95*(23), 13513–13518.

Myers, M. P., Stolarov, J. P., Eng, C., Li, J., Wang, S. I., Wigler, M. H., et al. (1997). P-TEN, the tumor suppressor from human chromosome 10q23, is a dual-specificity phosphatase. *Proceedings of the National Academy of Sciences of the United States of America*, *94*(17), 9052–9057.

Nguyen, H. N., Afkari, Y., Senoo, H., Sesaki, H., Devreotes, P. N., & Iijima, M. (2014). Mechanism of human PTEN localization revealed by heterologous expression in *Dictyostelium*. *Oncogene*, *33*(50), 5688–5696. https://doi.org/10.1038/onc.2013.507.

Nguyen, H. N., Yang, J. M., Afkari, Y., Park, B. H., Sesaki, H., Devreotes, P. N., et al. (2014). Engineering ePTEN, an enhanced PTEN with increased tumor suppressor activities. *Proceedings of the National Academy of Sciences of the United States of America*, *111*(26), E2684–E2693. https://doi.org/10.1073/pnas.1409433111.

Odriozola, L., Singh, G., Hoang, T., & Chan, A. M. (2007). Regulation of PTEN activity by its carboxyl-terminal autoinhibitory domain. *Journal of Biological Chemistry*, *282*(32), 23306–23315. https://doi.org/10.1074/jbc.M611240200.

Okumura, K., Mendoza, M., Bachoo, R. M., DePinho, R. A., Cavenee, W. K., & Furnari, F. B. (2006). PCAF modulates PTEN activity. *The Journal of Biological Chemistry*, *281*(36), 26562–26568. https://doi.org/10.1074/jbc.M605391200.

Qiao, Y., Molina, H., Pandey, A., Zhang, J., & Cole, P. A. (2006). Chemical rescue of a mutant enzyme in living cells. *Science*, *311*(5765), 1293–1297. https://doi.org/10.1126/science.1122224.

Rahdar, M., Inoue, T., Meyer, T., Zhang, J., Vazquez, F., & Devreotes, P. N. (2009). A phosphorylation-dependent intramolecular interaction regulates the membrane association and activity of the tumor suppressor PTEN. *Proceedings of the National Academy of Sciences of the United States of America*, *106*(2), 480–485. https://doi.org/10.1073/pnas.811212106.

Redfern, R. E., Redfern, D., Furgason, M. L., Munson, M., Ross, A. H., & Gericke, A. (2008). PTEN phosphatase selectively binds phosphoinositides and undergoes structural changes. *Biochemistry*, *47*(7), 2162–2171. https://doi.org/10.1021/bi702114w.

Schwarzer, D., & Cole, P. A. (2005). Protein semisynthesis and expressed protein ligation: Chasing a protein's tail. *Current Opinion in Chemical Biology*, *9*(6), 561–569. https://doi.org/10.1016/j.cbpa.2005.09.018.

Sun, H., Lesche, R., Li, D. M., Liliental, J., Zhang, H., Gao, J., et al. (1999). PTEN modulates cell cycle progression and cell survival by regulating phosphatidylinositol 3,4,5,-trisphosphate and Akt/protein kinase B signaling pathway. *Proceedings of the National Academy of Sciences of the United States of America*, *96*(11), 6199–6204.

Torres, J., & Pulido, R. (2001). The tumor suppressor PTEN is phosphorylated by the protein kinase CK2 at its C terminus. Implications for PTEN stability to proteasome-mediated degradation. *The Journal of Biological Chemistry*, *276*(2), 993–998. https://doi.org/10.1074/jbc.M009134200.

Van Veldhoven, P. P., & Mannaerts, G. P. (1987). Inorganic and organic phosphate measurements in the nanomolar range. *Analytical Biochemistry*, *161*(1), 45–48.

Vazquez, F., Ramaswamy, S., Nakamura, N., & Sellers, W. R. (2000). Phosphorylation of the PTEN tail regulates protein stability and function. *Molecular and Cellular Biology*, *20*(14), 5010–5018.

Vila-Perello, M., & Muir, T. W. (2010). Biological applications of protein splicing. *Cell, 143*(2), 191–200. https://doi.org/10.1016/j.cell.2010.09.031.

Wang, X., Trotman, L. C., Koppie, T., Alimonti, A., Chen, Z., Gao, Z., et al. (2007). NEDD4-1 is a proto-oncogenic ubiquitin ligase for PTEN. *Cell, 128*(1), 129–139. https://doi.org/10.1016/j.cell.2006.11.039.

Worby, C. A., & Dixon, J. E. (2014). PTEN. *Annual Review of Biochemistry, 83*, 641–669. https://doi.org/10.1146/annurev-biochem-082411-113907.

FURTHER READING

Vazquez, F., Grossman, S. R., Takahashi, Y., Rokas, M. V., Nakamura, N., & Sellers, W. R. (2001). Phosphorylation of the PTEN tail acts as an inhibitory switch by preventing its recruitment into a protein complex. *The Journal of Biological Chemistry, 276*(52), 48627–48630. https://doi.org/10.1074/jbc.C100556200.

Kolibri 360 GmbH
Ferdinand-Jühlke-Straße 7
99095 Erfurt, Deutschland
produktsicherheit@kolibri360.de